Construction Estimates and Costs

CONSTRUCTION ESTIMATES AND COSTS

FOURTH EDITION

H. E. Pulver, B.S., C.E.

Emeritus Professor of Civil and Structural Engineering, University of Wisconsin

McGRAW-HILL BOOK COMPANY

New York St. Louis San Francisco Düsseldorf Johannesburg
Kuala Lumpur London Mexico Montreal New Delhi
Panama Rio de Janeiro Singapore Sydney Toronto

CONSTRUCTION ESTIMATES AND COSTS

 Library of Congress Catalog Card Number 68-30565

07-050931-1

567890 MAMM 75432

Preface

This book has been written primarily for practical men who wish to study methods of estimating construction costs. The methods given in this book have been developed by the author as a result of many years of teaching estimating construction costs to hundreds of men residing in nearly every state of this country. Many illustrative estimates have been included so that the application of these methods can be thoroughly understood.

In this book, attention is called to variations in the time required by different men to do certain work as well as to variations in hourly wages.

The information given in regard to wages and prices is approximate only. This information is included so that the reader or student may have some general information in regard to wages and prices prevailing in 1967 when this revision was written.

Many tables have been included. Most of the tables relating to labor give both the time in hours required to do work and the amount of work done per hour.

Many diagrams have been prepared and included so that the preparation and use of simple diagrams for estimating purposes can be understood. Practically all the diagrams are straight-line diagrams. Such diagrams are easy to construct and to use.

In the preparation of this fourth edition, every effort has been made to bring the material up to date. Much obsolete material has been deleted, and considerable new material has been added. Many tables have been revised, half of the diagrams have been redrawn, and almost all the illustrative estimates have been revised.

Every effort has been made to make the tables and diagrams comprehensive, accurate, and useful. The author would appreciate receiving notice of any errors in these tables and diagrams.

H. E. Pulver

Abbreviations

aggr	aggregate	fph	feet per hour
ac	alternating current	fpm	feet per minute
amp	ampere	fps	feet per second
bbl	barrel	ft	foot
bhp	brake horsepower	ft-lb	foot-pound
Btu	British thermal unit	f.o.b.	freight on board
CL	carload	gal	gallon
cfm	cubic feet per minute	gph	gallons per hour
cfs	cubic feet per second	gpm	gallons per minute
cu ft	cubic foot	gas	gasoline
cu yd	cubic yard	hp	horsepower
cwt	hundred weight	hp-hr	horsepower-hour
deg	degree	hr	hour
diam	diameter	in.	inch
dsl	diesel	in.-lb	inch-pound
dc	direct current	kw	kilowatt
D & M	dressed and matched lumber	kwhr	kilowatthour
F	Fahrenheit	lb	pound
fbm	feet board measure	LCL	less than carload

lin	lineal
lin ft	lineal foot
M	thousand
mph	miles per hour
min	minute
No.	number
oz	ounce
pcf	pounds per cubic foot
psf	pounds per square foot
psi	pounds per square inch
pt	pint
qt	quart
sk	sack
sec	second
sq	square (= 100 square feet)
sq ft	square foot
sq in.	square inch
sq yd	square yard
S4S	surfaced four sides
S2S	surfaced two sides
tph	tons per hour
wt	weight
wk hr	working hour

Contents

TWO Handling and Transporting Materials 13

THREE Excavation 41

TWENTY Complete Estimates 440

TWENTY-ONE Approximate Estimates 445

TWENTY-TWO Construction Management 458

CHAPTER 8. WOOD CONSTRUCTION

CHAPTER 9. STRUCTURAL STEEL

CHAPTER 10. ROOFING AND FLASHING

CHAPTER 11. LATHING, PLASTERING, AND DRYWALL CONSTRUCTION

CHAPTER 12. HEATING, COOLING, AND AIR CONDITIONING

CHAPTER ONE

Estimating in General

Estimating of construction work may be defined as the process of calculating the quantities and costs of the various items entering into the work.

As the estimate is made before the work is done, the estimated cost is only the probable cost and is never the actual cost. The agreement of the estimated cost with the actual cost will depend upon the skill and judgment of the estimator. Skill implies the accurate use of good estimating methods, and judgment implies the correct visualization of the work as it will be done.

Kinds of Estimates An estimate of the cost of a construction job is the probable cost of that job as computed from the plans and specifications. There are several kinds of estimates. Some of the ones more commonly used are the following:

A *detailed* estimate is one which includes the quantities and costs of everything that a contractor is required to provide and do for the satisfactory completion of the work. A detailed estimate is the best and most reliable form of estimate that can be made. There are two ways of making a detailed estimate.

In the *unit-quantity* method, all the costs of a unit quantity of the item considered are found and summarized. Then the total cost for the item is found by multiplying the cost per unit quantity by the number of units. For example, on a simple concrete job, the total cost of 1 cu yd of concrete would be first found and then this unit cost multiplied by the number of cubic yards in the job in order to obtain the total estimate. This method presents the information in such a way that unit costs on various jobs can be readily compared and also that the total estimate can be easily corrected for variations in quantities.

In the *total-quantity* method, the total quantities of each kind or class of material and labor needed are first found, then these quantities are multiplied by their individual costs, and the results summed up. For example, on a simple concrete job, the costs of all the materials, all the labor, all the plant, etc., would be found separately and then added to obtain the total estimate.

An *approximate* or *short-cut* estimate is an approximate or rough estimate made to obtain an approximate cost in a short time or to check roughly the costs found by more accurate methods. Such an estimate is not very reliable. Examples are the square-foot and cubic-foot methods of estimating building costs.

A *complete* estimate is one that includes all costs relating to the work in addition to those included in the main contract and subcontracts. For example, a complete estimate of a large office building would include the following items: land, legal fees, architect's fees, main contract, subcontracts, extras, financing, interest, insurance, and taxes during construction.

A *quantity survey*, sometimes called a *quantity* estimate or materials *takeoff*, is a complete estimate of the quantities of all the materials required. In some localities, a quantity surveyor prepares a list of materials for any given building, and all the contractors use this list when preparing their bids. The contractors make their own estimates for material prices, labor, plant, overhead, and profit.

A *contractor's* estimate is made by the contractor for determining the price or prices to be bid. It is usually a carefully prepared detailed estimate.

An *architect's* or *engineer's* estimate is one made by the architect or engineer usually for the purposes of financing the work and for checking bids made by contractors.

Progress estimates are estimates made (by the engineer) at periodic intervals during the progress of the work for determining the amounts of partial payments to be made to the contractor. On large contracts, such estimates are commonly made each month and, hence, are frequently called *monthly* estimates. Usually 10 to 15 percent of the amount due the contractor at any payment is retained to ensure the satisfactory completion of the work by the contractor and also as a fund to cover errors and contingencies.

A *final* estimate is one made (by the engineer) at the completion of the contract. The purpose of this estimate is to determine the amount of final payment due the contractor and also to determine the exact quantities that have been used. Such an estimate is essential when there are extras on the work or when the contractor is paid according to the number of units of work, such as cubic yards of excavation or square yards of pavement.

Qualifications of an Estimator A good estimator of construction costs should possess the following qualifications:

1. A knowledge of the details of construction work
2. Experience in construction work
3. A fund of information relating to materials required, labor-hours required, plant needed, overhead problems, and costs of all kinds
4. Good judgment in regard to different localities, different jobs, and different workmen
5. A good method for preparing an estimate
6. Ability to do careful, thorough, painstaking, and accurate work
7. Ability to collect, classify, and evaluate data relating to estimating
8. Ability to visualize all the steps during the process of construction

Good instruction will help a man to become a good estimator. He can be instructed in a good method of preparing estimates, he can be taught how to make the needed computations, he can be shown what items to include in the estimate, and he can be cautioned concerning errors that he may make.

He cannot be taught experience or judgment. He must obtain his own practical experience and develop his own faculties of judgment. Some men learn rapidly by their own experiences and the observed experiences of others, and other men learn slowly. Some men develop good judgment in their lines of work in a comparatively short time, and others require more time.

Collection of Data The collection, classification, and evaluation of data relating to estimating are important. The estimator should take notes and keep cost records on as many jobs as he can. The notes and data should be complete, especially in regard to quantities and prices of materials; labor conditions, work, and wages; weather and job conditions; plant costs; delays and their causes; and conditions relating to overhead costs. The data collected should be arranged in an orderly manner and conveniently filed for reference.

Tables and Diagrams Nearly all estimators use tables and diagrams to simplify their work. Some estimators like tables, others like diagrams, and others like both tables and diagrams. The use of both tables and diagrams should be learned, as available cost data may be in either tabular or diagrammatic form. Almost any set of data can be expressed either in the form of a table or in the form of a diagram. Interpolations of values are often made more easily on diagrams than in tables. The selection of tables or diagrams to be used as an aid in estimating is a question for each individual estimator to decide for himself after he has studied both. Both tables and diagrams are used in this text, though more emphasis is placed on diagrams.

Data in tables should be arranged in an orderly manner and divided by lines or spaces so that the table will be easy to read. Some estimators like the tables to be arranged so that most of the reading is across the table, and others prefer to read up and down. Complicated tables and tables so large that they extend over several pages should be avoided.

When the data are presented in the form of diagrams, the diagrams should be simple and easy to read and use. In most instances, it will be found that the desired relations between the variables can be shown in the diagram by straight lines instead of curves. Straight lines are easier to compute, plot, and read than curves. Though only two points are needed to determine any straight line, at least three points should be computed and plotted. All three points should be in the same straight line. If all three points are not in the same straight line, there is at least one error in the computations. The straight lines or curves in a diagram should not be too close together for ease in reading. Also, when practical, the straight lines or curves should cross the horizontal and vertical "grid" lines of the diagram at an appreciable angle, say about 20 to 70 deg.

Material Diagrams In the material diagrams, the relations among quantities, prices, and costs are readily seen at a glance. In most of the material diagrams, the cost of the material is usually given along the horizontal scale, the cost per unit to be used in estimating is given along the vertical scale, and each straight-line "curve" represents a certain quantity per estimating unit.

A typical simple material-cost diagram is shown in Fig. 1-1. This diagram can be used for estimating the lumber cost per square foot of form surface for concrete. The horizontal scale gives the cost of lumber in dollars per 1,000 fbm. The vertical scale gives the lumber cost in dollars per square foot of form surface. Each straight line represents a certain amount of board feet of lumber required for 1 sq ft of form surface.

To compute three points or values for one of the straight lines, consider that 4 fbm of lumber is required for 1 sq ft of form surface. At $40 per 1,000 fbm, the cost per square foot of form surface will be $40/1{,}000 \times 4$, or $0.16. For lumber costs of $100 and $200 per 1,000 fbm, the costs per square foot of form surface will be 40 and 80 cents, respectively. All three of these values, when correctly plotted, are on the same straight line. Values for other straight lines in this diagram can be found in a like manner.

For an illustrative example of the use of this diagram, assuming that lumber costs $135 per 1,000 fbm delivered at the job and that 2.66 fbm of

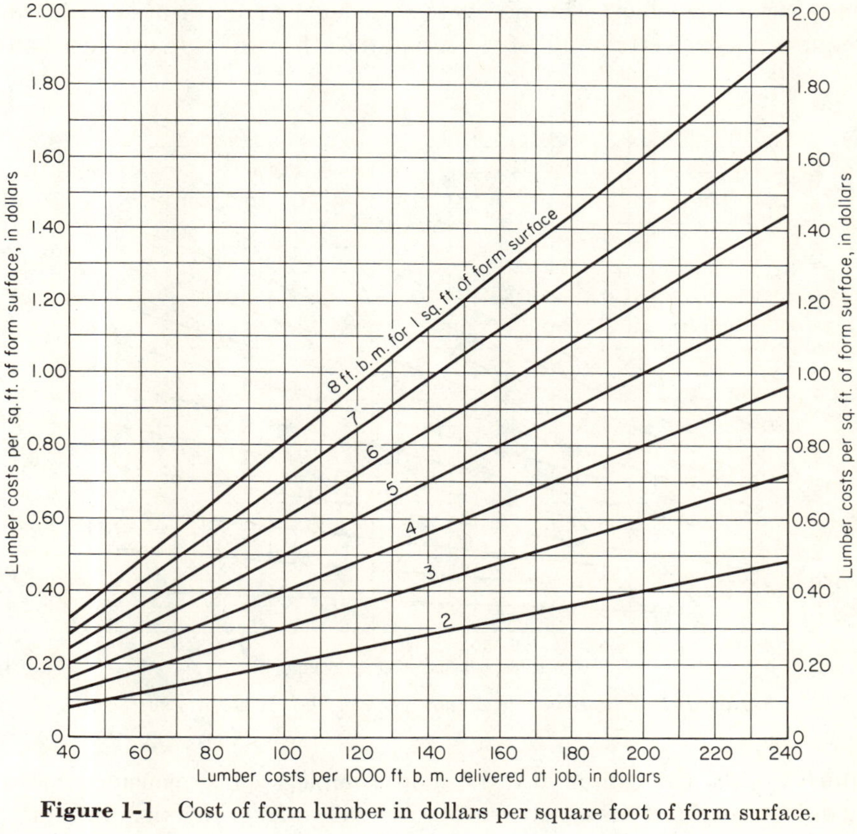

Figure 1-1 Cost of form lumber in dollars per square foot of form surface.

lumber is required for each square foot of form surface, the diagram shows that the lumber cost per square foot of form surface is 36 cents. If there were 380 sq ft of form surface, the cost of the lumber would be $0.36 × 380, or $137 (to the nearest dollar).

Labor Diagrams In the labor diagrams, the relations among wages, costs, and hours of labor required for certain items of work are shown. In this text, emphasis is placed upon wages per hour and hours of labor required, instead of upon wages per day and days of labor required. When the hour instead of the day is used as a base, the effects of workdays of varying lengths are eliminated.

When labor diagrams are prepared, the wages per hour are usually given along the horizontal scale, the unit costs along the vertical scale, and straight-line "curves" are drawn for the varying labor-hours required to do a unit of work or for the varying number of units of work done per labor-hour.

In the preparation of simple labor-cost diagrams, there are two ways of considering the labor. One way is to consider the number of labor-hours required to do a certain unit of work, and the other way is to consider how

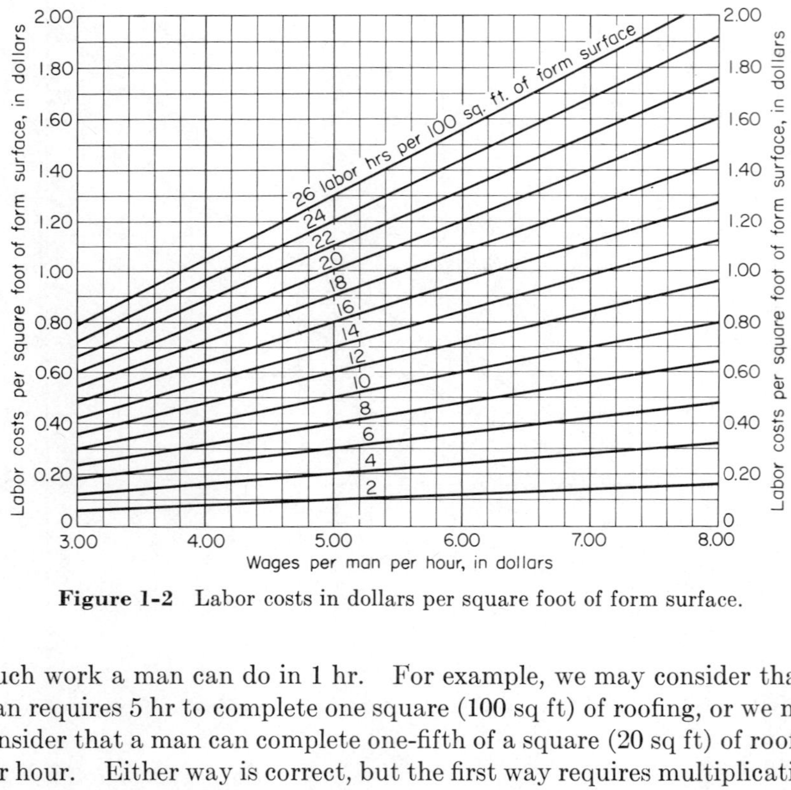

Figure 1-2 Labor costs in dollars per square foot of form surface.

much work a man can do in 1 hr. For example, we may consider that a man requires 5 hr to complete one square (100 sq ft) of roofing, or we may consider that a man can complete one-fifth of a square (20 sq ft) of roofing per hour. Either way is correct, but the first way requires multiplication, and the second requires division. For example, by using the rate of work previously mentioned, the labor-hours required for six squares (600 sq ft) of roofing would be $5 \times 6 = 30$ hr by the first way and $600 \div 20$, or $6 \div \frac{1}{5}, = 30$ hr by the second way. The first way is preferable because multplication is usually easier than division and also because most material quantities and costs are found by multiplication. If the numerical work is multiplication, there is less chance of error than when both multiplication and division have to be used.

A typical simple labor-cost diagram is shown in Fig. 1-2. This diagram can be used for estimating the labor costs for forms for concrete. The horizontal scale gives the wages per man per hour in dollars, and the vertical scale gives the labor cost per square foot of form surface in dollars. Each straight line represents a certain number of hours of work required for 100 sq ft of form surface. If desired, the straight lines in this diagram could be drawn to represent the number of square feet of form surface per hour.

To compute three points or values for one of the straight lines, assume that it takes 10 hr for a man to assemble and erect 100 sq ft of form surface. If his hourly wage is $4, the cost per square foot of form surface is $4 times 10 divided by 100, or 40 cents. For hourly wages of $6 and $8, the costs per square foot of form surface are 60 and 80 cents, respectively. When plotted correctly, all three of these values will be on the same straight line. Values for the other straight lines can be found in a like manner.

For an illustrative example of the use of this diagram, assuming that the hourly wage is $6 and the rate of work for assembling and erecting forms is 9 man-hours for 100 sq ft of form surface, the diagram shows that the labor cost is 54 cents per square foot of form surface. If there were 380 sq ft of form surface, the labor cost for assembling and erecting the forms would be 54 cents × 380, or $205 (to the nearest dollar).

Preliminary Investigations (Visiting the Site) Before preparing an estimate, the estimator should visit the site (or have a responsible party visit it for him) and make a study of the conditions there. For example, if the construction of a large building is planned, the estimator or his representative should visit the site; note the location of the proposed building; get all available data regarding the soil; make a sketch of the site showing all important details; decide on the best location for the plant and the materials storage yard; partly plan the construction plant (visualize it); obtain information concerning light, power, and water; obtain information regarding local ordinances and regulations; secure information concerning banking facilities; note conditions of roads and streets leading to railway yards and to material dealers; investigate cost of local haulage; consult with material dealers regarding prices and deliveries (if local dealers are to be used); study local labor conditions in regard to wages, supply, and general efficiency of local workmen; and investigate boarding and rooming places for workmen (if workers have to be imported).

Five Subdivisions of Estimating The work of estimating may usually be divided into five essential parts, or subdivisions, as follows:

1. *Materials.* Estimating quantities and costs
2. *Labor.* Estimating labor-hours and costs

3. *Plant.* Estimating amount and kind of equipment needed and costs
5. *Overhead.* Determining items included and estimating costs
5. *Profit.* Estimating percentage for time, place, and job

In some estimates, one of these five main items may be negligible, and some subitem may become large enough to be considered by itself. For example, in estimating excavation, the cost of the soil excavated may be zero, but the cost of transportation of this soil is large and may be considered as a main item.

Estimating Materials The preparation of a list of materials is known as the *takeoff*. The estimator takes off the various materials from the plans and specifications and tabulates on *quantity sheets* all these different materials as to kind, number, size, weight, volume, or other units used. If the job is a large one and many quantity sheets are required, the total quantities for the various items are brought forward in condensed form to summary sheets and totals obtained for each kind or class of material. These totals are used in preparing the final estimates.

In some places, a *quantity surveyor* prepares a complete estimate or list of the quantities of all the materials required (called a *quantity survey*), and the contractors use this list or estimate when preparing their bids.

When estimating material costs, the costs of the materials at the job are usually used. This cost includes first cost, freight, unloading, cartage, storage, inspection, testing, and insurance.

Estimating Labor When labor estimates are prepared, due allowances must be made for variations in wages, for variations in the length of time (hours) required to do a certain item of work, for variations in working conditions, and for variations in the classes (skilled and unskilled) of labor required for different kinds of work. Some allowance should also be made in regard to the probable labor supply and demand.

Several years ago, the day was usually selected as the unit of time and labor estimates were based on the amount of work done per day and the daily wages. At the present time, the hour is the most satisfactory unit of time because the effects of workdays of varying lengths (6, 7, 8, 9, or 10 hr) are eliminated and also because most wages are now expressed as hourly wages. The hour is used in this text as the unit of time for estimating labor.

The length of time required to do a certain piece of work may vary considerably owing to the skill and mental attitude of the workman and also to the working conditions on the particular job. In some localities, union rules and regulations tend to limit the hourly output of the worker. When work is plentiful, labor is scarce, and jobs are easy to get, the time required for a laborer to do a certain unit of work will probably be more than the average, and vice versa. However, when a contractor has had a number of workmen working for him for a few months, his experience should enable him to estimate quite closely the length of time that any of these workmen will need to do a certain job. These same considerations hold when estimating the length of time required to do a certain item of work by a gang of workers.

When estimating wages, consideration must be given to the different kinds of work involved, to the different classes (skilled and unskilled) of laborers needed, and to their prevailing hourly wages. Hourly wages of various classes of labor (electricians, carpenters, plumbers, skilled and unskilled) vary considerably in different parts of the country. Local customs or union regulations often require the use of certain classes of labor for certain work. The local demand for and supply of labor will probably cause some variations in wages. Hourly wages will also vary from year to year, and in some localities and for some trades there will be seasonable variations.

Some estimators like to estimate the labor costs of doing a piece of work by multiplying the units of work by the estimated labor cost per unit. They vary, from time to time in their own minds, the labor cost per unit according to the efficiency of their workmen and the prevailing labor wage. This method is simple but is usually not very accurate.

Other estimators like to estimate the labor-hours (or days) required per unit of work separate from the labor wage. Then, to obtain the estimated labor cost of the work, they multiply the labor-hours (or days) required per unit of work by the wage per hour (or per day) and then by the units of work. This is the better method. The work of computing can be lessened by the use of diagrams in which both variations in the lengths of time required to do a certain item of work and variations in the hourly wages are considered.

Estimating Plant A complete construction plant or equipment for any construction job will include all temporary buildings and structures, machinery, and tools necessary for the work. For example, a construction plant for a reinforced concrete building will include machinery and tools for making forms, for bending and cutting reinforcement, for storage and handling of materials, and for mixing, transporting, and placing concrete. All the construction plant may be located at the job, or part of it may be at the job and other parts at convenient places. In this text, the terms *plant* and *equipment* are used synonymously.

The selection of the plant will usually depend upon the machinery and tools that the contractor has available for the job. In some instances, new plant or equipment must be purchased. Some machines can be used for only one job, and other machines can be used over and over again on many jobs. Consequently, consideration should be given to the estimated "life" of any particular machine, the length of time it will be used on the job, and the amount of work to be done.

Plant costs usually include costs of ownership (or rentals), transportation, erection, moving, dismantling, and operation. These costs may include the labor cost of machine operators, such as hoisting engineers and foremen.

Unit costs may be based on time or on output. For example, the cost per hour can be computed by dividing the total plant costs by the number of hours that the plant is held available for the job or else by dividing by the number of hours that the plant is actually operated on the job. Some estimators use two plant-hour rates: one when the plant is operating and the other when the plant is idle. When based on output, the total estimated plant cost is divided by the total number of units produced. For example, in constructing a reinforced concrete bridge, the total plant cost would be divided by the total number of cubic yards of concrete to obtain the plant costs per cubic yard of concrete.

Information in regard to plant or equipment costs for different kinds of work is given in the various chapters of the text in which the estimates for these kinds of work are discussed. A general discussion of construction-plant economics, design, operation, and costs is given in Chap. 18 of this text. For ownership costs and rentals, see Appendix B and C.

Estimating Overhead. Overhead costs are usually divided into two classes: general overhead costs and job overhead costs.

General overhead costs include all costs that cannot readily be charged to any particular job. General overhead costs may include such items as office rent, office equipment and supplies, utilities, telephone, general insurance, taxes, interest, legal expenses, traveling, sundries, plans, and specifications. General overhead costs may also include the salaries (or parts of salaries) of general officers, managers, estimators, engineers, stenographers, clerks, draftsmen, etc., that cannot readily be charged to a particular job.

Job overhead costs may include all costs which can readily be charged to the job but which cannot be charged to materials, labor, or equipment. Some such items are job insurance, public liability insurance, job telephone, extra sets of plans and specifications, surveys, bonds, permits, inspection costs, architect's and engineer's fees, part of wages of superintendent or general foreman, etc.

Some estimators include payroll taxes and insurance in overhead costs, but most estimators add these items to the labor costs. These items usually include Social Security (FICA), Medicare, Federal Unemployment Insurance (FUTA), State Unemployment Insurance, Workmen's Compensation Insurance, Contractor's Liability Insurance, and Health and Welfare charges. The sum of these items may amount to 8 to 25 percent of the labor payroll, and they should be added to the labor costs.

Total overhead costs may vary from 8 to 30 percent of the sum of materials, labor, and equipment costs or from 12 to 50 percent of the labor costs, depending on the kind of job, locality, and items included.

Overhead expenses are sometimes apportioned to the several parts of a job according to the sum of the materials, plant, and labor costs of these parts. Another method is to apportion the overhead expense according to the labor-hours or the labor costs of the several parts of the job, though occasionally some of the overhead expense may be assigned to the materials or to the plant. On small jobs, the overhead expense is usually figured as a certain percentage of the labor costs or as a certain percentage of the sum of material, labor, and plant costs. Any reasonable method of computing and apportioning overhead expenses will be satisfactory if the method used is properly understood and correctly applied.

Some discussion of overhead costs is given in each of the following chapters that relate to the estimating of costs for particular types of work (concrete, masonry, steel, plumbing, etc.) A general discussion is included in Chap. 19.

Estimating Profit Last, but not least in importance, is the profit. The amount of profit is usually expressed as a percentage of the total estimated cost of the job. This percentage usually varies from about 8 to 15 percent, depending on the contractor's desire for the work, what he thinks is fair, and what he thinks he can get. For small jobs, 15 percent is a common figure for profit, with 12½ percent for medium jobs, 10 percent for large jobs, and possibly 8 percent for very large jobs. The percentage of profit added also depends, to some extent, on the risks and unforeseen difficulties of the job and on how often payments are to be made and in what amounts. On large jobs, payments up to about 85 percent of the total work done may be made each month.

Sources of Error There are many ways in which errors may enter into an estimate. Some of the more common errors are the following:

Errors in arithmetic. Addition, subtraction, multiplication, division, and decimal points.

Errors in copying items from one paper to another.

Omission of items of materials, labor, plant, or overhead. The items should be checked against a reminder list. For example, one contractor omitted one floor in a 10-story building estimate, one forgot the interest charge on the money he had to borrow, another omitted the accident and liability insurance, another forgot his foreman's wages.

Errors in estimating the lengths of time required to do certain items of work by different workmen and gangs.

Errors in estimating wages of men.

Errors in estimating quantities of materials.

Errors in estimating wastage of materials.

Errors in estimating prices of materials, including effects of rising, falling, and steady markets on prices.

Errors in estimating costs of transporting, handling, and storing materials.

Omission of plant charges.

No allowance for delays due to bad weather, labor troubles, lack of arrival of materials, breakdown of machinery, etc.

Forgetting to check all computations and results.

Omission of profit.

Checking the Estimate All details of an estimate should be carefully checked either by the estimator or preferably by another person. Any results that appear unreasonable should be recomputed. Omissions and errors in computations and copying should be watched for.

Many estimators use a reminder list, checking off each item as its cost is computed. Separate lists should be prepared for different kinds of jobs, say one list for a reinforced concrete building job, another list for a highway job, another list for a steel-bridge job, etc. These reminder lists are very helpful. However, it is practically impossible to prepare a reminder list that will be absolutely complete. The estimator should not depend entirely on a reminder list but should study the plans and specifications carefully for little details and inconspicuous clauses. A typical reminder list is given in Appendix A.

Summary After all the five subdivisions of the estimate have been carefully prepared and the computations and results in each of these subdivisions have been checked, the results can be summarized as follows:

Total estimated cost	
Materials.	$
Labor.	
Plant.	
Overhead.	
Profit.	
Total.	$

CHAPTER TWO

Handling and Transporting Materials

The Cost of Handling and Transporting Materials The cost of moving materials from one place to another is often a large part of the total cost of construction. The work required for moving materials from one place to another consists of handling (unloading, loading, piling, and repiling) and transporting the materials, as the case may be. This work may be done partly or wholly by hand labor, by machinery, or by combinations of hand labor and machinery.

The cost of handling, transporting, or combined handling and transporting of materials may be divided into four parts: labor, equipment or plant, overhead, and profit. In combined handling and transporting, each of these four items may be subdivided into handling and transporting if desired.

The cost of the materials is usually not included in handling and transporting estimates. If the cost of materials is to be included in the estimate, then this item must be added to the four items mentioned.

Labor Costs The cost of labor in the handling of materials usually consists of the labor costs of unloading, loading, and piling of the materials. When the labor gang is large enough so that a foreman is needed to supervise the work, his wages must be included in the labor costs.

Materials may be moved or transported from one place to another by carrying by hand; by hand trucks, carts, or barrows; by motor trucks, tractors, trailers, scrapers, etc.; by railways; by boats; by airplanes; or by combinations of various agencies.

The labor cost of transporting materials includes the cost of all labor (ordinary labor, machine operators, truck drivers, foremen, etc.) used in the work.

All payroll taxes and insurance, based on the labor wage, should be included in the labor costs. Payroll taxes and insurance, in the handling and transporting of materials, may vary from about 8 to 16 percent of the labor wage.

Equipment Costs Equipment or plant costs in the handling of materials include the costs of all machines and tools used for unloading, loading, and piling. The machines and tools may include such items as power shovels, cranes, tractors, conveyors, hoists, planks, runways, derricks, and hand and small tools such as gloves, shovels, picks, crowbars, pinch bars, hand hooks, saws, hammers, axes, and special small tools. Hand trucks, carts, or barrows may be needed on some jobs.

Ordinary hand tools rarely last the contractor more than one year because of being worn out, lost, or "missing." Hand trucks, barrows, and like tools may last up to two or three years, depending upon the amount and kind of use.

For transporting materials, the contractor may have and use his own equipment, he may rent equipment, or he may contract with others to do the work.

When the materials are transported by freight or express (by railways, boats, trucks, airplanes, etc.), such costs are usually included in the estimate as one item. However, when the contractor uses his own equipment, his transportation costs may be divided into labor, equipment, overhead, and profit if he desires.

When equipment or plant costs are computed, such items as the costs of ownership, operation, erection and dismantling, and transportation to and from the job should be included. Ownership and operation costs are discussed in Chap. 18 and Appendix B.

Overhead and Profit Overhead costs should include all such items as are mentioned in Chap. 1, page 10, that apply to the job for which the estimate is being prepared. Overhead costs may be based on the labor costs, on the equipment costs, or on the sum of the labor and equipment costs. When based on the labor costs alone, the overhead costs may vary from about 10 to 30 percent of the labor costs. When based on the sum of the labor and equipment costs, the overhead costs may vary from about 6 to 20 percent of this sum.

Profit is usually based on the sum of all other costs. The percentage allowed for profit may vary from about 8 to 15 percent.

Summary The total costs of handling and transporting materials may be summarized as follows. If unit costs (per cubic yard, per ton, per 1,000 fbm, etc.) are desired, these unit costs can be found by dividing the total costs by the corresponding number of units. If desired, the labor and equipment costs may be subdivided into handling and transporting costs.

Item	Unit cost	Total cost
Labor...............	$	$
Equipment..........		
Overhead...........		
Profit..............		
Total.............	$	$

Unloading, Loading, and Piling by Hand Labor Many construction materials are unloaded, loaded, and piled by hand. In this article, only a few materials will be considered, but the principles involved are the same for various other materials.

The time rate of work may be given in terms of the hours of labor per 100 units (sacks or block), per ton, per 1,000 fbm, per 100 lb, or per whatever unit is customary or convenient. If desired, the time rate of work may be given as the number of units (sacks, tons, cubic yards, etc.) per hour. The time required for a man to pick up a load and to lay it down will depend on several things such as the weight of the load, the size of the load, the ease of taking hold of the load, and the kind of material, as well as on the skill of the workman, his rate of work, and the ability of the foreman. Note that certain materials have to be handled more carefully than others. One-half minute is usually taken as the time required for a man to pick up and to lay down a load, and 1 min to pick up and to pile a load. These times are fair average values but may be too long or too short depending on the material and the working conditions.

A reasonably heavy load for the average workman is about 100 lb. Heavier loads may be handled by two or more workmen. Awkward loads, such as large light boxes and long pieces of lumber, may often be more efficiently handled by two workmen.

Perhaps the simplest way of preparing tables or diagrams for estimating the costs of unloading, loading, and piling materials is (1) to estimate the quantity (sacks of cement, number of bricks, tile, or board feet of lumber) or weight (pounds) handled at one time in one load, (2) to estimate the time required for handling one load, (3) to compute the time required for handling a given quantity or weight (say 1,000 articles or 1 ton), and (4) to compute the corresponding labor costs. Varying rates of work must be considered as well as varying hourly wages.

Table 2-1 gives the approximate times in hours required for the picking up and laying down or piling of some materials of construction. The unit selected may be sacks, pieces, board feet, or pounds as the case may be.

These time estimates are for such work as moving material from car to truck, from truck to car, from car to ground, from truck to ground, from ground to truck, or from one place to another on about the same level and where the distance traveled is small, say not more than about 10 ft. Materials may be laid down or piled.

Diagrams. For convenience in preparing estimates of labor costs of picking up, taking a few steps, and laying down or piling various materials, diagrams can be constructed (see Diagrams 2-1 and 2-2 in Appendix F). These diagrams give the labor costs for 1,000 units of any material when the hourly wage is known and the time in hours required to handle 1,000 units can be reasonably assumed (see Table 2-1). The material units may be brick, pounds, pieces, sacks, board feet, etc., as the case may be. If desired, diagrams could be constructed using units or articles per hour instead of hours per 1,000 units.

Table 2-1 *Time Required to Pick Up Materials, Take a Few Steps, and Lay Down or Pile Materials*

Material	One-man load	Approximate number of articles per hr		Approximate time in hr per 1,000 articles	
		Pick up and lay down	Pick up and pile	Pick up and lay down	Pick up and pile
Cement	1 sack	100–170	70–125	6–10	8–14
Block and tile:					
Large	1 tile	100–170	70–125	6–10	8–14
Small	2 tile	140–250	115–200	4–7	5–9
Brick:					
Hands	2–4	400–700	300–500	1.5–2.5	2.0–3.5
Tongs	4–10	700–1,200	500–1,000	0.8–1.4	1.0–2.0
Steel bars:					
Bundled	50–120 lb	3.500–10,000	2,000–5,000	0.1–0.3	0.2–0.5
Loose	40–100 lb	2,500–5,000	1,500–3,500	0.2–0.4	0.3–0.7
Lumber	20–40 fbm	1,200–3,500	800–2,500	0.3–0.8	0.4–1.2

Illustrative Estimates The three illustrative estimates in this article show how estimates may be prepared for handling different kinds of materials.

Illustrative estimate. Prepare an estimate of the labor costs of unloading 800 sk of cement from a freight car and piling these sacks in a storehouse. Labor wage is $2 per hour. The freight car is "spotted" on a side track in front of the storehouse door. The average distance of carrying the cement from the car to the storehouse is about 15 ft. Payroll taxes and insurance are 10 percent of the labor wage.

By assuming good workers, the time required may be taken as about 12 hr per 1,000 sk (see Table 2-1). Then the time required for 800 sk would be $(12 \div 1{,}000) \times 800$, or 9.6 hr. The labor costs would be $\$2 \times 1.10 \times 9.6$, or $21.12.

By using Diagram 2-2, the labor costs per 1,000 unit or sacks are found to be $26.50, or $0.0265 per bag. Then the labor costs for 800 sk would be $\$0.0265 \times 800$, or $21.20.

Illustrative estimate. Prepare an estimate of the labor costs of unloading 8,500 face brick from freight cars to trucks. Trucks can back up to car doors so that the brick can be carried from the cars and piled on the trucks. Brick tongs will be used, and 6 brick will be handled at a time. Labor wage is $3.30 per hour, including payroll taxes and insurance.

The time required to pick up the brick, take a few steps, pile on the truck, and return may be taken at about 1.5 hr per 1,000 brick, fair workmen and no delays being assumed. The time required for 8,500 brick would be $(1.5 \div 1{,}000) \times 8{,}500$, or 12.75 hr. The labor costs would be

$$\$3.30 \times 12.75 = \$42$$

By using Diagram 2-2, the labor costs per 1,000 units or brick are found to be $5.00. The labor costs for 8,500 brick (8.5 thousand) would be $\$5.00 \times 8.5$, or $42.50.

Illustrative estimate. Prepare an estimate of the labor costs of unloading 68,000 fbm of lumber from flatcars and piling it on ground near the track. The lumber is in the form of 1-in. boards, 2 by 4s, and 2 by 6s, varying in length from 12 to 16 ft. Labor wage is $3.30 per hour, including payroll taxes and insurance.

On the assumption that the men work in pairs and that they are average workmen, the time (labor-hours) required to unload and pile 1,000 fbm may be taken as about 0.8 hr (i.e., 2 men working together would require 0.4 hr for 1,000 fbm). Total time (labor-hours) required would be $(68{,}000 \div 1{,}000) \times 0.8$, or 54.4 hr. The labor costs would be $\$3.30 \times 54.4$, or $180.

By using Diagram 2-1, the labor costs per 1,000 fbm are found to be $2.65. The labor costs for 68,000 fbm would be $\$2.65 \times 68$, or $180.

Shoveling The cost of shoveling depends upon the kind of material (heavy, medium, or light), condition of material (loose or compact, dry or wet), vertical distance or lift, kind of shovel used, as well as upon the skill and inclination of the workmen. Working conditions, such as sureness of footing and freedom of movement, also affect the rate of work.

The presence of a foreman is usually desirable even on small jobs so that the work will be properly supervised and the men kept busy. With small gangs, say about five men or less, the foreman can work with the men. The amount of shoveling that the foreman can do with a small gang will depend largely upon the time required of him to lay out and direct the work. However, with larger gangs, the work will usually proceed more

Table 2-2 *Shoveling Loose Materials*

Material	Work	Cu yd		Tons	
		Cu yd, per hr, per man	Labor-hr per cu yd	Tons per hr per man	Labor-hr per ton
Sand	From car or truck to ground, from car to truck, or from pile to pile at same elevation	1.50–3.00	0.35–0.70	2.00–4.00	0.25–0.50
Crushed stone		1.00–2.50	0.40–1.00	1.25–3.00	0.35–0.80
Gravel		1.00–2.50	0.40–1.00	1.25–3.25	0.30–0.80
Cinders		2.00–4.00	0.25–0.50	1.75–3.50	0.30–0.60
Loam		1.25–2.75	0.35–0.80	1.65–3.50	0.30–0.60
Sand	From ground to car, truck, or wagon where lift is 4 ft or less	1.25–2.50	0.40–0.80	1.65–3.50	0.30–0.60
Crushed stone		0.75–2.25	0.45–1.35	0.90–2.75	0.35–1.10
Gravel		0.75–2.25	0.45–1.35	0.95–3.00	0.35–1.05
Cinders		1.50–3.50	0.30–0.65	1.25–3.25	0.30–0.80
Loam		1.00–2.50	0.40–1.00	1.35–3.35	0.30–0.75
Medium soil		0.75–2.00	0.50–1.35	1.00–2.65	0.40–1.00
Hard soil or clay		0.60–1.50	0.65–1.65	0.80–2.00	0.50–1.25
Hard pan		0.50–1.25	0.80–2.00	0.70–1.65	0.60–1.45
Sand	From ground to car, truck, or wagon where lift is 4 to 6 ft	0.75–2.00	0.50–1.35	1.00–2.75	0.35–1.00
Crushed stone		0.50–1.75	0.60–2.00	0.65–2.25	0.45–1.50
Gravel		0.50–1.75	0.60–2.00	0.65–2.50	0.40–1.50
Cinders		1.00–3.00	0.35–1.00	0.85–2.75	0.35–1.20
Loam		0.65–2.00	0.50–1.55	0.85–2.75	0.35–1.20
Medium soil		0.50–1.50	0.65–2.00	0.65–2.00	0.50–1.55
Hard soil or clay		0.40–1.25	0.80–2.50	0.55–1.65	0.60–1.80
Hard pan		0.30–1.00	1.00–3.35	0.40–1.35	0.75–2.50

rapidly if the foreman directs the work and keeps the men busy but does no shoveling himself.

There is a great variation in the rates at which workmen will work. It is easier to shovel from smooth ground or a smooth platform than from a rough or muddy surface. The proper size and kind of shovels or forks should be provided. The shovel should be of such size and shape that it will hold a load of about 22 lb. There is a knack to shoveling which a workman must learn (or be taught) before he can become efficient.

If there is much shoveling to be done on a construction job, the foreman in charge of the shoveling gang should select the shovels to be used for different materials and show how these shovels should be handled for the best results. In some cases, it may pay to make time and motion studies of shoveling.

The unit of work is usually the cubic yard with the materials measured in a loosened condition, though the materials can be measured in a compact form in some cases. However, when materials are purchased by weight instead of volume, the ton of 2,000 lb can be used as the unit of measurement.

Table 2-2 shows variations in shoveling rates under different working conditions and using men of varying efficiency.

When the total lift is over 6 ft, platforms can be used so that any one lift or stage will not exceed 5 or 6 ft.

In Table 2-3, the approximate weights per cubic yard of some of the materials are given. The weights per cubic yard will vary with different kinds of materials and will depend on their looseness or compactness and also on their wetness or dryness.

Table 2-3 *Weight per Cubic Yard*

Material	*Weight per cu yd, lb*
Sand	2,200–3,000
Gravel (screened)	2,200–2,800
Gravel (bank run)	2,400–3,200
Crushed stone	2,000–2,700
Cinders (very variable)	1,600–2,500
Loam or dirt	2,200–2,800

Diagram. Diagram 2-3 (Appendix F) can be used for finding the cost per cubic yard or per ton of shoveling materials when the hourly wage is known and the labor-hours required to shovel 1 cu yd or ton of material can be assumed correctly.

Illustrative estimate. Prepare an estimate of the labor cost of unloading a railway car containing 45 tons of gravel. The gravel is to be shoveled from the car to trucks. The labor wage is $3 per hour, including payroll taxes and insurance.

On the assumption that there would be no delays on account of the trucks and that the men of the gang of shovelers are good workmen, about 0.75 labor-hour will be required for shoveling 1 ton of gravel into the trucks. Then the cost of unloading 45 tons will be $\$3 \times 0.75 \times 45$, or $101.

With the use of Diagram 2-3, the labor cost of unloading 1 ton of gravel is found to be \$2.25. The labor cost of unloading 45 tons would be \$2.25 × 45, or \$101.

Unloading and Loading Materials with Power Equipment There are many types of power equipment available for handling materials. These types may be classified according to kind such as belt or bucket loaders and conveyors, small power-driven loaders such as tractors with front-end loader attachments, cranes, power shovels, derricks, and draglines.

In this book, no attempt will be made to describe all the various types of power equipment used for handling materials. For such descriptions, the reader is referred to books on construction equipment and to the catalogues of the various manufacturers.

Perhaps the best classification is the capacity of the machine in hours per unit of work or work units per hour. In the absence of better data, the manufacturer's rated capacity can be used and modified as required by the particular job. The manufacturer's rated capacity is the hours required to do a certain amount of work or the amount of work that the machine may be reasonably expected to do per hour when operated by a good workman and under average working conditions. On most jobs, the actual working time may vary from about one-fourth to three-fourths of the working hours because of necessary machine movements and delays caused by repairs, waits (as for trucks), and other reasons.

When estimating the time in hours required to move a certain quantity of a material, it is necessary to know the load per movement of the machine and the number of movements per minute or per hour. It is difficult to make an accurate estimate unless the estimator knows the machines, crew, and materials and can make an accurate guess as to working conditions.

Belt conveyors have belts 18, 24, or 30 in. wide and of varying lengths.

Bucket loaders are of different sizes with capacities ranging from 1 to 8 cu yd per minute.

Four-wheel tractors, with front-end loaders, may have buckets or scoops of ⅓- to 2-cu-yd capacity. Crawler tractors, with front-end loaders, may have buckets or scoops of ½- to 9-cu-yd or more capacity, depending upon the size and power of the tractor. Four-wheel tractors with fork lifts may have capacities ranging from about 1,000 to 15,000 lb or more; lifts of about 10, 15, or 20 ft; and speeds up to 20 mph.

Truck-mounted shovels may have dippers varying from ⅜ to 1 cu yd in size. Crawler-mounted shovels may have dippers varying in size from ⅜ to 5 cu yd and more.

Truck-mounted cranes may have capacities up to 25 tons with a 10 ft radius. Crawler-type cranes may have capacities up to 25 tons or more with a radius of 10 to 45 ft.

Clamshell and orange-peel buckets may vary in size from 3/8 to about 4 cu yd.

Derricks are of various sizes with capacities varying from about 2 to 50 tons, with booms from 30 to 100 ft and masts from 30 to 60 ft.

Crawler draglines may have buckets up to 4 cu yd or more in size and booms varying in length up to 75 ft or more.

When handling loose material, like sand, gravel, or crushed rock, most tractor loaders, shovels, and cranes may handle from one to four dipper, bucket, or scoop loads per minute. The smaller sizes of power equipment will usually operate at a faster rate than the larger sizes. The number of loads per minute will also vary with the movements required of a tractor loader and with the angle of swing of a shovel or crane.

Table 2-4 *Approximate Outputs of Power Shovels and Tractor Loaders Handling Loose Materials*

Dipper capacity, cu yd	Cu yd per hr	Hr per 1,000 cu yd	Dipper capacity, cu yd	Cu yd per hr	Hr per 1,000 cu yd
1/2	30–100	33.3–10.0	1 3/4	105–245	9.5–4.1
5/8	38–115	26.7–8.7	2	120–270	8.3–3.7
3/4	45–130	22.2–7.7	2 1/2	145–320	6.9–3.1
1	60–160	16.7–6.2	3	170–370	5.9–2.7
1 1/4	75–190	13.3–5.3	4	210–460	4.8–2.2
1 1/2	90–225	11.1–4.5	5	250–550	4.0–1.8

Almost every machine will require the services of a laborer to keep it supplied with material and to keep it operating efficiently. A large power shovel will need an operator and perhaps one or more laborers to assist in shovel movements and to do the needed hand shoveling, etc.

When the material (such as bricks, blocks, sacks, cartons, etc.) can be stacked on suitable pallets, the use of pallets and fork-lift trucks will often result in a saving of labor, time, and expense in the loading, unloading, piling, and moving of the materials in the materials yard or on the job.

Table 2-5 gives information about several types of equipment used in handling and transporting materials.

Information on various types of power equipment used for excavation is given in Chap. 3 on Excavation.

Table 2-4 gives approximate outputs of power shovels and tractor loaders handling loose materials.

The costs of power equipment used for the handling of materials includes the costs of ownership (or rentals), operation, transportation to and from the job, and sometimes erection and dismantling. Perhaps the best way of estimating these equipment costs is first to compute the ownership and operating costs per working hour as explained in Chap. 18 on Construction Plant and Equipment and in Appendix B on Equipment Costs and then add transportation, erection, and dismantling costs.

Diagrams. Simple diagrams can be prepared showing the relations among total hourly costs, hours required to do certain amounts of work or amounts of work per hour, and unit costs. Diagrams 2-4 and 2-5 (Appendix F) show the costs of handling (loading or unloading) 1 cu yd or 1 ton of material (such as coal or sand) by power equipment (such as power shovels or cranes and buckets). Diagram 2-6 (Appendix F) shows the cost of handling 1,000 units (such as sacks or block) by power equipment (such as a belt or other type of conveyor).

Illustrative estimate. Estimate the cost of loading 840 cu yd of gravel from a pile to trucks of 4-cu-yd capacity by a diesel crawler shovel with a 1-cu-yd dipper. The shovel is at the job, so there are no transportation costs for it.

Under good working conditions this shovel should handle 3 dipper loads per minute, thus requiring 1.33 min to load a truck. Assuming that it takes about 0.67 min for a loaded truck to move out and an empty truck to move in and that about 0.5 min will be lost in delays, the total time required for loading a 4-cu-yd truck will be 2.5 min. That is, 24 trucks (60 ÷ 2.5) will be loaded each hour.

Hours required for the job = 840 ÷ (24 × 4) = 8.75

Labor cost: With the operator's wages at $3.30 per hour, including payroll taxes and insurance, the labor cost is

Labor cost	= $3.30 × 8.75	= $29

Equipment cost: The capital investment in the shovel is $24,000. Appendix B gives the ownership cost per hour as 0.0244 percent of the capital investment, and an operating cost of $0.85 per hour.

Hourly cost	= $24,000 × 0.0244% + $0.85	= $6.71
Equipment cost	= $6.71 × 8.75	= $59
Overhead, assume 17 percent of the labor cost		= $5
Profit at 10 percent of all other costs		= $9
Total cost		= $102
Cost per cubic yard		= $0.12 (about)

Transporting Materials by Carrying Transporting materials by carrying is usually not very efficient. However, carrying is used on some jobs where the amount of materials is comparatively small or where the nature of the work is such that other methods cannot be readily used.

The amount of material that a man can carry will depend upon the kind of material and the condition of the ground or surface over which he must walk. An average man can carry a load of about 100 lb at a speed of about 100 fpm (a little over 1 mph). If the load is awkward or bulky or hard to hold, the weight carried at one time may be less. Loads heavier than 100 lb and loads awkward for one man to handle may be carried by two or more men. If the surface is hard to walk over or if the grade is steep, a man will either carry a lighter load or walk slower. Under very favorable conditions, a man will walk faster than 100 fpm. He may walk as fast as about 200 fpm with a load and as fast as about 250 fpm without a load. Roustabouts loading or unloading river steamers have often averaged better than 300 fpm by going at a dogtrot. However, 100 fpm is usually taken as the average walking speed as this rate allows for minor delays.

Diagram. The cost of carrying materials by men can be shown in the form of a diagram, as in Diagram 2-7 (Appendix F). When this diagram was prepared, a man was assumed to take 1 min to pick up and lay down his load and 1 min to walk 100 ft. These assumptions allow for small delays. Hence, the time required to transport one load 200 ft and return would be $1 + 2 + 2$, or 5 min. Like reasoning gives 3 min for 100 ft, 7 min for 300 ft, 11 min for 500 ft, etc.

Illustrative estimate. Estimate the cost of carrying 220 sk of cement (1 sk weighs 94 lb) 160 ft. Assume that each man carries 1 sk each trip. Men receive $3.10 per hour, including payroll taxes and insurance.

Time required for one trip $= 1 + 1.6 + 1.6 = 4.2$ min
Time required for 220 trips $= 4.2 \times 220 = 924$ min $= 15.4$ hr
Cost $= \$3.10 \times 15.4 = \$48$
With the use of Diagram 2-7, cost of carrying 1 sk is about $0.22. Cost of carrying 220 sk $= \$0.22 \times 220 = \$48$.

Transporting Materials by Vehicles The vehicles ordinarily used by a contractor for transporting materials are wheelbarrows, two-wheel handcarts, scrapers, tractors, dump wagons, and autotrucks. Transportation by freight and express is considered in later articles.

The cost of transporting materials by a vehicle depends upon several things such as the cost of owning and operating the vehicle, the capacity, the load, the time required for loading and unloading, the speed of travel, and delays. The actual capacity of a vehicle is usually taken as about 80 percent of the measured or rated capacity to allow for variations in materials and loadings. The time required for loading and unloading will vary on different jobs and for different vehicles, different materials, and different methods of loading. The travel speed will depend upon the particular vehicle, the road or path, and traffic and other conditions. Hence, for accurate estimating the estimator must have fairly exact information in regard to the variables present.

In Table 2-5, information is given in regard to various vehicles used for transporting materials. This information is approximate only. Further information concerning vehicles used in excavation is given in Chap. 3, Table 3-7.

Table 2-5 *Vehicular Data (Approximate)*

Vehicle	Capacity, cu yd or tons	Economical haul, ft	Time, min		Rate of travel, mph	
			Load	Unload	Loaded	Empty
Wheelbarrows*	0.07–0.15	Up to 150	1.0–3.0	0.2–0.4	70–125	100–175
Handcarts, two-wheel*	0.07–0.15	Up to 150	1.0–3.0	0.2–0.4	70–125	100–175
Front end loaders:						
Four-wheel	0.33–2.00	Up to 1,500	0.5–1.0	0.2–0.5	4–15	6–20
Crawler type	0.33–9.00	Up to 1,500	0.5–1.3	0.2–0.7	3–12	4–15
Fork lift†	0.50–8.00	Up to 1,500	0.3–1.0	0.3–1.0	3–15	5–20
Autotrucks‡	2–20	Over 500	1.0–3.0	0.5–2.0	10–45	15–60
Truck-trailer unit‡	3–25	Over 1,000	2.0–6.0	1.0–3.0	10–40	15–45
Truck-wagon unit‡	3–40	Over 1,000	2.0–6.0	1.0–3.0	10–35	15–45
Tractor-drawn wagons‡	3–25	Up to 3,000	1.0–3.0	0.3–1.0	4–12	5–15

* Rate of travel of wheelbarrows and handcarts is in feet per minute. Wheelbarrows and handcarts are hand loaded. Other vehicles are power loaded.

† Capacities of fork lifts are in tons.

‡ Tractors and trucks may draw more than one vehicle.

Computation of Transportation Costs. The computation of transportation costs by autotrucks, for example, can be divided into four steps.

1. Compute total cost of truck per hour, including such items as ownership and operating costs, labor (driver), and overhead.

2. Compute time in hours required for one round trip. This time includes time for loading, unloading, delays, traveling loaded, and traveling unloaded. Traveling time depends upon speed and speed depends upon the truck, the road, the driver, and the traffic.

3. Compute cost per trip. This equals cost per hour times hours for trip.

4. Compute unit costs by dividing cost per trip by number of units of material carried (cubic yards, tons, sacks, etc.).

A beginning estimator is apt to estimate the average truck speed at too high a value and to estimate the time lost in delays at too low a value. Plenty of time should be allowed for loading and dumping. Time lost in delays may be caused by waiting to load, waiting to unload, punctures and other tire trouble, engine trouble, and accidents. Delays may average 10 min or more per trip.

Diagrams. Diagrams can be prepared showing the relations among the different variables such as distance or length of haul, speed of trucks, time for trip, time for loading and delays, cost per hour of truck, cost per trip, capacity of truck, and unit costs. As the relations among only three variables can be shown in a simple diagram, several diagrams are necessary.

Diagram 2-8 (Appendix F) shows the relations among length of haul loaded, time per trip (trucks hand loaded), and speed of trucks. When computing the time per trip, 1 min is allowed for dumping or unloading, 3 min for delays, and 20 min for hand loading, giving a total of 24 min, or 0.40 hr. Note that total distance is twice the length of haul loaded.

Diagram 2-9 (Appendix F) is similar to Diagram 2-8 except that the trucks are power loaded. In Diagram 2-9, the loading time is taken as 3 min, unloading 1 min, and delays 5 min, giving a total of 9 min, or 0.15 hr. Other diagrams can be prepared, using other time allowances for loading, unloading, and delays.

Diagram 2-10 (Appendix F) shows the relations among cost per hour of truck, time required for trip, and cost of trip. Unit costs can be found by dividing the trip cost by the capacity of the truck (cubic yards, tons, sacks, etc.).

Diagram 2-11 (Appendix F) shows the relations among cost per trip, capacity of truck, and unit costs per cubic yard or per ton for trucks of varying capacities.

Illustrative Estimates The following two illustrative estimates show how transportation estimates may be prepared.

Illustrative estimate. Estimate the cost of transporting cement per sack from a railway siding to the job, assuming the following data: Truck capacity is 120 sk of cement. Cost of truck and driver is \$6.25 per hour. Average speed is 24 mph. Length of haul is 8 miles. Time to load is 12 min. Time to unload is 12 min. Delays getting into position, for waiting turn, and for all other causes are 16 min per trip.

One round trip = 2 × 8 = 16 miles
Time to travel 16 miles at 24 mph = 40 min
Time for one round trip = 12(load) + 12(unload) + 16(delays) + 40(travel)
= 80 min = 1.33 hr
Cost of one round trip = \$6.25 × 1.33 = \$8.33
Cost per sack of cement = \$8.33 ÷ 120 = \$0.07

Illustrative estimate. Estimate the cost of transporting 120 cu yd of gravel from a material yard to a job 6 miles away. Truck capacity is 6 cu yd. Cost of truck and driver is \$7.25 per hour. Average speed is 30 mph. Truck is power loaded at material yard, and material is dumped at job. Time to travel 6 miles and return at 30 mph is 24 min. Assume time required for loading 3 min, unloading 2 min, and delays 4 min.

Total time for one round trip = 33 min = 0.55 hr
Cost of one round trip = \$7.25 × 0.55 = \$4

Number of trips = 120 ÷ 6 = 20
Total cost = \$4 × 20 = \$80
Cost per cubic yard = \$80 ÷ 120, or \$4 ÷ 6 = \$0.67

If desired, this estimate can be prepared by the use of Diagrams 2-9, 2-10, and 2-11. Although the time allowed for unloading is 1 min more and the time for delays is 1 min less than the time assumed when preparing Diagram 2-9, the total trip time will be the same. For a length of haul 6 miles (round trip of 12 miles) and a speed of 30 mph, Diagram 2-9 indicates that about 0.55 hr will be required. Diagram 2-10 shows that the cost of one trip will be about \$4, and Diagram 2-11 gives the cost per cubic yard to be about 70 cents.

Total cost = \$0.70 × 120 = \$84

Vertical Transportation of Materials Among the ways of moving materials vertically are carrying by men up inclines, stairs, or ladders; wheeling in barrows or carts up inclines; use of simple block and tackle and other types of hand hoists; and use of power hoists such as derricks, cranes, block and tackle with power vehicles, and elevators. The selection of the way to be used will depend on the equipment available, the job conditions, and the amount and kind of material to be elevated.

Carrying. A workman will carry loads up to about 100 lb up (and down) inclines at a speed of about 80 to 100 fpm. The steeper the incline, the slower the speed and the smaller the load. The vertical distance, however, will usually be less than half the inclined distance; hence, the vertical rate may be about 25 to 40 fpm. The rate of carrying loads up or down stairs may vary from about 40 to 70 fpm on an inclined line or from about 20 to 35 fpm vertically. A man will carry a load up or down a ladder at about 20 to 30 fpm. The cost of transporting materials vertically by carrying can be estimated by the methods given on page 22 of this chapter, using vertical rates of speeds of from 20 to 40 fpm as the case may be.

Hand Wheeling. A man can push a wheelbarrow or a two-wheeled cart up an incline at an average rate of 70 to 100 fpm measured on the incline or about 20 to 40 fpm vertically. The amount of material (load) carried by the barrow or cart will be less than that carried on the level; the steeper the incline, the less the load.

Hand Hoists. Two or more workmen with block and tackle or other types of hand hoists or hand elevators can be used for elevating materials on small jobs where the use of power equipment is not justified. The number of pounds of material that a workman can raise a certain number of feet in an hour will depend upon his ability, the efficiency of the machine, the kind of material, the dead load or tare, and the working conditions.

On the assumption that 1 manpower is the equivalent of ⅙ hp, one man should be able to do 33,000 ÷ 6, or 5,500 ft-lb of work in 1 min, or 330,000 ft-lb of work in 1 hr. On a hand-hoisting job, this energy can be used for loading, hoisting, and unloading. Of the energy used for hoisting, part will be used in overcoming friction and other machine losses, part in hoisting the container of the materials (dead load or tare), and part in hoisting the materials. Consequently, the part of the energy of the man used for actually hoisting the materials may vary from about 25 to 75 percent of his total energy. This will amount to about 80,000 to 250,000 ft-lb of work per hour. That is, one man should be able to elevate 4,000 to 12,500 lb of material a distance of 20 ft in 1 hr, depending on conditions.

The hourly cost of elevating materials by hand power will depend upon the hourly wages of the men, the cost per hour of the equipment used, and in some cases, the overhead costs. Hand-equipment costs will usually be quite small.

Power Hoists. The amount of materials that can be raised vertically by machine power will depend on the net load the machine can handle at one time, the distance the load is raised, the speed at which the machine operates, and the time required for loading and unloading. The machine or equipment costs should include such items as ownership (or rental), transportation to and from the job, erection and takedown, and operation. When the hoist is used for hoisting several kinds of materials (as during the construction of a several-storied building), the costs can be apportioned among the kinds of materials hoisted, preferably according to the number of hours that the machine is used for hoisting each kind of material. The labor costs may include the labor costs of loading and unloading materials on and off the machine, cost of machine operation (if not included in machine costs) and costs of erecting and taking down the machine (if not cared for otherwise). All overhead costs relating to the work should be included.

Because of the different types of hoisting machines, various kinds of materials, and different jobs and working conditions, it is impractical to attempt to prepare cost tables and diagrams. For example, two workmen with a small power chain hod elevator can elevate 2,000 to 3,000 brick a vertical distance of about 30 ft in 1 hr. A power-driven two-barrow elevator with an operator (and necessary workmen for loading and unloading) can hoist 8,000 to 12,000 brick a vertical distance of about 30 ft in 1 hr.

When preparing estimates of costs of hoisting materials by power equipment, the following procedure is suggested:

1. Estimate the hours that the hoist will be operated.
2. Compute the hourly cost of ownership and operation.
3. Compute total cost of ownership and operation for the job.

4. Compute costs of transportation, erection, and dismantling.
5. Compute all labor costs, including payroll taxes and insurance.
6. Compute overhead costs.
7. Compute profit if this is to be included.
8. Sum all costs.
9. Compute unit costs by dividing the sum of all costs by the total number of units of material hoisted.

Transporting Materials by Express The cost of transporting an article by express depends upon the distance transported, the weight of the package or shipment, the kind or class of material (ordinary, fragile, perishable, livestock, bulky, etc.), and the way it is packed or boxed.

First-class express rates are based primarily on distance. For determining the rates, the country is divided into blocks or areas and rates are established between blocks. The rates are not directly proportional to distance as they include terminal costs (pickup and delivery) as well as transportation costs. For any one package, the terminal costs would be practically constant and the transportation costs would vary with the distance traveled.

The first-class merchandise rate between two blocks, for example, is the rate in dollars per 100 lb for packages or shipments of 100 lb or more, and for ordinary materials so packed that the packages can be readily handled without damage. Materials requiring special care in handling are charged higher rates which may vary from 1.5 to 3 or 4 times the first-class rates. In a few instances, there are special rates less than the first-class rates.

For packages weighing less than 100 lb, the charge is less than that for 100 lb, but not in proportion. For packages or shipments weighing over 100 lb, the cost will be directly proportional to the cost of shipping 100 lb. For example, the express charges on a 250-lb shipment will be 2.5 times that on a 100-lb shipment.

Express shipments may be transported by ground (train, truck, etc.), by air (airplane), or by combined air and ground (airplane and truck, etc.).

Ground rates (1967) per 100 lb for first-class express shipments may be approximately equal to

$$\$3.00 + \$1.00 \text{ per } 100 \text{ miles}$$

For example, the approximate charge for a 220-lb shipment of first-class express transported 400 miles would be about

$$(\$3.00 + \$1.00 \times 4.00) \times 2.20 = \$15.40$$

Air transportation rates for first-class express shipments are about two to three times the corresponding ground transportation rates.

All express shipments are subject to minimum charges, which may vary at different times, in different states, and by different companies. The minimum charge (1967) is about $3 for a ground shipment, and about $5 or $6 for an air shipment.

As the regulations and rates for express shipments vary from time to time, the local express company should be consulted to obtain accurate and up-to-date information.

Transporting Materials by Parcel Post Packages may be shipped by parcel post subject to the regulations of the U.S. Postal Department. The parcel post rates are based primarily on the weight, dimensions, distance, and method of transportation. The transportation may be by ground (train, truck, etc.), by air (airplane), or by combined ground and air methods.

For the determination of the rates for distances transported, the country around any post office is divided into circular concentric zones whose center is the local post office. The local zone usually includes all territory covered by the carriers of that local city or village post office. Zones 1 and 2 combined have a radius of not over 150 miles from the local post office. The radii and distances of the parcel post zones are given in Table 2-6.

Table 2-6 *Zone Radii and Distances*

Zone	*Distance, miles*
1 and 2	Up to 150
3	150–300
4	300–600
5	600–1,000
6	1,000–1,400
7	1,400–1,800
8	Over 1,800

Each zone is divided into sections and zip-code areas to aid in the correct determination of the correct parcel post charges at the local post office.

The maximum weights and dimensions for ground transportation are: For the local zone and Zones 1 and 2, 40 lb and 100 in. in combined length and girth (measured at the widest point). For Zones 3 to 8, inclusive, 20 lb and 72 in. in combined length and girth (measured at the widest point).

The maximum weight and dimensions for air transportation are 70 lb and 100 in. in combined length and girth (measured at the widest point).

The rates for air transportation (or combined air and ground transportation) are two or more times the corresponding rates for ground transportation.

Parcel post regulations and rates are subject to change from time to time. Consequently the local post office should be consulted for up-to-date information when making a parcel post shipment.

Transporting Materials by Air Freight Air freight rates are based mainly on distances transported from airport to airport, weight of shipments, dimensions, kind of commodity, and value. There are additional charges for pick up, for delivery, and for some special services.

One of the larger air freight companies classifies its rates and charges somewhat as follows:

General Commodity Rates. These rates apply to all shipments except special commodities and commodity exceptions. The minimum charge is $6 or the charge for minimum weight, whichever charge is greater. General commodity rates vary from about $6 to $30 per 100 lb depending upon the distance from air port to air port.

Special Commodity Rates. Some commodities are subject to special rates, which may be more or less than the general commodity rates. The minimum charge is $6 or the charge for minimum weight, whichever charge is greater.

Exception Rates. These are rates which apply to certain shipments such as livestock, uncrated vehicles, human remains, etc.

Excess Value Rates. An additional rate is charged on a shipment when the declared value exceeds 50 cents per lb. This excess value rate is about 10 cents per $100 of excess valuation.

Volume Rates. Extra charges are made for bulky lightweight shipments which weigh less than about 7 lb per cu ft.

Pickup and Delivery Charges. Extra charges are made for either pickup or delivery services. These charges may range from about $0.50 to $1.25 per 100 lb. The minimum charge for any one shipment may range from about $1.25 to $2.50.

Additional Services. Most air freight companies offer additional services such as combined air and truck service, air and express service, air and parcel post service, ocean and air service, etc.

As air freight regulations, rates, flights, and other services change from time to time, the air freight company should be consulted before making a shipment so that up-to-date information may be obtained. The air freight company should also be consulted in regard to maximum weights, maximum dimensions, and other information in regard to air freight shipments.

Transporting Materials by Railway Freight Freight rates are based on the class (kind of material and way it is packed and shipped), quantity of material in a shipment (carload or less than carload), and the distance transported. All freight classifications and rates are controlled by the Interstate Commerce Commission.

For the purpose of determining freight rates, the railroads of the United States have been combined into three large groups. Each of these groups has adopted certain rates to be charged for all materials shipped in their territory. The Official (or Eastern) group includes all territory east of Lake Michigan and the eastern boundary of Illinois and north of the northern boundaries of Kentucky and North Carolina and, in some instances, the state of Illinois and the southern part of the state of Wisconsin. The Southern group includes all territory east of the Mississippi and south of the northern boundaries of Kentucky and North Carolina. The Western group includes all territory west of the Official and Southern groups.

Each of these groups has also adopted certain classifications or ratings for all the different kinds of freight. However, the class or rating may be different in different groups. The classifications in use in the three groups are called Official Classification, Southern Classification, and Western Classification. There is also a fourth classification, the Uniform Freight Classification, which is used when a shipment is transported over more than one group area.

In classifying freight, certain abbreviations and symbols have been used. A list of these follows.

Characters Appearing in Territorial Rating Columns

1	= first class	D1	= double first class
2	= second class	2½t1	= 2.5 times first class
3	= third class	3t1	= 3 times first class
4	= fourth class	3½t1	= 3.5 times first class
5	= fifth class	4t1	= 4 times first class
6	= sixth class	A	= class A
R25	= rule 25 class	B	= class B
R26	= rule 26 class	C	= class C
1¼	= 1.25 times first class	D	= class D
1½	= 1.50 times first class	E	= class E

Abbreviations Appearing in Descriptions of Articles

LCL	= less than carload	NOIBN	= not otherwise indexed by name
CL	= carload	Lb	= pounds
SU	= set up	Min wt	= minimum weight
KD	= knocked down		

An examination of the preceding list of characters shows that freight ratings are given by number, rule, and letter. In general, the higher the numerical rating, the lower the freight rate. Freight shipped in carload lots is given a higher numerical rating than freight shipped in less than carload lots. In addition to the many classes listed and their accompanying rates, there are many special commodity rates applying to certain kinds of freight and between different stations.

The reader is referred to his local freight agent for a list of articles (materials) and the classification or rating for each article.

All railway freight shipments are subject to minimum weights and minimum charges, which may vary for different companies. The minimum weight for any one shipment may be 4,000, 5,000 or 6,000 lb. The minimum charge for any one shipment is the charge for the minimum weight, but not less than some minimum amount (say from about $2 to $6).

For example, if the minimum weight is 5,000 lb and the minimum charge is $4, the charge for a 1,200-lb shipment with a rate of 35 cents per 100 lb would be

$$\$0.35 \times 50.00 = \$17.50 \qquad \text{(and not } \$0.35 \times 12.00\text{)}$$

Some railway companies do not like to handle small shipments and refer the shipper to a freight forwarding company.

As freight regulations, rates, and classifications vary from time to time, the estimater should always consult the local railway agent for up-to-date information.

Whenever practical, shipments should be made in carload lots, as the carload rates are much less than the less-than-carload (LCL) rates. The minimum weight for a carload of materials varies with the materials and also if the shipment originates in or out of the state. The maximum weight depends upon the weight or volumetric capacity of the car. Some approximate minimum weights for carload lots of some materials are given in Table 2-7.

Table 2-7

Material	*Minimum weight, lb*
Asphalt, pitch, tar	40,000
Brick	50,000–60,000
Cement, lime, in bulk	50,000
Cement, lime, in sacks, barrels, boxes	40,000
Concrete or clay block or tile	50,000–60,000
Coal in open cars	80,000
Iron and steel products, reinforcing steel and rolled-steel shapes	36,000–40,000
Lumber and lumber products	36,000
Timber piles	36,000
Plumbers' goods	30,000
Sand, gravel, rock, in open cars	80,000
Sand, gravel, rock, in sacks	60,000

When materials are shipped by freight over railroads, there may be other expenses in addition to the freight rates. In a few localities, the railroad will collect and deliver package freight. In most localities, the shipper or consignor must meet the expense of delivering the freight to the railroad company, and in some cases he must load the freight on the cars. At the destination, the consignee must pay for the haulage from the railroad, and in some instances he is expected to unload the freight. There may be switching charges when freight is loaded or unloaded from cars spotted or switched on sidings not located in the freightyards. Also, if a car is kept by the shipper (consignor) or consignee for more than a certain length of time (usually two days), demurrage charges must be paid. These charges are usually $4 per car per day after the two free days, or 48 hr.

The discussion so far has applied to freight shipped over railroads. Freight may also be shipped by boat. Freight rates by boat are often less than railroad rates, but the time required to transport a shipment a given distance is usually appreciably longer. When shipping by boat, the consignor usually delivers the freight to the wharf and the consignee takes the freight from the wharf at the end of the trip.

Transporting Materials by Truck Freight Less-than-carload freight can often be conveniently shipped by truck. The time required by trucks is sometimes less than the time required by railroads.

There appear to be two kinds of truck freight companies; highway common carriers and highway contract carriers. Highway common carriers usually have regular schedules and routes, and maintain a pickup and delivery service. Highway contract carriers usually do not have regular schedules and routes but transport freight by the truckload when and where there is a demand. They may or may not provide a pickup and delivery service.

Truck freight rates are about the same LCL railways freight rates for like goods, weights, valuations of shipments, distances transported, and other services rendered. Truck freight rates are approved by the Interstate Commerce Commission or by a state commission as the case may be.

As truck freight rates and classifications may vary in different parts of the country, the estimator should consult the local trucking company agent for correct information.

Freight and Express Forwarding Companies There are many freight and express forwarding companies which will take complete charge of a shipment from the time the shipment is given them by the shipper until the time the shipment is received by the consignee. Some forwarding companies specialize in small shipments (LCL and package shipments), others in large shipments (CL), and others in both small and large shipments.

The charge made by the forwarding company may include such items as pickup, sorting, loading, insurance, transportation, terminals, unloading, distribution or delivery, etc.

The estimator should consult the forwarding company for up-to-date information in regard to the services provided, transportation methods, time required, and costs.

Combined Handling and Transportation of Materials The movement of materials from one place to another usually includes both handling and transportation. For economy, these two types of work should be combined so that the total cost of doing the work will be a minimum. In general, the use of more machinery will result in a saving of labor costs, and vice versa. However, the total combined costs of labor, equipment or plant, and overhead should govern.

The fundamental law is to select, arrange, and use the men and machinery so that the work will be done at a minimum cost. In some instances, the selection, arrangement, and use of machinery is limited to the equipment that the contractor has available for the job. In other instances, it may be advisable to rent or purchase additional equipment. Of equal importance is the selection of good workmen, their arrangement in gangs of the right number of men, and the use or supervision of the gangs so that the men will work efficiently and economically, especially when their work is combined with that of machines. Whenever practical, all men and all machines should be worked to their full capacity. The time lost by delays due to any cause should be reduced whenever possible. There is no special rule whereby the best combination of men and machines can always be determined by one computation. Often several combinations must be tried and computations made in detail before the most economical combination is found. A further discussion of the economical operation of men and equipment is given in Chap. 22 on Construction Management.

When estimating the cost of combined handling and transportation of materials by autotrucks, the hours required per truck load (or loads per hour) will depend upon the number of trucks loaded and unloaded per hour. For example, if 20 min are required for loading and 15 min for unloading one truck, only three truckloads can be handled per hour. The unloading gang will have a 5-min wait between trucks. Also, if four trucks are provided per hour and only three can be handled per hour, each truck will have a wait of 20 min per trip on an average (the loading time required for one truck).

When considering gang times, the times of the gangs for loading and unloading should be taken as equal. For example, if the loading time is 20 min and the unloading time is 15 min, only three trucks can be loaded per hour. Hence, only three trucks per hour will arrive at the unloading platform so that the unloading gang will have a 5-min wait between trucks and will be idle 5 × 3, or 15 min, out of every hour.

When the trip time required for a truck is computed, this time will equal the loading time plus travel time loaded plus unloading time plus travel time empty plus delays. No two of these times have to be taken equal. The number of trucks needed per hour will equal 60 min divided by the loading time in minutes (or divided by the unloading time if that is larger than the loading time).

In the preparation of any estimate for the cost of handling and transportation of materials, the four items mentioned in the article on the cost of handling and transporting materials and in the summary article in this chapter must be considered.

Illustrative Estimates on Combined Handling and Transportation of Materials The following four illustrative estimates show how cost estimates may be prepared for the handling and transportation of materials.

Illustrative estimate. Estimate the cost of unloading 226,000 fbm of lumber from freight cars, transporting 1 mile by truck, and piling in yard. Contractor has a gang of 6 men with 1 truck and driver. Capacity of truck is 4,000 fbm and average speed (allowing for small delays) is 16 mph. The men ride back and forth on the truck. Assume cost of truck, driver, and 6 men to be $32 per hour, including 15 percent for overhead charges.

Solution: The 6 men working together would require about 0.45 hr to load the truck if it is assumed that the truck can back up to the car and perhaps an equal time to unload and pile in the yard if it is assumed that the lumber can be piled directly from the truck (see Table 2-1).

Time required for one trip = 0.45 (load) + 0.45 (unload) + 0.0625 (travel loaded) + 0.0625 (travel empty) = 1.025 hr
Cost per hour of truck, driver, and gang, including overhead = $32
Cost of one trip = $32 × 1.025 = $32.80
Cost per 1,000 fbm = $32.80 ÷ 4 = $8.20
Cost for 226,000 fbm = $8.20 × 226 × $1,853

NOTE: As the time required for loading and unloading is large when compared with the travel time, the costs would be reduced by increasing the number of men in the gang, it being assumed that the men would not interfere with each other in their work.

Illustrative estimate. A certain contractor has a quantity of face brick to unload from freight cars, transport 2 miles to the job, and stack at the job. He has one truck available which can haul 1,000 brick per load and can maintain an average speed of 20 mph with an allowance for short delays. The cost of this truck with driver is $5.65 per hour, and wages of an ordinary laborer are $2.80 per hour. The overhead charge is 20 cents per hour per laborer and 35 cents per hour for the truck. Brick tongs will be used for loading and unloading. The laborers will ride from freightyard to job and back again with the truck. The driver does not assist in loading or unloading. Assume that 1 man will load 600 brick from car to truck per hour or from truck to stack; 2 men, at rate of 1,200 brick per hour; 3 men, at rate of 1,700 brick per hour; 4 men, at the rate of 2,100 brick per hour; and 5 men, at the rate of 2,400 brick per hour. More than 2 men will interfere with each other to some extent. What size of gang would be the most economical, and what should be the estimated cost per 1,000 brick?

Solution:

One-man gang:

One round trip will care for 1,000 brick.
Total truck cost = \$5.65 + \$0.35 = \$6 per hour, including overhead
Total wage per man per hour = \$2.80 + \$0.20 = \$3, including overhead
Load = 1.67 hr Unload = 1.67 hr Travel 4 miles = 0.20 hr
Total = 3.53 hr for 1 trip
Cost per 1,000 brick = 3.53(\$6 + \$3) = \$31.77

Two-man gang:

Load = 0.833 hr Unload = 0.833 hr Travel = 0.20 hr
Total = 1.87 hr for 1 trip
Cost per 1,000 brick = 1.87(\$6 + 2 × \$3) = \$22.44

Three-man gang:

Load = 0.59 hr Unload = 0.59 hr Travel = 0.20 hr
Total = 1.38 hr for 1 trip
Cost per 1,000 brick = 1.38(\$6 + 3 × \$3) = \$20.70

Four-man gang:

Load = 0.476 hr Unload = 0.476 hr Travel = 0.20 hr
Total = 1.15 hr for 1 trip
Cost per 1,000 brick = 1.15(\$6 + 4 × \$3) = \$20.70

Five-man gang:

Load = 0.417 hr Unload = 0.417 hr Travel = 0.20 hr
Total = 1.034 hr for 1 trip
Cost per 1,000 brick = 1.034 (\$6 + 5 × \$3) = \$21.71

According to the data given, the truck and driver with a 3- or 4-man gang would be the most economical combination in regard to cost.

Illustrative estimate. A certain contractor has a large quantity of cement in sacks to unload from freight cars, transport 3 miles to the job, and pile in a cement house. He has 3 trucks available for this purpose, each truck capable of hauling 60 sacks of cement per load. Each truck can maintain an average speed of 20 mph when traveling, with allowance for small delays. Cost per hour (including overhead) of a truck and driver is estimated at \$7 per hour. Ordinary labor can be secured at a cost (including overhead) of \$4 per hour. It takes 1 man 1 min to carry a sack of cement from a freight car and pile it in a truck. It takes 1 man $1\frac{1}{4}$ min to carry a sack of cement from the truck and pile it in the cement shed. One gang will be used for loading and another gang for unloading. One man in each gang will be the straw boss of that gang, and his wage is \$5 per hour, including overhead. The straw bosses will work with the men. The truck drivers will not help load or unload the trucks. Determine the most economical size of each gang, and compute the cost of handling the cement per sack.

Solution: This estimate is similar to the preceding estimate in that varying sizes of gangs may be tried, but it differs in that two gangs will be needed and the gangs will not travel on trucks. It may be that the gangs will be idle some of the time. No doubt it would be advisable to keep trucks busy all the time, even if one or both gangs are idle a little of the time. Loading times should equal unloading times (approximately) so as to reduce the amount of time that any gang is idle. The unloading gang should be about 1.25 as large as the loading gang, to keep these times about equal.

Try 3 men loading and 4 men unloading:

$$\begin{aligned}
&\text{Time to load a truck} &&= \frac{60 \times 1}{3} = 20 \text{ min}\\
&\text{Time to travel 3 miles} &&= 9 \text{ min}\\
&\text{Time to unload} &&= \frac{60 \times 1.25}{4} = 18.75 \text{ min}\\
&\text{Time to return 3 miles} &&= 9 \text{ min}\\
&\text{Time for one trip} &&= 56.75 \text{ min if no delays due to gangs}
\end{aligned}$$

However, it takes 20 min to load 1 truck, or 60 min to load 3 trucks. Hence, a truck would be delayed 60 − 56.75, or 3.25, min per trip by the loading gang. The unloading gang would have 1.25 min for idleness between trucks.

$$\begin{aligned}
&\text{Time for one trip} = 56.75 + 3.25 = 60 \text{ min} = 1 \text{ hr}\\
&\text{Trucks loaded and unloaded} = 3 \text{ in } 60 \text{ min} = 1 \text{ in } 20 \text{ min}\\
&\text{Truck cost one trip} = \$7 \times 1 = \$7\\
&\text{Cost of loading (one man is straw boss)} = (\$5 + 2 \times \$4) \times {}^{20}\!/_{60} = \$4.33\\
&\text{Cost of unloading (one man is straw boss)} = (\$5 + 3 \times \$4) \times \frac{18.75 + 1.25}{60}\\
&\qquad = \$5.67\\
&\text{Total cost of one trip} = \$7.00 + \$4.33 + \$5.67 = \$17.00\\
&\text{Cost per sack} = \$17.00 \div 60 = \$0.283
\end{aligned}$$

Evidently these gangs are too small, as the trucks are delayed by the gangs.

Try 4 men loading and 5 men unloading:

$$\begin{aligned}
&\text{Time to load 1 truck} &&= \frac{60 \times 1}{4} = 15 \text{ min}\\
&\text{Time to unload 1 truck} &&= \frac{60 \times 1.25}{5} = 15 \text{ min}\\
&\text{Time to travel 3 miles and return} &&= 9 + 9 = 18 \text{ min}\\
&\text{Time for one trip} &&= 48 \text{ min if no delays}
\end{aligned}$$

However, it takes 15 min to load (or unload) one truck; hence, 4 trucks could be loaded per hour if trucks arrived that fast (i.e., every 15 min).

Trucks will arrive every 48 ÷ 3, or 16, min; hence, each gang will have a 1-min wait between trucks. There will be no truck delays.

$$\text{Truck cost per trip} = \$7 \times 48/60 = \$5.60$$

$$\text{Cost of loading} = (\$5 + 3 \times \$4) \times \frac{15 + 1}{60} = \$4.53$$

$$\text{Cost of unloading} = (\$5 + 4 \times \$4) \times \frac{15 + 1}{60} = \$5.60$$

$$\text{Total cost of one trip} = \$15.73$$

$$\text{Cost per sack} = \$15.73 \div 60 = \$0.262$$

Here the gangs are balanced, and each gang gets a minute's rest between trucks. A good arrangement.

Try 5 men loading and 6 men unloading:

$$\text{Time to load 1 truck} = \frac{60 \times 1}{5} = 12 \text{ min}$$

$$\text{Time to unload 1 truck} = \frac{60 \times 1.25}{6} = 12.5 \text{ min}$$

Time to travel 3 miles and return = 18 min
Total time for one trip = 42.5 min if no delays

Trucks will arrive every 42.5 ÷ 3 = 14.2 min, so that loading gang will have a 2.2-min wait between trucks and the unloading gang a 1.7-min wait.

$$\text{Truck cost per trip} = \$7 \times \frac{42.5}{60} = \$\ 4.96$$

$$\text{Cost of loading} = (\$5 + 4 \times \$4) \times \frac{12 + 2.2}{60} = \$\ 4.97$$

$$\text{Cost of unloading} = (\$5 + 5 \times \$4) \times \frac{12.5 + 1.7}{60} = \$\ 5.92$$

$$\text{Total cost per trip} = \$15.85$$

$$\text{Cost per sack} = \$15.85 \div 60 = \$0.264$$

Try 6 men loading and 8 men unloading: By following the same process of analysis, the total time per trip is found to be 37.4 min, the cost per trip to be $16.31, and the cost per sack to be $0.272.

Try 7 men loading and 9 men unloading: By following the same process of analysis, the time per trip is found to be 34.9 min, the cost per trip to be $16.87, and the cost per sack to be $0.281.

Try 8 men loading and 10 men unloading: By following the same process of analysis, the time per trip is found to be 33 min, the cost per trip to be $17.42, and the cost per sack to be $0.290. Another example of balanced gangs. Each gang has an idle period of 3.5 min between trucks.

According to these results, a gang of 4 men for loading and one of 5 men for unloading would give the least cost per sack for handling and transporting the cement under the conditions given in the problem. In actual practice, it would be of doubtful value to carry the estimates out to so many decimal places. For this problem, the minimum cost per sack would be taken as 26¼ cents, and gangs selected that would give about this cost.

Illustrative estimate. A highway contractor has 2 miles of gravel road to construct, requiring a total of 7,800 cu yd of gravel. Two sources of gravel are available. A material supply company owning a gravel pit will supply suitable bank-run gravel at the job for \$1.80 per cubic yard. The other source is a gravel pit located about 8 miles (average haul 8 miles) from the job. Gravel may be purchased in the pit for 25 cents per cubic yard, but the buyer must load and transport the gravel from the pit to the job. Roads between this pit and the job are paved. The highway contractor has no equipment available for loading and transporting this gravel, but he may rent a gasoline shovel and trucks from two sources, Company *A* and Company *B*, all equipment being new and in good condition.

Company *A* has a 1-cu-yd gasoline shovel and a fleet of twelve 5-cu-yd trucks which it will rent for the following hourly prices, for a minimum working day of 6 hr: Shovel, \$13 per hour, including fuel, 1 operator, repairs, and all overhead costs except transportation of shovel to and from the pit. Cost of transporting shovel to and from pit is estimated at \$70. Trucks, \$9 per hour, including driver, oil, fuels, tires, and all costs.

Company *B* has a 2-cu-yd gasoline shovel and a fleet of fourteen 9-cu-yd trucks which it will rent for the following hourly prices for a minimum working day of 6 hr: Shovel, \$18 per hour, including 1 operator, fuel, repairs, and all overhead except transportation to and from the pit. Cost of transporting shovel to and from the pit is estimated at \$90. Trucks, \$13 per hour, including driver, oil, fuel, tires, and all costs.

Contractor will need a superintendent with a small car at an estimated total cost of \$7 per hour, and 2 shovelers for the 1-cu-yd shovel, or 3 shovelers for the 2-cu-yd shovel. Wage of shovelers, including all overhead is \$4 per hour per man.

Estimate the costs of supplying gravel by each of these methods, assuming average truck speeds of 30 mph for the 5-cu-yd trucks and 28 mph for the 9-cu-yd trucks.

Solution: 1. Estimated cost of gravel from material supply company, \$1.80 (per cubic yard) × 7,800 (cubic yard) = \$14,040.

2. Estimated cost of gravel purchased from pit and using rented equipment from Company *A*: Assume trucks to average 30 mph.

Time for one trip = 3 min (get in position and load) + 16 min (travel loaded) + 1 min (dump) + 16 min (return) + 4 min (delays) = 40 min = 0.667 hr

Then 12 ÷ 0.667 gives 18 trucks per hour at shovel. Shovel could load 20 trucks per hour if they were available.

Gravel loaded per hour = 18 × 5 = 90 cu yd
Hours required for 7,800 cu yd = 7,800 ÷ 90 = 87 hr

Gravel cost	= \$0.25 × 7,800 cu yd	= \$ 1,950
Shovel cost	= \$70 + \$13 × 87 hr	= 1,201
Truck cost	= \$9 × 12 trucks × 87 hr	= 9,396
Superintendent and shovelers	= (\$7 + 2 × \$4) × 87 hr	= 1,305
Total cost		= \$13,852

Cost per cubic yard = \$13,852 ÷ 7,800 = \$1.78
The use of 13 trucks instead of 12 would reduce costs a little.

3. Estimated cost of gravel purchased from the pit and using rented equipment from Company *B*: Assume trucks to average 28 mph.

Time for one trip = 3 min (get in position and load) + 17 min (travel loaded) + 1 min (dump) + 17 min (return) + 4 min (delays) = 42 min = 0.70 hr

Then 14 ÷ 0.70 gives 20 trucks per hour at shovel. Shovel can load 20 trucks per hour.

Gravel loaded per hour = 20 × 9 = 180 cu yd
Hours required for 7,800 cu yd = 7,800 ÷ 180 = 43.3, say 44 hr

Gravel cost	= \$0.25 × 7,800	= \$ 1,950
Shovel cost	= \$90 + \$18 × 44	= 882
Truck cost	= \$13 × 14 × 44	= 8,008
Superintendent and 3 shovelers	= \$(7 + 3 × \$4)44	= 836
Total cost		= \$11,676

Cost per cubic yard = \$11,676 ÷ 7,800 = \$1.50

It would be cheaper for the highway contractor to rent equipment from Company *B* and to haul the gravel, instead of buying the gravel delivered at the job from the material supply company.

CHAPTER THREE

Excavation

Excavation Excavation deals primarily with the handling and transportation of the soil, though in a complete excavation job many other items may need to be included. For example, the excavation of a basement for a large building may include such items as clearing the site, removing old buildings, loosening and removing soil, breaking and removing rock, filling and grading, bracing and shoring, sheet piling, pumping water, digging trenches, and protecting adjacent property.

The costs of excavation and the estimates of these costs depend upon several things such as class of excavation (special and general); quantity or amount of excavation; kind of soil; disposal of excavated materials; other items of work required in connection with the particular job; methods used for loosening and digging, transporting, and disposing of the materials; and delays caused by weather, machinery breakdowns, labor troubles, etc.

Classes of Excavation Excavation may be divided into two classes:

1. General excavation, such as excavation for a large basement or for a large cut for a railway or highway. General excavation can be done by

almost any method such as tractors, scrapers, or power shovels and trucks, for example.

2. Special excavation, such as digging trenches or footing holes for foundations. Handwork is nearly always required for special excavation.

Kinds of Soil Soil may be divided into five kinds according to the difficulty experienced in excavating it. Soils vary greatly in character, and no two are exactly alike.

1. *Light soil.* Requires no loosening. Can be shoveled easily. Sand is an example.

2. *Medium or ordinary soil.* Easily loosened by pick and plow. Preliminary loosening is rarely required when power excavating equipment such as shovels and dragline scrapers are used. Ordinary soil and loam are examples.

3. *Heavy or hard soil.* Can be loosened by picks. Hard to loosen with an ordinary plow. May be dug by sturdy power shovels without preliminary loosening. Stiff clay, compact gravel, and a hard and compacted loam containing gravel and small stones and boulders are examples.

4. *Hard pan or shale.* A soil that is partly hardened and is difficult to loosen. Hard to loosen with picks. Light blasting is often required when power excavating equipment is used.

5. *Rock.* Requires blasting before handling and transporting. Rock may be subdivided into different kinds or grades (such as soft, medium, and hard) according to the difficulty of drilling and breaking it.

Most soils expand when loosened. For example, 1 cu yd of earth measured in excavation may measure up to 1.25 cu yd when loosened and loaded into a truck. Some soils expand more than others, the rate being about 10 to 25 percent. This is a reason why the capacities of dump wagons and trucks are often taken at about 75 or 80 percent of their rated or measured capacity. Rock, when blasted and broken, will expand more than soils. This expansion may be as much as 40 or 50 percent.

Soil, when deposited and compacted as in a fill or a trench, may shrink up to 10 to 15 percent as compared with its original volume. The reason for this is that ordinary soil in the ground usually contains some voids, such as may be caused by the passage of water through the soil. Very loose soil will compact more than soil not so loose. Loose rock usually will not compact.

Some soils are firm so that the sides of the excavation will remain vertical and will not cave in. Other soils are not firm, and the sides of the excavation will require bracing to keep the soil in place. Bracing and shoring should be provided when the excavation is deep, when it may remain open for some time, or when it may be exposed to rain. Sometimes, when the excavation is not deep, it is cheaper to slope the banks instead of bracing them. A discussion of the cost of bracing is given in the following chapter on Piling and Bracing.

Angles of Repose. Table 3-1 gives the angles of repose, or the angles determining the slopes, for some kinds of soil.

Table 3-1 *Approximate Angles of Repose*

Material	Angle with horizontal, deg		
	Dry	Moist	Wet
Sand	20–35	30–45	20–40
Earth	20–45	24–45	25–30
Gravel	30–50		
Gravel, sand, and clay	20–35		

Computing Quantities The unit used when computing quantities of excavation is usually the *cubic yard.* The cubic foot is sometimes used when the quantity considered is quite small.

In the takeoff, each item of general and special excavation should be listed separately, with dimensions, quantities, and the kind of soil. It should be noted if the earth is firm and will retain its vertical position without side supports or bracing or if the earth wall will require shoring or bracing. Sometimes when excavating with power equipment it is more economical to omit the shoring or bracing and to slope the banks.

When computing quantities for excavation, the estimator must be careful to include the dimensions from outside to outside of all footings and forms and from the top of the grade to the lowest point of excavation. If the surface of the ground is sloping or uneven, the horizontal projection can be divided into a number of convenient squares and rectangles and the average height of each section estimated.

Methods Used for Excavating There are several methods that can be used in excavating soil. For example, the soil can be loosened by picks, team and plow, tractor and plow, or blasting. The material can be dug by hand shovels, tractor excavators, or tractor and scraper, or by a power excavator such as a shovel or dragline scraper. The transporting can be done by wheelbarrows, scrapers, dump wagons, autotrucks, crawler wagons, or railway dump cars. The material can be disposed of by merely dumping or by dumping and leveling by hand, scrapers, graders, or bulldozers, as the case may be. A typical excavating job will usually require loosening and digging, transportation, and disposal.

Factors Affecting Excavation Costs The estimator must remember to include every cost item. No item should be omitted. It is not so important as to what part of the cost classification any item is placed under, provided that the estimator knows and understands the classification used and classifies the same item in the same manner each time. When not so classified, the exception should be noted.

When excavation costs are estimated, several items must be considered, some of which may not always be apparent. For example, when estimating trenches, the following questions must be answered:

Will the sides of the trench remain vertical, or will they cave in?

Will bracing be required?

Will the work be done by hand, or will a trenching machine be used?

Will picks be needed to loosen the soil?

Will soil have to be handled more than once on account of the depth of the trench?

What kinds of soil will be encountered?

Will the soil be dry or wet?

Will any pumping be required?

Will drilling or blasting be required?

Will part or all of the soil have to be transported away and dumped?

If so, what transportation methods will be used?

Will backfilling be required?

What is the probable weather forecast?

What are the local wages for labor?

Are permits needed, and what do they cost?

Are there any special state laws or local ordinances to be complied with? If so, what are the requirements?

Must protection be provided for sidewalks and pavements?

Are lights needed, and what will they cost?

What materials and equipment are needed, and what is their cost?

What overhead expenses are to be charged to the job?

Is any profit to be included in the estimate?

When general excavation is estimated, somewhat similar questions will need to be answered. In addition, some other questions, such as the following, will also need consideration:

Will a power excavator (shovel or dragline excavator) be used?

What is its capacity and cost?

Will tractors and scrapers be needed? If so, how many?

Will a tractor and a plow be required?

If transportation is required, what vehicles will be used?

If dump wagons are to be used, how many are needed?

If trucks are to be used, how many are required and of what capacity?

In what manner will the excavated earth be disposed of?

Will extra laborers be required for any purpose? If so, how many?

What are the hourly costs of the various items of equipment which are to be used?

The fundamental law for securing low costs is "to select, arrange, and use the men and equipment so that the total cost of the job will be a minimum." In general, all men and machines should be worked at full capacity during working hours. Delays should be kept at a minimum. If some of the men or machines must be idle at times, the most expensive should be kept as busy as possible. For example, in an excavation job the power shovel may be the most expensive piece of equipment, and it should be kept busy during all the working hours. Sufficient trucks should be provided to keep the shovel busy, even if a truck should have to wait a little occasionally. Likewise, if the use of a few shovelers will speed up the work of the shovel at the excavation, maintain ramps in good order, or reduce the time required by the trucks at the dumping ground, these shovelers should be provided even if they may be idle a few minutes out of every hour on the average. Sometimes more than one shift can be used to speed up production and reduce overhead.

Methods of Estimating Excavation Costs There are two ways of estimating excavating costs. One way is to consider each class of excavation and each kind of soil and prepare separate estimates (labor, equipment, overhead, and profit) for each. The other way is to estimate the total cost of handling of all excavation and the total cost of transportation of all excavation. The first way is usually preferred.

There are also two methods of preparing cost estimates for any one kind or class of excavation. One method is to obtain the total costs of each of the four main items (the total-cost method) and then find the sum. Unit costs are found by dividing the total costs by the number of cubic yards of excavation. The other method (the unit-cost method) is to estimate the cost per cubic yard of handling, transporting, overhead, profit, and total, and then get the cost of the job by multiplying the unit cost by the number of cubic yards. Either method is satisfactory if no costs are omitted.

Preparation of Excavation-cost Estimates After the takeoff has been completed showing all kinds and quantities of excavation, and after the methods and equipment to be used have been decided upon, a cost estimate can be prepared. This estimate can be divided as follows:

Item	Unit cost per cu yd	Total cost
Labor..............................	$	$
Equipment..........................		
Overhead...........................		
Profit (if any is to be included)........		
Total..............................	$	$

No allowance is usually made for the cost of the material to be excavated. If the material must be purchased or if a price must be paid for a place to dispose of surplus material, such costs should be included as a separate item in the estimate.

Labor Costs. The labor costs include the wages of all labor on the job, plus all payroll taxes, insurance, etc. The crew may include a foreman, machine operators, truck drivers, common laborers, etc., as may be needed. Wages per hour (1967) are about as follows: foremen $4 to $8, machine operators and truck drivers $3 to $6, and common labor $2.50 to $4.

Payroll taxes and insurance include such items as social security, Federal unemployment, state unemployment, workmen's compensation, and contractor's liability, which are based on the labor payroll. The sum of these items may vary from about 8 to 18 percent of the labor wage.

Table 3-2 *Loosening Soil (Approximate)*

Method	Cu yd per hr			Hr per cu yd		
	Medium soil	Heavy soil	Hard pan	Medium soil	Heavy soil	Hard pan
Man and pick.....	2–4	1–3	0.5–1.5	0.25–0.50	0.33–1.00	0.67–2.00
Team and plow....	25–50	15–30	....	0.02–0.04	0.03–0.07	
Tractor and one plow............	40–70	25–50	5–20	0.01–0.03	0.02–0.04	0.05–0.20
Tractor and 2-gang plow............	50–100	40–70	....	0.01–0.02	0.01–0.03	

Equipment Costs. These include costs of ownership, rentals, operation, transportation to and from the job, and sometimes erection and dismantling. A method of estimating equipment costs is given in Chap. 18, Construction Plant and Equipment, and in Appendix B, Equipment Costs.

The cost of small tools, such as picks, shovels, barrows, etc., is usually estimated as a lump sum for any job, including transportation.

Overhead. Overhead costs on an excavation job may vary from about 5 to 15 percent of the sum of the labor and equipment costs or from about 8 to 25 percent of the labor costs alone.

Profit. If the excavation job is a separate job, some allowance, say 5 to 12 percent, should be allowed for profit.

Loosening Soil Table 3-2 gives the approximate times required for loosening various soils with different equipment. Light soil requires no preliminary loosening.

The cost of loosening rock by drilling and blasting will depend upon the kind of rock, ease of drilling, use of hand or machine drills, skill and wages of workmen, and amount and cost of explosives used. The general subject of rock excavation will be considered in a later article.

Digging Soil by Hand Labor Digging soil by hand shoveling is perhaps the oldest and simplest method. The amount of work done per man per hour varies greatly with the skill of the laborer, his inclination to work, the condition of the soil (dry or wet), the kind of soil, the height or lift, and the working conditions on the job. A further discussion on excavation by hand shoveling is given in the articles on Backfilling, Trenching with Hand Labor, and Excavation by Hand.

Table 3-3 gives the approximate outputs for shoveling loosened material into wagons or trucks. An increase in the total lift up to about 6 ft will decrease the hourly output about 5 to 10 percent for each foot of increase. Likewise, a decrease in the total lift will increase the hourly output about 5 to 10 percent per foot.

Table 3-3 *Shoveling Loosened Soil into Wagons or Trucks (Total lift about 4 ft)*

Soil	Cu yd per labor-hr	Hr per cu yd
Light, from ground	1.20–2.50	0.40–0.85
Medium, from ground	1.00–2.00	0.50–1.00
Hard or heavy, from ground	0.80–1.50	0.65–1.25
Hard pan, from ground	0.70–1.25	0.80–1.40
Light, from excavation	1.10–2.30	0.45–0.90
Medium, from excavation	0.90–1.80	0.55–1.10
Hard or heavy, from excavation	0.70–1.35	0.75–1.45
Hard pan, from excavation	0.60–1.10	0.90–1.65

Table 3-4 gives the approximate outputs per labor-hour for shoveling in excavation, including some loosening with a pick, and for a total lift not over 6 ft. In general, the output increases as the total lift decreases, and this must be considered when selecting values from the table.

When the excavation is more than 5 ft in depth, the soil must be rehandled by extra men using stages or platforms. For every two or three diggers, an extra man is required for each rehandling or for each stage. Note that the amount of excavation per labor-hour will be reduced as extra men are used to reshovel the soil.

Diagram 3-1 (Appendix F) can be used for estimating the costs of hand shoveling when the labor wage is known and the labor output can be reasonably assumed.

Digging Soil with Power Equipment Some kind of power equipment is usually used for digging the soil except on very small jobs or on

Table 3-4 *Shoveling Soil into Wagons and Trucks from Excavation, Including Some Loosening with Picks (Total lift not over* 6 *ft)*

Soil	Excavation	Cu yd per labor-hr	Hr per cu yd
Light...............	General, dry	1.00–1.75	0.55–1.00
	General, wet	0.70–1.30	0.75–1.45
	Special, dry	0.90–1.50	0.65–1.10
Medium............	General, dry	0.80–1.40	0.70–1.25
	General, wet	0.55–1.00	1.00–1.80
	Special, dry	0.70–1.20	0.85–1.45
Hard or heavy......	General, dry	0.60–1.15	0.85–1.70
	General, wet	0.35–0.65	1.55–2.85
	Special, dry	0.45–0.80	1.25–2.25
Hard pan...........	General, dry	0.50–1.00	1.00–2.00
	General, wet	0.25–0.50	2.00–4.00
	Special, dry	0.35–0.65	1.55–2.85

special excavation. Types of equipment commonly used are power shovels, draglines, bulldozers, tower excavators, cranes and derricks with booms and buckets, trenching machines, tractor-drawn scrapers, elevating graders, and various types of special equipment. For detailed descriptions of different types of power equipment, the reader is referred to books on construction equipment and to the catalogues of the manufacturers. The discussion given in the preceding chapter in the article on Unloading and Loading Materials with Power Equipment may be reviewed at this time.

The output of a power excavator, such as a power shovel or dragline excavator, depends upon the size of the bucket, dipper, drag bucket or scraper; the number of loads handled per minute or per hour; the kind of soil, weather, and job conditions; amount of moving required; delays of various kinds; and the skill of the operator.

A power excavator should average 1 to 3 bucket, dipper, or scraper loads per minute during the working day. The number of loads per minute will usually vary inversely with the amount of the load, the reach or length of boom, and the distance through which the load is swung or carried. A larger machine usually operates more slowly than a smaller one, and more time is required to operate a long boom than a short one.

Each type of power excavator has its advantages and disadvantages. Some types of machines work more efficiently on some kinds of jobs than on others. For example, a power shovel or dragline scraper may be the most suitable for the excavation of a basement for a large building, a derrick or crane with a bucket for deep excavation, and a tractor-scraper for a cut on a new highway.

Table 3-5 gives the approximate capacities of power excavators such as shovels and dragline scrapers.

When trucks or wagons are used with a power excavator, the capacity of a truck or wagon should be equal to one or more times the load handled at one time by the excavator (dipper, bucket, or scraper load). There should be a sufficient number of trucks or wagons available to keep the excavator busy during all working hours.

Table 3-5 *Approximate Outputs of Power Excavators Used for Digging Soil*

Bucket, dipper, or scraper capacity, cu yd	Machine with a short reach*		Machine with a longer reach†	
	Cu yd per hr	Hr per 1,000 cu yd	Cu yd per hr	Hr per 1,000 cu yd
½	30–100	10.0–33.3	25–75	13.3–40.0
¾	45–130	7.7–22.2	40–100	10.0–25.0
1	60–160	6.3–16.7	55–125	8.0–18.2
1¼	75–190	5.3–13.3	70–150	6.7–14.3
1½	90–220	4.6–11.1	80–175	5.7–12.5
1¾	105–245	4.1–9.5	90–200	5.0–11.1
2	120–270	3.7–8.3	100–220	4.6–10.0
2½	145–320	3.1–6.9	120–260	3.9–8.3
3	170–370	2.7–5.9	140–300	3.3–7.1
3½	190–420	2.4–5.3	160–330	3.0–6.3
4	210–460	2.2–4.8	175–350	2.9–5.7
5	250–550	1.8–4.0		
6	285–630	1.6–3.5		

* Such as a power shovel with short boom and dipper.

† Such as a dragline scraper with a long boom, or a crane with a long boom and bucket.

The cost of power excavators includes such items as costs of ownership, rentals, operation, transportation to and from the job, erection, and dismantling. See Chap. 18 and Appendixes B and C for equipment costs.

There are several ways of computing costs of power equipment. Almost any way is satisfactory provided that it is understood and that all cost items are included. Perhaps the best way is to compute the hourly cost per working hour because most costs vary with the time the equipment is in use. This hourly cost should include allowances for all such items as are mentioned in the preceding paragraph.

Backfilling Backfilling may be done by hand shoveling, by scrapers, or by bulldozers (tractors with a scraper fastened to the front). The time required for backfilling by hand will depend on the ease with which the material can be shoveled, and on the temping or compacting required.

Table 3-6 gives approximate rates of work.

Table 3-6 *Backfilling with Hand Labor*

Soil	Shovel only		Shovel and tamp	
	Cu yd per hr	Hr per cu yd	Cu yd per hr	Hr per cu yd
Light................	1.5–3.0	0.35–0.65	0.8–2.2	0.45–1.25
Medium..............	1.3–2.5	0.40–0.75	0.7–1.8	0.55–1.45
Hard or heavy........	1.0–2.0	0.50–1.00	0.6–1.5	0.65–1.65

Tamping time is assumed to vary from about one-fourth to one-half of the total time.

When the backfill material must be wheeled, the time required will be increased proportionately.

The amount of backfilling accomplished by a bulldozer and operator will vary with the material, the size of the machine, and the skill of the operator. The quantity of earth handled per hour may vary from about 3 to 30 cu yd.

Spreading Soil One man with a shovel should be able to spread and level 1 to 3 cu yd per hour of light or medium soil. If the soil must be wheeled, these quantities will be reduced proportionately. A bulldozer should be able to spread and level 3 to 30 cu yd per hour or more, depending upon the requirements of the particular job.

Pumping An estimate of the pumping requirements and costs cannot be readily made unless there is some way of determining in advance the probable amount of water that will need to be handled and the distance it must be pumped. If the amount of water is small, one or two hand pumps or a small power pump will answer the purpose and the cost will not be large. However, if the amount of water is large, one or more large power pumps will be required.

No worthwhile detailed information can be given in regard to estimating pumping costs on excavation jobs. The estimator must rely upon his own judgment and experience. Each job is a problem by itself, depending on the kind of work, the location of the job, the general topography of the locality, the season of the year, and the weather.

Transporting Soil The excavated material may be transported by such vehicles as wheelbarrows, scrapers, four-wheel or crawler front-end loaders with scoops or buckets, dump wagons, tractor-drawn crawler wagons, autotrucks, and railway dump cars. The methods of estimating the transportation of materials by vehicles as described in the preceding chapter in the article on Transporting Materials by Vehicles apply equally well to the transportation of soil. The use of light railways will not be discussed in this text.

Table 3-7 shows approximate capacities, economical hauls, loading and unloading times, and rates of travel for various vehicles.

Table 3-7 *Vehicular Data (Approximate)*

Vehicle	Capacity, cu yd	Economical haul, ft	Time, min		Rate of travel, mph	
			Load	Unload	Loaded	Empty
Wheelbarrows*	0.07–0.15	Up to 150	1.0–3.0	0.2–0.4	70–125	100–175
Handcarts, two wheel*	0.07–0.20	Up to 150	1.0–3.0	0.2–0.4	70–125	100–175
Front-end loaders:						
Four wheel	0.33–2.00	Up to 1,500	0.5–1.0	0.2–0.5	4–15	6–20
Crawler type	0.33–9.00	Up to 1,500	0.5–1.3	0.2–0.7	3–12	4–15
Tractor-drawn wagons†	3–25	Up to 2,500	1.0–3.0	0.3–1.0	3–10	4–12
Tractor-scraper units:‡						
Crawler type	5–30	Up to 2,500	1.0–2.0	0.3–1.0	3–7	4–10
Rubber tires	5–30	Up to 5,000	1.0–2.0	0.3–1.0	10–20	15–30
Auto dump trucks‡	2–20	Over 500	1.0–3.0	0.5–2.0	10–45	15–60

* Rate of travel of wheelbarrows and handcarts is in feet per minute.

† Tractors may draw more than one vehicle.

‡ Special wheel and carrying scrapers, tractor-scraper units, crawler wagons, and auto dump trucks can be obtained in larger sizes.

Trenching with Hand Labor The items affecting the cost of trenching with hand labor have been stated in the article on Factors Affecting Excavation Costs. The main items are the length, width, and depth of the trench; kind and condition of the soil; bracing and pumping if required; backfill and removal of surplus earth; and the wages and skill (output) of the workmen.

When a trench is more than about 5 ft in depth, the soil must be rehandled once for depths of 6 to 10 ft, twice for depths of 11 to 15 ft, etc. One man will be required for each rehandling for every two or three diggers. The rehandling is usually done by men on stages or platforms, the stages and platforms being moved as required. Note that, for a trench over 5 ft in depth, rehandling is not required for the first 5 ft and that one rehandling is required for each additional 5 ft or fractional part. Hence, no rehandlers will be needed for the first 5 ft.

Considering the output per hour per man for a 3-ft trench as 100 percent, the relative outputs per man (with allowances for rehandling) will be about as follows:

Depth of trench, ft	3	5	8	10	12	15
Percentage of output	100	95	86	81	71	67

The labor required for digging various kinds of soil in trenches can be found from Table 3-8. The values given in the table are approximate

Table 3-8 *Labor Required for Shoveling Soil in Trenches (Including loosening with picks as required)*

Soil	Depth of trench, ft					
	3	5	8	10	12	15
	Output, cu yd per man per hour					
Light	1.00–1.80	0.95–1.70	0.85–1.55	0.80–1.45	0.70–1.30	0.65–1.20
Medium	0.85–1.60	0.80–1.50	0.75–1.40	0.70–1.30	0.60–1.15	0.55–1.05
Heavy	0.65–1.25	0.62–1.20	0.55–1.05	0.53–1.00	0.45–0.90	0.43–0.85
Hard pan	0.50–1.00	0.47–0.95	0.43–0.85	0.40–0.80	0.35–0.70	0.33–0.76
	Labor-hours per cu yd					
Light	0.55–1.00	0.60–1.05	0.65–1.15	0.70–1.25	0.75–1.40	0.80–1.50
Medium	0.65–1.20	0.70–1.25	0.75–1.40	0.80–1.50	0.90–1.70	0.95–1.80
Heavy	0.80–1.55	0.85–1.65	0.95–1.80	1.00–1.90	1.15–2.20	1.20–2.30
Hard pan	1.00–2.00	1.05–2.10	1.15–2.35	1.25–2.50	1.40–2.80	1.50–3.00

and are for soils that are reasonably dry (not too wet). When the soil is quite wet, the labor output will be reduced perhaps as much as half in some cases.

The question of bracing or shoring (see Chap. 4 on Piling and Bracing) should be considered separately.

If water must be cared for, it can be drained to a low part of the trench and removed by hand or power pumps.

The minimum width of a trench that men can work in is about 20 to 24 in. The trench width required for pipe laying is equal to the width of the pipe plus 16 to 20 in. About 8 to 10 in. is required on each side of a pipe for lowering, placing, and joining or calking. Hence, the width required for a 4-in. pipe would be about 4 plus 16, or 20 in., and that for a 30-in. pipe would be about 30 plus 20, or 50 in. When the pipe sections can be joined together on the ground and then lowered into the trench, the required trench width will be less.

Perhaps the simplest way of estimating trenching costs by hand labor is to (1) compute the quantity of each kind of soil in cubic yards; (2) knowing the depth of trench, estimate the hours required for a man to excavate 1 cu yd of each kind of soil; (3) knowing the hourly wage, compute the cost per cubic yard for each kind of soil; (5) compute the total cost for each kind of soil; and (6) sum up the costs.

Equipment costs will be the wear on tools (picks, shovels, etc.), their loss, and the expense of transporting them to and from the job. These costs may vary from a few dollars up to $100 or more, depending on the particular job. On some jobs, allowances must be made for bracing and shoring, for pumping, and for stages and platforms for rehandling the soil.

General overhead costs may vary from about 8 to 20 percent of the labor costs.

Profit may vary from 5 to 15 percent.

Diagram 3-1 (Appendix F) can be used to find the labor costs per cubic yard for excavating and backfilling trenches by hand labor when the hourly wages are known and the hourly outputs for the laborers may be reasonably assumed.

Illustrative Estimates The following two estimates relate to trenching with hand labor.

Illustrative estimate. Estimate the cost of excavating and backfilling a trench 240 ft long, 4.5 ft deep, and 2 ft wide. Soil is ordinary loam (medium soil) and is firm enough so that no bracing is required. Labor wage is $3.15 per hour plus 11 percent for payroll taxes and insurance. Allow $9 for equipment costs. Overhead is 9 percent of the labor costs, and profit is 10 percent of all other costs.

Assume a rate of 0.90 hr per cubic yard (1.10 cu yd per hour) for digging, and 0.45 hr per cubic yard (2.20 cu yd per hour) for backfilling with a little tamping. See Tables 3-6 and 3-8.

$$\text{Cubic yards in trench} = \frac{240(4.5)2}{27} = 80$$

Labor cost per hour per man = ($3.15)1.11 = $3.50		
Labor cost, excavation = $3.50(0.90)80 =	$252	
Labor cost, backfill = $3.50(0.45)80 =	126	
Labor cost total		= $378
Equipment		= 9
Overhead, 9% of labor cost		= 34
Profit, 10% of all other costs		= 42
Total		= $463

Cost per cubic yard = $463 ÷ 80 = $5.79

The labor cost can be found by Diagram 3-1. At $3.50 per hour and 0.90 plus 0.45, or 1.35 hr per cubic yard, the labor cost per cubic yard is $4.75.

Total labor cost = $4.75 × 80 = $380

Illustrative estimate. Estimate the cost of excavating and backfilling a trench 140 ft long, 8.5 ft deep, and 2 ft wide. Soil is heavy, requiring some loosening with picks. Soil is firm so that no bracing is required. Backfill requires tamping. Labor wage is \$3.15 per hour plus 11 percent for payroll taxes and insurance. Equipment allowance is \$35 for picks, shovels, stages, or platforms and for transportation to and from the job. Overhead is 12 percent of the labor cost. Profit is 10 percent of all other costs.

Assume a rate of 1.45 hr per cubic yard (0.69 cu yd per hour) for digging and 0.60 hr per cubic yard (1.33 cu yd per hour) for backfill and tamping. See Tables 3-6 and 3-8.

$$\text{Cubic yards in trench} = \frac{140(8.5)2}{27} = 88.15$$

Labor cost per hour per man = (\$3.15)1.11 = \$3.50			
Labor cost, excavation	= \$3.50(1.45)88.15 =	\$448	
Labor cost, backfill and tamping	= \$3.50(0.60)88.15 =	185	
Labor cost total			= \$633
Equipment			= 35
Overhead, 12% of labor cost			= 76
Profit, 10% of all other costs			= 74
Total			= \$818
Cost per cubic yard = \$818 ÷ 88.15 = \$9.28			

The labor cost can be found by Diagram 3-1. At \$3.50 per hour and 1.45 plus 0.60, or 2.05 hr per cubic yard, the labor cost per cubic yard is \$7.20.

Total labor cost = \$7.20 × 88.15 = \$635

Trenching with Machines There are several types of trenching machines. A satisfactory trencher is a power shovel with a long handle or dipper arm. Another kind is the *ladder* trencher consisting of two endless chains with small buckets. The chains are mounted on a ladder which can be lowered into the trench. This machine is power driven. A third kind is a wheel type in which small buckets are mounted on a large wheel.

Trenchers work best in firm soils that are not too hard or heavy, that do not contain boulders, and that do not require bracing.

The ladder and shovel types will dig trenches up to 52 in. wide and 8 to 26 ft deep. The wheel trencher can be used for trenches about 1 to 2 ft wide and up to about 7 or 8 ft deep. The output of these machines may range from about 20 to 250 cu yd per hour, depending on the capacity or size of the machine, the kind and condition of the soil, and the skill of the operator. An operator and a helper are required to operate most trenchers. In addition, a trencher must be kept supplied with fuel and oil.

The cost of excavating trenches by machines will depend upon the kind and condition of the soil, the size and capacity of the machine, the hourly output of the machine, the skill of the operator, the expense of moving the trencher to and from the job, and delays due to various causes. The hourly cost of operating a trencher may vary from about 50 cents an hour for a small machine up to about \$2 an hour or more for a large machine. An operator and one or more laborers will be required, depending on the size of the machine, the kind of soil, and the difficulties encountered on the job. For example, in a soil containing boulders, the trencher will be delayed while the boulders are removed by the laborers. Trenchers are suitable for use in ordinary soils which do not require bracing and which do not contain boulders.

The backfilling can be readily done by bulldozers. The hourly cost of a bulldozer will vary from about \$2 to \$10. This cost does not include the cost of moving the machine to and from the job. Usually, one to three shovelers will be required for cleaning up and general help. If tamping or wetting down of the trench is required, laborers will be needed for this purpose.

Diagram 3-2 (Appendix F) shows the relations between the total hourly operating costs of the machine and crew, the output of the machine, and the cost per cubic yard of soil. In this diagram, the curves are drawn for various outputs in hours per cubic yard, with cubic yards per hour in parenthesis. Many manufacturers rate their machines in cubic yards per hour. The cost of moving the machine to and from the job is not included and, hence, must be added as an extra item. This diagram can be used in determining the costs per cubic yard for excavating with a trenching machine and also for backfilling with a bulldozer when the total hourly operating costs are known and the output can be reasonably assumed.

Illustrative estimate. Estimate the cost of excavating with a ladder trencher and backfilling with a bulldozer a trench 1,620 ft long, 5 ft deep, and 2.5 ft wide in medium soil which is firm enough so that no bracing is required. The trencher cost \$15,200, has diesel power and a rated capacity of 50 cu yd per hour in medium soil. The bulldozer cost \$7,400 and has diesel power (40 draw bar hp). Costs of moving trencher and bulldozer to and from the job are \$42 and \$12, respectively. The contractor's records show that the ownership and operating costs per working hour are \$5.15 for the trencher and \$2.75 for the bulldozer. These costs were computed by the method given in Chap. 18 and Appendix B. Labor wage is \$5 per hour for machine operators and \$3 per hour for laborers. Payroll taxes and insurance are 11 percent of the labor wage. Overhead is 12 percent of the labor cost. Profit is 10 percent of all other costs.

Work Plan: The trencher crew will be 1 operator and 2 laborers. The bulldozer crew will be 1 operator and 1 laborer. To allow for minor delays, etc., an average output of 37.5 cu yd per hour (0.0267 hr per cubic yard) will be assumed for the trencher and 25 cu yd per hour (0.04 hr per cubic yard) for the bulldozer.

$$\text{Cubic yards in trench} = \frac{1{,}620(5)2.5}{27} = 750$$

Amount of backfill will be the same
Trencher crew-hours = 750 ÷ 37.5 = 20
Bulldozer crew-hours = 750 ÷ 25 = 30

Labor:

Trencher crew cost per hour = ($5 + $3 × 2)1.11 = $12.21
Trencher crew cost = $12.21 × 20 = $244
Bulldozer crew cost per hour = ($5 + $3)1.11 = $8.88
Bulldozer crew cost = $8.88 × 30 = $266
Total labor cost = $244 + $266 = $510

Equipment:

Trencher cost = $5.15 × 20 + $42 =	$145	
Bulldozer cost = $2.75 × 30 + $12 =	95	
Small tools, etc., allow =	10	
Total equipment cost		= 250
Overhead at 12% of labor cost		= 61
Profit at 10% of all other costs		= 82
Total cost		= $903

Cost per cubic yard = $903 ÷ 750 = $1.20

Excavation with Hand Labor The cost of hand excavation depends in general upon the class of excavation, kind and condition of soil, the hourly wages and hourly output of the shovelers. The work will consist of loosening the soil, digging, and backfilling or spreading. This work has been discussed in the articles on Loosening Soil, on Digging Soil by Hand, on Backfilling, on Spreading, and on Trenching with Hand Labor. Tables 3-2, 3-3, 3-4, and 3-8 can be consulted when estimating the output. Then, when the hourly wage is known, the cost per cubic yard can be found by means of Diagram 3-1 or computed by arithmetic, as desired.

Illustrative estimate. Estimate the cost of 23 cu yd of special excavation and 15 cu yd of tamped backfill of ordinary loam (medium soil). This special excavation is required in the construction of the foundation of a large building. Hand labor is to be used at $4 per hour plus 10 percent for payroll taxes and insurance. Allow $6 for equipment, 11 percent of the labor cost for overhead, and 8 percent of all other costs for profit.

From Table 3-4, about 1.30 labor-hours per cubic yard can be assumed for digging, and from Table 3-6 about 1.10 labor-hours per cubic yard can be assumed for backfilling and tamping.

Labor cost for excavation	= \$4(1.10)(1.30)23 =	\$132
Labor cost for backfill and tamp	= \$4(1.10)(1.10)15 =	72
Equipment cost	=	6
Overhead, 11% of \$204	=	22
Profit, 8% of \$232	=	19
Total cost	=	\$251

Excavation with Hand Labor and Wheelbarrows The cost of excavating by hand labor and wheelbarrows will depend upon such items as the kind and condition of the soil, the labor output per hour loosening and shoveling, the capacity of the barrows and their loading, the difficulties of wheeling (steepness of incline and roughness of path), the distance wheeled, the weather, the skill of the workmen, and their wages. The output of workmen loosening and shoveling soil has been discussed in the articles on Loosening Soil and on Digging Soil by Hand Labor, and transportation by wheelbarrows in the article on Transporting Soil. The output with wet soil may be 20 to 40 percent less than with dry soil.

The rated capacities of wheelbarrows vary from about 2 to 5 cu ft with an average of about 3.5 cu ft. Because of the expansion of the soil when loosened, the difficulties of wheeling and the inclination of workmen not to load the barrows to capacity, the actual capacity of a barrow will vary from about 2 to 4 cu ft with an average load of about 2.5 cu ft. Whenever practical, the path for wheeling should be made smooth so that the barrows may be fully loaded.

The limit of economical haul for wheelbarrows is about 150 ft, though in some cases barrows have been used up to 200 or even 300 ft. However, when the length of haul is over 100 ft, the cost of transporting with wheelbarrows should be compared with the cost of transporting with other vehicles. Excavation by hand labor and barrows is costly.

The number of barrows and laborers to be used on a job will depend to some extent upon the size of the job, the number of barrows and men available, and the number that can be used without interference with one another. For example, there may be one laborer with each barrow, or there may be more laborers than barrows with the extra laborers being used to help in the loading. The size of gang and the work of the men in the gang should be such that all the men can be kept busy all the time during working hours.

When rough estimates of time (and cost) are prepared, the time required for one man to load a barrow is usually taken as 3 min (two men, 1.5 min; three men, 1 min), and a man is assumed to travel at a rate of 100 fpm. Time for dumping is very small and can be included in either the loading or traveling time.

However, the total cost of loosening, digging, and moving earth by hand shoveling and wheelbarrows depends upon the time required for one man to loosen, dig, and move 1 cu yd and upon the labor, equipment, and overhead costs. Labor wages may vary from about $2.50 to $4.50 per hour, and overhead costs may vary from 8 to 20 percent of the labor costs. Small tool and barrow costs will not usually be over a few cents an hour per laborer for a pick, shovel, barrow, and runways plus costs of transportation to and from the job. Equipment costs (including transportation), overhead, and profit should be included as separate items.

For convenience, the data in regard to digging and transporting soil by barrows have been summarized in Table 3-9. Values are based upon 2.0 cu ft per load and travel at 100 fpm and include loosening, transporting, and dumping soil.

Table 3-9 *Excavation by Hand Labor with Wheelbarrows (Average barrow load assumed as 2 cu ft)*

Soil	Loosen and load, min	Cu yd per man per hr			Labor-hours per cu yd		
		Haul 50 ft	Haul 100 ft	Haul 150 ft	Haul 50 ft	Haul 100 ft	Haul 150 ft
Light (sand)	1.5–3.0	1.12–1.80	0.90–1.28	0.75–1.00	0.56–0.89	0.78–1.11	1.00–1.33
Medium (loam)	2.0–4.0	0.90–1.50	0.75–1.12	0.64–0.90	0.67–1.11	0.89–1.33	1.11–1.56
Heavy (clay)	2.5–5.0	0.75–1.28	0.64–1.00	0.56–0.82	0.78–1.33	1.00–1.56	1.22–1.78
Hard pan	3.0–6.0	0.64–1.12	0.56–0.90	0.50–0.75	0.89–1.56	1.11–1.78	1.33–2.00

The length of haul (50, 100, or 150 ft) is the length of haul loaded. The trip length will be twice the length of haul.

For other barrow capacities, the outputs will be in proportion. For example, for an average barrow load of 1.5 cu ft, the cubic yards per man per hour will be 1.5 ÷ 2.0, or 0.75 of those given in the table, and the labor-hours required per cubic yard will be 2.0 ÷ 1.5, or 1.33 times the labor-hours given in the table.

Other tables can be prepared for other barrow capacities and loading and wheeling times.

Diagram 3-1 gives the cost per cubic yard of digging and wheeling soil in barrows when the hourly wage is known and the hours required per cubic yard of excavation (digging and wheeling) can be reasonably assumed (as in Table 3-9).

Illustrative estimate. Estimate the cost of excavating 90 cu yd of loam. Weather is good, and soil is fairly dry. Soil will be dug by hand and wheeled in barrows for an average distance of 75 ft. Wage of laborers is $3 per hour, plus 10 percent for payroll taxes and insurance. Equipment costs are assumed as

$8, including transportation. Overhead is 12 percent of the labor cost, and profit is 8 percent of all other costs.

From Table 3-9, it may be assumed that 1 hr will be required to dig and wheel 1 cu yd of soil with barrow loads averaging 2.25 cu ft.

Labor cost = $3(1.10)90	= $297
Equipment cost	= 8
Overhead, 12% of labor cost	= 36
Profit, 8% of all other costs	= 27
Total cost	= $368
Cost per cubic yard = $4.09	

Excavation with Hand Labor and Auto Dump Trucks The method of preparing estimates for excavation with hand labor and auto dump trucks is like that discussed in the preceding article, and is similar to the estimates prepared in the article in Chap. 2 on Combined Handling and Transportation of Materials.

Since the cost per hour of a motor truck and driver is large when compared with the cost of a shoveler, as many shovelers should be used to load each truck as can work conveniently together. This will tend to reduce the time (and cost) required for a trip and, consequently, will reduce the cost of transporting 1 cu yd of soil.

When an estimate is prepared, the following procedure can be used.

1. Compute number of cubic yards of each class and kind of soil to be excavated.
2. Determine number of men in loading gang.
3. Estimate time required for loading one truck.
4. Decide on number of loading gangs.
5. Estimate time for one trip by truck.
6. Estimate number of trucks required. Number of trucks required for each loading gang will be equal to trip time divided by loading time.
7. If the answer is fractional, the size of gang or rate of work or both should be changed, as a fractional part of a truck cannot be provided.
8. Allow for loosening soil.
9. Allow for spreading material at dump (hand labor or bulldozer).
10. Compute hourly cost of entire equipment and crew, including foremen and extra labor. Allow for overhead and profit if these are to be included.
11. Compute hourly output of crew in cubic yards (or compute time required per cubic yard). The cubic yards per hour will equal the average truck load times the number of trucks loaded per hour.
12. Compute cost per cubic yard of excavation. This cost may be itemized if desired.
13. Compute total cost of excavation. This may be itemized if desired.

Diagram 3-3 may be used for finding the cost per cubic yard of excavation after the total hourly cost of the crew and trucks has been computed and the hourly output (or time per cubic yard) has been reasonably assumed.

Some estimators prefer the procedure given in the article on Excavation with Power Excavators and Movers.

Illustrative estimate. Estimate the cost of excavating a basement for a store building. The basement is 50 by 90 ft in size and averages 5.1 ft in depth. The soil is a fairly dry loam and does not require lossening. The length of haul from the basement to the dump is about 0.75 mile. Overhead is 7 percent of the labor and equipment costs. Profit is 8 percent of all other costs. Payroll taxes and insurance are 10 percent of the labor wage. Prevailing hourly wages are foreman $4, truck drivers $3.50, and laborers $3. Contractor has dump trucks of 2.5 cu yd rated capacity each.

Assume average truck speed of 15 mph for the 0.75-mile haul, and average load as 2 cu yd as measured in the ground (or 2.5 cu yd measured loose). Truck ownership and operating cost is $2.10 per hour. Allow 40 cents per hour for small tools.

Total yardage = 90 × 50 × 5.1 ÷ 27 = 850 cu yd

Try a loading crew of 10 shovelers (about as many as can work around one truck). Time to load one truck, assuming 0.85 man-hour to shovel 1 cu yd, will be

2 cu yd × 0.85 ÷ 10 men = 0.17 hr, or 10 min

The basement is large enough so that two shoveling crews can be used.

Time for 1 truck trip = 1 min position + 10 min load + 3 min travel + 2 min unload + 3 min travel + 1 min delays = 20 min

Each gang will need 20 ÷ 10 or 2 trucks, and 2 gangs will need 4 trucks. Two laborers will be needed for leveling dirt at the dump.

The total crew will be 1 foreman, 4 truck drivers, 20 shovelers, and 2 men at the dump.

Labor cost per hour, including payroll taxes and insurance	= ($4 + $3.50 × 4 + $3 × 20 + $3 × 2)1.10	= $ 92.40
Equipment cost per hour	= $2.10 × 4 (trucks) + $0.40 (tools)	= $ 8.80
Overhead cost per hour	= ($92.40 + $8.80) × 7%	= $ 7.08
Profit per hour	= ($92.40 + $8.80 + $7.08) × 8%	= $ 8.66
Total cost per hour	= $116.94, say	$117

Cubic yards excavated per hour with each crew loading 2 cu yd in 10 min (12 cu yd per hour) = 24 cu yd for 2 crews

Hours for the job = 850 ÷ 24 = 35.4, say 36 hr

Total cost of the job = $117 × 36 = $4,212

Cost per cubic yard for 850 cu yd = $4.96

Diagram 3-3 gives a cost of $5.00 per cubic yard.

On this job a small power loader would be cheaper than hand labor.

Excavation with Power Excavators and Movers The power excavators include such machines as power shovels, cranes with buckets, draglines, excavators, tower excavators, and elevating graders. The material may be transported by auto dump trucks or by tractors with crawler wagons. A tractor pulling one or two crawler wagons of 15 cu yd or more capacity each is an economical arrangement when the quantity of earth to be moved is large, the haul comparatively short, and the road difficult. The large quantity handled per trip offsets the slow speed. Auto dump trucks are more economical for long hauls over good highways. Each of the power excavators mentioned has its advantages and disadvantages, which determine its suitability for the particular work.

Excavation with power excavators and movers is a problem in the combined handling and transportation of material such as was discussed in the article on Combined Handling and Transporting of Materials, in Chap. 2. The items of cost are the same as those mentioned there. For economy of operation and of costs, the types of power excavators and movers selected will depend primarily on the characteristics of the job. Enough movers should be available to keep the excavator working full time with no waits on the part either of the excavator or of the movers.

The sizes and capacities of power shovels and dragline excavators are given in Table 3-5. Elevating graders vary somewhat in size and capacity. Their capacity may range from 70 to 150 cu yd per hour. The capacity of a crane and bucket will depend (as in the case of a power shovel) upon the size of the bucket and number of bucket loads handled per minute or per hour. The hourly costs of operating shovels, dragline scrapers, and cranes are approximately the same for machines with the same size of dipper, scraper, or bucket. In general, a power shovel will handle a few more cubic yards of material per hour than a crane or a dragline scraper.

The rated capacities of an auto dump truck vary from 2 to 20 cu yd or more. The actual loaded capacity with loose material will be about 75 or 80 percent of the rated capacity.

Small power loaders, such as tractors with front-end loaders, have bucket capacities varying from ⅜ to 3 cu yd or more. Their capacities per hour are usually a little less than corresponding shovel capacities, but the small power loaders are more mobile, cost less to move from job to job, and are easy to operate. They are preferred for small excavation jobs.

The output of a back hoe is usually less than that of a shovel with the same size dipper, say from 15 to 30 percent less.

Crawler-tractor-scraper units are used for moving large quantities of soil for short distances, say up to about 1,500 ft. Their loading ability is greater than that of wheel-tractor-scraper units, but their speed is low (about 7 mph or less). Crawler-tractor-scraper units come in several sizes. The scraper sizes may vary from about 5 to 20 cu yd or more.

Wheel-tractor-scraper units are usually more economical than crawler-tractor-scraper units for hauls of more than about 1,000 ft because of their greater speed (up to about 30 mph). Their capacities are about the same as the capacities of the crawler-type units. Usually another tractor is needed to assist a wheel-tractor-scraper unit when loading. When the length of haul is a mile or more, a shovel and dump trucks may be more economical.

Diagrams 2-8 to 2-11, 3-2, and 3-4 to 3-9 may be helpful when preparing and checking excavation estimates.

Before starting to prepare an estimate, the articles on Factors Affecting Excavation Costs and on Preparation of Excavation Cost Estimates should be reviewed.

The procedure for preparing an estimate is about the same as that outlined in preceding articles.

Some estimators prefer to consider the entire crew of men and machines as a unit and to compute the hourly cost of the unit and the amount of work done, in cubic yards per hour, by the unit. The procedure is about as follows:

Compute the number of cubic yards of excavation.

Select equipment and crew.

Compute output of equipment and crew in cubic yards per hour.

Compute number of working hours required for the job.

Compute cost per hour. Sum of costs of labor, equipment, overhead, and profit.

Compute total cost of the job.

Compute unit costs per cubic yard as desired.

This procedure is followed in the following illustrative estimate.

Illustrative estimate. Excavation with wheel-tractor-scraper units. Estimate the cost of excavating, transporting, and dumping 43,500 cu yd of loam in a highway-cut job. Soil is waste material, and no compacting is required. Length of haul averages 4,200 ft.

Contractor will use 5 diesel-powered wheel-tractor-scraper units of 13-cu-yd capacity each (heaped) and one 115-hp diesel tractor to assist in loading the scrapers. Haul road is fairly level so that a speed of 12 mph loaded and 18 mph empty can be attained. Pay load of a scraper will be assumed as 10 cu yd bank measure (about 12.5 cu yd loose and heaped).

Trip time will be approximately 2 min load, 4 min travel loaded, 1 min unload and turn, 2.7 min travel empty, and 2.3 min delays, giving a total of 12 min.

Each scraper will haul 5 loads of 10 cu yd each per hour, or 50 cu yd per hour. Five scrapers will haul 250 cu yd per hour.

$$\text{Hours required for job} = 43{,}500 \div 250 = 174$$

Labor: Crew will be 1 foreman, 6 tractor operators, and 2 laborers. Labor cost per hour, including payroll taxes and insurance:

\$6 (foreman) + \$5 × 6 (operators) + \$4 × 2 (laborers) = \$44

Equipment: See Chap. 18 and Appendix B for method of computing costs.
Capital investment = tractor \$16,000, tractor-scrapers \$20,000 each
Ownership expense per working hour = 0.024% for the tractor and 0.0245% for the tractor-scrapers
Ownership cost, tractor = \$16,000 × 0.024% = \$3.84 per hr
Ownership cost, tactor-scraper = \$20,000 × 0.245% = \$4.90 per hr
Operating cost per hour = \$0.010 per horsepower for tractor and \$0.0125 per horsepower for a tractor-scraper
Operating cost, tractor = \$0.010 × 115 = \$1.15 per hour
Operating cost, tractor scraper = \$0.0125 × 115 = \$1.44 per hour

Transportation costs, estimated, are \$20 for the tractor and \$55 for each tractor-scraper unit. Hourly costs, based on 174 hr, are tractor 12 cents and each tractor-scraper 32 cents.

Equipment costs per hour are:

Tractor = \$3.84 + \$1.15 + \$0.12 = \$5.11
Each tractor-scraper = \$4.90 + \$1.44 + \$0.32 = \$6.66
Total equipment cost per hour = \$5.11 + \$6.66 × 5 = \$38.41
Overhead at 6% of labor and equipment costs = \$4.94 per hour
Profit at 8% of all other costs = \$6.99 per hour
Total cost per hour = \$44 + \$38.41 + \$4.94 + \$6.99 = \$94.34
Cost per cubic yard = \$94.34 ÷ 250 (cubic yards per hour) = \$0.38
Total cost of job = \$94.34 × 174 hr = \$16,415

Illustrative estimate. Excavation with tractor loaders and auto dump trucks. Estimate the cost of excavating the basement for a store building. The basement is 100 by 90 ft in size and averages 5.1 ft in depth. The soil is a fairly dry loam. The length of haul from basement to dump is about 0.75 mile. Overhead is 7 percent of labor and equipment costs. Profit is 8 percent of all other costs. Payroll taxes and insurance are 10 percent of the labor wage. Hourly labor wages are tractor operator and truck drivers \$4.50 and laborers \$3. Tractor operator will act as a straw boss. Contractor has a 40-hp diesel tractor with a front-end loader of 1-cu-yd capacity and four 3.5-cu-yd gasoline dump trucks. Allow \$12 for transportation cost of tractor.

Yardage = 100 × 90 × 5.1 ÷ 27 = 1,700 cu yd

Assume truck speed of 20 mph. Truck load is 3 cu yd bank measurement. Truck trip time equals 2.5 min load, 2.25 min travel, 1 min dump, 2.25 min travel, and 2 min delays, giving a total trip time of 10 min. Each truck will make 6 trips per hour and haul 18 cu yd per hour. Four trucks will haul 72 cu yd per hour.

Hours for the job = 1,700 ÷ 72 = 23.6, say 24 hr

Labor: Crew will be 1 tractor operator, 4 truck drivers, and 1 laborer.

Crew cost per hour = ($4.50 + $4.50 × 4 + $3) × 1.10 = $28.05

Equipment (see Appendix B):

Tractor loader, hourly ownership cost = 0.024%		
Ownership cost	= $10,800 (capital investment) × 0.024%	= $2.60 per hour
Operating cost	= $0.0125 × 40 hp	= $0.50 per hour
Transportation cost	= $12 ÷ 24	= $0.50 per hour
Total tractor cost		= $3.60 per hour

Trucks. Capital investment is $7,800 each, and hourly ownership expense is 0.026%.

Ownership cost per working hour = $7,800 × 0.026% = $2.03
Operating cost = $1.27 per hour
Total truck cost = $3.30 per hour each
Total equipment cost per hour = $3.60 + $3.30 × 4 = $16.80
Overhead at 7% of labor and equipment costs = $3.14 per hour
Profit at 8% of all other costs = $3.84 per hour
Total cost per hour = $51.83
Cost per cubic yard = $51.83 ÷ 72 (cubic yard per hour) = $0.72
Total cost of job = $51.83 × 24 hr = $1,244

Illustrative estimate. Excavation with a power shovel and auto dump trucks. Estimate the cost of excavating the surplus soil on a building lot 150 by 225 ft in size, hauling it 3 miles in auto dump trucks, and dumping. Surplus soil averages 8.4 ft in thickness over the entire lot. Soil is a sandy loam that is easy to handle. Contractor will use a diesel crawler shovel with a 2.5-cu-yd dipper and heavy-duty diesel dump trucks of 12-cu-yd capacity. Overhead costs are 7 percent of labor and equipment costs, and profit is 8 percent of all other costs.

Yardage = 150 × 225 × 8.4 ÷ 27 = 10,500 cu yd

Assume that truck capacity is 10 cu yd bank measurement and that the shovel can dig 2 full dipper loads per minute. Assume trucks to average 24 mph loaded and 30 mph empty.

Truck trip time equals 2.5 min position and load, 7.5 min travel loaded, 1 min dump, 6 min travel empty, and 3 min delays, giving a total trip time of 20 min.

Each truck can average 3 trips per hour, and 10 cu yd per trip, giving 30 cu yd per hour.

Shovel can load 60 ÷ (0.5 + 2) = 24 trucks per hour
Trucks required = 24 ÷ 3 = 8
Cubic yards per hour = 30 × 8 trucks = 240
Hours required for the job = 10,500 ÷ 240 = 43.75, say 44 hours

Labor: Crew will be 1 foreman at $4.75 per hour, 1 shovel operator at $4 per hour, 8 truck drivers at $3.75 per hour each, and 2 shovelers at $3 per hour each. Payroll taxes and insurance are 10 percent of the labor wage.

Labor cost per hour = (\$4.75 + \$4 + \$3.75 × 8 + \$3 × 2) × 1.10 = \$49.23

Equipment: See Appendix B. Capital investment is \$68,000 for the shovel, and \$23,000 for each truck. Ownership expense for the shovel is 0.0238 percent of the capital investment, per working hour.

Ownership cost of shovel = \$68,000 × 0.0238% = \$16.20 per hour
Ownership expense for a truck = 0.0213 percent of the capital investment per working hour
Ownership cost of a truck = \$23,000 × 0.0213% = \$4.90 per hour
Operating cost of shovel = \$1.90 per hour
Operating cost of truck = \$1.40 per hour
Transportation cost for the shovel = \$60, or \$60 ÷ 44 = \$1.36 per hour
Cost of shovel = \$16.20 + \$1.90 + \$1.36 = \$19.46 per hour
Cost of one truck = \$4.90 + \$1.40 = \$6.30 per hour
Equipment hourly cost = \$19.46 + \$6.30 × 8 = \$69.86
Overhead hourly cost at 7% of labor and equipment costs = \$8.34
Profit per hour at 8% of all other costs = \$10.19
Total hourly costs = \$137.62
Cost per cubic yard = \$137.62 ÷ 240 cu yd per hour = \$0.57
Total cost for job = \$137.62 × 44 hr = \$6,056

Illustrative estimate. Excavation with a dragline. Estimate the cost of excavating 4,000 cu yd of gravel from a gravel pit and piling the gravel on the bank by the dragline. The dragline is a crawler type with a diesel engine, a 3-cu-yd bucket, and a 60-ft boom. The dragline is located on the bank of the gravel pit, so there are no transportation costs. Dragline operator wage is \$4.50 per hour plus 10 percent for payroll taxes and insurance. Overhead is 6 percent of labor and equipment costs. Profit is 8 percent of all other costs. Capital investment is \$77,000.

Equipment: Ownership cost per hour is 0.024 percent of the capital investment.

Ownership cost per hour = \$77,000 × 0.024%	= \$18.48	
Operating cost per hour	= 2.30	
Total equipment cost per hour		= \$20.78
Labor cost per hour = \$4.50 × 1.10		= 4.95
Overhead cost per hour = 6% of labor and equipment costs		= 1.54
Profit per hour at 8% of all other costs		= 2.18
Total cost per hour		= \$29.45

Assume dragline to dig and dump on pile (90-deg swing) 3 bucket loads every 2 min, or 270 cu yd per hour.

Cost per cubic yard = \$29.45 ÷ 270 = \$0.11
Hours required for job = 4,000 ÷ 270 = 14.83, say 15
Total cost of job = \$29.45 × 15 = \$442.

If a 3-cu-yd clamshell bucket was used instead of a drag bucket, the cost would be about the same as a clamshell bucket works well in an open gravel pit.

Spreading and Compacting Soil On many jobs involving earth fill, the contractor is required to spread the soil in layers and to compact each layer to a uniform thickness. The soil must be moist (not too wet or too dry) to compact readily. The desired moisture content is usually about 12 percent by weight. If the soil is too dry, water must be added, usually by a sprinkler truck. The desired compacted thickness of one layer is usually 6 in. The compacting may be done by crawler tractors, pneumatic tires, smooth rollers of varying weights, or sheep's-foot rollers. The spreading may be done by bulldozers or motor graders.

The specifications usually state the required moisture content by weight, the thickness of a compacted layer, and the method of compacting. From 6 to 15 passes with a sheep's-foot roller may be required. The speed of the roller or compactor should be slow, probably not more than 3 or 4 mph.

The contractor must have sufficient equipment to care for the soil as it is received. He may need one or more motor graders for spreading the soil uniformly, one or more sprinkling trucks for adding water, and one or more crawler tractors each pulling two or more sheep's-foot roller drums.

Illustrative estimate. Spreading and compacting soil. Estimate the cost of spreading and compacting 43,500 cu yd of loam (bank measure) in a highway fill. Soil will be delivered by wheel-tractor-scraper units at the rate of 250 cu yd per hour. The soil has a moisture content of 8 percent, and 12 percent is required. Each layer, when compacted, is to be approximately 6 in. thick. The compacted soil weighs 90 pcf. Specifications require spreading with a motor grader and compacting with 12 passes with sheep's-foot rollers. (This illustrative estimate deals with the spreading and compacting of the soil excavated in the illustrative estimate on Excavation with Wheel-tractor-scraper Units.)

Hours needed for the job = 43,500 ÷ 250 = 174

Two hundred and fifty cubic yards of earth per hour spread 6 in. thick will cover an area of $250 \times 27 \times {}^{12}\!/_{6}$, or 13,500 sq ft per hour.

Spreading: A 75-hp gasoline grader with an 8-ft blade (5 ft effective width) and an average speed of 2 mph, including turns and delays, will cover an area of $2 \times 5{,}280 \times 5$, or 52,800 sq ft per hour per pass. This motor grader should be able to make 3 to 4 passes per hour over an area of 13,500 sq ft. Usually 3 passes are sufficient.

Sprinkling: The water required will be 12 minus 8, or 4 percent, by weight of the soil spread.

Soil spread per hour = $250 \times 27 \times 90$ = 607,500 lb
Water required per hour = $607{,}500 \times 4\%$ = 24,300 lb or 2,930 gal
One gallon of water = 8.33 lb
Contractor has a 2,000-gal gasoline sprinkler truck. He will need 1.5 truck

loads per hr (3 loads every 2 hr). Water costs \$1 per load. Truck has to go 3 miles for water at a speed of 18 mph. Truck trip time is 10 min travel, 5 min load, 10 min travel, 12 min sprinkle, and 3 min delays, giving a total of 40 min. Sprinkler truck can make 3 trips in 2 hr.

Compacting: Area to be covered in one pass is 13,500 sq ft.

Equivalent area for 12 passes = 12 × 13,500 = 162,000 sq ft per hour
Will use sheep's-foot rollers with drums 4 or 6 ft wide
Area covered per hour by one 4-ft roller traveling at 2.75 mph
= 2.75 × 5,280 × 4 = 58,000 sq ft
Rollers needed = 162,000 ÷ 58,000 = 2.8, say 3 rollers 4 ft wide
Two rollers 6 ft wide could be used if desired

Contractor will use a 75-hp diesel crawler tractor to pull the 3 rollers.

Labor: The crew will be 1 foreman at \$5.50 per hour, 1 grader operator at \$5 per hour, 1 tractor operator at \$5 per hour, and 1 truck driver at \$4.50 per hour. Payroll taxes and insurance are 10 percent of the labor wage.

Labor cost per hour = (\$5.50 + \$5 + \$5 + \$4.50) × 1.10 = \$22.00

Equipment: The contractor's records show the following ownership and operating costs per working hour: motor grader \$2.75, sprinkler truck \$6.35, diesel tractor \$3.10, and 3 sheep's-foot rollers \$0.90. Transportation costs for equipment are \$87, or \$87 ÷ 174 gives \$0.50 per hour.

Total equipment cost per hour = \$2.75 + \$6.35 + \$3.10 + \$0.90 + \$0.50
= \$13.60

Water: Three loads in 2 hr at \$1 per load gives an hourly cost of \$1.50.

Overhead at 6% of labor and equipment costs = \$2.14 per hour
Profit at 8% of all other costs = \$3.02 per hour
Total cost per hour = \$40.76
Cost per cubic yard = \$40.76 ÷ 250 (cubic yards per hour) = \$0.163
Total cost of job = \$40.76 × 174 hr = \$7,092

Excavation of Rock The excavation of rock is a problem in the handling and transportation of materials. However, the work of excavating rock is somewhat different from that of excavating ordinary soil. Rock may be classed as soft, medium, or hard, according to difficulty of drilling and blasting it.

For computing yardage, measurements are taken of the rock as in its original position and before loosening. The volume will be larger after loosening and breaking; 1 cu yd of solid rock may be equal to 1.25 to 1.50 cu yd measured loose.

For more complete information in regard to rock excavation, the reader is referred to books on the excavation of rock, rock quarrying, and mining.

The excavation of rock will include the loosening by drilling and blasting and the breaking of larger pieces into smaller ones; the loading of the rock into vehicles for transporting, the transportation of the rock, and the disposal at the end of the transportion route.

The rock is usually drilled by machine drills and broken by blasting. Large pieces may be broken by sledges, falling weights, or blasting or by drilling and blasting.

Drills. There are several kinds of machine drills which are used for drilling rock. Some of the more common are listed below.

Jackhammers are air-operated drills which are small enough to be held and operated by a man. These drills weigh from about 30 to 70 lb each and will drill holes up to about 2.5 in. in diameter and up to about 20 ft in depth.

Drifters are like jackhammers but are larger and heavier, weighing up to about 200 lb, and drilling holes up to about 4.5 in. in diameter and up to 40 ft or more in depth. These drills need to be mounted on masts or tripods. When the mast is mounted on wheels, this drill is called a wagon drill. Drifters and jackhammers are percussion drills; i.e., they drill by breaking the rock into fine pieces by repeated blows.

Rotary and abrasion drills grind the rock by the abrasive action of a rotating bit. Diamond and shot drills are examples of this type.

Core drilling is done by a rotating steel pipe drill, often of the diamond-drill type.

Bits. There are several kinds and types of bits. They may be solid bits of the percussion or rotating types or hollow (core) bits of the rotating type.

A detachable bit is one that can be removed from the drill.

A carbide bit is a detachable bit with cutting edges made of tungsten carbide embedded in a steel base.

A diamond bit is a detachable bit with cutting edges made of small diamonds in a metal base.

Table 3-10 gives approximate rates of drilling holes in rock.

Blasting. The explosives used in blasting may be a slow explosive such as blasting powder or a fast explosive such as dynamite, depending upon the results desired. The amount of explosive required per cubic yard of rock will depend primarily upon the hardness of the rock and may vary from about 0.5 to 2 lb per cubic yard of rock loosened in open cut. Rock in tunnels may require up to 10 lb of explosives per cubic yard of rock loosened. In addition to the explosives, other materials are required, such as blasting caps, primers, fuses, lead wires, and blasting machines.

Other equipment required for loosening rock will include various small tools, sledges, machine drills, bits, masts, tripods, air compressors with power, forges, and apparatus needed for sharpening the drills.

The cost of loosening rock will depend upon the labor wage and output and the amount and cost of the drilling and blasting materials and equipment needed. The cost of power drills, explosives, supplies, and equipment may vary from about 50 to 100 percent of the labor costs. The cost of rock excavation in tunnels will be more than that in open cuts because more explosives will be required and the labor output will be less owing to restricted space, inconvenient working conditions, and longer waits and delays.

Labor. The labor required for loosening 1 cu yd of rock will depend upon the hardness of the rock and the working conditions as well as upon

Table 3-10 *Approximate Drilling Rates in Rock, Feet per Hour*

Drill	Rock	Diameter of hole, in.					
		1.5	2	2.5	3	4	6
Jackhammer	Soft	15–20	12–20	10–15			
	Medium	10–15	8–12	7–10			
	Hard	7–10	5–9	4–8			
Drifter or wagon	Soft	30–50	30–50	30–50	20–50	12–25	
	Medium	25–35	20–35	20–30	15–30	7–15	
	Hard	15–30	12–30	10–25	5–20	3–8	
Diamond	Soft	5–12	5–10	4–9	4–8	3–7	3–6
	Medium	4–10	4–8	3–7	3–6	2–5	2–4
	Hard	3–8	3–7	3–6	2–5	1–4	1–3

the skill and inclination of the workers. Open-cut rock excavation usually requires less labor than tunneling.

The broken rock may be loaded into vehicles (barrows, carts, wagons, trucks, cars on rails, etc.) by hand labor or power equipment, depending upon the amount of material and the working conditions. Power equipment such as power shovels should be used when practical. The cost of loading will depend upon the hourly output in cubic yards and the hourly costs of labor and equipment.

When the labor required on smaller jobs is estimated, the drilling, blasting, and loading are often considered together. Table 3-11 gives the approximate labor required. However, when estimates are prepared for larger jobs, the estimate should be divided into loosening (drilling and blasting) and loading.

Diagram 3-10 (Appendix F) shows the labor costs of loosening (by drilling and blasting) and loading (by hand labor) rock in excavation. The rock is considered as measured in its original position in the ground. From this diagram, the labor costs per cubic yard can be found when the labor-hours per cubic yard and hourly wage are known or the total cost per cubic yard can be found when the labor-hours per cubic yard and the total cost per labor-hour (labor, equipment, overhead, and profit) are known.

Table 3-11 *Labor Required for Drilling, Blasting, and Loading Rock*

Operation	Cu yd per labor-hour			Labor-hours per cu yd		
	Soft	Medium	Hard	Soft	Medium	Hard
Hand drill, plug and feathers	0.05–0.09	0.03–0.06	0.02–0.04	11–20	17–33	25–50
Hand drill, blasting	0.06–0.11	0.04–0.08	0.03–0.06	9–17	12–25	17–33
Machine drill, plug and feathers	0.10–0.18	0.08–0.13	0.06–0.10	5–10	8–13	10–17
Machine drill, blasting	0.20–0.35	0.15–0.25	0.12–0.18	3–5	4–7	5–9

Quantities measured are of rock in place in ground.

The expense of dumping or spreading, as the case may be, will depend upon the work to be done, whether hand or machine labor is to be used, and their costs. The costs of leveling and spreading broken rock will usually be higher than those for leveling and spreading ordinary soil, as the comparative output per hour will probably be less.

The *transportation* costs will depend upon the time required for one trip, the load carried per trip, and the cost per hour or per trip. The time lost owing to delays may be quite large if the transportation equipment must wait for drilling and blasting operations. The work should be planned so that the delays will be a minimum. The transportation equipment used may be barrows or carts propelled by men; cars on rails and propelled by men, horses, motor equipment, or small engines; wagons drawn by horses; wagons drawn by tractors; or trucks. The methods used for estimating transportation costs as explained in other articles of this chapter and of Chap. 2 can be used for estimating transportation costs of rock.

The *overhead* costs of rock excavation will be comparatively large because of the hazards of the occupation, special regulations to be met, etc. Overhead costs may range from 10 to 20 percent of all labor and equipment costs, or from 15 to 35 percent of the labor costs.

The percentage allowed for *profit* should be somewhat higher than that for other excavation jobs because of the greater hazards and uncertainties. Profit may range from 10 to 25 percent of all other estimated costs.

The costs of rock excavation can be summarized as follows: A separate estimate should be prepared for each kind of rock (soft, medium, and hard) and for excavation in open cut and tunneling.

Item	Cost per cu yd	Total cost
Labor	$	$
Equipment		
Overhead		
Profit		
Total	$	$

Factors Affecting Rock-excavation Costs. When cost estimates for rock excavation are prepared, the following items need consideration:

Kind of rock, soft, medium, and hard.
Amount of each kind of rock in cubic yards.
Open-cut excavation or tunneling.
Will bracing or timbering be needed?
Will water have to be cared for?
Loosening method to be used (i.e., machine drilling and blasting).
Materials and equipment needed for drilling and blasting.
Delays in drilling and blasting.
Loading by hand labor or power equipment.
Length of haul and kind of road for transporting broken rock.
Transportation methods (barrows, trucks, etc.).
Equipment needed for transportation.
Transportation delays, especially those caused by drilling and blasting.
Disposal, dumping, leveling, spreading, etc.
Special rules, regulations, and safety requirements to be complied with.
Overhead expenses to be cared for.
Profit to be charged.

Illustrative estimate. Rock excavation. Estimate the costs of drilling, blasting, and loading into autotrucks of 43 cu yd of medium rock and 27 cu yd of hard rock in open cut. Labor wage is $6 per hour, including payroll taxes and insurance. Assume cost of drills and blasting to be $11 per cubic yard for medium rock and $18 per cubic yard for hard rock. Overhead is 22 percent of the labor cost, and profit is 15 percent of all other costs.

From Table 3-11, assume 5 labor-hours per cubic yard for medium rock and 6.5 labor-hours per cubic yard for hard rock.

Item	Cost per cu yd	
	Medium	Hard
Labor	$30.00	$39.00
Equipment	11.00	18.00
Overhead	6.60	8.58
Profit	7.14	9.84
Total per cu yd	$54.74	$75.42

Cost of 43 cu yd of medium rock at $54.74 = $2,354
Cost of 27 cu yd of hard rock at $75.42 = 2,036
Total cost = $4,390

If desired, the labor costs can be checked by Diagram 3-10.

A more detailed and accurate estimate should be made for a larger job. The labor for drilling, blasting, and loading should be listed in detail, and the cost computed. Likewise the materials and equipment needed for drilling, blasting, and loading should be itemized and the costs computed. If the job requires transporting, dumping, and spreading the broken rock, these costs must be included. Such an estimate is much more detailed and accurate than the illustrative estimate given here.

CHAPTER FOUR

Piling and Bracing

General Discussion

Kinds of Piles The two types of piles in general use are sheet and bearing piles. Sheet piles are used for bracing in trenches and excavation, retaining walls, and bulkheads. Their main purpose is to retain earth in place or to keep out water or earth as the case may be. Bearing piles are used for supporting loads. The materials commonly used for piles are wood, steel, and concrete.

Placing and Driving Piles of all kinds may be placed by hand, gin pole, special derrick, special tractor crane, or pile-driver derrick or crane, depending on the size and weight of the piles and the equipment available. Machines should be used for placing large and heavy piles.

Piles may be driven by hand labor and mauls, by drop hammer, by power hammer, or by jetting. Only the smaller and shorter piles can be economically driven by hand.

Drop hammers with pile-driver frames, guides or leads, winches, and engines were used considerably in the past. The drop hammers vary in weight from about 500 to 5,000 lb, and the fall from a few feet up to 20 ft or so. The weight of hammer used depends upon the length and weight of the piles and the character of the soil. Drop hammers require the use of leads or guides and operate at 4 to 10 blows per minute.

Power hammers suspended by means of a boom connected with a derrick or crane are now frequently used for pile driving. The crane may have hanging leads or guides or no leads at all if a template is used. The hammers may be driven by steam or compressed air and may be single, double, or differential acting. In a single-acting hammer, the ram is raised by steam or air and allowed to fall by gravity. In the double-acting hammer, steam or air is used to assist gravity on the fall or downstroke. In a differential acting hammer there are two pistons. Steam or air is used to raise the ram and then to accelerate it on the downstroke. A diesel hammer is a self-contained and self-activated unit using diesel fuel for power. More information on pile hammers is given in the section of this chapter on pile-driving equipment.

There are several formulas for finding the safe static load on a bearing pile. As an estimator rarely has to compute such loads, only one formula will be discussed in this chapter.

The *Engineering News-Record* formula has been long used in connection with pile driving by drop and power hammers.

For a drop hammer, this formula is

$$R = \frac{2Wh}{S + 1}$$

For a single acting air or steam hammer, and for some diesel hammers, this formula is

$$R = \frac{2Wh}{S + 0.1} \qquad \text{or} \qquad R = \frac{20WhB}{B + 10}$$

For a double-acting or differential-acting air or steam hammer, this formula is

$$R = \frac{2E}{S + 0.1} \qquad \text{or} \qquad R = \frac{20EB}{B + 10}$$

where R = safe static load on the pile, lb

W = weight of ram or drop hammer or weight of ram piston for a single-acting power hammer, lb

h = free fall of a drop hammer ram or stroke of a single-acting power hammer, ft

E = energy of the ram piston of a double-acting or a differential-acting power hammer, ft-lb per blow

S = average penetration for last few strokes or blows, in.

B = number of strokes or blows for the last inch of penetration

$S = 1/B$ (approximately)

It is often more convenient to use B instead of S when using power hammers.

Diagram 4-9 shows the relations among the safe bearing load (tons), the energy per blow (Wh or E), and the number of blows (B) for the last inch driven.

The formulas will give fairly good results for piles driven in soils that possess high internal friction without cohesion, such as sand, gravel, and permeable fills. The formulas may not give good results for piles driven in soils possessing cohesion, such as fine-grained silts and soft clays.

Jetting or a combination of jetting with a light power hammer can be advantageously used for driving piles into sand, gravel, or soft-packed clay. The jet water, under pressure, loosens the material and carries it to the surface, as well as lubricating the surface of the pile. If possible, two jets should be used, one on each side of the pile. If one jet is used, it should be used first on one side and then on the other. The jet pipes should not be fastened to the pile but should be kept moving up and down as desired.

Alignment of piles can be secured by fixed leads or guides on the pile-driver derrick, by hanging leads on a crane, by templates (usually of timber), or by other forms of timber frame guides. Leads are not needed when templates are used.

Pulling Piles may be pulled by hand power with a lever arrangement, by jacks, by a line run through a crane or derrick, or by a power pile extractor. Levers and jacks are suitable for small piles in soft soil, but power equipment is usually required for larger and longer piles and in medium or hard soils. Jetting around a pile often helps in loosening the pile while it is being pulled upon.

Other Operations Other operations connected with pile driving are protection of the tops when driving, use of followers, splicing, and cutting off tops.

Piles should be pointed for ease in driving and protection of the lower ends. Special caps can be used if desired. Metal straps over the lower ends of timber piles are a protection. The top of the pile can be protected from the impact of the hammer by using a cushion of timber, old rope, old belting, etc. The cushion should protect the pile head but not absorb too much of the energy of the blow. A steel ring placed around the top of a timber pile will help to prevent brooming. In many cases, caps of metal are placed over the tops of piles to protect the pile heads from being shattered by the blows of the hammer.

A follower can be used when the pile is to be driven below the guides or below the ground or water surface. It may be a short pile or a cylindrical casting of some type placed on the top of the pile to transmit the force of the blow from the hammer to the pile. A follower protects the top of the pile to some extent.

Splicing of piles is necessary when extra-long piles are required or when there is a lack of headroom for driving. The piles can be spliced by placing a metal sleeve over the joint or by bolting fishplates on the sides. Lap joints are rarely satisfactory.

Piles may require cutting off for several reasons. When the top of a timber pile is broomed, the broomed part must be removed. When piles cannot be driven so far as desired, the tops must be cut off to the proper height. Timber can be sawed, steel can be sawed or cut with a flame or torch, and concrete can be cut with a chisel and hammer or a power tool.

Factors Affecting Piling Costs The more important factors affecting the cost of piling are as follows:

Kind of soil. Piles can be driven more easily in some kinds of soils than in others. Driving is easy in mud, silt, or soft loam. Soils like coarse sand, gravel, and hard pan offer much more resistance to driving.

Cost of piles and waling delivered at the job.

Size and weight of pile. The greater the size and weight of the pile, the more the cost of driving.

Depth driven. The greater the depth, the greater the cost.

Spacing of piles. The spacing often determines the amount of moving of the pile-driving equipment.

Method selected. The method chosen for placing and driving the piles should be suitable for the kind of soil encountered and the kind, size, weight, and length of piles used.

Equipment used. The plant or equipment used should be suitable for the soil encountered and the piles used.

Total ownership and operating costs of the equipment.

Amount of pulling to be done.

Amount of other operations (splicing, straightening, cutting off tops, etc.) required.

Skill and wages of foreman and crew.

Special laws and regulations to be complied with.

Overhead and profit.

Preparation of Estimates The cost estimate for piling is usually divided into materials, labor, plant or equipment, overhead, and profit.

The cost of materials is the cost of the piles and waling delivered at the job. This cost includes the first cost plus the transportation to the job plus all incidental expenses connected with the first cost and transportation. If the piles are to be pulled, then the cost of transportation from the job to the storage yard must be added. Also if the piles are to be pulled and used again, then the salvage value should be deducted.

The cost of labor will include the wages of superintendents, foremen, inspectors, and all skilled and unskilled labor used on the job, plus payroll taxes and insurance.

The cost of the plant will be the cost of ownership (or rental cost, if rented), cost of transportation to the job, erection costs, operating costs (including fuel, grease, power, minor repairs, and moving), takedown or dismantling costs, and costs of transportation from the job. Operating costs may or may not include cost of operator.

Overhead costs include such items as have been previously mentioned in preceding chapters. Overhead costs may be based on labor alone, on labor plus plant, or on labor plus plant plus material as the case may be. The percentages usually allowed for overhead will be comparatively high, say from about 10 to 35 when based on labor alone.

Profit is based on the sum of all other costs, the percentage often ranging from 10 to 25.

Perhaps the best way of estimating piling costs is to

1. Compute all material quantities and costs.
2. Decide on crew, and compute hourly cost of crew.
3. Decide on plant, and compute hourly cost of plant.
4. Estimate number of piles handled per hour (placed, driven, pulled, cut off, etc.), hours required per pile, lineal feet per hour, or square feet of wall surface per hour as desired. Then compute total hours required for the job.
5. Estimate amount of waling handled per hour (placed, removed), and compute hours required.
6. Compute total labor and plant costs.
7. Compute overhead costs.
8. Compute profit.
9. Obtain total cost for the job.

Unit costs per pile, per lineal foot of piling, per square foot of piling surface, per lineal foot of waling, etc., can be computed as desired.

Some estimators prefer to make complete estimates for each class of work such as material costs, placing and driving, cutting off, and pulling for piling and material costs, placing, and removal costs for waling. Then the estimates for each class of work are summed up to obtain the total estimate.

The estimate can be summarized about as follows:

Item	Unit cost per pile or per sq ft of wall surface or per lin ft of pile	Total cost
Materials.	$	$
Labor.		
Plant. . .		
Overhead.		
Profit.	________	________
Total.	$	$

*Pile-driving Equipment**

Compressed-air and Steam Power Hammers In a single-acting hammer the weight of the ram may vary from a few hundred pounds up to 20,000 lb or more. The weight of the ram is usually a little less than half of the weight of the hammer.

The double-acting hammer, compared with the single-acting hammer, has a lighter ram, a shorter stroke, a greater striking velocity, and about double the number of blows per minute. The weight of the ram varies from about one-seventh of the weight of a small hammer to about one-third of the weight of a large hammer.

A differential-acting hammer is somewhat like a modified double-acting hammer and has about the same number of blows per minute, though the weight and free fall of the ram are more like that of a single-acting hammer. A differential-acting hammer will drive a pile much faster than a single-acting hammer of the same size.

The data given in Table 4-1 are approximate and do not include all sizes and makes of air and steam hammers. The manufacturers' catalogues should be consulted for more accurate information. The smaller hammers usually operate at more blows per minute than the larger hammers.

Table 4-1 *Approximate Data on Compressed-air and Steam Pile Hammers*

Item	Single acting	Double acting	Differential acting
Rated energy, ft-lb per blow	800–60,000	1,000–55,000	3,600–50,000
Blows per minute	80–50	300–85	150–100
Weight of ram, lb	550–20,000	200–6,000	1,800–20,000
Weight of hammer, lb	1,400–35,000	1,500–21,000	4,100–39,000
Approx. stroke, in	21–40	7–36	10–16
Air 100 psi, cfm	65–2,100	175–2,400	300–1,800
Approx. boiler hp	5–150	15–160	25–120

Diesel Power Hammers Diesel power hammers seem to be economical, light, safe, and effective for hard driving. The blows per minute are less than those of a double-acting air or steam hammer. Diesel power hammers are somewhat similar to single-acting air or steam hammers in operation. The overall length of a diesel hammer is greater than that of an air or steam hammer of corresponding size. The overall length of a diesel hammer may be 13 ft or more. The weight of the ram is

* Drop hammers are discussed in the article on Placing and Driving.

usually about 30 to 40 percent of the weight of the hammer. The ratings (energy per blow) are computed differently by different manufacturers.

The data in Table 4-2 are approximate and do not include all sizes and makes of hammers.

Table 4-2 *Approximate Data for Diesel Power Hammers*

Rated energy per blow, ft-lb	4,500–45,000
Blows per minute	40–106
Weight of ram, lb	1,500–5,000
Weight of hammer, lb	3,000–13,000
Diesel fuel, gph	0.80–3.50
Lubricating oil, gph	0.12–0.50

The gallons of diesel fuel required per hour are approximately equal to

$$\frac{\text{Energy of blow, ft-lb} \times \text{blows per min}}{500{,}000}$$

Air Compressors Most of the air compressors used with air power hammers are of the portable type. Free air is usually delivered at 65, 100, or 125 psi. Compressors may be single-stage or two-stage, have capacities from about 100 to 1,100 cfm, and be gasoline or diesel powered. Portable air compressors may be wheel mounted (four-wheel trailer) or truck or tractor mounted. Very large sizes may be mounted on skids.

The amount of air needed for any air hammer is given in the manufacturers' ratings. However, these ratings do not always allow for leakage and waste, so the compressor selected should be able to deliver 10 or 15 percent more air than the ratings call for. The smaller hammers require more air proportionally.

An approximate formula for finding the amount of air required, with an allowance for leakage and waste, is

$$\text{Air, cfm} = \frac{\text{energy per blow, ft-lb} \times \text{blows per min}}{K}$$

where K = 1,000 for smaller single-acting hammers
= 1,250 for larger single-acting hammers
= 1,500 for smaller double-acting hammers
= 1,700 for larger double-acting hammers
= 1,600 for smaller differential-acting hammers
= 2,000 for larger differential-acting hammers

Cranes The crane selected for pile driving should be of adequate size and capacity to lift and place the piles and to handle the hammer.

That is, the crane capacity in tons should be a little greater than the weights to be handled, and the boom should be longer than the length of the pile plus the length of the hammer. The crane may be truck, crawler, or skid mounted. Capacities may range from 4 to 40 tons or more with booms up to 100 ft in length. Cranes are usually diesel or gasoline powered.

Accessories There are many accessories required in pile-driving work. Some of the accessories required are pile points, driving caps, cushions, leads, guides, templates, air or steam hose and fittings, followers, small tools, and spare parts. A small pickup truck is useful for hauling fuel and supplies.

Boilers are required for steam hammers. Vertical-type boilers may vary from 20 to 90 bhp, and locomotive-type boilers from 25 to 175 bhp, with steam pressure at 150 psi. The smaller boilers may be mounted on wheels, and the larger ones on skids.

Energy Required for Driving Piles The energy required to drive a pile is best expressed in foot-pounds. The amount of energy required to drive a pile depends upon the size and weight of the pile, the depth driven, the kind of soil, the friction between the soil and the pile surface, and various energy losses such as impact, the temporary compression of the pile head and cap, and the internal friction and other losses in the hammer, etc.

These energy losses may be classified as hammer loss, impact loss, temporary compression loss, and soil friction loss. The energy provided by the hammer must be great enough to care for these four losses and have some energy left to push the pile into the ground.

Selecting a Pile Hammer The hammer selected must be able to provide enough energy to care for the losses mentioned in the preceding article and have some energy left to push the pile into the ground. It is good practice to select the largest hammer that can be used without overstressing or damaging the pile.

Experiments have shown that a heavy ram with a short stroke is better than a light ram with a long stroke but delivering the same amount of energy per blow. It is thought that a hammer with a large number of strokes per minute tends to reduce the soil friction by keeping this friction from setting up between strokes and developing its full value.

The tops of piles can be protected by using proper caps. Brooming of the tops of timber piles can be lessened by placing an iron band around the top. The tops of concrete piles can be further protected by "cushions." Pointing or points may be needed in hard soil.

The theoretical energy per blow of a single-acting power hammer is equal to the weight of the ram times the length of stroke. For a double-acting or a differential-acting power hammer, the theoretical energy per blow is equal to the weight of the ram times the length of stroke plus the added effect of the air or steam on the downward stroke.

There is no single factor on which to base the selection of a hammer. For example, the hammer losses are independent of the pile, the impact loss depends on the weight and stroke of the ram and the weight of the pile, the temporary compression loss depends on the material and the force of the blow, and the friction loss depends on the soil and the pile surface.

The hammer efficiency should be high to lessen hammer losses.

Larger and heavier piles require larger and more powerful hammers.

Harder driving requires larger and more powerful hammers.

The weight of the ram should equal or more than equal the weight of the pile so as to reduce impact losses.

The minimum size (foot-pounds per blow) of the hammer needed will vary with the kind of soil, depth driven, and the kind, size, and weight of the pile. Consequently, the values given in Table 4-3 are approximate. In some instances the minimum size of the needed hammer may be less or more than the size suggested in the table.

Table 4-3 *Approximate Sizes of Hammers for Driving Piles from Half to Full Penetration [Size of hammer (ft-lb per blow) = constant in table × wt of pile (lb)]*

Kind and weight of pile	Length of pile, ft	Medium-driving constant	Hard-driving constant
Wood sheet piles, 10–20 psf	15	6	7
	25	5	6
Wood bearing piles, 25–35 lb per lin ft	25	5	6
	50	4	5
	80	3	4
Steel sheet piles, 15–40 psf or 12–60 lb per lin ft	20	3.5	4.0
	40	3.0	3.5
	60	2.5	3.0
Steel bearing piles, H sections, 40–300 lb per lin ft	30	3.0	3.5
	60	2.5	3.0
	90	2.0	2.5
Concrete sheet piles, 75–150 psf or 75–375 lb per lin ft	20	2.5	3.0
	40	2.3	2.7
	60	2.0	2.3
Concrete bearing piles, 100–400 lb per lin ft	30	2.5	3.0
	60	2.0	2.5
	90	1.5	2.0

Sheet Piling

Materials for Sheet Piling Wood, steel, and concrete piling can be used. Wooden sheet piling is used up to lengths of about 20 or 30 ft, steel sheet piling up to 40 or 50 ft or more as it may be spliced, and concrete sheet piling up to about 100 ft. Waling, bracing, and shoring materials are usually of wood, though other materials can be used. Wooden waling materials may vary from 2-in. plank up to 12 by 12 timbers.

Wooden sheet piling may cost \$95 to \$250 per 1,000 fbm f.o.b. plus transportation to the job. Wooden waling and shoring prices are about the same. The salvage of wooden sheet piling and waling may range from about 40 to 90 percent. Thicknesses of wooden sheet piling are 2, 3, or 4 in. with widths of 4, 6, 8, 10, or 12 in. Common sizes of wooden sheet piling are 2 by 6, 2 by 8, and 3 by 8. Lengths are usually available up to 20 ft. The piles may have plain edges or may be tongued and grooved. Wakefield sheet piles are made by nailing or bolting three planks together so that the middle plank forms a tongue and groove. These piles may be 5, 8, or 11 in. thick and 6, 8, 10, or 12 in. wide, depending on sizes of plank used.

Diagram 4-1 (Appendix F) can be used for estimating the cost of wooden sheet piling.

Steel sheet piling costs about \$6 to \$8 per 100 lb (base price) at the mills plus transportation to the job. Costs at the job may range from about \$7 to \$12 per 100 lb. The salvage of steel sheet piling may range from 70 to 95 percent of the number used. Steel sheet piling may vary in net width from about 7 to 20 in., about 15 or 16 in. being common. Lengths up to about 50 ft are usually available. Weights may vary from about 12 to 60 lb per lineal foot or from 15 to 40 lb per square foot of wall. Corrugated sheet steel piling can be obtained in lighter weights. Steel waling and accessories for steel sheet piling may cost \$6 to \$10 per 100 lb (base price) plus transportation to and from the job.

Diagram 4-2 (Appendix F) can be used for estimating the cost of steel sheet piling.

Reinforced concrete sheet piling may cost about \$35 to \$90 per cubic yard (say from \$9 to \$23 per 1,000 lb), plus transportation from casting yard or station to job. Reinforced concrete piling will weigh about 4,000 lb per cubic yard or 150 pcf. This piling is usually priced at so much per pile, with dimensions, mix, reinforcement, etc., as specified. Concrete sheet piling is rarely used when it is to be pulled later. Concrete sheet piles are usually rectangular in shape with tongue and groove joints. Grooves can be grouted after driving to improve watertightness. Thickness may vary from about 6 to 12 in., and widths from about 12 to 30 in. Concrete piles can be made in almost any length, the length being limited by the facilities for transporting, handling, and driving. Concrete sheet piling is economical as to cost when it is to be left in place and when the loads to be restrained are comparatively large.

Whenever possible, the sheet piling and bracing should be designed by a competent engineer. Then the amounts and kinds of the materials can be more accurately computed.

Labor for Sheet Piling The size of the crew used in sheet-piling work may vary from 3 or 4 men with one workman acting as straw boss or foreman up to 10 or 12 men under a foreman or superintendent, depending upon the particular job, size and weight of piles, and the plant used.

For hand driving with mauls or a lightweight (20 or 25 lb) power hammer, a crew of about three or four men is usually satisfactory. For driving with a medium-weight power hammer, a crew of four to six men may be satisfactory. A larger crew is usually required for long heavy piles and heavy equipment.

For pulling, a crew of three or four men may be used with hand equipment such as levers and simple block and tackle. A crew of four or five men or more may be needed with a power hoist and a pile extractor, depending on size and weight of the piles.

For waling, a crew of two or more men should be used, depending on the amount of waling to be done and the size of the timbers used. The amount of waling cut, framed, and placed per labor-hour may vary from about 12 to 75 fbm, or from 13 to 80 labor-hours may be required per 1,000 fbm. About 150 to 400 fbm may be removed per labor-hour, or 2.5 to 7 labor-hours will be needed for removing 1,000 fbm.

The amount of work done per man per hour will vary greatly depending on the various factors mentioned in the article on Factors Affecting Piling Costs. Perhaps the best way is to base the estimate on the number of square feet of wall surface per labor-hour with kind of soil, depth driven, and equipment used as variables. Note that the kind of soil does not affect labor for placing. For sheet piling, approximate labor outputs are given in Table 4-4.

Table 4-4 *Approximate Labor Required for Sheet Piling*

Operation	Square feet of surface* per labor-hour	Labor-hours per 100 sq ft of surface
Placing or setting piles	15–30	3–7
Driving piles	10–40	2.5–10
Pulling piles	15–50	2–7

* This refers to the surface area of the wall or of one side of the pile.

Diagram 4-3 (Appendix F) can be used for estimating the labor costs for sheet piling.

Equipment or Plant for Sheet Piling The amount, kind, and size of equipment used (and the costs also) will vary with the kind of soil and the size, weight, and length of the piles.

For placing light (and comparatively short) wooden and steel sheet piling, no equipment is required as the placing can be readily done by hand. The placing of longer and heavier piles can be done by the pile-driving equipment or by separate equipment such as gin poles, derrick, or cranes. A crawler crane is very satisfactory.

The equipment used for driving may be hand mauls and sledges, light power hammers handled by one or two men, drop hammers with frames and hoisting equipment, medium power hammers suspended by a crane or derrick, or heavy power hammers with heavy frames and other equipment. When practical, the equipment should be mounted so that it can be readily moved as required.

The equipment used for pulling may be rope and hand lever of timber, block and tackle, jacks, hand hoists, power hoists, or power pile extractors as the case may be, depending on the difficulties encountered in starting and pulling the piles, such as kind of soil, depth driven, and weight of pile.

The cost of the equipment used will vary from a few cents per hour for simple hand equipment up to several dollars per hour for the larger types of power equipment. Costs of transportation of equipment to and from the job and of erection and dismantling must be included. Operating costs may or may not include wages of operator and helpers. For example, the hourly operating cost of a small power hammer (25- to 100-lb ram) with an air compressor (including fuel and compressor operation) will be about $2 to $5 per hour. A crawler crane for handling and placing piles may cost $3 to $10 per hour. A medium-weight power hammer with a crane and power equipment may cost about $7 to $15 per hour. Larger power hammers and equipment may cost about $10 to $25 per hour. Drop hammers with frames and hoists may cost about $10 to $25 per hour, depending on weight of hammer and size of frame.

Diagram 4-4 (Appendix F) can be used for estimating the plant or equipment costs for sheet piling.

Overhead and profit Owing to the hazards accompanying sheet-piling work, the percentages assigned to overhead and profit are comparatively high. As stated in the article on Preparation of Estimates, percentages for overhead may range from 15 to 35 percent when based on labor costs alone or from 10 to 25 percent when based on labor and plant costs. Profit usually ranges from 10 to 35 percent of all other costs.

Wooden-sheet-piling Estimates Planks 2 in. thick are commonly used for wooden sheet piling. However, when the depth is fairly great, the load restrained is fairly large, or the wales are spaced some distance apart, thicker piles, say 3 or 4 in., should be used. In general, 2-in. plank is suitable for depths up to 10 or 12 ft, 3-in. plank for depths from 10 to 20 ft, and 4-in. plank for depths from 15 to 25 ft, depending on conditions. Wooden sheet piling is rarely used for depths exceeding 20 ft.

Rangers should be spaced apart horizontally from 3 to 6 ft. The spacing should be closer near the bottom of the excavation. For excavation up to 10 ft deep, 4 by 4 or 4 by 6 rangers spaced not over 4 or 5 ft apart with braces not over 6 or 8 ft apart will usually be satisfactory. For excavations up to about 15 ft, 6 by 6 or 6 by 8 rangers may be needed. Excavation over 15 ft in depth may require 8 by 8 or larger timbers.

Table 4-5 *Approximate Labor Output for Installing Wooden Sheet Piling*

		Sq ft of surface per hr per man		Labor-hr per 100 sq ft of surface	
Depth driving, ft	Method of driving	Place, drive, and brace	Place and drive only	Place, drive, and brace	Place and drive only
5	By hand with maul	10–16	14–25	6.0–10.0	4.0–7.0
10		9–14	12–20	7.0–11.0	5.0–8.5
15		8–12	10–16	8.0–12.5	6.0–10.0
10	By light power hammer	12–20	16–30	5.0–8.0	3.3–6.5
15		10–18	14–27	5.5–10.0	3.7–7.0
20		8–16	13–24	6.0–12.5	4.1–7.5
25		7–15	12–22	6.5–14.0	4.5–8.5

Sizes of cross braces in trenches will vary from 4 by 4 for a trench width of 3 or 4 ft up to about 8 by 8 for a trench width of 12 ft or so.

Diagram 4-1 shows the cost per square foot of wooden sheet piling for varying thicknesses of piles and costs per 1,000 fbm.

Two-inch plank can be driven by hand in fairly soft soil up to 6 or 8 ft in depth. A light power hammer is usually required for greater depths and for harder soils. Wooden sheet piles can be driven all the way by hand when it is possible to keep the excavation at or near the foot of the piles.

The approximate labor output for man per hour for placing, driving, and bracing wooden sheet piling is given in Table 4-5. One gang is supposed to consist of two to four men.

The approximate labor output for removing braces and pulling piling for a gang of two to four men is given in Table 4-6.

Diagram 4-3 shows the relations among hourly labor wages, labor output per hour, and costs per square foot of surface. This diagram can be used either for placing and driving or for pulling the sheet piles when the hourly output per man per hour can be reasonably assumed.

Equipment costs will depend on equipment used. For hand driving by mauls and pulling by timber levers, an allowance of a cent or so per labor-hour will usually be satisfactory. A small power hammer (20- to 25-lb ram) with an air compressor (and no special operator) may cost about $2 an hour plus transportation to and from the job. Gin poles with block and tackle or other hand hoists may cost about 25 cents to $1 an hour plus transportation. A heavier power hammer (say with a ram from 100 to 500 lb) with a movable crane, compressor, fuel and no operator may cost about $6 to $15 per hour plus transportation. When the equipment must be erected and later dismantled, these costs must be included. Overhead and profit percentages will vary about as stated in the article on Overhead and Profit.

Table 4-6 *Approximate Labor Output for Removing Wooden Sheet Piles*

Depth pulled, ft	Method of pulling	Sq ft surface per hr per man: Piling and bracing	Sq ft surface per hr per man: Piling only	Labor-hr per 100 sq ft of surface: Piling and bracing	Labor-hr per 100 sq ft of surface: Piling only
5	By hand with timber lever or puller	20–40	25–50	2.5–5.0	2.0–4.0
10		18–35	23–45	3.0–5.5	2.2–4.5
15		15–30	20–40	3.5–6.5	2.5–5.0
10	By block and tackle and hand hoist on gin pole	18–35	23–45	2.8–5.5	2.2–4.4
15		16–30	20–40	3.3–6.2	2.5–5.0
20		14–26	18–35	3.8–7.1	2.8–5.6
25		12–23	16–30	4.3–8.3	3.3–6.2
10	Start with power extractor and pull by power hoist or crane	23–50	28–60	2.0–4.3	1.7–3.6
15		21–46	25–55	2.2–4.8	1.8–4.0
20		19–42	23–50	2.4–5.3	2.0–4.4
25		17–38	21–45	2.6–5.9	2.2–4.8

Illustrative estimate. Prepare a complete estimate of the cost of installing and removing wooden sheet piling and bracing for an excavation 60 ft wide, 100 ft long, and 10 ft deep in fairly soft soil. The sheet piles are 2 by 8 in. in size by 12 ft in length (the piles must be 2 ft longer than the excavation depth).

Wales are 2 lines of 6- by 6-in. timbers.
Braces are 3 by 8 in. in size, and there are required 34 each of lengths of 9 and 12 ft.
Stakes are 6 by 6 in. in size and 6 ft. long and 34 are needed.
Cost of lumber is \$130 per 1,000 fbm delivered at the job.
Salvage value is estimated at 60 percent.
Cost of trucking lumber from job to contractor's yard and piling is \$2.10 per 1,000 fbm.
Spikes (estimated at 20 lb per 1,000 fbm) cost \$0.20 per pound.
Labor wages are \$4.50 per hour for hammer operator and \$4 per hour for laborers, plus 15 percent for payroll taxes and insurance.
Piles are to be driven by a small power hammer and hand pulled by a timber lever.
All overhead costs assumed as 20 percent of the labor cost.
For profit, 10 percent of the total costs is to be allowed.

Materials Estimate:

$$\text{Sheet piling} = (100 + 100 + 60 + 60)12 \times 2 = 7{,}680 \text{ fbm}$$

$$\text{Wales} = (100 + 100 + 60 + 60)2 \times \frac{6 \times 6}{12} = 1{,}920 \text{ fbm}$$

$$\text{Braces} = 34\left(\frac{3 \times 8}{12}\right)(9 + 12) = 1{,}428 \text{ fbm}$$

$$\text{Stakes} = 34\left(\frac{6 \times 6}{12} \times 6\right) = 612 \text{ fbm}$$

Total lumber	= 11,640 fbm
Allow about 5 or 6% for waste, say	= 660 fbm
Total	= 12,300 fbm

Say 12,300 fbm at \$130 per 1,000	= \$1,599
Deduct salvage value of 60%	= 959
Net cost of lumber	= \$ 640
Spikes, 20 × 12 × \$0.20	= 48
Trucking costs for taking pulled piling and bracing to contractor's yard and piling = 12.3 × \$2.10	= 26
Total material costs	= \$ 714

Labor Estimate: Assume a crew of 1 hammer operator (and straw boss) and 4 laborers for placing and driving piles and installing bracing.

Hourly cost of crew = (\$4.50 + \$4 × 4) × 1.15 = \$23.58

Assume labor output at 16 sq ft of wall surface per hour, or 80 sq ft per crew hour.
Crew-hours = (100 + 100 + 60 + 60) × 12 ÷ 80 = 48
Labor cost of installing piles and bracing = \$23.58 × 48 = \$1,132
Assume a crew of 4 laborers (one as straw boss) for removing piling and bracing.

Hourly cost of this crew = \$4 × 4 × 1.15 = \$18.40

Assume a rate of 35 sq ft of wall surface per man-hour for removing and piling the piles and bracing, or 140 sq ft per crew-hour.
Crew-hours = (100 + 100 + 60 + 60) × 12 ÷ 140 = 27.4, say 28 hr
Labor cost of removing piles and bracing = \$18.40 × 28 = \$516
Total labor cost = \$1,648

Equipment Estimate: Small power hammer, air hose, and compressor cost \$2.10 per hour plus \$10 for transportation. Allow \$5 for timber pile puller.

Equipment cost = \$2.10 × 48 + \$10 + \$5	=	\$ 116
Overhead at 20% of the labor cost	=	330
Profit at 10% of all other costs	=	281
Total cost of job	=	\$3,089

Cost per square foot of wall surface = \$3,089 ÷ 3,840 = \$0.80

Steel-sheet-piling estimates Steel sheet piling varies considerably in net width and in weight per square foot of pile. Extreme ranges in width are from about 7 to 20 in., with 15 or 16 in. being common. Ranges in weights are from about 15 to 40 lb per square foot of wall, depending on the horizontal thrusts to be resisted by the piling and the bracing provided. Steel sheet piling should be used where watertightness of the wall is essential. As steel sheet piling can be pulled and used many times, from eight or ten to fifty, it is economical for use when compared with other materials. As steel piling can be readily spliced, it can be driven to a considerable depth. Lengths of piling (without splices) are available up to about 40 or 50 ft.

Bracing for steel sheet piling is often required and should be provided about as for wooden sheet piling. Sometimes the heavier weights do not require bracing, and in some instances the bracing timbers may be larger and spaced farther apart.

As steel sheet piling is sold by the pound, it is necessary to know the length of pile, net width of pile, and weight per foot of length or per square foot of surface when preparing estimates. As previously stated, the price of steel sheet piling is about \$6 to \$8 per 100 lb at the mill plus freight and other transportation charges to the job. The salvage value is very high, ranging from about 70 to 95 percent. Steel sheet piling can be rented at a cost of \$2 to \$5 per 100 lb plus transportation.

Diagram 4-2 shows the relation among the cost of steel sheet piling per 100 lb, the weight per square foot of wall surface, and the cost per square foot.

The approximate outputs per labor-hour for placing, driving, and bracing steel sheet piling and for removing the bracing and pulling the piling is about the same as that for wooden sheet piling for like depths. Tables 4-7 and 4-8 give approximate values.

Table 4-7 *Approximate Labor Output for Installing Steel Sheet Piling*

Depth driven, ft	Method of driving	Sq ft of surface per hr per man: Place, drive, and brace	Sq ft of surface per hr per man: Place and drive	Labor-hr per 100 sq ft of surface: Place, drive, and brace	Labor-hr per 100 sq ft of surface: Place and drive
5	By hand with maul	10–16	14–25	6.0–10.0	4.0–7.0
10		9–14	12–20	7.0–11.0	5.0–8.5
15		8–12	10–16	8.0–12.5	6.0–10.0
10	By light power hammer	12–20	16–30	5.0–8.0	3.3–6.5
15		10–18	14–27	5.5–10.0	3.7–7.0
20		8–16	13–24	6.0–12.5	4.1–7.5
25		7–15	12–22	6.5–14.0	4.5–8.5
10	By medium power hammer with crane or derrick	14–25	20–35	4.0–7.0	2.8–5.0
20		12–22	17–31	4.5–8.5	3.2–6.0
30		10–20	14–28	5.0–10.0	3.6–7.0
40		8–18	12–25	5.5–12.0	4.0–8.5
50		7–16	10–23	6.0–14.0	4.3–10.0

Table 4-8 *Approximate Labor Output for Removing Steel Sheet Piling*

Depth pulled, ft	Method of pulling	Sq ft of surface per hr per man: Piling and bracing	Sq ft of surface per hr per man: Piling only	Labor-hr per 100 sq ft of surface: Piling and bracing	Labor-hr per 100 sq ft of surface: Piling only
5	By hand with timber lever or puller	20–40	25–50	2.5–5.0	2.0–4.0
10		18–35	23–45	3.0–5.5	2.2–4.5
15		15–30	20–40	3.5–6.5	2.5–5.0
10	By block and tackle and hand hoist or gin pole	18–35	23–45	2.8–5.5	2.2–4.4
15		16–30	20–40	3.3–6.2	2.5–5.0
20		14–26	18–35	3.8–7.1	2.8–5.6
25		12–23	16–30	4.3–8.3	3.3–6.2
10	Start with power extractor and pull by power hoist	23–50	28–60	2.0–4.3	1.7–3.6
20		20–45	24–55	2.2–5.0	1.8–4.2
30		17–40	21–50	2.5–5.8	2.0–4.8
40		15–36	18–45	2.8–6.7	2.2–5.5
50		13–33	16–40	3.0–7.7	2.5–6.2

Diagram 4-3 can be used for estimating labor cost per square foot for either placing, driving, and bracing or removing bracing and pulling steel sheet piling.

Equipment costs will vary greatly depending upon equipment used. These will be about the same as those stated in the preceding article. For power hammers with heavier rams than those mentioned, the hourly costs will be greater. Transportation costs to and from the job and erection and takedown costs should not be omitted.

Overhead and profit percentages will be about as stated in the article on Overhead and Profit.

Contract figures on steel sheet piling work in 1967 gave prices ranging from \$3 to \$9 per square foot of wall surface, depending on weight per square foot of pile, soil, distance driven, and prices of materials and labor. One contract gave a price of about 15 cents per pound for piles in place.

Illustrative estimate. Prepare an estimate of the cost of installing and removing steel sheet piling for an excavation 60 ft wide, 100 ft long, and 10 ft deep in soft soil. The steel sheet piles available are 15 ft long and 16 in. in net width. They weigh 30.7 lb per lineal foot or 23 lb per square foot of wall surface. Piles cost \$7.40 per 100 lb delivered at the contractor's yard. Trucking costs to and from the job are \$1.70 per ton each way. Salvage value is assumed at 95 percent. No wales or braces are needed, as the piles will extend 5 ft below the excavation.

Labor wages per hour are foreman \$5, hammer and crane operators \$4.50, and laborers \$4. Payroll taxes and insurance are 15 percent of the labor wage.

Piles will be driven two at a time by a single-acting power hammer with a rated capacity of 4,100 ft-lb per blow. A portable air compressor with a capacity of 250 cfm at a pressure of 100 psi will be used. A crawler crane of 4-ton capacity and sufficient boom length will be used for handling the piles and hammer and also for pulling the piles.

The cost of transportation to and from the job is \$12 for the compressor and hammer and \$22 for the crane.

The installing crew will be 1 foreman, 1 hammer operator, 1 crane operator, and 3 laborers. This crew and equipment should place and drive 8 piles (2 at a time) per hour. This corresponds to about 27 sq ft of pile surface per man-hour.

Piles will be pulled and loaded on trucks by a crew of 1 crane operator and 2 laborers, using the crane. This crew should pull and load 12 piles per hour, as the piles are only 5 ft in the ground after the excavation.

Overhead is 20 percent of the labor cost.

Profit is 10 per cent of all other costs.

Material Estimate:

$$\text{Sheet piling} = (100 + 100 + 60 + 60) \times 15 \times 23 = 110{,}400 \text{ lb}$$

$$\text{Cost of piles, less salvage,} = \frac{110{,}400}{100} \times \$7.40 \times (1.00 - 0.95) = \$409$$

Cost of trucking $= \dfrac{110{,}400}{2{,}000} \times \$1.70 \times 2$ = 188

Total cost of steel piling = \$409 + \$188 = \$597

Labor Estimate:

Number of piles = (100 + 100 + 60 + 60) × 12 ÷ 16 = 240
Crew hours for installing piles = 240 ÷ 8 = 30
Crew cost per hour for installing = (\$5 + \$4.50 + \$4.50 + \$4 × 3)1.15
= \$29.90, say \$30.
Labor cost of installation = \$30 × 30 = \$900
Crew-hours for pulling and loading piles = 240 ÷ 12 = 20
Crew cost per hour for pulling and loading = (\$4.50 + \$4 × 2)1.15
= \$14.38
Labor cost of pulling and loading piles = \$14.38 × 20 = \$288
Total labor cost for job = \$900 + \$288 = \$1,188

Equipment Estimate: Contractor's records for ownership and operating costs give an hourly cost of \$1.13 for the pile hammer, \$2.48 for the air compressor and hose, and \$4.22 for the crane. Transportation costs are \$12 for the hammer and compressor and \$22 for the crane. Allow \$10 for the cost of small tools.

Cost of hammer, compressor, and hose	= (\$1.13 + \$2.48)30 + \$12	= \$120
Cost of crane	= \$4.22 × 50 hr + \$22	= 233
Cost of small tools		= 10
Total equipment cost	= \$120 + \$233 + \$10	= \$ 363
Overhead at 20% of labor cost		= 238
Profit at 10% of all other costs		= 239
Total cost of job		= \$2,625

Cost per square foot of pile surface = \$2,625 ÷ (320 × 15) = \$0.55
Cost per linear foot of pile = \$2,625 ÷ (240 × 15) = \$0.73

Concrete-sheet-piling Estimates Concrete sheet piling is often used when the piles are to remain in place. These piles are, in general, larger and heavier than most sheet piles of wood or steel. Consequently, power equipment is used for placing and driving. This piling is usually sold at so much per pile, with length, dimensions, and other details being specified. Approximate estimates of the cost per square foot of wall surface for reinforced concrete sheet piles will vary largely with the thickness of the pile, other factors being constant. This cost may range from about 85 cents to \$2.50 per square foot of wall surface for piling 6 in. thick to \$1.40 to \$4 per square foot for piling 12 in. thick. Concrete sheet piling can be designed to resist horizontal thrusts either with or without waling and bracing.

Labor and equipment costs of placing and driving a reinforced concrete sheet pile may be comparatively high because of the larger size and heavier weight when compared with wooden and steel sheet piles. Heavier and sturdier equipment must be provided. However, when the sheet piling is to be left in place, reinforced concrete sheet piling is sometimes economical when compared with the wood and steel piling of equal strength.

The same methods can be used for estimating the cost of concrete sheet piling as were used in the two preceding articles, due allowance being given to the heavier weight of the pile and the stronger equipment needed for handling and driving. The output per labor-hour will usually be a little less than the values given in Table 4-7.

When the estimates are prepared, allowances must be made for grouting, if this is required for making the joints watertight, and for cutting the tops of the piles to grade.

Trench-piling and -bracing Estimates The bracing required in trenches will vary from none, when the walls are firm and the trench is not deep, to skeleton bracing, when only a few braces are needed, to light sheet piling in soft soil, to fairly heavy piling and bracing in very soft and loose soil.

Skeleton piling and bracing consist of placing planks (usually 2 by 4, 2 by 6, or 2 by 8), either vertically or horizontally, every few feet on the opposite sides of the trench and using braces (2 by 4 or 4 by 4) between the planks to keep them in position. The planks can be placed closer together or farther apart as conditions require. The cost of materials for skeleton bracing is not great, as comparatively little material is required and this can be almost all salvaged. The labor cost is also small, as the planks and braces can be put in place by the diggers as needed without taking more than a few minutes of their time out of each hour. The cost of skeleton piling and bracing is sometimes included in the estimates of cost of digging and of tools, no separate estimate being made.

Light piling and bracing are usually composed of wood and can be put in place as the trench is dug. The planks used for sheet piling are placed vertically and close together. Wales and cross braces are used to keep the piling in place. The costs of such piling and bracing can be estimated according to the method given in the article on Wooden Sheet Piling Estimates and by using comparatively high labor outputs per hour for both placing and removing piling and bracing. The planks used for piling are usually 2 by 6 or 2 by 8 in size, the wales may be 2 by 6 or 4 by 4 in size, and the cross braces about 4 by 4.

Medium and heavy sheet piling and bracing for trenches may consist of either wood or steel sheet piling with wooden wales and wooden or steel cross braces. This sheet piling is usually placed and driven as the trench is dug. In very soft or loose soils, the bottoms of the piles should be kept a few inches or more below the bottom of the trench. With soils not quite so loose, the trench can be first dug a little below the bottoms of the piles and then the piles can be driven as far as the trench bottom. In this method, the driving is much easier than driving the piles into the soil in advance of the digging. The method given in the article on Wooden-sheet-piling Estimates can be used for estimating the costs of placing, driving, bracing, and removing the sheet piling and braces, due consideration being given the kind of work. The sheet piling may be a light steel piling if water is to be kept out or 2 by 6 or 2 by 8 planks if the trench is not too deep (say 8 or 10 ft or less) or the horizontal thrust of the soil too great or 3 by 6 or 3 by 8 for deeper trenches and greater soil pressure. Timber 4 in. thick or Wakefield piling can be used if needed. Wales are 4 by 4 or 4 by 6 in. in size and spaced from about 5 to 3 ft apart as the depth of the trench increases. Cross braces 4 by 4 or 4 by 6 in. in size should be placed in contact with the wales and about 8 to 5 ft apart; the deeper the trench, the closer the spacing.

Timber Bracing for Cofferdams Timber is generally used for the bracing in cofferdams. This bracing usually consists of horizontal rangers braced by horizontal struts extending across the cofferdam in one or both directions. In some instances, diagonal braces are used at the corners and at ends where the cofferdam is rectangular in shape, with horizontal braces in one direction. Vertical wales, struts, or posts may be required to keep the horizontal members in place.

The timbers used for cofferdam bracing will be comparatively large so as to carry the load, prevent buckling, and provide free spaces in which the men and machines can work without too much interference. The timbers used may vary from about 4 by 4 to 12 by 12 in. in size.

The amount of material required is usually expressed in feet board measure, with size and length given. To determine the correct amount, a complete takeoff should be made showing all sizes and lengths. Some allowance must be made for cutting and waste. This allowance may range from about 5 to 15 percent.

The cost of the material will be the first cost less salvage value plus transportation to and from the job as the case may be.

The labor required will vary with the skill of the crew, the size of the timbers to be handled, and the working conditions. Carpenters and handymen are usually satisfactory. Labor costs will depend upon the items mentioned and the labor wage. Timbers from 4 by 4 to 8 by 8 in. in size can be placed and removed by hand. For larger timbers, a derrick or crane will usually be needed. The labor required is given in Table 4-9. Installing includes cutting, framing, and erecting.

The equipment used will vary with the tools available, size of timbers, and number of them. The usual hand tools such as saws, hammers, axes, and bits and braces will always be needed. Power tools, such as portable power drills and power saws, should be provided when practicable, as they will save much hand labor. A stationary power saw is sometimes used to saw the timbers to exact lengths. A hand hoist, derrick, or crane is helpful in handling (placing and removing) medium- and large-sized timbers. For the larger jobs and larger sizes of timbers, a power hoist of some kind (derrick or crane) is desirable.

The cost of equipment will vary with the kind and size of tools and machines used. The cost of small hand tools may be a few cents per day

Table 4-9 *Labor Required for Installing and Removing Timber Bracing in Cofferdams*

Size of timber	Crane or derrick	Installing		Removing	
		Fbm per hr	Hr per 1,000 fbm	Fbm per hr	Hr per 1,000 fbm
4 by 4	None	20–50	20–50	250–400	2.5–4.0
6 by 6	None	16–40	25–65	200–350	3.0–5.0
8 by 8	None	12–30	35–85	150–300	3.5–6.5
8 by 8	Yes	20–40	25–50	200–350	3.0–5.0
10 by 10	Yes	18–35	30–55	175–300	3.5–6.0
12 by 12	Yes	15–30	35–70	150–250	4.0–7.0

per laborer. A hand hoist may cost up to 5 or 10 cents per hour. Small power tools such as drills and saws each cost 5 to 35 cents per hour. A hand derrick or crane may cost 20 to 50 cents per hour, and a small power derrick or crane may cost $1 up to $4 per hour with operator extra. These costs are approximate.

As this type of work is hazardous and often difficult to estimate accurately, the labor estimate should be liberal, the overhead fairly high (say 15 to 35 percent when based on labor costs), and the profit reasonably high (say 10 to 35 percent).

The items entering into a complete estimate will be materials, labor, plant or equipment, overhead, and profit.

Bearing Piles

Materials for Bearing Piles The materials commonly used for bearing piles are wood, steel, reinforced concrete, and combinations of steel and concrete such as a steel shell filled with concrete.

Wooden bearing piles are usually round and tapered in shape, about 6 in. in diameter at the small end and 12 to 20 in. in diameter near the butt, and vary in length from a few feet up to a maximum of about 90 ft.

Wooden bearing piles are usually priced at so much per lineal foot, this unit price varying with the size and length of the pile, the nearness to the source of supply, and the transportation to the job site.

Steel bearing piles, usually of IWF or H structural sections, have been used satisfactorily in many places, the H sections being preferred. According to the steel companies, load tests and life tests were very satisfactory. Steel piles are adaptable when they can be driven through loose, unstable soil or fill to a relatively hard stratum, when considerable penetration in hard driving material is required, when extremely great penetration is necessary to secure adequate bearing capacity, and where

Table 4-10 *Approximate* 1967 *Prices for Wood (Pine) Untreated Bearing Piles f.o.b. New York**

Dimensions			
Points, in.	2 ft from butt, in.	Length, ft	Approx. price per lin ft
6	12 at butt	30–39	$0.40–$0.50
6	12	40–49	0.70–1.00
6	12	50–59	1.00–1.10
6	14	60–69	1.15–1.25
6	14	70–79	1.35–1.45
5	14	80–89	1.55–1.70

* The 70- to 89-ft lengths are also available in fir. Treated wood piles may cost from about 30 to 80 cents more per lineal foot, depending on the size of the pile and the treatment.

piling may be susceptible to attack by borers or other insects. The H sections are usually available in the A36, A242, A441, and A442 grades of steel. The size of the H sections may vary from about 8 by 8⅛ in., area of 10.6 sq in., and a weight of 36 lb per lineal foot up to about 14¼ by 14⅞ in., area of 34.44 sq in., and a weight of 117 lb per lineal foot. The sections are readily obtainable in lengths up to about 40 ft and can be spliced. Prices are the same as those for structural shapes and are equal to mill base prices, plus extras, plus freight and other transportation. Mill prices usually range from $6 to $8 per 100 lb. Dealer prices in large cities may range from $8 to $10 per 100 lb. Transportation from dealer, railroad yard, or contractor's yard to the job will vary with the length of haul and other conditions. The resulting price of steel bearing piles at the job may vary from about $8 and up per 100 lb.

Reinforced concrete and combination steel and concrete bearing piles of various types have been designed, made, and used. Concrete bearing piles of the larger sizes and lengths (up to 100 ft long) are economical.

Prices are usually quoted at so much per pile f.o.b. at the casting yard, with dimensions, mix, reinforcement, and other details as specified. Transportation charges to the job must be added. Prices per cubic yard of reinforced concrete may range from $35 to $90 per cubic yard (say from $9 to $23 per 1,000 lb), depending on cost of concrete materials, reinforcement, forms, labor, etc.

Labor for Bearing Piles The labor required for placing, driving, and other operations connected with bearing piles will depend upon the kind of soil, size of pile, depth driven, and equipment used. These conditions vary greatly on different jobs.

Table 4-11 *Approximate Number of Bearing Piles Driven per Hour in Soils with Average Friction Resistance*

Kind of pile and approx. wt, lb per lin ft	Length of pile, ft						
	20	30	40	50	60	75	90
Wood	5	4	3	2.5	2	1.5	1
Steel:							
30	5	4	3				
50	4.5	3.5	2.5	2	1.5		
75	4	3	2.3	1.8	1.3		
100	3.5	2.5	2	1.5	1.2	0.9	0.6
150	3	2	1.7	1.3	1	0.7	0.4
200	2.5	1.8	1.3	1	0.8	0.5	0.3
Concrete:							
100	. . .	. . .	2	1.6	1.3		
150	. . .	. . .	1.6	1.3	1	0.7	0.4
200	. . .	. . .	1.3	1	0.8	0.5	0.3
300	. . .	. . .	1	0.8	0.6	0.4	0.2
400	. . .	. . .	0.7	0.5	0.4	0.2	0.1

Table 4-11 gives the approximate number of bearing piles that can be driven per hour by an adequate crew and equipment under normal working conditions. The rates given in the table are for soils with average friction resistance and medium driving. The rates should be reduced, say 25 percent or more, for soils with high friction resistance and hard driving conditions.

The crew required will vary greatly, depending on the particular job. For example, on a job using small piles, the crew may consist of one or two operators, a foreman, and one to three helpers. The operator may act as foreman in some cases. For medium-sized piles, there may be a foreman, two or three operators, depending on equipment used, and two to six helpers. With large heavy piles, there may be a foreman, two to four operators, and four to ten helpers.

Diagrams 4-5 and 4-6 (Appendix F) can be used for estimating labor costs for bearing piles.

Equipment or Plant for Bearing Piles The equipment or plant required for bearing piles will be about the same as that required for sheet piling except that heavier and larger equipment may be needed for the heavier and larger piles.

The time required to drive a bearing pile is usually greater than that required for a sheet pile, because whereas a sheet pile is driven to a certain depth, the bearing pile is driven until a hard stratum is reached or until the pile is judged capable of bearing the desired load. The time required to drive a bearing pile the last few feet is much greater in proportion. Note formulas given for bearing piles in the article on Placing and Driving.

The equipment used may be a derrick and drop hammer, a crane and power hammer, or a jet with a crane and light hammer. The pile-driving equipment may be used for picking up and placing the piles, or separate equipment (such as a crawler crane) may be used. Cushions or caps are required to protect the pile tops in driving. Followers may be needed when the piles are to be driven below the surface of the water or of the ground. Leads and guides are usually required unless a suitable template (say of timber) is provided for aligning and guiding the piles. Sometimes other equipment is required to straighten the piles as they are driven. The equipment selected should be suitable for the kind of soil and the size, weight, and length of the piles. Some equipment may be needed for cutting the tops of piles and capping them.

The cost of pile-driving equipment will vary greatly with the size and kind used. The cost is usually given as so much per hour for ownership and operation. Transportation to and from the job, erection, and dismantling are added as separate items. Approximate hourly operating costs for pile-driving equipment may vary from about $4 to $20 or more per hour. The cost of a crawler crane may vary from about $4 to $15 per hour. Jetting equipment with pumps and an operator may cost about $7 and up per hour. The costs of templates, guides, blocking, timbering, etc., must not be omitted.

Diagrams 4-7 and 4-8 (Appendix F) can be used for estimating plant or equipment costs for bearing piles.

For jetting operations, pumps of adequate capacity are needed because volume of water is more important than pressure. The size of pipe usually varies from 1.5 to 2.5 in., and the size of the nozzle from 0.5 in. up to the size of the pipe (i.e., no contraction). Two pipes and nozzles with hose and other connections should be provided. The maximum water pressure at the pump may vary from 100 to 225 psi, and the volume of water discharged from 50 to 500 gpm (say 50 to 250 gpm per pipe). A suitable water pump and equipment with power and no operator may cost $2 to $5 per hour depending on capacity of pump and number of jet pipes, plus transportation to and from the job.

The rate of driving piles may vary from 100 lin ft or more per hour for small piles in soft soil up to only a few feet per hour for the larger, heavier, and longer piles in hard ground. When estimates are prepared, the time required to drive one pile, the number of piles driven per hour, or the average number of lineal feet driven per hour may be chosen. The accuracy of the estimate will depend upon the estimator's judgment and his experience on somewhat similar jobs.

Some outputs are as follows:

220 wooden piles, 65 ft long, in 9 hr in *very soft* ground
100 to 200 piles, about 40 ft long, in 8 hr, in soft ground
20 to 60 piles (average 35), 35 to 40 ft long, in 8 hr, in medium soil
10 to 35 piles, about 50 ft long, in 9 hr, in fairly hard soil

Overhead and Profit As pile-driving operations are hazardous and difficult to estimate as to time and cost, the percentages allowed for overhead and profit should be comparatively large. For overhead, the percentage may range from 15 to 35 when based on labor costs alone or from 10 to 25 when based on cost of labor and equipment. The percentage charged for profit may range from 10 to 25 of the sum of all other costs.

Bearing-pile Estimates The different items affecting the costs of piling were discussed in the article on Factors Affecting Piling Costs. As stated in the article on Preparation of Estimates, the main items are materials, labor, equipment or plant, overhead, and profit. The method of estimating piling costs given in that article can be followed, using hours

Table 4-12

Piles driven in place	*Cost per lin ft*
Wood, untreated, 30–60 ft	$ 2.50–$ 3.00
Wood, treated, 30–60 ft	$ 3.00–$ 3.50
Steel H piles, 8 in	$ 6.00–$ 7.00
Steel H piles, 10 in	$ 9.00–$10.50
Steel H piles, 12 in	$11.00–$12.50
Reinforced concrete precast, 10 in	$ 6.00–$ 7.00
Reinforced concrete precast, 14 in	$ 7.50–$ 9.50
Reinforced concrete precast, 18 in	$ 9.00–$13.00

required per pile (or piles per hour) or hours per 100 lin ft (or lineal feet per hour) in step 4.

Some 1967 prices gleaned from various sources are given in Table 4-12.

Diagrams 4-5 to 4-8, inclusive, show relations among hourly costs of crew, hourly costs of equipment, hours per pile (piles per hour), hours per 100 lin ft (lineal feet per hour), costs per pile, and costs per lineal foot.

Illustrative estimate on wooden bearing piles. Estimate the cost in place of 288 treated-timber bearing piles, each 40 ft long and costing $1.55 per linear foot f.o.b. siding. Cost of transporting piles to the job is $1.45 per pile. Template cost is $65 for the job. Soil is medium hard with rock at 40 ft. Piles are to have metal caps costing 55 cents per pile.

Equipment will consist of a 6-ton diesel crawler crane with a 60-ft boom, a 7,200-ft-lb differential-acting power hammer, a 650-cfm diesel-powered portable air compressor with hose and connections, and the necessary small tools. Transportation costs are crane $46; hammer, air compressor, hose, small tools, etc., $22.

Labor crew will be 1 foreman at $5 per hour, 1 hammer operator at $4.50 per hour, 1 crane operator at $4.50 per hour, and 4 helpers at $4 per hour each. Payroll taxes and insurance are 15 percent of the labor wage. This crew and equipment should be able to place and drive 3 piles per hour.

Crew-hours for job = 288 ÷ 3 = 96

Overhead cost is 20 percent of the labor cost.
Profit is 10 percent of all other costs.

Materials:

Cost per pile = $1.55 × 40 + $1.45 + $0.55 = $64
Cost of 288 piles = $64 × 288 = $18,432

Labor:

Crew cost per hour = ($5 + $4.50 + $4.50 + $4 × 4)1.15 = $34.50
Labor cost for job = $34.50 × 96 = $3,312

Equipment: Contractor's books show the following costs per hour: hammer $1.80, air compressor and hose $5.55, and crane $4.65.

Total hourly costs = $12.00

Equipment costs for job, including transportation and template

	= $12 × 96 + $22 + $46 + $65	= $ 1,285
Overhead at 20% of labor cost		= 662
Profit at 10% of all other costs		= 2,369
Total cost of job		= $26,060

Cost per pile = $26,060 ÷ 288 = $90.50
Cost per lineal foot of pile = $90.50 ÷ 40 = $2.26

Illustrative estimate on steel bearing piles. Estimate the cost in place of 128 structural steel bearing piles (H section, 42 lb per lineal foot), each 40 ft long and costing \$157 each f.o.b. siding plus \$3 each for transportation to the job. Template cost is \$65 for the job. Piles are of A36 grade steel.

Equipment will consist of two 6-ton diesel crawler cranes, one with a 60-ft boom; a 7,200-ft-lb differential-acting power hammer; a 650-cfm air compressor with hose and connections; and the necessary small tools. The second crane will be used for handling piles at the job. Transportation costs are cranes \$80; hammer, air compressor, hose, small tools, etc., \$22. Soil is medium hard with rock at 40 ft.

Labor crew will be 1 foreman at \$5 per hour, 1 hammer operator and 2 crane operators at \$4.50 per hour each, and 6 helpers at \$4 per hour each. Payroll taxes and insurance are 15 percent of the labor wage.

Overhead is 20 percent of the labor costs.

Profit is 10 percent of all other costs.

Materials:

Cost per pile = \$157 + \$3 = \$160
Cost of 128 piles = \$160 × 128 = \$20,480

Labor: This crew and equipment should be able to place and drive about 2.75 piles per hour, say 48 hr for 128 piles.

Crew cost per hour = (\$5 + \$4.50 × 3 + \$4 × 6)1.15 = \$48.88
Labor cost for job = \$48.88 × 48 = \$ 2,346

Equipment: The equipment costs per hour will be equal to those in the preceding estimate plus \$4.65 per hour for the second crane, or \$12.00 + \$4.65 = \$16.65.

Equipment costs for job, including transportation and template
= \$16.65 × 48 + \$80 + \$22 + \$65 = \$ 966
Overhead at 20% of labor cost = 469
Profit at 10% of all other costs = 2,426
Total cost of job = \$26,687
Cost per pile = \$26,687 ÷ 128 = \$209
Cost per lineal foot, of pile = \$209 ÷ 40 = \$5.23

Illustrative estimate on reinforced concrete bearing piles. Estimate the cost in place of 128 reinforced concrete bearing piles. Each pile is 40 ft long, weighs 6,000 lb, and costs \$130 at the casting yard plus \$20 for transportation to the job. Soil is medium hard with rock at 40 ft.

This illustrative estimate is the same as the preceding illustrative estimate on steel bearing piles except that reinforced concrete bearing piles are used instead.

In order to drive a 6,000-lb reinforced concrete pile efficiently, a hammer of from 12,000 to 16,000 ft-lb of energy per blow is needed. A diesel hammer with a capacity of 15,000 ft-lb per blow and operating at 50 blows per minute will be used. This hammer will require about 2 gal of diesel fuel oil per hour.

A diesel crawler crane of 10-ton capacity at 10 ft radius and fitted with a 70-ft boom will be used for driving the piles. A smaller diesel crawler crane of about 6-ton capacity will be used for handling the piles at the job. The cost of transporting the two cranes, diesel hammer, and small tools, etc., to and from the job is $90. The cost of the template is $65.

The crew will be 1 foreman at $5 per hour, 1 hammer operator and 2 crane operators at $4.50 per hour each, and 5 helpers at $4 per hour each. Payroll taxes and insurance are 15 percent of the labor wage.

This crew and equipment should be able to place and drive 1.5 piles per hour, or 85 crew-hours for 128 miles.

Overhead is 20 percent of the labor cost.

Profit is 10 percent of all other costs.

Materials:

Cost per pile = $130 + $20 = $150
Cost of 128 piles = $150 × 128 = $19,200

Labor:

Crew cost per hour = ($5 + $4.50 × 3 + $4 × 5)1.15 = $44.28
Labor cost = $44.28 × 85 = $3,764

Equipment: See Appendix B for method of computation. Capital investments are diesel hammer $8,700, 10-ton crane $24,000, and 6-ton crane $15,500. Ownership costs:

Diesel hammer = $8,700 × 0.0321% = $2.80 per hour
10-ton crane = $24,000 × 0.0244% = $5.85 per hour
6-ton crane = $15,500 × 0.0244% = $3.80 per hour

Operating costs:

Diesel hammer = $0.50 per hour
10-ton crane = $0.80 per hour
6-ton crane = $0.55 per hour

Total ownership and operating cost per hour = $14.30
Transportation cost = $90
Template cost = $65

Total equipment costs = $14.30 × 85 hr + $90 + $65	= $1,371
Overhead at 20% of labor cost	= 753
Profit at 10% of all other costs	= 2,509
Total cost for job	= $27,597

Cost per pile = $27,597 ÷ 128 = $216
Cost per linear foot of pile = $216 ÷ 40 = $5.40

If desired, unit costs can be computed and the estimate summarized as follows:

Item	Cost per lin ft	Cost per pile	Total cost
Materials	$3.75	$150	$19,200
Labor	0.74	29	3,764
Equipment	0.27	11	1,371
Overhead	0.15	6	753
Profit	0.49	20	2,509
Total	$5.40	$216	$27,597

NOTE: The costs of the piles, labor wages, and equipment costs used in these three illustrative estimates on bearing piles were not based on actual jobs. The data assumed were used, not for the purpose of comparing the costs for these three kinds of piles, but for the purpose of illustrating the preparation of cost estimates for bearing piles.

CHAPTER FIVE

Concrete

Estimating Concrete Work in General A concrete construction job can be divided into the following divisions or parts in regard to quantities and units:

1. Excavation in cubic yards (See Chap. 3 for excavation estimates.)
2. Forms in square feet of form surface
3. Concrete in cubic yards or cubic feet
4. Reinforcing steel in pounds or tons
5. Finishing in square feet or square yards
6. Curing in cubic yards or cubic feet or in square yards or square feet

The plant, overhead, and profit items can be readily apportioned among these six divisions, though some estimators prefer separate divisions for overhead and profit. The cost of cleaning up should not be omitted.

Each of these six divisions can be subdivided as follows:

a. Materials in units suitable for each kind of material
b. Labor
c. Plant
d. Overhead
e. Profit

Most concrete work is estimated by the *unit-quantity* method instead of by the *total-quantity* method. Most form work is estimated in square feet of form surface, reinforcing steel in pounds, concrete in cubic yards, and finishing in square feet. Unit costs are first found, and then total costs are found for each item by multiplying unit costs by the number of units.

When it is desired to prepare estimates using the unit-quantity method and for different classes or kinds of concrete, the cost per cubic yard (or other suitable unit) of each class of concrete is computed. For example, in a reinforced concrete building the classes of concrete would be foundations and footings, columns, beams and girders, floor slabs, etc. Then when preparing an estimate for one class, such as the columns, the estimator would compute costs per cubic yard and total costs, including such items as forms, reinforcement, concrete, finishing (if any), and curing. This method is used when it is necessary to submit bids or estimates of different kinds of concrete in place.

A complete list of materials, known as the takeoff, is first prepared. The estimator tabulates all the different materials as to kind, number, size, weight, volume, or other units used. The selection of units for different materials will be explained in more detail in later paragraphs.

When material costs are estimated, the costs of the materials at the job are usually computed. This cost includes first cost, freight, unloading, cartage, storage, inspection, testing, and insurance.

When the quantity of labor required for any unit of work is estimated, the total number of labor-hours for each class of labor and each kind of work is listed. Total labor costs for any unit of work are found by multiplying the hours estimated for each class of labor by the corresponding wage rate per hour and then adding results. Payroll taxes, insurance, and other direct labor costs should be included.

The selection of the plant to be used on any particular concrete job often depends upon the machinery and tools that the contractor has available for that job. Sometimes new equipment must be purchased, and this should be considered when costs are estimated.

Plant costs usually include costs of ownership, rentals, transportation, erection, dismantling, and operation. The labor costs of machine operators, such as hoisting engineers, etc, are usually included in the labor costs. For concrete work, plant costs are usually computed for each unit quantity or work, say per square foot or per 100 sq ft of form surface for forms or per cubic yard of concrete, for example. Some estimators prefer hourly rates for the plant, based on the total time the plant is held available for the job, on the actual operating time, or on a combination of both.

Overhead costs include such general office and other labor costs that are not considered as direct productive labor on the job. Other costs included are insurance, rents, office stationery, expense of plans and specifications, interest, legal expenses, traveling expenses, sundries, etc. These overhead expenses are often apportioned to the several parts of the job according to the sum of the materials, labor, and plant costs or according to the labor costs of these parts. Overhead costs on concrete work may range from 5 to 15 percent of the sum of the materials, labor, and plant costs or from 10 to 30 percent of the labor costs alone.

Estimates for profit usually vary from 8 to 15 percent of the total costs.

The estimating of excavation was covered in detail in Chap. 3 and will not be considered in this chapter.

Estimating Forms

Units of Measurement The unit of measurement for forms should be the actual area in square feet of the surface of the concrete in contact

Table 5-1 *Units of Measurement for Forms*

Forms for	*How measured*
Floors	Total area, sq ft
Walls	Total area, sq ft. Forms can be placed on one or both sides
Columns	Circumference of the column times the net height from floor to floor, ft
Column caps, drops, bands, etc	Total area, sq ft
Roofs	Total area, sq ft. When the slope of the roof with the horizontal exceeds 25 deg, the upper side of the roof requires forms
Footings	Total area of concrete surface next to forms, sq ft
Beams and girders	Total area, sq ft. For a beam, this is equal to the net length between columns or supports times the sum of the breadth and twice the depth
Staging and bridging	No definite rules. Total number of 1,000 fbm required should be computed
Moldings and cornices	Total number of lin ft. Other dimensions should be noted
Window sills and copings	Total number of lin ft. Other dimensions should be noted
Stairs	Total area, sq ft, composed of area of the underside, areas of ends, and areas of risers

with the forms. The estimated materials for forms should include materials for struts, posts, bracing, bolts, wire, ties, oiling, cleaning, and repairing but should not include materials for staging and bridging. Forms for each different part of the structure should be listed and described separately. Forms for moldings, window sills, and copings are measured by the lineal foot. No deductions in form measurement are made for openings having an area of less than 25 sq ft, because the extra labor in forming around the openings will often cost more than the value of the lumber saved. No allowance is made for construction joints except in very large structures such as dams.

Materials for Wooden Forms *Quantities.* The approximate quantities of materials required for wooden forms of different types are given in Table 5-2.

Table 5-2 *Approximate Quantities of Materials Required for 100 Sq Ft of Form Surface of New Wooden Forms*

Kind of forms	Lumber, fbm	Nails, bolts, and wire, lb
Footings and piers	200–350	6–11
Walls and partitions	200–270	6–9
Floors	180–280	6–9
Roofs	200–300	6–10
Columns	190–320	6–11
Column caps	200–400	6–12
Beams and girders	300–700	8–16
Stairs	300–600	8–14
Molding and cornice*	200–800	6–15
Sills and lintels*	250–800	7–14

* Per 100 lin ft.

When the work permits the reuse of certain types of forms, perhaps 50 to 80 percent of these forms can be used again after repairs have been made. The extra lumber required for form repairs may vary from about 40 to 200 fbm per 100 sq ft of form surface.

On some jobs, the form lumber can be salvaged and used for other purposes. From 20 to 80 percent can be so used, depending upon the job and the damage done in stripping and cleaning.

If the form surfaces are to be oiled, the amount of oil required may be estimated at ½ to 1 gal per 100 sq ft of form surface.

Staging and bridging should be estimated separately for each job. No general values can be given for this work.

Costs. The cost of form lumber delivered at the job may vary from about $95 to $200 per 1,000 fbm. The cost of ordinary plywood, ⅝ or

¾ in. thick, may range from about $150 to $200 per 1,000 sq ft. Special concrete-form-grade fir plywood, 4- by 8-ft panels, five-ply, sanded two sides, water-resistant glue, carload lots, per 1,000 sq ft of surface may cost from about $175 to $250. Many contractors prefer this special-form plywood because of its higher salvage value, lower labor costs, and better surface when compared with ordinary lumber sheathing.

There is usually some salvage value of the old form lumber. When there is no salvage value, the costs will be comparatively high. The salvage value of the old form lumber may be 20 to 80 percent of its original cost, depending upon care exercised by the workmen and the possible uses to which the old lumber may be put. The salvage value of special concrete form plywood may range from 50 to 90 percent.

Nails, bolts, and wire will cost about 16 to 30 cents per pound on the average when purchased in quantity. The cost of nails, bolts, and wire will vary from about 75 cents to $2 per 100 sq ft of form surface.

Oil for oiling forms will cost a few cents per 100 sq ft of surface, say 7 to 15 cents, depending on kind and price of oil.

Material Diagram. Diagram 5-1 (Appendix F) can be used for estimating the materials cost of wooden forms when the cost per 1,000 fbm is known and the number of board feet required for 1 sq ft of form surface can be reasonably assumed.

Labor for Wooden Forms *Quantities.* The approximate amounts of labor required for various types of wooden forms are given in Table 5-3.

Table 5-3 *Approximate Labor-hours Required for 100 Sq Ft of Form Surface for Wooden Forms*

Kinds of forms	Labor-hours per 100 sq ft of form surface			
	Assemble	Erect	Strip and clean	Repair
Footings and piers	3–7	2–4	2–4	2–5 hr for all types
Walls and partitions	5–9	3–5	2–5	
Floors	3–8	2–4	2–4	
Roofs	3–9	2–5	2–4	
Columns	4–8	2–4	2–4	
Column caps	5–11	3–7	2–5	
Beams and girders	6–10	3–4	2–5	
Stairs	6–12	4–8	3–5	
Molding and cornice*	5–11	3–9	3–5	
Sills and lintels*	5–10	3–6	3–5	

* Per 100 lin ft.

In general, the labor-hours per 100 sq ft of form surface will be less when special concrete fir plywood is used than when plain boards or shiplap are used.

Oiling or wetting the form surface will require a few minutes to about an hour for each 100 sq ft of form surface.

Staging and bridging should be estimated separately for each job, and no general values for this work can be given.

Costs. Form work is often all done by carpenters, but when possible, a combination gang of carpenters, handy men, and ordinary laborers should be used to save expense. The proportion of the different classes of laboring men will vary in any gang, but one-third carpenters, one-third handy men or rough carpenters, and one-third laborers will not be far from the average.

The average cost of laborers will be about as follows: carpenters $4 to $7 per hour, handy men or rough carpenters $3 to $6 per hour, and laborers $2.50 to $4.50 per hour. Wages will vary at different times and in different localities. Labor costs per man-hour for a combination gang consisting of a foreman, carpenters, handy men, and laborers will range from about $3.50 to $6, plus payroll taxes and insurance.

Foreman will cost $5 to $8 per hour, and a good superintendent $6 to $9 per hour.

Labor Diagram. The labor costs for forms are shown in Diagram 5-2 (Appendix F). The diagram can be used for determining the labor cost per square foot for form surface for any kind of work relating to forms, such as assembling, erecting, stripping, cleaning, and repair or any combinations of these kinds of work. To use this diagram, the labor wage per hour and the labor-hours required per 100 sq ft of form surface must be known or assumed. For example, assume that the hourly wage is $4 and that the labor-hours required to assemble, erect, strip, and clean the forms for 100 sq ft of form surface are 6.5. Then, from the diagram, the labor cost per square foot of form surface is 26 cents.

Plant The plant used will depend on the size of the job and the equipment available. On comparatively small jobs, a few workbenches and the hand tools for the carpenters will be all the plant needed. On large jobs, sheds, more benches, power saws, etc., will be required in addition to the ordinary hand tools.

Plant costs will vary from a few cents per 100 sq ft of form surface where benches only and no power saws or other equipment are provided up to about $2, depending upon the amount of mechanical equipment provided and the total area of form surface. The use of power saws and other mechanical equipment will tend to reduce the labor costs.

Overhead and Profit Overhead costs, including general supervision, engineering, and general office expense, will vary from 5 to 15 percent of the total cost, with 10 percent as an average value. When based on labor costs alone, overhead costs will vary from 10 to 30 percent.

Profit on form work (when figured separately) will usually vary from 10 to 25 percent.

Summary The total and unit costs of form work can be summarized as follows:

Item	Unit costs per 100 sq ft or per sq ft	Total costs
Materials	$	$
Labor		
Plant		
Overhead		
Profit		
Total	$	$

Metal Forms A variety of metal forms have been satisfactorily used for concrete work for some years. The first cost of a metal form varies considerably with the weight of metal used, design, shape, and cost of fabrication. Usually the first cost of a metal form is much greater than that of a wooden form. However, when the metal form is rugged and can be used and reused many times (i.e., many more times than a corresponding wooden form), the cost per use may be reasonably low and may in some cases be less than the cost of wooden forms. When a metal form can be used many times, the cost per square foot of form surface may range from about 5 to 30 cents.

Table 5-4 *Approximate Labor Required for Metal Forms*

Item	Labor-hours per 100 sq ft of form surface	Sq ft of form surface per labor-hour
Erection (and assembling)	2–6	17–50
Strip and clean	1–4	25–100
Minor repairs	1–3	33–100
Total	4–13	8–25

The labor costs for erecting, cleaning, and stripping metal forms will usually be less than for assembling, erecting, stripping, and cleaning wooden forms that can be used for the same kind of concrete work. Assembling costs, if any, are usually included with the erection costs. The labor required for metal forms is shown in Table 5-4.

Diagram 5-2 can be used for estimating the labor costs of metal forms when the labor wage is known and the hourly output can be readily assumed.

In some cities, the metal-form work is let as a subcontract. The subcontractor supplies the forms and furnishes all the labor required for a complete job of forming.

In some localities, the general contractor or the concrete contractor may rent metal forms for use on any particular job. Rental prices vary considerably according to type and make of forms.

Because of the many varieties and designs of metal forms on the market, it is impractical in this text to give any estimates of the cost of metal forms.

Illustrative estimate for wooden forms. Make a complete estimate of the cost for the following concrete forms, including the form lumber required in board feet, pounds of nails and bolts, labor in hours, and costs of each for the following form surfaces (Table 5-5): foundations and footings 785 sq ft, walls 2,180 sq ft, columns 3,150 sq ft, column heads 1,780 sq ft, floors 14,200 sq ft, beams 3,160 sq ft, roofs 4,100 sq ft. Take unit prices as follows: form lumber $150 per 1,000 fbm delivered at the job, bolts and nails 20 cents per pound, 1 foreman at $6 per hour, 3 carpenters at $5 per hour, 3 handy men at $4.50 per hour, and 3 laborers at $4 per hour. Assume 8 percent of total cost for overhead and 10 percent for profit. Forms are to be made and erected and later stripped and cleaned. Assume salvage value of form lumber to be 60 percent. (Some estimators use different salvage values for different forms.) Payroll taxes and insurance, 12 percent of the labor wage.

Table 5-5 *Materials Estimate*

Item	Form surface, sq ft	Fbm per sq ft of form surface	Lumber, fbm
Foundations and footings	785	3.0	2,355
Walls	2,180	2.5	5,450
Columns	3,150	2.5	7,875
Column heads	1,780	3.0	5,340
Floors	14,200	2.3	32,660
Beams	3,160	5.0	15,800
Roofs	4,100	2.5	10,250
Total	29,355	...	79,730

About 80,000 fbm is required.

The values for number of board feet per square foot of form surface are estimated for the several items, and represent fair average values.

Cost of form lumber less salvage value is \$150 × (1.00 − 0.60), or \$60 per 1,000 fbm.

Cost of form lumber at \$60 per 1,000 fbm = \$60 × 80.000 = \$4,800
Cost of nails and bolts, etc., at 8 lb per 100 sq ft of form surface
and \$0.20 per pound = \$0.20 × 8 × 294 = 470
Total cost (assuming 60% salvage value) = \$5,270

If desired, the materials estimate can be prepared with the aid of Diagram 5-1. The cost of lumber per 1,000 fbm delivered at the job is \$150 less salvage value, or \$60. The cost per square foot of form surface is obtained from Diagram 5-1. Computations can be made by slide rule to the nearest dollar. This estimate can be tabulated as shown in Table 5-6.

The labor estimate can be prepared as in Table 5-7.

Labor-hours per 100 sq ft of form surface are average values for the different kinds of form work.

Total wages per hour for gang, including payroll taxes and insurance
= (\$6 + \$5 × 3 + \$4.50 × 3 + \$4 × 3) × 1.12 = \$52.08

Average wage per hour = \$5.21

Table 5-6 *Materials Estimate Using Diagram* 5-1

Item	Form surface, sq ft	Fbm per sq ft of form surface	Cost per sq ft of form surface	Cost
Foundations and footings	785	3.0	\$0.18	\$ 141
Walls	2,180	2.5	0.15	327
Columns	3,150	2.5	0.15	473
Column heads	1,780	3.0	0.18	320
Floors	14,200	2.3	0.14	1,988
Beams	3,160	5.0	0.30	948
Roofs	4,100	2.5	0.15	615
Total for lumber	29,355	...	.....	\$4,812
Add for nails and bolts, 2,350 lb at \$0.20				\$ 470
Total for lumber, nails, and bolts				\$5,282

In computing this average wage, it is assumed that the presence and ability of the foreman to direct the work of the men will be at least equivalent to the work of one man.

Labor cost for forms is \$5.21 × 3,340 = \$17,401

Table 5-7 *Labor Estimate*

Item	Form surface, sq ft	Labor-hours per 100 sq ft form surface: Assemble and erect	Labor-hours per 100 sq ft form surface: Strip and clean	Total labor-hours	Labor costs
Foundations and footings.	785	7	3	79	$ 412
Walls..................	2,180	10	3	284	1,480
Columns.................	3,150	8	4	378	1,969
Column heads...........	1,780	12	4	285	1,485
Floors..................	14,200	7	3	1,420	7,398
Beams..................	3,160	10	4	443	2,308
Roofs..................	4,100	8	3	451	2,350
Total.................	29,355	...	...	3,340	$17,402

If desired, the labor estimate can be prepared with the aid of Diagram 5-2. The average wage assumed for the 10-men gang is $5.21, as before. The estimate can be tabulated as shown in Table 5-8.

Labor-hours required = $17,390 ÷ $5.21 = 3,340

Table 5-8 *Labor Estimate Using Diagram* 5-2

Item	Form surface, sq ft	Labor-hours per 100 sq ft form surface*	Unit costs per sq ft of form surface†	Total labor costs
Foundations and footings..	785	10	$0.52	$ 410
Walls....................	2,180	13	0.68	1,480
Columns..................	3,150	12	0.63	1,990
Column heads............	1,780	16	0.83	1,480
Floors...................	14,200	10	0.52	7,380
Beams...................	3,160	14	0.73	2,310
Roofs...................	4,100	11	0.57	2,340
Total..................	29,355	...	.....	$17,390

* Assemble, erect, strip, and clean.
† From Diagram 5-2.

In this illustrative estimate, the materials and labor estimates have each been prepared with and without the use of diagrams. A slide rule was used to reduce the time required for multiplication and division.

Plant costs will be small and will be taken at about 50 cents per 100 sq ft of form surface, or about $150 for the job, assuming that very little mechanical equipment will be used.

Overhead costs will be 8 percent materials, labor, and plant costs, or 8 percent of $22,822, or $1,826.

Profit will be 10 percent of all other costs, or 10 percent of $24,648, or $2,465.

This estimate for forms may be summarized as follows:

Item	Foundations and footings	Walls	Columns	Column heads	Floors	Beams	Roofs	Totals
Materials	$155	$ 359	$ 520	$ 351	$ 2,155	$1,043	$ 678	$ 5,270
Labor	412	1,480	1,969	1,485	7,398	2,308	2,350	17,402
Plant	4	13	17	13	63	20	20	150
Overhead	46	148	200	148	770	270	244	1,826
Profit	62	200	271	200	1,039	364	329	2,465
Total	$679	$2,200	$2,977	$2,197	$11,424	$4,005	$3,621	$27,113
Cost per 100 sq ft	$86	$101	$94	$123	$81	$127	$88	$93

Estimating Concrete

Concrete Takeoff The first step in preparing an estimate for concrete is to takeoff or list all the quantities of concrete required for all the different parts of the work and for all the different mixes.

In preparing the list of concrete quantities, it is customary to use a sheet or page for each different mix and to arrange the columns on each page in about the following order:

Description (footing, beam, etc.)
Dimensions (of each unit in feet)
Volume (cubic feet or cubic yards)

Other columns can be added for unit and total prices, etc.

In taking off quantities of concrete, it is customary to begin at the bottom or one end of the structure and go over it systematically. In a concrete building, the order of takeoff would be about as follows:

Footings
Foundation walls

Columns (interior and exterior). (Ordinary column caps and brackets are usually included with columns.)

Floor and roof slabs
Drop panels
Beams and girders (exterior and interior)
Exterior walls
Partitions
Window sills and copings
Stairs
Miscellaneous
Sidewalks and drives

Units of Measurement All concrete is usually measured net, as fixed or placed in the structure. Units of measurement are cubic yards or cubic feet, cubic yards being the common unit. No deductions are made for steel beams and reinforcement in the concrete unless the steel has a cross-sectional area of more than 1 sq ft. No deductions are made for pipes or holes having a sectional area of less than 1 sq ft. Each mix of concrete should be measured and described separately, and the concrete in different members of the structure should be measured and described separately according to location or purpose of the work.

Many rules are given for measuring stairs, but the best rule is to compute the quantity of concrete required in cubic yards or cubic feet by some simple method.

Sidewalks and pavements should be measured by the square foot or square yard with the thickness and mix stated so that the number of cubic yards or cubic feet of each mix can be easily computed.

The unit of measure for precast concrete work is usually the cubic foot.

Kinds and Measurement of Materials The materials used in concrete are portland cement, water, and fine and coarse aggregates. They may be measured by weight or by volume.

The cement is portland cement Type I unless another kind or type is specified. Cement is usually measured by the sack. There are several types of portland cement.

The water should be free from injurious amounts of acids, alkalis, organic matter, or other injurious ingredients. A water suitable for drinking is usually satisfactory. Water is usually measured by the U.S. gallon.

The fine aggregate is usually a clean, well-graded sand up to $3/8$ in. in size.

The coarse aggregate is usually a well-graded clean gravel or crushed stone ranging in size from $1/4$ in. up to the maximum size permissible for the kind of work in question.

The aggregate may be measured by the pound, ton, cubic foot, or cubic yard.

Ready-mix concrete is usually measured and sold by the cubic yard with the proportions of the mix and other details stated.

The following information may be helpful.

Cement:

1 sack = 1 cu ft = 94 lb
1 barrel = 4 sacks = 4 cu ft = 376 lb

Water:

1 U.S. gal = 231 cu in. = 0.1337 cu ft = 8.35 lb
1 cu ft = 7.5 U.S. gal = 62.4 lb

Aggregate:

1 ton = 2,000 lb = 19 cu ft = 0.70 cu yd
1 cu yd = 27 cu ft = 2,800 lb = 1.40 ton

Ready mix concrete:

1 cu yd = 27 cu ft = 2 ton = 4,000 lb
1 ton = 2,000 lb = 0.50 cu yd = 13.5 cu ft

Proportioning of Materials The proportioning of the materials in a concrete mix is usually based on the water-cement ratio stated as U.S. gallons of water per sack of cement. The cement is usually measured in sacks, the water in U.S. gallons, and the aggregate in pounds or tons. Sometimes the aggregates are measured by volume in cubic feet or cubic yards.

Proportions of the aggregates by weight may be changed to proportions by volume by dividing the weights by the corresponding weights per cubic foot of the aggregates in question. For example, if the weight of sand in a one sack batch is 210 lb and the weight per cubic foot is 105 lb, the corresponding proportion by volume would be 210 ÷ 105 or 2 cu ft.

When proportioning and mixing the concrete in the field, the free water in the aggregates must be allowed for. The presence of a small amount of water in the sand often causes the sand to bulk or swell. For some sands this bulking effect may be as much as 20 to 30 percent, and usually occurs with about 6 or 7 percent of free water. The bulking effect is practically negligible when the sand is completely inundated. However, when estimating the quantities of aggregates required, it is rarely necessary to allow for the free water unless the sand is measured by volume. Aggregates are assumed to be surface dry when computing quantities in estimating.

Concrete mixes may be proportioned for strength (compressive or flexural), durability, and workability (consistency). The strength and durability are determined by the water-cement ratio and the workability by the slump.

Tables 5-9, 5-10, and 5-11 show the relations between strength and water-cement ratio, durability and water-cement ratio, and workability and slump. The values are approximate.

Table 5-9 *Water-cement Ratios and Approximate Strengths of Concrete Mixes*

Water-cement ratio U.S. gal per sack	Approximate strength, psi	
	Compressive	Flexural (bending)
4	6,000	700
5	5,000	625
6	4,000	550
7	3,300	500
8	2,800	450

Table 5-10 *Suggested Water-cement Ratios for Durability*

Type of construction	Water-cement ratio, U.S. gal per sack	
	Severe or moderate climate	Mild or warm climate
Massive (thick) sections	5.5–7.0	5.5–7.5
Pavement slabs	5.5–6.5	6.0–7.0
Medium sections	5.5–6.5	5.5–7.0
Thin sections	5.0–6.0	5.0–6.0

Table 5-11 *Suggested Slumps for Workability*

Type of construction	*Slump, in.*
Heavy mass construction	1–3
Reinforced footings, foundations, and walls	2–5
Plain footings, caissons, and substructure walls	1–4
Reinforced slabs, beams, and walls	3–6
Building columns	3–6
Pavement slabs	2–3

A stiff mix is one with a slump of about 2 in. or less.
A medium mix is one with a slump of about 2 to 4 in.
A wet mix is one with a slump of about 4 to 6 in.

Quantities of Materials After the takeoff has been completed and the total yardage of each kind of concrete found, the quantities of cement, water, and fine and coarse aggregates can be computed.

When the proportions of a concrete mix are specified, the required quantities of the materials per cubic yard of concrete may be computed by

Table 5-12 *Approximate Mixes for Concrete of Medium Consistency (3-in. slump) with Medium Sand*

	Per sack of cement			Per cubic yard of concrete			
Maximum size of aggregate, in.	Fine aggregate, lb	Coarse aggregate, lb	Yield, cu ft of concrete per sack	Cement, sacks	Water, U.S. gal	Fine aggregate, tons	Coarse aggregate, tons
Water = 5 U.S. gal per sack of cement. Compressive strength, 28 day = 5,000 psi							
0.75	180	220	3.57	7.6	38	0.685	0.835
1.00	165	250	3.67	7.4	37	0.610	0.925
1.50	160	290	3.91	7.0	35	0.560	1.015
2.00	160	325	4.14	6.6	33	0.530	1.070
Water = 6 U.S. gal per sack of cement. Compressive strength, 28 day = 4,000 psi							
0.75	235	265	4.29	6.3	38	0.740	0.835
1.00	215	295	4.36	6.2	37	0.665	0.915
1.50	210	345	4.66	5.8	35	0.610	1.000
2.00	210	390	4.96	5.5	33	0.580	1.070
Water = 7 U.S. gal per sack of cement. Compressive strength, 28 day = 3,300 psi							
0.75	290	305	4.98	5.45	38	0.790	0.830
1.00	270	340	5.08	5.35	37.5	0.725	0.915
1.50	265	395	5.42	5.0	35	0.660	0.985
2.00	265	450	5.78	4.7	33	0.625	1.060
Water = 8 U.S. gal per sack of cement. Compressive strength, 28 day = 2,800 psi							
0.75	345	330	5.57	4.75	38	0.820	0.790
1.00	330	385	5.84	4.65	37	0.770	0.895
1.50	320	440	6.14	4.40	35	0.705	0.965
2.00	325	510	6.64	4.10	33	0.670	1.045

the absolute volume method or by the simplified method explained in the article on Quantities by Weight by Absolute-volume Method and in the Illustrative Problem of this chapter. If ready-mix concrete is used, only the yardage of each kind of concrete is needed.

On some jobs the proportions of the mix are not given. Sometimes the strength, water-cement ratio, and (or) slump may be specified. Then the estimator will need to determine the proportions of the mix (or mixes) and compute the quantities required. Sometimes trial mixes may have to be made. Usually the estimator may obtain sufficient information for estimating purposes from data already found and tabulated, as in Table 5-12. The mixes listed in this table are those suggested by the Portland Cement Association. For more information on concrete mixtures, the reader is referred to the booklet "Design and Control of Concrete Mixtures" published by the Portland Cement Association, Chicago, Ill.

The materials referred to in Table 5-12 are portland cement Type I, suitable water, fine aggregate (a well graded, clean, surface-dry medium sand), and coarse aggregate (a well-graded, clean, surface-dry gravel or crushed stone). In Table 5-12, the quantities of materials required were computed by the absolute-volume method.

Note that on a job, extra water will be needed for washing and cleaning. The total amount of water required per cubic yard of concrete may vary from about 70 to 150 gal. About 100 gal of water per cubic yard of concrete is a fair value to use for estimating purposes.

Table 5-12 has been prepared for mixes having a slump of about 3 in. Approximate workable mixes, having other slumps, may be estimated as follows:

One-sack Batch. To increase the slump, decrease the weights of each of the fine and coarse aggregates by 3 percent for each increase of 1 in. in slump.

To decrease the slump, increase the weights of each of the fine and coarse aggregates by 3 percent for each decrease of 1 in. in slump.

One-cubic-yard Batch. To increase the slump, increase the quantities of cement and water by 3 percent for each increase of 1 in. in slump.

To decrease the slump, decrease the quantities of cement and water by 3 percent for each decrease of 1 in. in slump.

Quantities by Weight by the Absolute-volume Method For an accurate method of computing the sacks of cement, U.S. gallons of water, and the weights of the fine and coarse aggregates, the absolute-volume method is recommended. The proportions of a concrete mix are usually given by the cement in sacks, the water in U.S. gallons, and the fine and coarse aggregates in pounds or tons. The amount of concrete made by a one-sack batch is first computed and then the amounts of materials required for one cubic yard of concrete are computed.

When making computations by the absolute-volume method, the apparent specific gravities and the solid weights of the various materials are needed. The values given in Table 5-13 may be used for estimating purposes. However, the actual apparent specific gravities must be used for special aggregates, such as cinders and blast-furnace slag. The

Table 5-13

Material	Apparent specific gravity	Solid weight per cu ft
Water (fluid weight)	1.00	62.4
Portland cement	3.15	196.5
Average sand	2.65	165.5
Average gravel or coarse aggregate	2.65	165.5

amount of entrained air in regular concrete mixes is very small and may be neglected when computing quantities. One U.S. gallon of water = 8.35 lb = 231 cu in = 0.134 cu ft. One sack of portland cement, Type I, = 0.481 solid cu ft.

Illustrative problem. For an example of the absolute-volume method, compute the quantities of materials required for 1 cu yd of concrete. The stated proportions are: 1 sk of portland cement, Type I, to 5.5 U. S. gal of water to 185 lb of surface dry medium sand to 280 lb of surface-dry gravel.

Solution: First compute the yield or cubic feet of concrete made by a 1-sk batch.

$$\text{Cement, 1 sack} = 94 \text{ lb} = \frac{94}{196.5} = 0.480 \text{ cu ft}$$

$$\text{Water, 5.5 U.S. gal} = \frac{5.5 \times 8.35}{62.4} = 5.5 \times 0.134 = 0.735 \text{ cu ft}$$

$$\text{Sand, 185 lb} = \frac{185}{165.5} = 1.115 \text{ cu ft}$$

$$\text{Gravel, 280 lb} = \frac{280}{165.5} = 1.695 \text{ cu ft}$$

Absolute volume of a one sack batch = 4.025 cu ft of concrete.

That is, one sack of cement with the correct proportions of water and surface-dry aggregates will make 4.025 cu ft of concrete.

Quantities required for 1 cu yd of concrete are:

Cement = 27 ÷ 4.025 = 6.7 sk
Water = 6.7 × 5.5 = 36.85 or 37 U.S. gal
Sand = 6.7 × 185 = 1,240 lb = 0.62 tons
Gravel = 6.7 × 280 = 1,875 lb = 0.94 tons

The total weight of 1 cu yd of concrete is 4,055 lb

Simplified Solution: A variation of this method is to assume that 1 cu yd of concrete made with average aggregates, having an apparent specific gravity of 2.65, will weigh about 4,000 lb per cubic yard, or about 150 lb per cubic foot.

Then a 1-sk batch made with 1 sk of cement, 5.5 U.S. gal of water, 185 lb of surface-dry sand, and 280 lb of surface-dry gravel will weigh: 94 + (5.5 × 8.35) + 185 + 280 = 605 lb.

Cement, sacks per cubic yard of concrete = 4,000 ÷ 605 = 6.65 sk
Water, U.S. gal per cubic yard = 5.5 × 6.65 = 36.6 or 37 U.S. gal
Sand per cubic yard = 6.65 × 185 = 1.230 lb = 0.615 tons
Gravel per cubic yard = 6.65 × 280 = 1,860 lb = 0.93 tons

This gives a total weight per cubic yard of concrete of 4,025 lb.
These results agree closely with those obtained previously.

Costs of Materials The cost of cement delivered on the job is generally used, though on very large jobs the cost of cement will be the price at the mill plus costs of freight, unloading, trucking to job, storing, inspection and testing, and loss due to waste and spoiling.

Mill prices per barrel may be quoted for the cement in bulk or in paper bags and in carload lots f.o.b. the mill or in cars at a station near the job. The price of the cement (without sacks) may range from about $3.50 to $5 per barrel. Paper bags cost about 60 cents per barrel and are not returnable. Prices are higher for quantities less than carload lots. Testing may cost from 2 to 7 cents a sack. Unloading, trucking, and storing may cost from 7 to 50 cents per sack. The cost of cement at the job may range from about 90 cents to $1.75 a sack.

Diagram 5-3 can be used for determining the cost of cement per cubic yard of concrete when the number of sacks per cubic yard and the price per sack at the job are known.

The cost of sand, gravel, and crushed rock at the job will vary greatly in different localities. Prices may be given per cubic yard or per short ton f.o.b. cars at station nearest the job or per cubic yard or per short ton delivered at the job. Sand or gravel may cost from $1.50 to $4 per short ton of 2,000 lb or from $2 to $5 per cubic yard at the job. Crushed rock may cost from $2 to $5 per short ton or from $2.50 to $6 per cubic yard at the job. Sand or gravel may weigh from 2,300 to 3,000 lb per cubic yard, and crushed rock may weigh from 2,200 to 2,700 lb per cubic yard.

Diagram 5-4 (Appendix F) can be used for determining the costs of the fine and coarse aggregates per cubic yard of concrete when the quantities of aggregates (measured in cubic yards) required for 1 cu yd of concrete and the prices per cubic yard at the job are known.

Diagram 5-5 (Appendix F) can be used when the fine and coarse aggregates are measured and priced by weight in tons.

The cost of water can best be found by finding the local rate per 1,000 gal and multiplying this rate by the number of thousand gallons required. In most localities, there are also extra charges for setting the water meter and turning the water on and off. On some jobs, the water is pumped from a well or a stream. On such jobs, the cost of pumping and piping the water should be included in the estimate.

Ready-mix concrete is concrete proportioned at a central plant and mixed in mixer trucks as it is transported to the job, or it is concrete proportioned and mixed at a central plant and transported in agitator trucks to the job. Mixer trucks have capacities up to 12 cu yd, and agitator trucks have capacities up to 14 cu yd or more.

Prices of ready-mix concrete vary in different localities. Prices (April, 1967) range from about $12 to $18 (average about $14) per cubic yard for a 1-to-2-to-4 mix (or a 1-to-2½-to-3½ mix) delivered at the job for quantities of 50 cu yd or more and for hauls up to about 5 miles from the central plant. Prices will be a little lower for weaker mixes and higher for stronger or special mixes, longer hauls, lesser quantities, and delays.

Table 5-14 *Approximate Labor Required for Mixing, Placing, and Curing Concrete*

Kind of work	*Labor-hours per cu yd*
Hand mixing	1–2
Machine mixing	0.5–1.2
Machine mixing plus heating of water and aggregate	0.7–1.5
Placing in footings and abutments	1–4
Placing in columns and thin walls	2–5
Placing in thick walls	1–4
Placing in floors and slabs	1–4
Placing in stairs	3–6
Placing structural concrete (average)	1–4
Placing structural concrete in cold weather	2–5
Curing concrete in ordinary or warm weather	0.5–1
Curing concrete in cold weather including installation, maintenance, and removal of protective enclosures, and heating	1–5
Mix, place, and cure, ordinary weather	2–6
Mix, place, and cure, cold weather	3–10

Labor Quantities and Cost The number of labor-hours required for mixing, placing, and curing 1 cu yd of concrete varies greatly according to the nature of the job and conditions of the work, weather, kind of plant, and skill and inclination of the workers. The labor includes that required for mixing the concrete, transporting and placing it in the forms, and curing. If ready-mix concrete is used, no labor for mixing will be needed at the job.

The values in Table 5-14 are approximate and are for structural concrete.

A labor gang includes the foreman and laborers, both skilled and unskilled. With the foreman at $5 to $8 per hour, mixer operator and other skilled labor at $3.50 to $6 per hour, and ordinary labor at $2.50 to $5 per hour, the average cost of labor for mixing, placing, and curing

1 cu yd of concrete will vary from about \$11 to \$20. Payroll taxes and insurance may vary from about 7 to 16 percent of the labor wage.

Diagram 5-6 (Appendix F) can be used for determining the cost of labor per cubic yard of concrete when the hours of labor required for mixing, placing, and curing 1 cu yd of concrete and the average hourly wage are known or can be readily assumed.

Plant The plant required for a structural concrete job will vary (usually with the size of the job) from a smaller mixer with a few wheelbarrows, shovels, hand tools, runways, etc., to a large mixer with weighing devices and cranes or derricks with buckets, cableway and buckets, belt conveyors, concrete pumping equipment, or hoists with carts and runways. Equipment is needed for weighing, mixing, transporting, placing, and sometimes finishing (as for concrete floors) and for curing the concrete. If ready-mix concrete is used, no plant will be needed for storing, weighing, and mixing.

The equipment needed for curing concrete in warm weather will include hose and a water supply and perhaps canvas, burlap, straw, etc., for covering the surface of the concrete to assist in keeping the concrete surface from drying out. All the concreting plant may be located at the job, or the proportioning plant may be in another location. Sometimes ready-mixed concrete is purchased delivered at the job.

For cold-weather concreting, some equipment must be provided for heating the aggregates and water. This equipment may be quite simple, such as a tank for the water with a place for a fire or a firebox beneath it and a corrugated iron pipe over which the aggregate is piled and in which a coal or wood fire is built. Perhaps a more satisfactory heating plant would be a steam boiler furnishing steam to pipe coils in the water tank and to steam pipes in the piles of aggregate.

The plant required for curing concrete in cold weather will include materials for enclosing and protecting the concrete and provision for heating during the curing period. Materials for enclosing and protecting the concrete may be canvas, burlap, straw, etc. The forms may be warmed by means of a hose and live steam from a boiler. The heating may be done by steam pipes or by salamanders. One salamander is required for about every 300 sq ft of floor area. In addition, one or two extra salamanders may be required for each exterior column, depending upon the weather and wind. If a steam boiler is used, at least a 50-hp boiler should be provided, i.e., licensed to carry a pressure of at least 80 psi. The curing time may vary from about 4 to 8 days. The reader is referred to publications of the Portland Cement Association for information on making, placing and curing concrete in cold weather.

Mixers are usually rated according to the approximate number of cubic feet in one batch. Capacities of mixers used in construction work vary from about 3½ cu ft to 3 or 4 cu yd or more. The mixers are often designated as 10S, 28S, 56E, the number denoting capacity in cubic feet, the letter S denoting side dump common to construction mixers, and the letter E denoting end dump common to paving mixers. For example, 28S means a side-dump mixer of 28 cu ft (or 1 cu yd) capacity. The mixing cycle of a batch mixed includes charging the mixer, mixing the concrete, and discharging the concrete. The minimum time per batch is about 3 min, with 4 to 5 min as an average. The minimum mixing time is 1 min, and frequently 1½ min or more are specified.

A laborer with a wheelbarrow can wheel 25 to 40 cu ft of concrete a distance of 50 to 30 ft per hour. With a two-wheel cart, one man can move 40 to 65 cu ft per hour.

A two-barrow or cart elevator working continuously can hoist 2,500 to 3,300 cu ft per foot per hour for concrete in barrows and 7,000 to 9,000 cu ft per foot per hour for concrete in carts.

The transporting and placing equipment should be of such capacity that it can handle the output of the mixer.

The output per hour of the entire concrete plant is restricted to the output per hour of the part of the plant having the least capacity.

The plant cost of mixing and placing 1 cu yd of concrete will vary considerably, depending upon the kind and conditions of the work, arrangement of the plant, amount of plant, and efficiency of plant operation. In general, a large plant will result in less labor per cubic yard of concrete, and vice versa. In general, the unit plant costs on large jobs, say 8,000 cu yd or over, are less than those on smaller jobs ranging from 2,000 to 6,000 cu yd, other things being equal. An elaborate complete plant will tend to cost more per cubic yard of concrete than a simple plant will, but the saving on labor costs may more than offset the extra plant costs.

Plant costs include not only the mixer costs, but also the costs of barrows, shovels, concreting tools, carts, hoists, towers, bins, cranes, derricks, cableways, buckets, chutes, runways, belt conveyors, etc.

Plant costs for heating water and aggregates in cold weather and for protecting and heating the concrete while curing may vary from about $1 to $5 per cubic yard, depending upon the equipment used, the fuel burned, and the temperature. If desired, plant costs for cold-weather curing can be divided into costs of equipment and costs of plant material used. The equipment will include the boiler, pipes, hose, heaters, salamanders, burlap, and canvas that may be used on many jobs. The plant materials will include water, sand, straw, and fuel that are used up or consumed on the job.

Plant costs per cubic yard of concrete may range about as given in Table 5-15. These values are approximate.

Diagram 5-7 (Appendix F) can be used for determining the plant cost

Table 5-15 *Approximate Plant Costs for Concrete*

Item	*Cost per cu yd of concrete*
Storage and weighing	\$0.05–\$0.25
Mixing	0.25–1.00
Mixing plus heating of water and aggregate	0.30–1.50
Placing	0.50–1.25
Curing, ordinary or warm weather	0.05–0.20
Curing, cold weather, including fuel	0.50–4.00
Total plant costs, ordinary weather	\$1.00–\$4.00
Total plant costs, cold weather, including fuel	1.50–9.00

per cubic yard of concrete when the output and hourly costs of the plant are known or can be assumed.

Overhead and Profit Overhead costs for mixing and placing concrete include such items as are usually included under this heading. The cost of a foreman is usually included in the labor costs.

When based on total costs of materials, labor, and plant, the overhead will range from about 5 to 15 percent. When based on labor costs alone, the overhead may range from 10 to 30 percent.

When the mixing and placing of the concrete are considered as a separate job, a reasonable amount of profit should be allowed. The percentage of profit may vary from about 5 to 15 percent of the costs of materials, labor, plant, and overhead.

Summary The unit and total costs of each class or mix of concrete can be summarized as follows:

Item	Class *A*		Class *B*		Etc.
	Cost per cu yd	Total cost	Cost per cu yd	Total cost	
Materials	\$	\$	\$	\$	
Labor					
Plant					
Overhead					
Profit					
Total					

Illustrative estimate on concrete. Using the absolute-volume method, make a complete cost estimate for 1,650 cu yd of concrete, given the following:

Mix is 1 sk cement to 6.5 U.S. gal water to 216 lb sand to 338 lb gravel.

Cost of cement per sack at job	$1.10
Cost of sand per ton at job	$2.50
Cost of gravel per ton at job	$2.70
Cost of water per 1,000 gal	$0.16
Cost of plant per cubic yard of concrete	$1.00
Average wage of labor per hour	$4.40
Labor-hours per cubic yard of concrete	4.20 hr
Overhead expense, based on materials, labor, and plant	9%
Profit, based on materials, labor, plant, and overhead	8%

Labor wage includes 10 percent for payroll taxes and insurance.

Quantities of materials required per cubic yard will be computed first, using the absolute-volume method and assuming average aggregates and a concrete weighing 4,000 lb per cubic yard.

Assuming a 1-sk batch,

Cement		= 94 lb
Water	= 6.5 U.S. gal × 8.35	= 54 lb
Sand		= 216 lb
Gravel		= 338 lb
Total weight for a 1-sk batch		= 702 lb

Quantities per cubic yard of concrete:

Cement	= 4,000 ÷ 702	= 5.70 sk	= 536 lb
Water	= 5.70 × 6.5	= 37.00 U.S. gal	= 309 lb
Sand	= 5.7 × 216 ÷ 2,000	= 0.617 ton	= 1,233 lb
Gravel	= 5.7 × 338 ÷ 2,000	= 0.965 ton	= 1,930 lb
Total (for a check)			= 4,008 lb

Summary of quantities and costs:

	1 *cu yd*		1,650 *cu yd*	
Materials,				
Cement	5.70 sk	$ 6.27	9,400 sk	$10,340
Water (including waste)	100 gal	0.02	165,000 gal	33
Sand	0.617 ton	1.54	1,020 tons	2,550
Gravel	0.965 ton	2.60	1,590 tons	4,293
Material costs		$10.43		$17,216
Labor, $4.40 × 4.20 hr		18.48	6,930 hr	38,492
Plant		1.00		1,650
Overhead, 9% of materials, labor, plant		2.69		4,442
Profit, 8% of materials, labor, plant, overhead		2.61		4,304
Total estimate		$35.21		$58,104

Using Ready-mix Concrete. Recompute this estimate using ready-mix concrete costing $14.60 per cubic yard delivered at the job. Assume that 3.5 labor-hours per cubic yard are required for placing and curing and that the plant costs (without weigher and mixer) are 60 cents per cubic yard. Summarizing:

	1 *cu yd*	1,650 *cu yd*
Materials	$14.60	$24,090
Labor, $4.40 per hour × 3.5 hr per cubic yard	15.40	25,410
Plant	0.60	990
Overhead, 9% of materials, labor, and plant	2.75	4,544
Profit, 8% of all other costs	2.67	4,403
Total	$36.02	$59,437

Estimating Reinforcement

Quantities of Materials Steel for reinforcement is estimated in pounds, or in tons for large quantities. All reinforcing bars are usually listed by bar number, the number corresponding to the bar diameter in eighths of an inch. For example, a No. 5 bar is a bar ⅝ in. in diameter. All reinforcing bars are round deformed bars except No. 2 which is a ¼-in. plain round bar. The relations among bar numbers, sizes, areas, and weights are given in Table 5-16.

The kinds of steel are rail steel (A16), billet steel (A15), and some special steels. Bar Nos. 14 and 18 are usually rolled from special steel. The grades of rail and billet steels are structural, intermediate, and hard. The kind and grade of steel should always be noted. Chairs, ties, pipe sleeves, clamps, units, threaded ends, turnbuckles, etc., should be tabulated separately. Wire cloth, expanded metal, and similar steel fabric sold by the roll or sheet should be measured and described by the square foot, stating size of mesh and weight per square foot. No allowances should be made for waste and cutting unless the cutting is done at the job. Allowances must be made for laps and splices when called for in the plans and specifications.

The takeoff sheet for reinforcing steel should have columns for number of bar, number of pieces, length, and bends. The summary sheet should give number of bar, weight per lineal foot, total weight, bends, unit price, and total price. Separate sheets may be prepared for straight and bent bars, spirals, and stirrups, and for accessories such as chairs, ties, and clamps.

Steel reinforcing bars may be purchased from a mill, from a main warehouse, or from a local warehouse. Bars purchased at the mill or at a main warehouse are cut to length, bundled, and tied. Bars purchased at a local warehouse must be bought from lengths in stock and then cut to the correct lengths at the job. Consequently, when time permits, the bars should be purchased from the mill to avoid cutting and waste. However, if the time is short or if the amount of steel is small, it may be advisable to buy the bars from a local warehouse.

Table 5-16 *Numbers, Sizes, Areas, and Weights of Bars*

Bar number	Size, diam in.	Area, sq in.	Weight, lb per ft
2*	1/4	0.049	0.167
3	3/8	0.110	0.376
4	1/2	0.196	0.668
5	5/8	0.307	1.043
6	3/4	0.442	1.502
7	7/8	0.601	2.044
8	1	0.785	2.670
9†	1 1/8	1.000	3.40
10†	1 1/4	1.2656	4.30
11†	1 3/8	1.5625	5.31
14	1 3/4	2.405	8.18
18	2 1/4	3.976	13.52

* No. 2 bar is a plain round bar.

† Nos. 9, 10, and 11 correspond to the old 1-, 1 1/8-, and 1 1/4-in. square bars, and are equivalent to those three square bars in weights and nominal cross-sectional areas.

Costs of Materials The price of steel varies considerably from time to time and, consequently, must be checked about each time that an estimate is made. The price depends upon the quantity in pounds purchased, the sizes of the bars, and the number of bends and hooks. The *base* price is the price quoted on bars f.o.b. the mill or main warehouse. Extra prices are charged for quantities, for sizes, bends made, and for engineering services. In addition, freight must be paid from mill to the city in which the job is located and a truck delivery charge paid for delivering the bars from the railway station to the job. Hence, the price charged for bars of any size is composed of the sum of base price, grade extras, quantity extras, size extras, bending extras, milling extras, engineering extras, freight charges, and delivery charges.

The *base* price is the price per 100 lb of reinforcing bars f.o.b. mill or main warehouse and includes cutting to specified lengths, bundling, and tagging. In 1968, base prices for new billet and rail steel bars in different localities ranged from $7 to $10 per 100 lb.

In regard to chairs, spacers, ties, etc., the total cost of these will vary greatly, depending on the kind used and the number required. An allowance of 50 cents to $1.25 per 100 lb of reinforcing bars is usually satisfactory.

Extras for Reinforcing Bars
(Grade extras vary from about 15 *cents to* $1.25 *per* 100 *lb)*

Quantity extras per 100 lb (for LCL shipments only)

Pounds	
Under 1,000	$1.80
1,000–1,999	$1.50
2,000–4,999	$1.20
5,000–9,999	$0.90
10,000–19,999	$0.60
20,000–29,999	$0.30
30,000–39,999	$0.10
40,000 and up	None

Size extras per 100 lb

No. 2	$2.50
No. 3	$1.40
No. 4	$1.00
No. 5	$0.70
Nos. 6, 7, 8, 9, 10, 11	$0.50

Bending extras per 100 lb

Heavy bars (not more than 6 bends)	$0.90
Light bars (½ in. and less)	$2.20

Engineering extras per 100 lb

Engineering (details and placing plans from designs by others)

Tons	*Extra per* 100 *lb*
Over 150	$0.30
20–150	$0.50 (but not over $900)
Under 20	$1.00 (but not over $200)
Minimum charge	$25.00
Design only	$0.60

Freight per 100 lb

Actual freight charge (CL or LCL lots) from mill or main warehouse to destination.

Delivery per 100 lb

Delivery by truck from local warehouse or station to job site, not less than $0.10.

Labor Quantities and Costs The labor required will depend upon the amount of cutting and bending to be done at the job, upon the difficulty of placing the bars in the forms, and also upon the sizes and lengths of the different bars and the number of stirrups and column ties.

Cutting to specified length is usually done by the manufacturer, and consequently, only a little cutting, if any, will have to be done at the job. The time required for cutting will usually range from 1 to 3 hr per 100 cuts depending on bar sizes, tools available, and workmen. If the bars must be sawed, the labor will be increased greatly.

The time in hours required for making 100 bends or 100 standard hooks is given in Table 5-17.

Table 5-17 *Labor-hours Required for Making* 100 *Bends or* 100 *Hooks*

Size of bar, in.	By hand		By machine	
	Bends, hr	Hooks, hr	Bends, hr	Hooks, hr
½ or less	2–4	3–6	0.8–1.5	1.2–2.5
⅝, ¾, and ⅞	2.5–5	4–8	1.0–2.0	1.6–3.0
1 and 1⅛	3–6	5–10	1.2–2.5	2.0–4.0
1¼ and 1½	4–7	6–12	1.5–3.0	2.5–5.0

The time required for placing 100 bars is given in Table 5-18. This time includes placing chairs and spacers; placing bars, stirrups, and spirals; and wiring in position.

Table 5-18 *Labor-hours Required for Placing* 100 *Bars*

Size of bar, in.	Length of bar, ft		
	Under 10	10–20	20–30
½ or less	3.5–6.0	5.0–7.0	6.0–8.0
⅝, ¾, and ⅞	4.5–7.0	6.0–8.5	7.0–9.5
1 and 1⅛	5.5–8.0	7.0–10.0	8.5–11.5
1¼ and 1½	6.5–9.0	8.0–12.0	10.0–14.0

The bending and placing of steel reinforcement bars can be done by skilled men, at an hourly wage of $4.50 to $7, under the direction of a competent foreman (wage $5 to $8 per hour). Payroll taxes and insurance may vary from about 7 to 16 percent of the labor wage.

If the reinforcement has to be piled or repiled at the job, an allowance of 10 cents per 100 lb is usually enough to cover the cost of each piling or repiling. A laborer can pile or repile 1,000 to 2,000 fph or about 2,000 to 4,000 lb per hour.

Diagram 5-8 (Appendix F) gives the labor costs for making bends and hooks, and Diagram 5-9 (Appendix F) gives the labor costs for placing reinforcement bars.

Plant When the cutting and bending are done by the manufacturer, the plant required by the contractor is very little. Only hand tools will be required for placing and tying. The cost of these tools should range from about 1 to 3 cents per 100 lb of bars.

If the cutting and bending are to be done at the job, the contractor will need hand or power shears for cutting, one or more hand benders (depending on size of job), and perhaps a power bender also. Plant costs for cutting should range from 1 to 3 cents per 100 cuts, depending upon the particular job. Plant costs for bending with hand machines may vary from 0.3 to 2 cents per 100 bends. For bending with power machines, the plant costs per 100 bends will be about the same, as the power bender is faster, though more costly to buy and operate. Trucking costs to and from the job will be extra and should be allowed for.

Overhead and profit Overhead charges for bending and placing reinforcement are usually based on labor costs and will range from about 10 to 25 percent. In some instances on reinforced concrete work, overhead expenses are charged to the job as a whole and not apportioned to forms, concrete, and reinforcement.

On a reinforced concrete job, profit is often estimated on the job as a whole and is not apportioned to the several parts of the job. However, if the reinforcement is to be considered as a complete unit by itself, then an allowance should be made for profit.

Summary The complete estimate for the reinforcement on a reinforced concrete job would include costs of materials, labor, plant, overhead, and profit (if not otherwise apportioned). This summary can be tabulated as follows:

Item	Unit cost per 100 lb	Unit cost per ton	Total cost
Materials...........	$	$	
Labor..............			
Plant..............			
Overhead...........			
Profit..............			
Total............	$	$	

Illustrative estimate on reinforcement. Make a complete cost estimate of the reinforcement in place for the bars listed in Table 5-19. Compute the weights and costs of the steel delivered at the job, assuming a base price of $8 per 100 lb and a truck rate of 25 cents per 100 lb. Allow 60 cents per 100 lb of bars for chairs, spacers, and ties. Steel will be cut to length by the manufacturer, but all bends and hooks will be made at the job. All bars are of billet steel, structural grade, except the column bars which are of billet steel, hard grade.

On most jobs, the manufacturer will cut the bars to specified length and make all bends and hooks. He will also make all bends and hooks for stirrups and will make all column spirals and attach these spirals to spacers. It is usually advisable to have as much work as is practical done by the manufacturer, so that the work at the job will be less.

Table 5-19 *Reinforcing-bar List**

Size of bars			Number of bars				
Bar no.	Diam., in.	Lin ft	0–10 ft	10–20 ft	20–30 ft	Bends	Hooks
3	3/8	1,092	14	44	16	None	None
5	5/8	1,266	42	66	0	86	65
6	3/4	1,972	57	52	28	56	109
8	1	2,846	96	74	38	72	110
5	5/8 columns	152	. . .	14	. . .	None	None
9	1 1/8 columns	192	. . .	12	. . .	None	None
3	3/8 ties	714	102	. . .	. . .	408	None
3	3/8 stirrups	976	324	. . .	. . .	648	648

* All bars are deformed except the 3/8-in. round ties (for columns), which are plain.

Consider extras for quantities and sizes and truck delivery as previously given. There will be no extras for bending and for engineering services.

All work is to be done by a gang of 1 foreman ($6 per hour) and 3 men ($5 per hour each). Payroll taxes and insurance are 10 percent of the labor wage.

Allow 14 percent of labor cost for overhead and 9 percent of all costs for profit.

Compute the total cost of steel placed in the forms. Also compute cost of steel per 100 lb and per ton.

Materials: The weights and costs of the steel can be tabulated and summed up as in Table 5-20.

Labor: The labor estimate for bending and placing this steel can be computed and tabulated as in Tables 5-21 and 5-22. Various labor-hours selected

Table 5-20 *Cost of Steel*

Size of bars					Unit costs					
Bar no.	Diam., in.	Lin ft	Wt, lb per ft	Wt, lb	Base	Quantity	Size	Truck del.	Total	Total cost
3	3/8	1,092	0.376	411	$8.00	$0.60	$1.40	$0.25	$10.25	$ 42
5	5/8	1,266	1.04	1,317	8.00	0.60	0.70	0.25	9.55	126
6	3/8	1,972	1.50	2,960	8.00	0.60	0.50	0.25	9.35	277
8	1	2,846	2.67	7,600	8.00	0.60	0.50	0.25	9.35	711
5	5/8 columns	186	0.85	158	8.00	0.60	0.70	0.25	9.55	15
9	1 1/8 columns	192	3.40	653	8.00	0.60	0.50	0.25	9.35	61
3	3/8 ties	714	0.376	268	8.00	0.60	1.40	0.25	10.25	28
3	3/8 stirrups	976	0.376	367	8.00	0.60	1.40	0.25	10.25	38
	Total for bars	.....	.....	13,734	.....	.....	.....	.....	9.45	$1,298
Allowance for spacers and wire at $0.60 per 100 lb of bars										$ 82
Total for all steel delivered at job site										$1,380

for bending and placing 100 bars of different sizes are fair average values for good workmen under fair working conditions.

Average hourly wage per man for the gang of 1 foreman and 3 men (foreman to work with the men) is $\frac{1}{4}(\$6 + \$5 \times 3)1.10 = \$5.78$.

Diagram 5-8 can be used for computing costs of bends and hooks, and Diagram 5-9 for computing cost of placing reinforcement.

Table 5-21 *Labor Estimate for Bending*

Size of bars								
Bar no.	Diam, in.	Bends	Labor-hr assumed per 100 bends	Labor-hr for bends	Hooks	Labor-hr assumed per 100 hooks	Labor-hr for hooks
3	3/8	None	...	.....	None		
5	5/8	86	1.1	0.95	65	2.0	1.30
6	3/4	56	1.3	0.73	109	2.3	2.51
8	1	72	1.5	1.08	110	2.5	2.75
5	5/8 columns	None	...	.....	None		
9	1 1/8 columns	None	...	.....	None		
3	3/8 ties	408	1.0	4.08	None		
3	3/8 stirrups	648	1.0	6.48	648	1.6	10.37
	Total	1,270	...	13.32	932	...	16.93

Labor cost of bends = \$5.78 × 13.32 = \$ 77
Labor cost of hooks = \$5.78 × 16.93 = 98
Labor cost of bends and hooks = \$175

Table 5-22 *Labor for Placing Bars*

Size of bars		Length 0–10 ft			Length 10–20 ft			Length 20–30 ft		
Bar no.	Diam, in.	No. of bars	Labor-hr per 100 bars	Labor-hr	No. of bars	Labor-hr per 100 bars	Labor-hr	No. of bars	Labor-hr per 100 bars	Labor-hr
3	3/8	14	4.0	0.56	44	5.5	2.42	16	6.5	1.04
5	5/8	42	5.0	2.10	66	6.5	4.29	0		
6	3/4	57	5.5	3.14	52	7.0	3.64	28	8.0	2.24
8	1	96	6.0	5.76	74	7.5	5.55	38	9.0	3.42
5	5/8	0	...	.....	14	6.5	0.91	0		
9	1 1/8	0	...	.....	12	9.0	1.08	0		
3	3/8 ties	102	...	.....	0	...	.....	0		
3	3/8 stirrups	324	5.0	16.20	0	...	.....	0		
	Total	...	...	27.76	...	...	17.89	...	...	6.70

Total labor-hours = 27.76 + 17.89 + 6.70 = 52.35
Labor cost for placing = \$5.78 × 52.35 = \$303

No labor costs are apportioned to the ⅜-in. round ties for columns, as the cost of securing these in place is included with the labor cost of placing the ⅝-in. and 1⅛-in. column bars.

On a job of this size, many estimators, perhaps, would not go into so much detail in preparing the labor estimate. In this particular problem, the labor estimate is worked out in detail so that the use of the method will be understood.

Total labor costs, bending and placing = \$175 + \$303 = \$478

Plant: The plant required for this job would be 1 machine bender, about 2 benches, and several hand tools. Plant charge can be estimated at about 1.50 cents per 100 lb of bars, or about \$2. To this must be added the costs of trucking bender, benches and tool box to and from the job, say \$8.

Total plant = \$10

Overhead: This will be 14 percent of the labor cost, and equals 14 percent of \$478, or \$67.

Profit: This will be 9 percent of all of the other costs, or 9 percent of \$1,935 equals \$174.

Summary: The total estimate, on the assumption that this job is a complete job by itself, including overhead and profit, will be summarized as follows:

Item	Unit cost per 100 lb	Unit cost per ton	Total cost
Materials	\$10.05	\$201	\$1,380
Labor	\$ 3.48	\$ 69	\$ 478
Plant	\$ 0.07	\$ 2	\$ 10
Overhead	\$ 0.49	\$ 10	\$ 67
Profit	\$ 1.26	\$ 25	\$ 174
Total	\$15.35	\$307	\$2,109

Estimating Surface Finish

Units of Measurement Surface finish of concrete surfaces is usually measured by the square foot, except for such work as finishing coves and molding, when the lineal foot is used.

Some allowance should be made for going over the surfaces of concrete work after the removal of forms, patching up voids and stone pockets if any are present, removing fins, etc. On some jobs, this work is included with the labor required for stripping and cleaning of forms. However, on other jobs, the contractors consider the work of surface cleaning or finishing as a separate item and prepare their estimates accordingly.

All different types of surface finish should be measured and described separately. That is, the kinds of finish (troweling, rubbing, tooling, scrubbing, washing, sandblasting, terrazzo, granolithic, etc.) should be listed separately with enough explanation so that a reasonable estimate of the required materials, labor, and plant can be prepared.

Materials No materials are required for such types of finishing as simple troweling, washing, and rubbing, but materials are needed for a cement-mortar finish and for terrazzo and granolithic finishes.

The quantities of cement and aggregate required for 100 sq ft of surfacing 1 in. thick will depend upon the mix. Table 5-23 gives the quantities required for various mixes. Quantities required for other thicknesses are readily computed from the values given. The aggregate may be all sand, part sand and part fine crushed stone, sand with stone (granite or marble) chips for surfacing, or all fine crushed stone.

Table 5-23 *Approximate Quantities Required for 100 Sq Ft of Surface Finish 1 In. Thick*

Material	Mix by volume				
	1–1	1–1½	1–2	1–2½	1–3
Cement, sack	6.0	4.8	4.0	3.5	3.0
Aggregate, cu ft	6.0	7.2	8.0	8.5	9.0
	Mix by weight				
	1–1	1–1½	1–2	1–2½	1–3
Cement, lb	600	480	400	345	300
Aggregate, lb	600	720	800	860	900

An allowance of water for all purposes of about 75 to 150 gal per 100 sq ft of surface is usually satisfactory.

The price of cement at the job may vary from about $1 to $3 per sack. The aggregates used are often special, such as washed and graded sand, graded stone chips, granite chips, or marble chips. These materials are often delivered in sacks weighing 100 to 200 lb. The price of aggregate at the job varies considerably and may range from about 50 cents to $3 per 100 lb.

Labor The approximate amounts of labor required for various types of surface finishing is given in Table 5-24.

Skilled labor is required for surface finishing, though some unskilled labor may be used for helping in some instances. Prices of skilled labor for finishing may range from about $4 to $6 per hour.

Diagram 5-10 (Appendix F) can be used for estimating labor costs of surface finishing when the hourly wage and labor-hours required per 100 sq ft of surface are known or assumed.

Plant The plant required will vary from a few hand tools required for troweling to a combination of hand tools, small mixers and accompanying equipment, and machine grinders required for granolithic or terrazzo finish. Sandblasting requires a machine, hose, etc., and a supply of fine sand. Scaffolding of some sort must be provided when walls and ceilings are to be finished.

Table 5-24 *Approximate Labor-hours Required for Finishing Concrete Surfaces*

Kind of work	*Labor-hours*
Troweling floor, walls, sidewalks, etc. (100 sq ft)	1–4
Troweling plain base, cove, etc. (100 lin ft)	1–4
Troweling fancy base, cove, etc. (100 lin ft)	3–7
Carborundum rubbing of floor and wall surfaces (100 sq ft)	4–12
Carborundum rubbing of window sills, base, cove, etc. (100 lin ft)	4–14
Machine grinding (100 sq ft)	3–12
Ornamental tooling, various kinds (100 sq ft)	8–20
1-in. granolithic or terrazzo finish laid after concrete has hardened, including mixing and placing (100 sq ft)	7–15
1-in. granolithic or terrazzo finish laid integral with the concrete, including mixing and placing (100 sq ft)	5–12
Removing fins, patching pockets, etc. (100 sq ft)	1–4
Scrubbing surface (100 sq ft)	2–5
Washing surface with acid (100 sq ft)	2–5
Sandblasting surface (100 sq ft)	3–6
Cement or other surface work per coat, (100 sq ft)	2–5

It is practically impossible to give any accurate plant costs in general because the requirements of the individual jobs vary so greatly. Plant costs may range from a few cents to several dollars per 100 sq ft of surface.

Overhead and Profit Overhead costs of surface finishing are usually based on labor costs and may range from about 10 to 25 percent of the cost of labor.

The profit may range from about 5 to 20 percent, depending upon circumstances.

Summary The total estimate for a surface-finishing job will include materials (if any required), labor, plant, overhead, and profit (if a separate job).

Illustrative estimate on surface finish. Estimate the cost of 100 sq ft of terrazzo surface finish 2½ in. thick and placed in two courses. The first course is 1½ in. thick and is a mortar course composed of 1 part of portland cement to 3 parts by weight of sand. The second course is 1 in. thick and is composed of 1 part of portland cement to 2 parts by weight of marble chips. Prices of materials at the job are cement $1.20 per sack, sand $2.50 per ton, and marble chips 75 cents per 100 lb. Labor gang will consist of 1 foreman at \$6 per hour, 2 finishers or grinders at \$5 per hour, and 3 laborers at \$4 per hour. Allow 10 percent of the labor wage for payroll taxes and insurance.

Estimate for first or base course. Mix is 1 to 3. Course is 1.5 in. thick (see Table 5-23).

Cement, pounds = 300 × 1.5 in. = 450 lb = 4.8 sk at \$1.20	= \$ 5.76
Sand, pounds = 900 × 1.5 = 1,350 lb at \$2.50 a ton	= 1.69
Water, estimated at	= 0.15
Materials	= \$ 7.60

Labor, mixing, placing, striking off, say 2.50 hr, at average wage of

$$\left(\frac{\$6 + \$10 + \$12}{6}\right) \times 1.10 \text{ or } \$5.13 \text{ per hr} \qquad = \$12.83$$

Plant, say	= 2.00
Overhead, say 12% of labor	= 1.54
Profit, say 10% of other costs	= 2.40
Total for first course	= \$26.37

Estimate for second or wearing course. Mix is 1 to 2. Course is 1 in. thick (see Table 5-23).

Cement, pounds = 400 = 4.25 sk at \$1.20	= \$ 5.10
Marble chips, pounds = 800 at \$0.75 per 100 lb	= 6.00
Water, estimated	= 0.15
Materials	= \$11.25
Labor, mixing, placing, troweling, and grinding, say 7 hr at an average wage of \$5.13 per hr	= \$35.91
Plant, mixing and grinding: say about	= 3.50
Overhead, 12% of labor	= 4.31
Profit, 10% of other costs	= 5.50
Total for second or wearing course	= \$60.47
Total for 100 sq ft both courses (or about \$0.87 a square foot)	= \$86.84

Complete Estimates

Complete Estimates A complete estimate of a structural concrete job will include costs of forms, concrete, reinforcement, and surface finishing (if required). Curing costs are usually included in the concrete estimate, though many estimators add a separate item for curing in cold weather.

On almost every large concrete job, there are a number of miscellaneous items which need consideration. Most estimators include these items or allow for them in the overhead charges apportioned to the different parts of the whole job. The percentages charged to overhead should be large enough to include all general overhead, cleaning up, and miscellaneous items. Cleaning up may cost about 1/10 of 1 percent to 1/2 of 1 percent of the total costs. Miscellaneous and sundry expenses and contingencies may vary from about 2 to 5 percent of the total cost. Such expenses should preferably be included in the allowance for overhead.

Some estimators use different percentages for overhead for the different parts of the work (forms, concrete, etc.) and include overhead and profit with the estimates for each kind of work. Other estimators use the same percentage for overhead for all parts of the work and then include overhead and profit as separate items. The first method seems preferable.

Summary The complete estimate for the entire job can be divided and itemized and summarized as desired. On unit-price jobs, the cost per cubic yard of each class of concrete is tabulated.

When overhead and profit are apportioned among the different parts of the work, the estimate can be summarized as follows:

Item	Unit cost per cu yd	Total cost
Forms	$	$
Concrete		
Reinforcement		
Surface finishing		
Curing (if separate)		
Total estimate	$	$

If it is not desirable to apportion the overhead and profit, the summary can be as follows:

Item	Unit cost per cu yd	Total cost
Forms (materials, labor, and plant)	$	$
Concrete, (materials, labor, and plant)		
Reinforcement, (materials, labor, and plant)		
Surface finishing, (materials, labor, and plant)		
Curing (if separate) (materials, labor, and plant)		
Overhead		
Profit		
Total estimate	$	$

Estimates for unit price jobs can be tabulated as follows for each class of concrete. Each item listed includes its proportion of overhead and profit.

Item	Class A cost per cu yd	Class B cost per cu yd	Etc.
Forms	$	$	$
Concrete			
Reinforcement			
Curing (if separate)			
Surface finishing (if any)			
Total costs per cubic yard	$	$	$

When the overhead and profit are included as separate items, the summary can be as follows:

Item	Class A cost per cu yd	Class B cost per cu yd	Etc.
Forms	$	$	$
Concrete			
Reinforcement			
Curing (if separate)			
Surface finishing (if any)			
Overhead			
Profit			
Total costs per cubic yard	$	$	$

Total costs for any class of concrete would equal the unit cost multiplied by the yardage for that class. The total cost of the job would equal the sum of the total costs of the different classes of concrete.

Estimating Concrete Walks, Driveways, and Garage Floors

Items Included Concrete walks, driveways, and garage floors are usually built on the ground. They are all constructed in the same manner. The building of a concrete walk, driveway, or garage floor on the ground may include the following items:

1. Excavation or fill
2. Leveling and compacting the soil
3. Placing, compacting, and leveling the subbase
4. Placing screeds or forms where required
5. Mixing and placing the concrete
6. Placing the reinforcement
7. Finishing the concrete surface and joints
8. Curing the concrete
9. Removing forms and cleaning up

Units of Measurement Concrete walks, driveways, and garage floors are usually measured by the square foot or square yard, with the thickness, mix, and other details given. The thickness may range from 3 to 8 in. Concrete walks, driveways, and garage floors are usually constructed in one course. The cement is measured in sacks, water in U.S. gallons, aggregates in pounds or tons (or cubic feet or cubic yards), reinforcing rods in pounds, and wire mesh in square feet or square yards.

Materials The subject of excavation and fill has been covered in Chap. 3. The reader is referred to that chapter for information on estimating excavation and fill.

Subbase. Well-compacted cinders, sand, gravel, or crushed rock is recommended. The price of material delivered at the job may range from about $2 and up per cubic yard, or from about $1.50 and up per ton. The quantities of materials required for 100 sq ft of subbase are given in Table 5-25.

Table 5-25 *Materials Required per* 100 *Sq Ft of Subbase*

	Compacted thickness, in.				
	2	3	4	5	6
Cu yd per 100 sq ft of base	0.70	1.00	1.35	1.65	2.00
Tons per 100 sq ft of base	0.90	1.30	1.75	2.15	2.60

Joint Filler. The joint filler should be premolded strips of bituminous filled fiber or mineral aggregate, 0.50 in. thick, as wide as the thickness of the concrete, and as long as the width of the walk or driveway.

Reinforcement. The reinforcement may be small steel rods (measured in pounds) or some form of wire mesh (measured in square feet or square yards).

Concrete Materials. These materials should be portland cement, good water, well-graded ⅜-in. sand, and well-graded ¼- to 1-in. gravel or crushed stone.

Approximate prices of these materials delivered at the job are: portland cement $1 to $2 per sack, sand and, gravel and crushed stone $1.50 to $5 per ton or $2 to $6 per cubic yard.

Concrete Mixes. The proportions of a concrete mix should be given by portland cement in sacks, to water in U.S. gallons, to fine aggregate (sand) in pounds or tons, to coarse aggregate (gravel or crushed stone) in pounds or tons.

The recommended slump is 3 in.

The recommended water-cement ratios are: 6 U.S. gal per sack of cement in cold and moderate climates, and 6.5 U.S. gal per sack of cement in mild and warm climates.

The following are suggested trial mixes for 1-sk batches and 1-cu-yd batches intended to give a slump of 3 in. with 6.0 and 6.5 U.S. gal per sack of cement with clean, surface-dry ⅜-in. sand and 1-in. gravel. If these proportions do not give the desired slump, the amounts of sand and gravel may be varied slightly to give the desired results.

Suggested Trial Mixes for a 3-in. Slump Using ⅜-in. Sand and 1-in. Gravel

	1-sack batch		1-cu-yd batch	
Cement, sacks	1.00	1.00	6.2	5.7
Water, U.S. gal	6.00	6.50	37.0	37.0
Sand, lb	215	240	1,340	1,370
Gravel, lb	295	320	1,830	1,820
Yield, cu ft	4.36	4.74	27	27

The proportions of a corresponding mix by volume may be found by dividing the weights of the fine and coarse aggregates by their corresponding weights per cubic foot. The water-cement ratio should not be changed.

Suggested minimum thicknesses of concrete are:

Residential walks, driveways, and garage floors	4 in.
Commercial walks	6 in.
Commercial driveways	8 in.
Commercial garage floors, medium trucks	6 in.
Commercial garage floors, very heavy trucks	8 in.

Table 5-26 gives approximate quantities of cement, water, and fine and coarse aggregates required per 100 sq ft of surface area for concrete walks,

Table 5-26 *Materials Required per 100 Sq Ft; Slump of 3 In.*

Item	Thickness, in.						Quantities per cu yd
	3	4	5	6	7	8	
	Water-cement ratio, 6.0 U.S. gal per sack						
Cement, sacks	5.75	7.70	9.60	11.50	13.45	15.35	6.20
Water, U.S. gal	34.5	46.0	57.5	69.0	80.5	92.0	37.0
Sand, tons	0.63	0.83	1.04	1.25	1.46	1.66	0.670
Gravel, tons	0.85	1.13	1.42	1.70	1.98	2.27	0.915
	Water-cement ratio, 6.5 U.S. gal per sack						
Cement, sacks	5.30	7.05	8.85	10.60	12.35	14.15	5.70
Water, U.S. gal	34.5	46.0	57.5	69.0	80.5	92.0	37.0
Sand, tons	0.64	0.85	1.06	1.27	1.49	1.70	0.685
Gravel, tons	0.85	1.13	1.41	1.70	1.98	2.26	0.910
Concrete, cu yd	0.93	1.24	1.55	1.86	2.17	2.48	1.00

driveways, and garage floors of different thicknesses. The quantities have been computed by the absolute-volume method.

Ready-mix Concrete. Ready-mix concrete is usually measured and sold by the cubic yard, with details given as to mix (sacks of cement per cubic yard, water-cement ratio, slump, and amounts, kinds, and sizes of the fine and coarse aggregates). About 0.31 cu yd of concrete is required per 100 sq ft for each inch in thickness. Concrete weighs about 4,000 lb (2 tons) per cubic yard. Ready-mix concrete may cost from about $12 to $18 per cubic yard.

Labor The labor required for constructing 100 sq ft concrete sidewalk will depend upon the particular job, the skill of the gang, and the distances that the materials and concrete must be transported.

The cement and aggregate piles should be placed as close to the mixer as practicable, and the mixer should be located so that the distance that the concrete must be transported is a minimum. With a small mixer (about a one-half bag), a concrete crew of about 3 to 6 men is satisfactory. For a one-bag mixer, a crew of 5 to 10 men may be needed.

If ready-mix concrete is used, no space for the mixer and for the aggregate storage piles will be needed at the job and no mixing labor will be required.

Table 5-27 *Approximate Labor Required for Concrete Walks, Driveways, and Garage Floors*

Item	*Labor-hours per* 100 *sq ft of walk*
Leveling and tamping soil	0.5–2.0
Placing, leveling, and tamping subbase	0.5–2.0
Placing forms and screeds	0.5–2.0
Mixing and placing concrete	1.5–3.0
Finishing	1.0–3.0
Curing, removing forms, cleaning up	0.5–2.0
Total	4.5–14.0

For good curing, a covering (canvas, burlap, sand, straw, etc.) should be placed over the fresh concrete as soon as the surface has hardened sufficiently, and this covering should be thoroughly wet at least twice a day on dry days for a week.

The approximate labor required for the construction of one-course concrete sidewalks is given in Table 5-27.

Labor wages vary at different times and in different localities. Hourly wages are approximately as follows: foremen $4 to $7; finishers, mixer operators, carpenters $3.50 to $6; and ordinary labor $2.50 to $4.50.

Forms Either lumber or metal forms can be used. Lumber forms are usually 2 in. thick and of a width equal to the depth of the concrete. Metal forms should be strong. One stake is required for about every 3 to 5 ft of form length. The lumber cost per lineal foot of forms will vary from about 10 to 25 cents for first cost. As the salvage value is high and the lumber can be used many times, the average cost (assuming salvage and reuse) will be about 3 to 10 cents per lineal foot. Metal forms will cost about the same or perhaps a little more. Some estimators consider the forms as a part of the plant and include the form costs with the plant costs.

Plant The plant and equipment required for the construction of concrete walks will include a small mixer, hose, water-measuring device, scale, barrows or carts, plank for runways, shovels, tampers, and hand tools for form builders and finishers. If ready-mix concrete is used, the mixer, water-measuring device, and scales are not needed.

The plant cost per cubic yard of concrete for a walk may range from about 35 cents to $3 per cubic yard of concrete or from about 50 cents to $4 for 100 sq ft of walk, or about 30 cents to $1 per hr that the mixer is operated.

Overhead and Profit Overhead costs will probably range from about 10 to 20 percent of the labor costs.

Profit may range from about 5 to 15 percent, with about 10 percent as a fair average value.

Summary The estimate for concrete walks can be summarized as follows:

Item	Unit cost per sq ft	Total cost
Materials	$	$
Labor		
Forms		
Plant		
Overhead		
Profit		
Total	$	$

Illustrative estimate for concrete walks. An estimate is required for a one-course concrete sidewalk 5 ft wide and 620 ft long. The walk is to be 4 in. thick and is to have a 2-in. sand base. Concrete mix is by weight with 1-in. aggregate. Land surface conforms to grade, so that there will be but little excavating and leveling. Prices of materials delivered at the job are cement $1.20 per sack net, sand $2 per ton, and 1-in. gravel $2.25 per ton. Cost of water is nominal and may be taken at 20 cents per 100 sq ft of walk to allow for wetting down, concrete mixing, curing, and waste. Hourly wages of laborers and number of men in the crew are 1 foreman at $5, 1 mixer operator at $4, 1 finisher at $4, and 5 laborers at $3 each. Allow 10 percent of the labor wage for payroll taxes and insurance. Plant costs are 50 cents per working hour. Allow $10 for transportation. Overhead is 9 percent of the labor cost. Profit is 10 percent of all other costs. Take totals to nearest dollar.

Concrete mix is 1 sk cement to 6 U.S. gal water to 215 lb sand to 295 lb gravel.

Area = 620 × 5 = 3,100 sq ft or 31 squares

Materials (Tables 5-25 and 5-26):

Sand base, 0.90 ton × 31 sq × $2.00	= $ 56	
Cement, 7.70 sk × 31 sq × $1.20	= 287	
Sand, 0.83 ton × 31 sq × $2.00	= 52	
Gravel, 1.13 tons × 31 sq × $2.25	= 79	
Water, allow	= 7	
Expansion-joint material	= 16	
Total materials		= $ 497

Labor (Table 5-27), 31 sq:

Leveling and tamping soil, 0.5 hr per square	= 16 hr	
Preparing sand subbase, 0.75 hr per square	= 23 hr	
Forms, 0.75 hr per square	= 23 hr	
Mixing and placing concrete, 2.25 hr per square	= 70 hr	
Finishing, 1.20 hr per square	= 37 hr	
Curing and cleaning up, 1 hr per square	= 31 hr	
Total labor-hours	= 200 hr	
Gang hours = 200 ÷ 8 = 25		
Hourly wage of gang = ($5 + $4 + $4 + $3 × 5)1.10 = $30.80		
Labor cost = $30.80 × 25		= $ 770
Forms (say about 1,240 ft of 2 by 4s and 250 stakes) With lumber at $120 per 1,000 fbm and salvage of 80 percent, cost of forms will be about		= $ 25
Plant costs = $0.50 × 25 = $12.50 + $10 transportation		= $ 23
Overhead at 9% of labor cost		= $ 70
Profit at 10% of all other costs		= $ 139
Total estimated cost		= $1,524

Unit cost per square foot $0.49 (about)

Using Ready-mix Concrete: Recompute this estimate using ready-mix concrete costing $16 per cu yd delivered at the job and assuming that concrete can be chuted into the forms so that only 1 labor-hour per square will be needed for placing, that plant costs will be 10 cents per working hour; that no mixer operator will be needed, and that the other costs will be unchanged.

Materials: 31 sq:

Sand base, 0.90 ton per square at \$2.00 per ton	= \$ 56	
Ready-mix concrete, $\frac{3{,}100 \times 4}{27 \times 12} \times \$16$	= 613	
Expansion-joint material	= 16	
Total materials		= \$ 685
Labor: 31 sq:		
Preparing soil as before	= 16 hr	
Preparing sand base as before	= 23 hr	
Forms as before	= 23 hr	
Placing concrete at 1 hr per square	= 31 hr	
Finishing as before	= 37 hr	
Curing and cleaning up as before	= 31 hr	
Total labor-hours	= 161 hr	
Gang-hours = 161 ÷ 7 = 23		
Hourly wage of gang = (\$5 + \$4 + \$3 × 5)1.10 = \$26.40		
Labor cost = \$26.40 × 23		= \$ 607
Forms as before		= \$ 25
Plant = \$0.10 × 23 = \$2.10, plus \$10 transportation		= \$ 12
Overhead at 9% of labor cost		= \$ 55
Profit at 10% of all other costs		= \$ 139
Total estimated cost		= \$1,523
Unit cost per square foot = \$0.49 (about)		

Estimating Concrete Roads and Pavements

Items Included in Estimate The items included in an estimate for a concrete highway would be materials, labor, plant, overhead, and profit. However, on most jobs it is often convenient to regroup these items according to the different divisions of the work. Such a regrouping may give:

1. *Materials costs.* Cost of cement, fine and coarse aggregates, water, reinforcement, and miscellaneous materials delivered to the contractor.
2. *Materials yard.* Cost of labor and equipment for handling, storing, and measuring materials at the materials yard (see Chap. 2 on Handling and Transporting Materials).
3. *Transportation of materials.* Costs of transporting materials to job (see Chap. 2 on Handling and Transporting Materials).
4. *Construction of pavement.* Costs of labor and equipment required for constructing the concrete pavement. This will include
   a. Excavation and fill. Usually the excavation and fill are constructed by a different contractor, and the roadway base is at grade. This item will rarely appear in concrete paving estimates (see Chap. 3 on Excavation).

b. Leveling subgrade. Occasionally some equipment and labor may be required for bringing the subgrade to the exact grade desired.
c. Providing water supply for mixer and for curing.
d. Forms. Placing and removing as work progresses.
e. Placing reinforcement, usually placed with concrete.
f. Mixing and placing concrete.
g. Finishing surface.
h. Curing.
i. Cleaning up (if considered as a separate item).

Units of Measurement The common unit of measurement for concrete roads and pavements is the square yard of pavement surface, with thickness, reinforcement, concrete mix, and other details stated. Sometimes other units, such as 100 sq ft, 100 sq yd, cubic yards per mile, cubic yards per station, foot-mile (1 ft wide and 1 mile long), etc., have been used.

The materials that the contractor uses will be priced in other units, e.g.,

Cement, sacks, barrels, 100 lb, or tons
Fine aggregate (sand), cubic yards or tons
Coarse aggregate (gravel or crushed stone), cubic yards or tons
Reinforcement, pounds or tons
Water, gallons, 1,000 gal, etc.

Materials *Concrete Materials.* The materials required for concrete are cement, water, and fine and coarse aggregates. The amounts of the different materials required per square yard of pavement will depend upon the proportion of the mix and the average thickness of the pavement section. Pavements vary from about 5 to 12 in. in average thickness, values of 6 to 9 in. being common. The amount of concrete required per square yard of pavement surface is given in Table 5-28.

Table 5-28

Thickness, in.	*Cu yd of concrete per sq yd of surface*
5	0.139
6	0.167
7	0.195
8	0.222
9	0.250
10	0.278
11	0.306
12	0.333

Many pavements have thickened edges, and the thickness consequently varies from one side to the other of the pavement. When estimating the quantity of concrete in cubic yards per square yard of pavement surface required for such pavements, the estimator first assumes a convenient length of pavement, then finds the cross-sectional area and multiplies this area by the assumed length to obtain the volume. Dividing this volume in cubic yards by the area of this portion of the pavement in square yards gives the average number of cubic yards of concrete required per square yard of pavement.

The consistency of the mix is very stiff. A slump of about 1 to 2 in. is common with machine finishing, and a slump of about 2 to 3 in. with hand finishing.

The proportions are practically always given by weight, and may vary from about a 1-2-3.5 mix to a 1-3-4.5 mix, with water varying from 5 to 6 gal (41.7 to 50 lb) per 100 lb of cement. The cement may vary from about 500 to 700 lb per cubic yard of concrete. The mix is usually designed to give a 28-day modulus of rupture (cross-bending strength) of from 600 to 800 psi and a corresponding compressive strength of 4,500 to 6,500 psi.

When the proportions are given, the estimator can compute the quantities of materials per cubic yard of concrete according to the methods explained in the section on Estimating Concrete. Methods of computing costs per cubic yard are also given in that section. After the quantities and costs per cubic yard are found, they can be reduced to quantities and costs per square yard of pavement surface.

Reinforcement. The reinforcement used in concrete pavements may be wire mesh or bars (such as tie bars, dowels, shear reinforcement, and corner reinforcement). The amount of reinforcement may vary from an average of a few pounds up to about 10 or 12 lb per square yard in some instances.

When a reinforcement takeoff is prepared, a section of the pavement between expansion joints may be considered. All the reinforcement in this section of pavement should be listed as to kind, size, and length and the total weight found. Then the pounds of reinforcement per square yard of surface can be found by dividing the weight of steel for this section by the area in square yards. Information in regard to weights of bars, prices, and price extras is given in the section on Estimating Reinforcements.

Miscellaneous Materials. All miscellaneous materials required should be listed and priced. Some of these materials may be joint filler for expansion joints, sleeves or tubes for steel bars at expansion joints, admixtures, coloring matter, etc.

Materials Yard The equipment at a materials yard may include unloading machines, elevating machinery, office, storage bins or sheds for cement, storage sheds for reinforcement and miscellaneous materials and supplies, aggregate bins, and batcher plants with weighing devices. The minimum material yard equipment for one 27E or 34E single-drum mixer would include an aggregate bin (or bins) of 75 tons or more capacity, a crane with a 1-cu-yd clamshell bucket and a 40- to 50-ft boom. One 34E double-drum mixer would need an aggregate bin (or bins) of 100 tons or more capacity and a crane with a 1.5- or 2-cu-yd clamshell bucket and a 40- to 60-ft boom. If the cement is in bulk, a cement storage bin of 100 bbl or more capacity should be provided.

The labor will be the labor required to care for the yard and operate the various pieces of equipment.

Perhaps the best unit for unit costs is the ton, though other units such as sacks or cubic yards may be selected as desired.

To find the unit costs per ton, the total cost of the yard for a length of time, such as a month or year, is found, and this cost divided by the tonnage cared for during that time. If the yard is a temporary one to be used for one job (or a few jobs), the costs of handling and transporting equipment from the contractor's main storage yard to this yard, costs of erection, costs of dismantling, and costs of transportation and handling from this yard to the main storage yard must be included.

On many jobs it is desirable to estimate the materials yard costs per working hour. The equipment costs include ownership and operating costs per working hour (see Appendix B) plus a proportional part of the transportation, erection, and dismantling costs. The labor costs per working hour include the hourly wages of the men plus payroll taxes and insurance. The materials yard cost per working hour is the sum of the hourly equipment and labor costs.

The materials yard costs may range from a few cents up to $1 per ton.

Materials Transportation The concrete materials are usually transported from the materials yard to the mixer by dump trucks. The body of a dump truck may be divided into two, three, or four compartments, each compartment holding enough material for one batch.

The transportation costs can be estimated by:

1. Computing the hourly cost of truck and driver
2. Computing time for one round trip in hours (load, travel loaded, unload, return travel, delays, etc.)
3. Computing cost per round trip
4. Computing cost per batch, per cubic yard of concrete, or per square yard of pavement, as desired.

Pavement Construction The items mentioned in the article on Items Included in Estimate will be discussed in order.

Excavation and Fill. This work is usually done by another contractor and will not be considered here.

Leveling Subgrade. On almost every job, a little leveling of the subgrade will need to be done. An allowance of 35 cents to $1 per 100 sq yd of surface will be sufficient when the work of excavation and fill has been well done. A fine grader may be needed.

Providing Water Supply. On some jobs, the water supply can be secured from a city or village. Then the plant needed will be pipe and hose to carry the water from the nearest hydrant to the job. The total cost of water, including allowance for pipe and hose, may range from 10 to 25 cents per 1,000 gal.

On some jobs, the water may have to be hauled in tanks mounted on trucks. Here the transportation costs are comparatively large. The total cost of the water delivered at the job may range from about 50 cents to $3 or more per 1,000 gal, depending on distance transported.

On other jobs, the contractor may need to install pumping equipment, pipe, and hose and pump the water from a stream or well. On such jobs, the cost of water delivered at the job may range from about 25 cents to $1 or more per 1,000 gal.

Forms. Forms for concrete highway pavements are usually made of steel, and because they can be used and reused many times, their cost per square yard of pavement is quite low. The total cost of the forms should include the proportional cost of the forms, transportation to and from the job, and labor costs for placing and removing. Labor required for placing 100 lin ft forms may range from about 0.5 to 2 labor-hours. Labor required for removing forms may vary from about 0.2 to 0.8 labor-hours per 100 lin ft. A flat-bed truck may be needed for transporting the forms on the job.

Perhaps the best way of computing form costs is to find the costs per 100 lin ft, and then reduce these lineal feet costs to costs per square yard of pavement. About 100 lin ft of form work is required for 100 sq yd of pavement. Approximate costs may range from 5 to 20 cents per square yard of pavement.

Placing Reinforcement. The labor costs of placing the reinforcement at the job will vary with the number and sizes of pieces placed and the skill and wages of the workmen. Approximate costs of placing reinforcement, per square yard of pavement, may range from about 5 to 20 cents. On many jobs the cost of placing reinforcement is included with the cost of mixing and placing the concrete.

Mixing, Placing, and Finishing Concrete. The plant required for mixing, placing, and finishing the concrete will be a mixer and a spreader, vibrator, and finishing machine. The amount of concrete mixed and placed per hour depends upon the size of the mixer and the number of batches mixed per hour. Standard sizes of mixers are the 16E double drum, the 27E single drum, and the 34E single or double drum. The number 16, 27, or 34 refers to the size of the batch in cubic feet, and the E means an end discharge. Most standard mixers have a capacity of 10 to 20 percent more than their rated capacity.

The minimum mixing time is about 1 min for the smaller mixers and about 1.25 min for the larger sizes. The minimum batch cycle time (load, mix, and discharge) may be about 1.5 to 2 min for a single-drum mixer and about 1 to 1.5 min for a double-drum mixer. The average batch cycle time on the job will be greater.

Machine finishing costs (plant and labor) may vary from about 5 to 25 cents per square yard. Hand-finishing costs may range from about 25 to 50 cents per square yard.

Curing. Costs of curing concrete pavements will vary somewhat with the method used, such as ponding; covering with straw, sand, or dirt; or covering with burlap or canvas. The costs will include the covering material, water, and labor. About 0.05 to 0.25 labor-hour may be required per square yard of surface. Approximate costs per square yard of surface, assuming a covering material such as canvas or burlap that can be used several times, may be about 10 to 40 cents per square yard.

Cleaning Up. The cost of cleaning up will not be large and may be less than 1 percent of the sum of the other costs. The cost may range from about 0.3 to 2 cents per square yard.

Summary The cost items for a concrete-paving estimate can be tabulated in different ways. Most estimators consider overhead and profit as separate items. Overhead costs may range from 5 to 15 percent of the sum of the materials, labor, and equipment costs or from 10 to 30 percent of the labor costs alone. Payroll taxes and insurance are usually included in the labor costs. Profit may range from 5 to 20 percent of the sum of all other costs.

The estimate can be tabulated as follows:

Item	Cost per sq yd	Total cost
Materials, delivered to contractor	$	$
Labor		
Materials yard		
Transportation		
Construction		
Plant or equipment		
Materials yard		
Transportation		
Construction		
Overhead		
Profit		
Total estimate	$	$

Illustrative estimate for a concrete pavement. Prepare an estimate of the total cost and the cost per square yard for a concrete highway 2.6 miles long. The concrete pavement is 22 ft wide and 7.5 in. thick (average). Subgrade has been prepared previously so that only a little fine grading is needed. The specifications require 6.25 sk of cement, 1,100 lb of sand, and 2,100 lb of gravel per cubic yard. Slump is 1 in., and water not over 5 gal per sack of cement.

The materials yard may be located near a railway siding in a village. Water can be purchased in the village at 10 cents per 1,000 gal. The contractor has a 34E single-drum concrete mixer, six 5-cu-yd diesel heavy-duty dump trucks, a 100-ton portable batcher plant with scales, a crane, and other needed equipment. Average length of haul to the job is 3.5 miles over good roads.

$$\text{Cubic yards of concrete required} = \frac{2.6 \times 5{,}280 \times 22 \times 7.5}{27 \times 12} = 7{,}000$$

Allow 200 cu yd for waste, etc., giving a total of 7,200 cu yd. A 34E mixer, with a 20 percent overload, will mix a batch of 40.8 cu ft, say 1.5 cu yd. Assuming an average output of 24 batches per hour, the hourly production will be 36 cu yd.

Mixer hours for the job = 7,200 ÷ 36 = 200

Materials estimate:

Cement = 6.25 × 7,200 = 45,000 sk at $1.20	= $54,000	
Sand = 1,100 × 7.200 ÷ 2,000 = 3,960 tons at $2.00 =	7,920	
Gravel = 2,100 × 7,200 ÷ 2,000 = 7,560 tons at $2.30 =	17,388	
Water = 75 × 7,200 = 540,000 gal at $0.10 per 1,000 =	54	
Dowel bars and joint fillers for the job =	1,650	
Total materials		= $ 81,012

Labor estimate:

Item	*Wage per hr*
1 superintendent	$ 7.50
Materials yard crew:	
1 batcher operator	4.75
1 crane operator	4.75
3 laborers, handling cement, etc., at $4	12.00
Forms and fine-grading crew:	
1 assistant foreman	5.25
1 fine-grader operator	4.75
4 laborers at $4	16.00
Mixer crew:	
1 foreman	6.00
1 mixer operator	4.75
3 spreader and vibrator men at $4.25	12.75
2 finishers at $4.50	9.00
5 laborers at $4	20.00
Transportation crew:	
6 dump truck drivers at $4.50	27.00
1 water truck driver at $4.50	4.50
Curing crew, 2 laborers at $4	8.00
Labor wage per hour (33 men)	$147.00
Payroll taxes and insurance, 10%	14.70
Labor cost per hour	$161.70

Total labor cost = $161.70 × 200 hr = $ 32,340

Equipment estimate: All hourly costs for equipment were computed as explained in Appendix B, Equipment Costs.

Item	*Cost per hr*
1 car or pickup truck for superintendent	$ 1.75
1 portable office and tool shed	0.10
1 cement storage shed	0.20
1 100-ton portable batcher plant and scales	2.20
1 crawler crane with 1.5-cu-yd clamshell bucket and 45-ft boom	4.60
3 hand trucks for handling cement	0.10
1 diesel flat-bed truck for forms	2.00
1 diesel powered fine grader	3.00
1 diesel tank truck (2,000 gal) with pump	3.30
2,500 lin ft steel forms with pins, etc., estimated at $330 for the job	1.65
1 34E single-drum mixer	9.20
1 spreader and vibrator machine	2.55
1 finisher machine	2.20
6 5-cu-yd heavy-duty diesel dump trucks at $3.30	19.80
Burlap mats for curing, $240 for job	1.20
Various hand tools, $60 for job	0.30
Equipment cost per hour	$54.15

Equipment cost for 200 hr = $54.15 × 200 = $ 10,830
Transportation, erection, and dismantling costs = 3,680
Total equipment cost = $ 14,510

Overhead estimate:
Allow 7 percent of materials, labor, and equipment costs = $127,862 × 7% = $ 8,950

Profit estimate:
Allow 10 percent of all other costs = $136,812 × 10% = $ 13,681

Total estimate: = $150,493

Square yards of pavement = (2.6 × 5,280 × 22) ÷ 9 = 33,557
Cost per square yard = $150,493 ÷ 33,557 = $4.49

Summary:

Item	Cost per sq yd	Total cost
Materials	$2.42	$ 81,012
Labor	0.96	32,340
Equipment	0.43	14,510
Overhead	0.27	8,950
Profit	0.41	13,681
Total	$4.49	$150,493

CHAPTER SIX

Masonry

Masonry includes structures or parts of structures that are built by a mason of such materials as stone, brick, block, and tile.

Brick Masonry

Brick Materials Brick may be classified as to manufacture (burning, casting), materials (clay, cement, sand lime), kind (ordinary or common, face brick), and size. For estimating purposes, the quantity, kind, cost, and size of brick and thickness of joint are important. Brick sizes range from 2 to 3 in. in thickness, $3\frac{1}{4}$ to $4\frac{1}{2}$ in. in width, $7\frac{1}{2}$ to 9 in. in length, and about 60 to 76 cu in. in volume. The standard size is $2\frac{1}{4}$ by $3\frac{3}{4}$ by 8 in., with a volume of 67.5 cu in. and a wall surface area (when laid as a stretcher) of 18 sq in. The thickness of the joints may vary from $\frac{3}{16}$ to $\frac{3}{4}$ in., from $\frac{1}{4}$ to $\frac{1}{2}$ in. being common.

Many estimators base their estimates for the number of brick on the number of square feet of wall surface, with walls 1, 2, 3, or 4 brick thick as the case may be. Other estimators base their estimates on the number of cubic feet of wall.

Table 6-1 gives the approximate number of brick required per cubic foot of wall for varying sizes of brick and thicknesses of mortar joints.

A simple method for finding the number of brick required for 1 sq ft of wall of a thickness equal to the width of one brick is to lay out on a drawing board to scale a certain number of square feet of wall surface. Then the brick required for this area are counted and this number is divided by the square feet of wall surface to get the brick per square foot. This method is sufficiently correct for practical purposes.

Table 6-1 *Approximate Number of Brick Required for 1 Cu Ft of Wall*

Average volume of one brick, cu in.	Thickness of mortar joint, in.			
	¼	5/16	⅜	½
	Number of brick			
60	23.7	22.4	21.1	18.6
62	23.3	21.9	20.4	18.2
64	22.4	21.3	19.9	17.8
66	21.9	20.8	19.6	17.4
67.5*	21.4	20.4	19.1	17.1
68	21.3	20.3	19.0	17.0
70	20.6	19.7	18.6	16.6
72	20.3	19.2	18.2	16.2
74	19.8	18.7	17.7	15.8
76	19.1	18.2	17.3	15.4

* Standard size.

Table 6-2 gives the number of brick required per square foot of wall surface for brick of different sizes and for varying thicknesses of joints and for a wall one brick thick.

The number of brick for a two-brick wall will be twice the values given in the table; for a three-brick wall, three times the values; and for a four-brick wall, four times the values.

Diagram 6-1 (Appendix F) shows the relations among wall area of brick, thicknesses of joints, and the number of brick required per 100 sq ft of wall surface.

Different procedures are followed when openings are considered. All large openings should be allowed for. Small openings (say about 2 sq ft or less) can be neglected unless there are many of them. Some estimators allow for all openings regardless of size, and other estimators allow for the larger openings only.

Some allowance must be made for breakage and waste. This allowance may vary from 2 to 5 percent for common brick and from 1 to 3 percent for face brick and special brick of various kinds.

The number of brick required for walls constructed of face brick backed with common brick can be estimated by making two takeoffs, one for the face brick and the other for the common brick. Allowance should be

Table 6-2 *Approximate Number of Brick Required per Square Foot of Wall Surface*

Wall area of brick, sq in.	Thickness of joint, in.					
	¼	5⁄16	⅜	½	⅝	¾
	Number of brick					
16	7.75	7.50	7.25	6.80	6.35	6.00
17	7.35	7.10	6.90	6.45	6.05	5.75
18*	7.00	6.75	6.55	6.15	5.80	5.50
19	6.65	6.45	6.25	5.90	5.55	5.30
20	6.35	6.15	6.00	5.65	5.35	5.10
21	6.10	5.90	5.75	5.40	5.15	4.90
22	5.85	5.65	5.50	5.20	4.95	4.70
23	5.60	5.45	5.30	5.00	4.75	4.55

* Standard size.

made for headers when face brick are so used. For a check, the entire wall may be considered as being made of one kind of brick, and the total number of brick estimated. This total number should equal the sum of the face and the common brick.

When estimating walls of brick veneer, usually consisting of a single thickness of stretcher courses, it is customary to include the metal ties for bonding the brick to the wall. The number of brick and ties per square foot of wall and for common, English, Flemish, or other bonds can be found by the method of sketching previously mentioned.

Brick are usually sold by the thousand; hence, the estimator in the takeoff should tabulate the number of each kind of brick in thousands. The price per 1,000 brick varies in different localities and for different kinds of brick. Prices may range from about $25 to $80 per 1,000 for common brick up to $160 or more per 1,000 for the better and more costly kinds of face brick. The price of brick may be quoted as f.o.b. manufacturers' or dealers' yard, f.o.b. nearest railway siding, or delivered at the job. If the price is not quoted as delivered at the job, the added expense of loading, transporting, and piling must be considered. See Chap. 2 on Handling and Transporting of Materials for ways of estimating handling and transportation costs per brick.

Mortar Materials The materials required for mortar per 1,000 brick will depend upon the size of the brick, the thickness of the joints, and the proportions of the mix. Most mortar mixes include cement, lime, sand, and water. Sometimes another ingredient, such as coloring matter, is added. The cement improves the strength, and the lime improves the workability. Enough water is used to give a workable mix. The sand should be clean and hard and be fairly fine and uniform in grading. Cement is usually purchased by the sack or barrel, hydrated lime by the ton in 40- or 50-lb sacks, and sand by the cubic yard or ton. Cement weighs 94 pcf, hydrated lime about 40 pcf, and sand 2,200 to 2,600 lb per cubic yard.

The amount of mortar required for 1,000 brick of standard size laid in a three-brick wall (12 in.) is approximately as listed in Table 6-3.

Table 6-3 *Mortar for* 1,000 *Standard Brick in a Three-brick Wall*

Thickness of joint, in.	1/4	5/16	3/8	7/16	1/2	9/16	5/8	11/16	3/4
Cu yd of mortar	0.32	0.38	0.44	0.50	0.56	0.62	0.68	0.74	0.80
Cu ft of mortar	8.65	10.25	11.90	13.50	15.10	16.75	18.35	20.00	21.60

These quantities are approximately correct but will vary in practice because of variations in brick dimensions, thicknesses of joints, voids in sand, and completeness of filling of joints.

A thick wall usually requires a little more mortar per 1,000 brick than a thin wall. Considering the mortar required for a 12-in. (three-brick) wall as 100 percent, the variation is about as follows:

Thickness of wall, in.	8	12	16	20	24
Percentage	95	100	103	106	108

The proportions of the mix are usually given by volume, though weight can be used. In stating the proportions, the proportion of cement is usually given first, lime next, and then sand.

Table 6-4 gives the quantities of cement, lime, and sand required for a cubic yard of mortar. Proportions are given by volume and by weight. When the tabular values were computed, 1 sk of cement was taken as 1 cu ft, 40 lb of hydrated lime as 1 cu ft, and 2,500 lb of sand as 1 cu yd. Approximately 1 cu yd or 2,500 lb (1.25 tons) of sand will be required for 1 cu yd of mortar.

Table 6-4 *Materials Required for 1 Cu Yd of Mortar for Brick Masonry*

Mix by volume, cement–lime–sand	Cement, sacks	Lime, lb	Sand, cu yd	Mix by weight, cement–lime–sand	Cement, lb	Lime, lb	Sand, lb
1–0.05–2	13.00	26	0.96	1–0.05–2	1,200	60	2,400
1–0.05–3	9.00	18	1.00	1–0.05–3	835	42	2,500
1–0.05–4	6.75	14	1.00	1–0.05–4	625	31	2,500
1–0.10–2	13.00	52	0.96	1–0.10–2	1,200	120	2,400
1–0.10–3	9.00	36	1.00	1–0.10–3	835	84	2,500
1–0.10–4	6.75	27	1.00	1–0.10–4	625	63	2,500
1–0.25–2	12.70	127	0.94	1–0.25–2	1,150	290	2,300
1–0.25–3	9.00	90	1.00	1–0.25–3	835	210	2,450
1–0.25–4	6.75	67	1.00	1–0.25–4	625	155	2,500
1–0.50–2	12.40	250	0.92	1–0.50–2	1,150	575	2,300
1–0.50–3	8.80	175	0.98	1–0.50–3	800	400	2,400
1–0.50–4	6.75	135	1.00	1–0.50–4	615	305	2,450
1–0.50–5	5.40	110	1.00	1–0.50–5	500	250	2,500
1–1–3	8.60	345	0.95	1–1–3	770	770	2,300
1–1–4	6.60	270	0.98	1–1–4	600	600	2,400
1–1–5	5.40	210	1.00	1–1–5	490	490	2,450
1–1–6	4.50	180	1.00	1–1–6	420	420	2,500
1–1.5–3	8.10	485	0.90	1–1.5–4	575	865	2,300
1–1.5–4	6.35	380	0.94	1–1.5–5	480	720	2,400
1–1.5–5	5.30	320	0.98	1–1.5–6	420	630	2,500
1–1.5–6	4.50	270	1.00	1–1.5–7	355	535	2,500
1–1.5–7	3.85	230	1.00	1–1.5–8	315	470	2,500
1–1.5–8	3.40	205	1.00	1–1.5–9	280	420	2,500
1–2–4	6.10	490	0.90	1–2–5	460	920	2,300
1–2–5	5.10	410	0.94	1–2–6	400	800	2,400
1–2–6	4.40	350	0.98	1–2–7	360	720	2,500
1–2–7	3.85	310	1.00	1–2–8	315	625	2,500
1–2–8	3.40	270	1.00	1–2–9	280	560	2,500
1–2–9	3.00	240	1.00				

Diagrams 6-2, 6-3, and 6-4 (Appendix F) can be used for finding the costs of the cement, lime, and sand per cubic yard of mortar when the proportions of the mix and the unit costs of the materials are known.

Diagram 6-5 (Appendix F) can be used for finding the cost of the mortar for 1,000 brick when the quantity required and the cost per cubic yard are known. The cost of 1 cu yd of mortar is equal to the sum of the costs of the cement, lime, sand, and water. Approximately 1 cu yd (2,500 lb) of sand is required for 1 cu yd of mortar.

The amount of water used per 1,000 brick may vary from about 30 to 100 gal, depending on the amounts used for mortar, washing, and waste. The cost of the water may vary considerably in different places. From 2 to 5 cents per 1,000 brick can be assumed as the cost of water.

For cleaning brick walls, a mixture of 1 part muriatic acid to about 20 parts water is commonly used. One gallon of muriatic acid weighs about 10 lb and (when mixed with water) will clean about 1,000 sq ft of brick surface. The cost of 1 gal of muriatic acid is usually less than $2; hence, the cost for cleaning purposes will usually be less than $2 per 1,000 sq ft of brick surface.

The costs of the materials will vary considerably at different times and in different localities. For example, the price of cement may vary from about $1 to $1.50 per sack for carload lots f.o.b. mills plus transportation to job, or the price may range from $1.20 to $2 per sack delivered at the job. The price of hydrated lime may vary from $20 to $45 per ton f.o.b. railway siding plus transportation to the job. Sand may cost $1 to $4 a ton or $1.25 to $5 per cubic yard f.o.b. railway siding, at dealer's yard, or delivered at the job.

Labor There are two classes of labor used in brick masonry, the masons (bricklayers) and the helpers. The masons lay the brick and point the joints. The helpers mix the mortar, keep the masons supplied with brick and mortar, and erect, move, and dismantle scaffolding as required. Cleaning of the brick wall may be done by masons, helpers, or both. In cold weather, the work of heating and protection is usually done by the helpers.

Usually the number of the helpers will be equal to the number of bricklayers. The proportion of helpers to masons will depend upon the work and may vary from one helper to two bricklayers to three helpers to two bricklayers, according to the difficulty of keeping the bricklayers busy.

When brick and mortar must be transported vertically for one or more stories a hoist should be used in preference to carrying the brick and mortar in hods up and down ladders. A hand hoist is suitable on small jobs. On large jobs, a power hoist will prove economical. See the article on Equipment for costs of hoists.

A mortar mixer is generally used on all jobs.

The wages of a bricklayer may range from $4.50 to $7 per hour, depending upon the locality and other considerations. Corresponding wages of helpers may range from $3.50 to $6 per hour. Payroll taxes and insurance may vary from 8 to 18 percent of the labor wage.

The number of brick laid per hour by a bricklayer will depend upon several factors, such as kind of brick (face or common), kind of joints, kind of mortar, thickness of wall, number of openings, pilasters, corners, paneling, working conditions, union and other regulations, and skill and inclination of the workers. Table 6-5 gives approximate rates of work for laying brick.

The labor required for pointing will depend primarily upon the type of joint. Ordinary joints such as struck, plain cut, rodded, or V joints can be made as the brick are laid and require no extra time. Special joints, where some of the mortar is raked out and replaced by special mortar that is struck, rodded, beaded, etc., require extra labor which must be allowed for. The cost of special pointing is usually based on the square foot of brick surface. One mason should be able to point 10 to 40 sq ft of surface per hour, depending upon the kind of joint.

Table 6-5 *Approximate Rates of Laying Brick*

Description	Brick per hr	Hr per 1,000 brick
Common brick		
Joints struck on one side of wall:		
2-brick wall	70–125	8.0–14.3
3-brick wall	80–150	6.7–12.5
4-brick wall	90–175	5.7–11.1
5-brick wall	100–200	5.0–10.0
6-brick wall	110–225	4.5– 9.1
Common brick		
Joints struck on two sides of wall:		
2-brick wall	60–110	9.1–16.7
3-brick wall	70–135	7.4–14.3
4-brick wall	80–160	6.3–12.5
5-brick wall	90–180	5.5–11.1
6-brick wall	100–200	5.0–10.0
Face brick		
Joints struck on two sides of wall:		
Common bond	50–100	10.0–20.0
Special bonds	40–80	12.5–25.0
Boiler settings	40–80	12.5–25.0
Fireplaces	30–50	20.0–33.3

The amount of work done per hour by the helper will vary greatly. The following are approximate rates of work. A good helper can in 1 hr do one of the following items of work:

Mix 1 to 2 cu yd of mortar with a mortar mixer.

Load and wheel (with a barrow) 700 to 1,200 brick a distance of 40 to 50 ft and dump them on a scaffold.

Carry 400 to 600 brick in a hod up to a height of 10 ft.

Cull 200 to 400 brick.

Load and wheel (with a barrow) 0.7 to 1.2 cu yd of mortar a distance of 40 to 50 ft.

Handle from truck and pile 300 to 600 brick.

Two good helpers can, in 1 hr, do one of the following:

Handle 12 to 20 lin ft of pole staging.

Take down and move horse scaffolds for about 10 to 12 bricklayers working on thick walls, say 2 ft or more thick.

Take down and move horse scaffolds for about 5 to 8 bricklayers working on 8- or 12-in. walls.

There are several ways of estimating the amount of work to be done. One method is to use the entire wall surface, with no allowance for openings, and the thicknesses of the wall (2 or 3 brick, etc.). The reason for this is that the extra labor required for laying brick around the opening is about balanced by the labor saved by the openings. Another method is not to allow for any openings less than about 4 or 6 sq ft and to allow half for the larger openings. A third, and perhaps better, way is to base the labor on the actual number of brick laid, giving consideration to the extra work required as around openings, in panels, for pilasters, for arches, and for corners.

When labor estimates of brickwork are prepared, the number of bricklayers and helpers in the gang should be determined, the total hourly wages found, and the average cost per bricklayer (including proportionate cost of helper) computed. The rate of work should then be carefully assumed, and the labor cost per 1,000 brick figured. The total labor cost will equal the cost per 1,000 brick multiplied by the number of brick in thousands. The cost should also include the expense of pointing and cleaning when this work is to be done.

Diagram 6-6 (Appendix F) may be used for estimating the labor cost per 1,000 brick and Diagram 6-7 may be used for estimating the labor cost per square foot of surface area when the hourly wage of the bricklayer and helper is known.

Equipment The equipment required on a bricklaying job may include the following:

- Bricklayer's hand tools
- Mortar boxes
- Mortar mixer
- Sand screen
- Shovels
- Hoes
- Hose
- Water barrel
- Wheelbarrows
- Mortar tubs or boards
- Brick and mortar hods
- Scaffolding
- Hand hoist
- Power hoist
- Elevator tower and requisite hoisting equipment

The cost of this equipment should include such items as ownership or rentals, transportation to and from job, erection and dismantling, and operation (fuel, oil, grease, etc.).

For the various hand tools used, an allowance of 25 cents to $1 per 1,000 brick is usually adequate.

A small mortar mixer with a gasoline engine will cost 20 to 50 cents per hour of operation plus transportation and repairs.

The cost of a hand hoist will be a few cents per hour plus transportation to and from the job.

The cost of a power hoist will depend upon size and capacity of the hoist and amount and kind of other equipment such as elevator tower. Operating costs with no operator may range from 40 cents to $2 per hour plus transportation, plus erection and dismantling, plus repairs. A power hoist may be capable of elevating 2,000 to 10,000 brick a vertical distance up to 30 ft in 1 hr.

The cost of scaffolding will vary considerably with the amount required and the type used. Ordinary horse scaffolds will cost but a few cents per 1,000 brick (assuming moving to be done by helpers) plus transportation. Special scaffold machines (such as may be used on a steel building three or more stories high) can be rented for $8 to $30 per week per machine, including transportation, erection, and dismantling. Pole or post scaffolds, such as are ordinarily used on the outside of a building, will cost 50 cents to $1 per 1,000 brick, on the assumption that the erection and dismantling will be done by the helpers and that most of the lumber can be salvaged. If carpenters have to be employed for the erection and dismantling of these scaffolds, the cost will be considerably higher and may vary from $1.50 to $7 per 1,000 brick.

The total cost of equipment will be the sum of the costs of all the types and kinds of equipment used on the job.

Brick Masonry Estimates The cost estimate for brick masonry can be summarized as follows for each kind of brick and type of work.

Item	Cost per 1,000 brick	Total cost
Materials (brick and mortar)	$	$
Labor		
Equipment		
Overhead		
Profit		
Total	$	$

The costs of materials, labor, and equipment or plant can be found as directed in the preceding articles.

Overhead may be based on the sum of the costs of materials, labor, and equipment or on the cost of labor alone. The percentage allowed for overhead may range from about 5 to 12 percent when based on the sum of the costs of materials, labor, and equipment and from about 8 to 20 percent when based on labor costs alone.

Profit may range from about 8 to 15 percent of the sum of all other costs.

Illustrative estimate. The walls of a building 60 ft wide, 150 ft long, and 12 ft high are to be constructed of brick of standard sizes, using face brick on the outside and common brick for backing. The walls are 12 in. (3 brick) thick. The building is to have 36 steel-sash windows 6 ft by 8 ft 6 in., 2 doors 10 by 10 ft, and 3 doors 3 by 7 ft in size. Brick are to be laid in common bond. Joints are ¼ in. and are to be struck joints, weather joints on outside and flush joints inside of walls. Walls are to be brushed. Horse scaffolds will be used. Brick and mortar will be placed on scaffolds by hand labor. Mortar will be mixed with a small mortar mixer. Mortar mix is to be 1 to 1 to 3 by volume. Cost of materials delivered at the job are

Common brick, per 1,000	$40.00
Face brick, per 1,000	75.00
Sand, per cubic yard	4.00
Cement, net per sack of 94 lb	1.40
Hydrated lime, in sacks, per ton	35.00
Labor wages:	
Bricklayers, per hour	5.50
Helpers, per hour	4.50

Mortar mixer $0.40 per hour plus $4 transportation
Payroll taxes and insurance, 10 percent of the labor wage.

Materials:

Wall area, total	= (150 + 150 + 60 + 60)12		= 5,040 sq ft
Deductions	= (10 × 10)2	= 200 sq ft	
	= (6 × 8.5)36	= 1,836 sq ft	
	= (3 × 7)3	= 63 sq ft	
Total		= 2,099, say	2,100 sq ft
Net wall surface (outside)			= 2,940 sq ft

Number of standard brick with ¼-in. joints per square foot of wall surface is 7 (Table 6-2 or Diagram 6-1) for a 1-brick (4-in.) wall or 21 for a 3-brick (12-in.) wall.

Total number of brick (no waste) = 2,940 × 21 = 61,740

The number of face brick will probably be a little more than one-third of the total because of openings and headers.

One-third of the total number = 20,580

NOTE: An allowance of 1.5 brick per lineal foot of opening, except at bottom of doors, will be satisfactory here, by assuming that an 8-in. depth of face brick is enough at the openings. If the face brick at openings were to extend clear through the wall, 3 brick extra per lineal foot would be required.

Extra face brick for openings will be [(30)2 doors + (17)3 doors + (29)36 windows] 1.5	= 1,735
For common bond, 1/6 of the brick will be headers, and the number of extra face brick will be 1/2 × 1/6 × 20,580	= 1,715
Net number of face brick	= 24,030
Allow about 1.5% of face brick for waste, say	= 370
Total number of face brick required	= 24,400
Net number of common brick = 61,740 − 24,030	= 37,710
Allow about 3% for waste	= 1,140
Total number of common brick required	= 38,850
Contractor should order 39,000 common brick	
Cost of face brick at job = $75 × 24.4	= $1,830
Cost of common brick at job = $40 × 39	= 1,560
Total cost of brick	= $3,390

Mortar mix is 1 to 1 to 3 of cement lime, and sand. Net brick in walls is 61,740, or 61.74 thousand. Mortar materials for 1/4-in. joints with 1 to 1 to 3 mix by volume:

Mortar per 1,000 brick (Table 6-3) = 0.32 cu yd
Mortar for 61.74 thousand brick = 0.32 × 61.74 = 19.80 cu yd (about)
By using Table 6-3 and increasing quantities about 5 percent to allow for waste

Cement = 8.60 sk × 19.80 cu yd × 1.05 = 179 sk at $1.40	= $ 251
Lime = 345 lb × 19.80 cu yd × 1.05 = 7200 lb at $35 per ton	= 126
Sand = 0.95 × 19.80 cu yd × 1.05 = 19.80 cu yd at $4	= 80
Water, say	= 10
Total cost of mortar materials	= $ 467
Total cost of brick and mortar materials	= $3,857

If desired, the unit costs of mortar materials can be found from Diagrams 6-2, 6-3, and 6-4. These unit costs can be multiplied by 19.80 cu yd to obtain total costs.

Labor: The gang for this brickwork will be assumed to consist of 3 bricklayers and 3 helpers. One bricklayer will act as straw boss. Note that a mortar mixer is to be used but no hoist; also that the helpers must supply the masons by hand labor.

Gang cost per hour, including payroll taxes and insurance
= ($5.50 × 3 + $4.50 × 3)1.10 = $33.00

This is a straight 12-in. (3-brick) wall with openings and the face brick will be built up with the common brick. Hence, the hourly output for all brickwork can be reasonably assumed at more than that for face brick and a little less than that for common brick for a wall of this type. An hourly output of 80 to 100 brick per hour per mason should be a fair estimate. Will use an output of 90 brick per hour or 11 hr per 1,000 brick for this estimate. Hours required for the job = 11 × 61.74, or 679, hr for one bricklayer or 679 ÷ 3 or about 227 hr for the gang for laying the brick. No extra time will be required for pointing, as the joints can be struck when the brick are laid. No extra time need be allowed for wire brushing the walls if the helpers will go over the walls near the close of each day's work.

Labor cost = $33 × 227 = $7,491

Equipment:

Cost of mortar boxes, hoes, barrows, hods, etc., = about $1 per 1,000 brick for a job of this kind, or $1 × 61.74		= $ 62
Cost of mortar mixer	= $0.40 × 227 + $4	= 95
Cost of horse scaffolds, including transportation = about $1 per 1,000 brick, or $1 × 61.74		= 62
Total cost of equipment		= $219

Overhead: No overhead cost is given in the problem. Perhaps a fair percentage would be about 8 based on other costs or 14 based on labor costs alone. Using 14 percent of labor costs,

Overhead cost = $7,491 × 0.14 = $1,049

Profit: Profit will be assumed as 10 percent of all other costs.

$12,616 × 0.10 = $1,262

Summary: The cost per 1,000 brick is based on 61.74 thousand.

Item	Cost per 1,000 brick	Total cost
Materials	$ 63	$ 3,857
Labor	121	7,491
Equipment	4	219
Overhead	17	1,049
Profit	20	1,262
Total	$225	$13,878

Stonemasonry

Stonemasonry Classification Stonemasonry consists of walls or other structures built of stone usually bound together by mortar. A classification of stonemasonry is as follows:

1. Dry masonry in which no mortar is used. Slope walls, stone paving, and riprap are examples.

2. Wet or mortar masonry in which mortar is used.
   a. Rubble masonry, which is composed of rough unsquared stone as it comes from the field or quarry.
   b. Squared-stone masonry in which the stones are roughly squared and dressed on beds and joints. The thickness of the mortar required in the joints is more than ½ in. The stones may be laid in courses (range), broken courses (broken range), or no courses (random).
   c. Cut-stone or ashlar masonry in which the stone is cut and the thickness of the mortar in the joints is less than ½ in. The ashlar masonry may or may not be laid in courses.

Stone veneer consists of a thin layer of cut stone on the face of the wall, and tied or bonded to the backing masonry of brick or stone.

Stone trim consists of cut stone used for such purposes as water tables, lintels, copings, and ornamental courses in buildings.

Table 6-6 gives some of the different kinds of stone used for masonry and the approximate weights per cubic foot.

Table 6-6

Stone	*Weight per cu ft*
Granite	165–170
Sandstone	145–150
Marble	165–170
Limestone	140–160
Slate	170–175
Trap	180–185

Estimating Stone The units of measurement may be the cubic yard or cubic foot for rubble and squared-stone masonry, the cubic foot for cut-stone work, and the cubic foot or the lineal foot for stone trim. Stone veneer may be estimated by the cubic foot or by the square foot of surface with the thickness being given.

When walls are measured, the length is usually measured along the outside and then multiplied by the height and thickness. Allowance should be made for all openings when figuring quantities of materials, though no deductions are usually made for corners.

The volume of the stone required will be less than the volume of the wall because of the space occupied by the mortar. Mortar may form 15 to 40 percent of the volume in rubble masonry, 10 to 25 percent in squared-stone masonry, and 4 to 10 percent in ashlar masonry. The waste of stone may vary from practically nothing, when cut stone is purchased already dressed, up to as much as 20 percent when the cutting and dressing are done at the job.

The cost of the stone delivered at the job may include cost of stone at quarry (rough or dressed), freight from quarry, and transportation from railway siding to job.

The cost of the stone at the quarry may vary about as shown in Table 6-7.

Table 6-7

Kind of stone	*Cost per cu ft*
Rough stone	\$ 0.50–\$ 1.00
Special stone (such as special limestones)	\$ 1.30–\$ 4.00
Special stone, squared on two sides	\$ 2.00–\$ 5.00
Special stone, squared on four sides	\$ 3.00–\$ 7.00
Special stone, cut, finished, and crated	\$ 6.00–\$10.00
Special stone, carved and crated	\$10.00–\$16.00

These prices are approximate only. Consequently, the estimator should, whenever possible, secure prices from the quarry or the dealer for the stone f.o.b. quarry, f.o.b. siding nearest the job, or delivered at the job.

Anchors and dowels are required on many jobs. Roughly one anchor or dowel, costing 10 to 30 cents, will be required for each cubic foot of masonry.

Mortar Materials The materials required for mortar for stone masonry are cement, lime, and sand plus such special ingredients as are required. The proportions for the mortar may vary from 1 to 2 up to 1 to 4 for portland cement and sand plus small amounts of lime; 1 to 1 to 3, 1 to 1 to 4, 1 to 1 to 5, 1 to 1 to 6, etc., for cement, lime, and sand; and 1 to 2 to 4, 1 to 2 to 5, 1 to 2 to 6, 1 to 2 to 8, 1 to 2 to 9 when a larger proportion of lime is desired. Proportions are usually by volume, though proportions by weight are coming into use.

The amount of mortar required will depend upon the size of the individual stones and the thickness of the joints. Only approximate amounts may be given.

The cubic yards of mortar required for 1 cu yd of stonemasonry may range about as follows:

Kind of masonry	*Cubic yards of mortar*
Rubble masonry	0.15–0.40 (average 0.25)
Squared-stone masonry	0.10–0.25 (average 0.15)
Cut-stone or ashlar masonry	0.04–0.10 (average 0.07)

Table 6-8 gives the quantities of materials required per cubic yard of mortar of varying proportions. When these values were computed, one sack of cement was taken as 1 cu ft, 40 lb hydrated lime as 1 cu ft, and 2,500 lb of sand as 1 cu yd.

The cost of mortar materials delivered at the job will range about as follows: cement $1.20 to $2 a sack, hydrated lime $25 to $45 per ton, and sand $1 to $4 per ton or $1.25 to $5 per cubic yard.

The water required may vary from about 100 to 200 gal per cubic yard

Table 6-8 *Materials Required for* 1 *Cu Yd of Mortar for Stonemasonry*

Mix by volume cement–lime–sand	Cement, sacks	Lime, lb	Sand, cu yd	Mix by weight cement–lime–sand	Cement, lb	Lime, lb	Sand, lb
1–0.10–2	13.00	52	0.96	1–0.05–2	1,200	60	2,400
1–0.10–3	9.00	36	1.00	1–0.05–3	835	42	2,500
1–0.10–4	6.75	27	1.00	1–0.05–4	625	31	2,500
1–0.25–2	12–70	127	0.94	1–0.10–2	1,200	120	2,400
1–0.25–3	9.00	90	1.00	1–0.10–3	835	84	2,500
1–0.25–4	6.75	67	1.00	1–0.10–4	625	63	2,500
1–0.50–2	12.40	250	0.92	1–0.25–2	1,150	290	2,300
1–0.50–3	8.80	175	0.98	1–0.25–3	835	210	2,450
1–0.50–4	6.75	135	1.00	1–0.25–4	625	155	2,500
1–0.50–5	5.40	110	1.00				
				1–0.50–3	800	400	2,400
1–1–3	8.60	345	0.95	1–0.50–4	615	305	2,450
1–1–4	6.70	270	0.98	1–0.50–5	500	250	2,500
1–1–5	5.40	210	1.00	1–0.50–6	420	210	2,500
1–1–6	4.50	180	1.00				
				1–1–4	600	600	2,400
1–1.5–3	8.10	485	0.90	1–1–5	490	490	2,450
1–1.5–4	6.35	380	0.94	1–1–6	420	420	2,500
1–1.5–5	5.30	320	0.98	1–1–7	355	355	2,500
1–1.5–6	4.50	270	1.00	1–1–8	315	315	2,500
1–1.5–7	3.85	230	1.00				
1–1.5–8	3.40	205	1.00	1–1.5–5	480	720	2,400
				1–1.5–6	420	630	2,500
1–2–4	6.10	490	0.90	1–1.5–7	355	535	2,500
1–2–5	5.10	410	0.94	1–1.5–8	315	470	2,500
1–2–6	4.40	350	0.98	1–1.5–9	280	420	2,500
1–2–7	3.85	310	1.00				
1–2–8	3.40	270	1.00				
1–2–9	3.00	240	1.00				

of mortar, depending on the amounts wasted and used for washing, etc. An allowance of 6 to 15 cents per cubic yard of mortar is usually satisfactory.

Diagrams 6-2, 6-3, 6-4, and 6-8 (Appendix F) can be used for estimating cost of materials for mortar for stonemasonry.

Labor The labor required for stonemasonry may vary considerably according to the kind of masonry, amount of cutting and dressing required, working conditions, and skill and inclination of the men. Approximate labor outputs are given in Table 6-9.

Table 6-9 *Labor Required for Stonemasonry*

Kind of work	Crew, approximate	Output of crew	
		Per hr	Per unit of work
Preparing stone:	1 cutter		
Rough squaring		2–10 cu ft	2.5–13 hr per cu yd
Smoothing beds and builds		1.5–4 sq ft	0.25–0.7 hr per sq ft
Hammering		0.8–2 sq ft	0.5–1.2 hr per sq ft
Fine hammering		0.3–0.7 sq ft	1.5–3 hr per sq ft
Laying stone by hand:	1 mason, 1 to 3		
Rubble	helpers as	4–11 cu ft	2.5–7 hr per cu yd
Rough squared	needed	4–11 cu ft	2.5–7 hr per cu yd
Cut stone or ashlar		3–8 cu ft	3.5–9 hr per cu yd
Cut-stone trim		3–7 cu ft	4–9 hr per cu yd
Cut-stone veneer (about 4 or 6 in. thick)		5–10 sq ft	1–2 hr per sq yd
Laying stone with aid of hand hoist or derrick:	1 mason, 1 to 3		
All kinds of masonry	helpers as	6–15 cu ft	2–4.5 hr per cu yd
Cut-stone trim	needed	4–10 cu ft	3–7 hr per cu yd
Cut-stone veneer		7–15 sq ft	0.6–1.3 hr per sq yd
Laying stone with aid of hand hoist or derrick:	1 mason, 3 to 6		
Masonry	helpers as	7–18 cu ft	1.3–4 hr per cu yd
Cut-stone trim	needed	5–12 cu ft	2.3–5.5 hr per cu yd
Cut-stone veneer		9–18 sq ft	0.5–1 hr per sq yd
Laying stone with aid of power hoist or derrick:	1 mason, 1 operator, 3		
Masonry	to 6 helpers as	13–27 cu ft	1–2 hr per cu yd
Cut-stone trim	needed	9–21 cu ft	1.3–3 hr per cu yd
Cut-stone veneer		15–30 sq ft	0.3–0.6 hr per sq yd
Laying stone with aid of power hoist or derrick:	1 operator, 2 or 3 masons, 8 to		
Heavy foundation	12 helpers as	27–54 cu ft	0.5–1 hr per cu yd
Medium masonry	needed	18–40 cu ft	0.7–1.5 hr per cu yd
Cut-stone trim		12–30 cu ft	0.9–2.2 hr per cu yd
Cut-stone veneer		25–60 sq ft	0.15–0.35 hr per sq yd
Pointing:	1 mason, ⅓ to		
Simple	1 helper as	20–50 sq ft	0.2–0.5 hr per sq yd
Special	needed	10–30 sq ft	0.3–0.9 hr per sq yd
Cleaning	1 man	20–70 sq ft	0.15–0.5 hr per sq yd

If the individual stones are not too heavy or if the distance to be carried is not too great, all the work can be done by hand labor. Hand or power hoists or derricks should be provided for handling the heavier stones and for raising the stones vertically if the lift is about 10 ft or more.

The problem in stonework is to keep the mason supplied with stone and mortar so that he will be busy all the working hours. All manual labor should be performed by the less skilled men whenever practical. A mortar mixer should be used on all jobs.

In many localities, the masonwork is let by the job at so much per cubic yard or per cubic foot. When the number of cubic feet of masonry are computed as a basis for labor, most localities do not allow any deductions for openings, and the minimum thickness of wall may be taken as 18 in. even if the actual thickness is less. In other localities, part or all openings above a certain size are deducted. The estimator must be informed in regard to local customs before preparing labor estimates.

On many jobs, a certain amount of cutting, trimming, fitting, and hammering may be required even though the cut stones are supposed to be cut to the correct dimensions before being shipped to the job. The amount of labor to allow for such extra work is difficult to estimate in advance. Some estimators allow for this extra work when estimating the hourly output of the crew. Other estimators figure this extra work as an extra item of cost. One cutter or fitter may be needed full time on a medium-sized job and more on large jobs to care for this extra work. The cost of such work must not be neglected.

Labor wages of stonemasons (layers and cutters) may vary from about $4.50 to $7 per hour, power-hoist operator from $3.50 to $5 per hour, helpers from $3 to $6 per hour. Payroll taxes and insurance may vary from about 8 to 18 percent of the labor wage.

Diagrams 6-9 and 6-10 can be used for estimating labor costs of laying stonemasonry when the hourly wage of the mason and helpers or the crew is known and the output can be reasonably assumed.

Equipment The equipment required for stonemasonry work may include the following:

Mason's hand tools, trowels, hammers, sledges, bars, etc.

Mortar box	Water barrel
Screens	Hand hoist or derrick
Mortar mixer	Power hoist or derrick
Mortar hods	Mortar tubs or boards
Mixing tools	Barrows
Hose	Scaffolding

For the various hand tools, an allowance of 5 to 40 cents per hour per mason will be sufficient depending upon the size of the crew and number of tools. Transportation costs to and from the job are extra.

Small power tools (such as hammers) may cost a few cents up to about 30 cents per hour.

A small mortar mixer, engine or motor driven, may cost 20 cents to $1 per hour of operation plus operator plus transportation.

The cost of a hand hoist or derrick may vary from about 5 to 20 cents per hour depending upon size, plus transportation, erection, and dismantling expenses.

The cost of a power hoist or derrick with no operator may vary from about $1 to $3 per hour, plus transportation, erection, and dismantling expenses.

The cost of scaffolding will vary with the amount required and the type used. Ordinary horse scaffolds (moved by helpers) will cost but a few cents per hour per mason, plus transportation. The cost of pole or post scaffolds, erected and dismantled by helpers, will be the cost of the lumber less salvage value and plus transportation. The cost of post scaffolds will be considerably greater if carpenters are required for erection and dismantling. Cost of post scaffolds may vary from about $1 to $5 per cubic yard of masonry. Special machine scaffolds can be rented for $8 to $30 per week per machine including transportation, erection, and dismantling.

The total cost of equipment is equal to the sum of the costs of all kinds and types of equipment used on the job.

Stonemasonry Estimates The cost estimate for stonemasonry can be summarized as follows for each kind of stone and type of work. Unit costs may be per cubic yard, per cubic foot, per square foot or square yard, or per lineal foot as desired.

Item	Unit cost	Total cost
Materials (stone and mortar)	$	$
Labor		
Equipment		
Overhead		
Profit		
Total	$	$

The costs of materials, labor, and equipment can be found as directed in the preceding articles. Overhead may be based on the sum of material, labor, and equipment costs or based on labor costs alone. The percentage allowed for overhead may range from about 5 to 15 percent when based on the sum of the materials, labor, and equipment costs, and from about 10 to 25 percent when based on labor costs alone. The percentage allowed for profit may vary from 8 to 15 percent of the sum of all other costs (materials, labor, equipment, and overhead).

Illustrative estimate. The foundation for a church building 80 ft long and 40 ft wide is to be of stonemasonry walls 2 ft thick and 10 ft high. The lower 6 ft of the wall (from bottom to grade) is to be of ordinary flat-stone work, and the upper 4 ft (from grade to top) is to be of coursed granite ashlar. The outside walls above grade are to be pointed smooth, and the inside walls are to be pointed smooth. Prepare a cost estimate for this job, assuming the following:

Ordinary flat stone, per cubic yard = \$16.00
Coursed granite ashlar, per cubic foot = \$4.00
Mortar mix = 1 to 1 to 3 by volume
Mortar materials, cement per sack = \$1.40
  Hydrated lime per ton = \$33.00
  Sand per cubic yard = \$3.50
Mason's wage, per hour = \$5.50
Helper's wage, per hour = \$4.25
Payroll taxes and insurance = 10% of labor wage
Prices given for stone and mortar materials are for these materials delivered at the job.

Materials: Coursed ashlar will be assumed 1 ft thick and backed by 1 ft of rough masonry.

Coursed ashlar (80 + 80 + 40 + 40) × 4 × 1 = 960 cu ft, or 35.5 cu yd
Assume small waste to equal space of joints.
Then 960 cu ft at \$4 = \$3,840
Flat stone (80 + 80 + 40 + 40) × 6 × 2 + (80 + 80 + 40 + 40) × 4 × 1 = 2880 + 960 = 3,840 cu ft, or 143 cu yd
Of this quantity, about 75 or 80 percent will be stone and 20 to 25 percent mortar. Allow 75 percent stone and 5 percent waste.
Yardage of flat stone = 143 × 80% = 115
Then 115 cu yd at \$16 = \$1,840
Total cost of stone = \$5,680

Mortar required for rough masonry will be about 25 percent of 143 cu yd, or 35.8 cu yd. Mortar for ashlar will be about 5 cu ft per cubic yard of masonry, or about 18.5 percent.

960 cu ft × 18.5% = 177.5 cu ft, or 6.6 cu yd

Total amount of mortar = 42.4 cu yd, say 42.5 cu yd
For a 1-to-1-to-3 mix by volume, this will require (Table 6-8):

0.95 × 42.5 = 40.4 cu yd sand at \$3.50	= \$ 142	
345 × 42.5 ÷ 2,000 = 7.33 tons lime at \$33	= 242	
8.60 × 42.5 = 365 sk cement at \$1.40	= 512	
Water, say	= 4	
Cost of mortar materials		= \$ 900
Cost of stone and mortar		= \$6,580

Labor: As the walls are not high and stones can be readily handled by hand, all handwork will be assumed. A crew of 3 masons (one acting as straw boss) and 4 helpers will be assumed.

From Table 6-9, this crew should be able to lay about 3 × 9, or 27, cu ft of flat-stone masonry per hour (say 1 cu yd per hour, or 1 hr per cubic yard). For ashlar, the crew rate is about 3 × 6, or 18, cu ft per hour (0.67 cu yd per hour, or 1.5 hr per cubic yard).

Hourly wage of crew = (\$5.50 × 3 + \$4.25 × 4)1.10 = \$36.85
Hours for laying stone = 143 × 1 + 35.5 × 1.5 = 143 + 53 = 196

Allow about 12 hr for this crew for cleaning the walls (about 240 × 10 + 240 × 4, or 3,400 sq ft) and for miscellaneous work. No extra time needs to be allowed for pointing as the joints are struck as the stone is laid.

Total crew time = 208 hr (624 hr for masons and 832 hr for helpers)
Total labor cost = \$36.85 × 208 = \$7,665

This can be divided among flat and ashlar masonry according to the hours required for laying, giving about \$2,070 for the ashlar and \$5,595 for the flat stone, with unit costs of \$59 per cubic yard for the ashlar and \$39 per cubic yard for the flat stone, based on wall yardage.

Equipment: The equipment, except mortar mixer and scaffolding, will be estimated at 5 cents per hour per mason, plus about \$10 for transportation, or \$0.05 × 3 × 196 + \$10 = \$40. Some horse scaffolding may be needed on the inside of the wall. This can be estimated roughly at about \$20 for the job, including transportation. Mortar mixer at 50 cents per hour for 208 hr plus \$8 for transportation will cost \$112.

Total equipment costs = \$112 + \$40 + \$20 = \$172
Overhead, say 11% of labor costs = \$7,665 × 11% = \$843
Profit, say 10% of sum of all other costs, or 10% of \$15,260 = \$1,526

Summary:

Item	Cost per cu yd	Total cost
Materials	\$37	\$ 6,580
Labor	\$43	\$ 7,665
Equipment	\$ 1	\$ 172
Overhead	\$ 5	\$ 843
Profit	\$ 8	\$ 1,526
Total	\$94	\$16,786

If desired, costs per cubic foot can be computed or costs per cubic foot or per cubic yard for the ashlar and for the rough stone masonry can be determined.

Concrete Block, Brick, and Tile Masonry

Concrete Block, Brick, and Tile Concrete block may vary from 4 to 12 in. in height, from 6 to 12 in. in thickness, and from 12 to 32 in. in length. The most common size is $7\frac{3}{4}$ in. high, 8 in. thick, and $15\frac{3}{4}$ in. long (sometimes called 8 by 8 by 16). In addition to the regular or "stretcher" block, there are half blocks, jamb blocks (full and half size), and corner blocks. Concrete block are usually hollow with one or more vertical openings. The common size weighs about 50 to 60 lb each. Tolerance in dimensions is usually $\frac{1}{8}$ in. in length, $\frac{1}{8}$ in. in height, and $\frac{1}{4}$ in. in width. Concrete block are divided into three classes according to their strength, namely, heavy-load-bearing, medium-load-bearing, and non-load-bearing.

Concrete tile are usually hollow and vary in height from 3 to 12 in. and in width from $3\frac{3}{4}$ to 12 in. Their length is usually 12 in. They are classed according to their strength as heavy-, medium-, and non-load-bearing. The non-load-bearing tile can be used for non-load-bearing partitions.

Concrete brick are usually made of standard size ($2\frac{1}{4}$ by $3\frac{3}{4}$ by 8 in.), are rarely hollow; and can be used in place of the clay brick.

Concrete-trim stone, lintels, and specially molded ornamental work are also made. If desired, the concrete units (block, tile, brick, trim, etc.) may have special facings.

The laying of concrete block and tile will be considered in this section. Concrete or cement brick are laid in the same manner as standard clay brick. Concrete-trim stone and ornamental work are laid in the same manner as ornamental and trim stone and architectural terra cotta.

The mortar used in concrete masonry work of these types is usually a 1-to-3 portland cement and sand mortar, with about 5 to 10 lb of hydrated lime per sack of cement added to improve the workability.

The joint thickness is usually $\frac{1}{4}$, $\frac{3}{8}$, or $\frac{1}{2}$ in.

The price of concrete block and tile will vary considerably because of the costs of materials, the mix used, the quantity of concrete per unit, the labor wage, and the transportation costs from manufacturer's yard to the job. The price of concrete block may range from about 17 to 30 cents each for 8- by 8- by 16-in. block and from 12 to 25 cents each for 6- by 8- by 12-in. tile.

In the material takeoff, the number of each kind (load- or non-load-bearing), type (regular, corner, jamb), and size should be listed.

Mortar Materials A mortar mix of 1 part portland cement to 3 parts sand with 5 to 10 lb of hydrated lime is often used, though a 1-to-2, 1-to-$2\frac{1}{2}$, 1-to-3, 1-to-$3\frac{1}{2}$, or 1-to-4 mix may be selected. The proportions may be by volume or by weight, with weight preferred.

The amount of mortar required for 100 concrete block or tile will depend on the size of the unit, the net-sectional area of the unit, and the thickness of the joints. From 0.15 to 0.33 cu yd of mortar may be needed, with an average of about 0.20 to 0.25 cu yd.

The cost of the mortar materials may range from $1.20 to $2 per sack of cement, from $25 to $45 per ton for hydrated lime, and from $1 to $4 per ton or from $1.25 to $5 per cubic yard for sand.

Table 6-10 gives the quantities of cement, lime, and sand required for 1 cu yd of mortar.

Table 6-10 *Mortar Materials Required for 1 Cu Yd of Mortar with 10 Lb of Hydrated Lime per Sack of Cement*

Mix, cement and sand	Proportions by volume			Proportions by weight		
	Cement, sacks	Lime, lb	Sand, cu yd	Cement, lb	Lime, lb	Sand, lb
1–2	13.00	130	0.96	1,200	130	2,400
1–2.5	10.50	105	0.98	980	105	2,450
1–3	9.00	90	1.00	835	90	2,500
1–3.5	7.75	80	1.00	715	80	2,500
1–4	6.75	70	1.00	625	70	2,500

* Sand is estimated at 2,500 lb per cubic yard.

Approximately 1 cu yd, or 1.25 tons (2,500 lb), of sand is required for 1 cu yd of mortar.

Diagrams 6-2, 6-3, and 6-4 can be used for estimating the cost of mortar materials. Diagram 6-11 (Appendix F) can be used for estimating the cost of mortar per 100 block or tile.

The amount of water required per cubic yard of mortar may vary from about 100 to 200 gal. An allowance of a few cents (2 to 5 cents) per 100 block or tile, or 6 to 15 cents per cubic yard of mortar, is usually sufficient.

Labor The mason labor required for laying 100 concrete block or tile will vary with the size of the unit and the number of openings, jogs, corners, etc., in the wall. The mason labor required for pointing the joints will depend primarily on the kind of joints.

The helper labor required per 100 concrete block and tile will depend upon the number laid per hour by the mason and the distance the block and tile have to be moved horizontally and vertically. If the number of units required per hour is quite large and the vertical distance 10 ft or more, some form of hoists (either hand or power) may be used. Labor will be required for scaffolding. This labor will vary with height of wall, kind of scaffolds used, and the number of movements, etc., required. Helper labor for pointing will be from ¼ to 1 times the mason labor for pointing.

Table 6-11 gives the labor required for 100 block or tile. The values are approximate.

Labor required per 100 sq ft of wall will range from 10 to 30 percent more than that required for 100 block. Twice as much time is needed if both wall faces are to be pointed or cleaned.

Table 6-11 *Labor Required for Laying Concrete Block or Tile*

Kind of work	Hr per 100 block or tile		Block or tile per hr	
	Mason	Helpers	Mason	Helpers
Foundation, few openings or corners 8 by 8 by 16 block (approx.)	2.5–5.0	2.5–5.0	20–40	20–40
Superstructure, few openings or corners 8 by 8 by 16 block (approx.)	2.8–5.5	2.8–6.5	18–35	16–35
Superstructure, several openings and corners 8 by 8 by 16 tile (approx.)	3.3–6.7	3.3–7.0	15–30	14–30
Partition, few openings 6 by 8 by 12	2.5–4.0	2.5–5.0	25–40	20–40
Partition, several openings 6 by 8 by 12	2.8–5.5	2.8–6.0	18–35	17–35
Pointing and cleaning concrete block and tile				
Pointing one side:				
Simple joints	1.7–5.0	0.25–0.50	20–60	100–200
Special joints	3.3–10.0	2.5–6.7	10–30	15–40
Cleaning one side	1.25–4.0	....	25–80	

Labor wages of masons range from about \$4.50 to \$7 per hour and of helpers from about \$3.50 to \$6 per hour. Payroll taxes and insurance may vary from about 8 to 18 percent of the labor wage.

Diagram 6-12 (Appendix F) can be used for estimating labor costs for block and tile.

Equipment The equipment required for laying concrete block and tile masonry is practically the same as that for brick masonry. Such equipment may consist of

Mortar boxes, screens, hoes, hose, water barrel, sand screen, mortar hods, mortar tubs and boards, and barrows
Small power mortar mixer
Hand hoist or derrick
Power hoist or derrick
Scaffolding

The cost of such equipment is given in the articles on Equipment of the sections on Brick Masonry and Stonemasonry. Roughly the cost of equipment and scaffolding per 100 concrete block or tile is about the same as the cost per 1,000 brick or the cost per 100 cu ft of stone.

Mortar boxes, screens, hoes, etc.	\$0.20–\$1.00 per 100 block
Scaffolding, horse	\$0.05–\$0.25 per 100 block
Scaffolding, pole or post	\$0.50–\$1.00 per 100 block
Scaffold machines	\$0.20–\$0.40 per hour
Mortar mixer with engine	\$0.30–\$1.00 per hour
Hand hoist or derrick	\$0.05–\$0.20 per hour
Power hoist or derrick	\$0.40–\$2.00 per hour

These hourly costs are approximate ownership and operating costs per working hour.

The cost of a single tower, 25 to 100 ft in height, may vary from 20 to 60 cents per working hour. The cost of a double tower, 25 to 100 ft in height, may vary from 35 to 85 cents per working hour.

Costs of transportation to and from the job and erection and dismantling, if required, are in addition to these costs.

Concrete Block and Tile Masonry Estimates The cost estimates for concrete block and tile masonry can be summarized as follows for each kind of block and later totaled as desired. Unit costs are usually expressed in costs per 100 block or tile, with the size of the regular or "stretcher" block being given.

Item	Cost per 100 block or tile	Total cost
Materials (block or tile and mortar)	\$	\$
Labor		
Equipment		
Overhead		
Profit		
Total	\$	\$

The costs of materials, labor, and equipment can be found as directed in the preceding articles. Overhead may range from 8 to 25 percent based on labor costs alone or from 5 to 15 percent based on the sum of materials, labor, and equipment costs. Profit may vary from 8 to 15 percent of the sum of all other costs.

Illustrative estimate. The walls of a building 60 ft wide, 150 ft long, and 12 ft high are to be constructed of standard-size concrete block 7¾ in. high, 15¾ in. long, and 8 in. thick. The walls are 8 in. thick. The building is to have 36 steel-sash windows 6 ft by 8 ft 6 in., 2 doors 10 by 10 ft, and 3 doors 3 ft by 7 ft 4 in. Joints are to be ¼ in. thick and are to be struck flush on both outside and inside. Walls are to be brushed. Horse scaffolds are to be used. Block and mortar are to be placed on scaffolds by hand labor. Mortar will be mixed by mixer. Mortar mix is 1 part portland cement to 3 parts sand by weight. Hydrated lime to the amount of 10 percent of the weight of the cement is to be added.

Cost of materials delivered at the job:

Regular or stretcher block, each	$0.22
Corner block	0.25
Jamb block, same as regular block	0.22
Half regular and half jamb block	0.14
Lintels, per foot of length	0.20
Sand, per ton	3.00
Cement, per sack	1.40
Hydrated lime in sacks, per ton	35.00

Labor wages:

Masons, per hour	$5.50
Helpers, per hour	$4.50

Payroll taxes and insurance, 10 percent of the labor wage.

Materials:

Wall area = (150 + 150 + 60 + 60)12 = 5,040 sq ft

Openings (6 × 8.5)36 = 1,836

(10 × 10)2 = 200

(3 × 7⅓)3 = 66

Total = 2,102, say = 2,100 sq ft

Net wall area (outside) = 2,940 sq ft

Corner block = 4 × 12 × 12 ÷ 8 = 72 at $0.25 = $ 18

Jamb block, doors = 2 × 2 × 10 × 12 ÷ 8 = 60

Doors = 3 × 2 × 7⅓ × 12 ÷ 8 = 66

Windows = 36 × 2 × 6 × 12 ÷ 8 = 648

Total jamb block = 774

Full-length jamb block = 387 at $0.22 = $ 85

Half-length jamb block = 387 at $0.14 = $ 54

Half-length regular block, estimated = 387 at $0.14 = $ 54

$$\text{Regular block} = \frac{2{,}940 \times 144}{8 \times 16} - 72 - 387 - \frac{387}{2} - \frac{387}{2} - \frac{366 \times 12}{16}$$

= 3,308 − 1,121 = 2,187 at $0.22 = $481

Lintels = 2 × 11.33 + 3 × 4.33 + 36 × 9.17

= 336 ft at $0.20 = $ 73

Total cost of block and lintels = $765

For mortar, allow 580 lb of sand per 100 block

$$\text{Mortar sand} = \frac{580 \times 3{,}700}{2{,}000 \times 100} = 11 \text{ tons at } \$3 = \$\ 33$$

Cement = ⅓ × 11 × 2,000 = 7,333 lb

= 78 sk at $1.40 = 109

Lime = 7,333 × 10% = 733 lb at $35 per ton = 13

Water, say = 3

Total cost of mortar materials = $158

Total cost of all materials = $923

Labor: Mason's time for laying, pointing, and cleaning wall is assumed at 4 hr per 100 block, as the joints are simple struck joints and cleaning can be by brushing.

Hours for 3,420 block	= 137		
Hours for 366 ft of lintels at 6.25 hr per 100 ft	= 23		
Total mason time	= 160		
Mason cost	= 160 hr at $6.05	= $968	
Helpers' time is assumed same as masons' time and includes handling materials, mortar, and scaffolds.			
Helpers' cost	= 160 hr at $4.95	= $792	
Total labor			= $1,760

Crew will consist of 4 masons and 4 helpers, with one mason acting as straw boss. Total crew time is 40 hr.

Equipment: Equipment needed, except mixer and scaffolds, will be assumed at 55 cents per 100 block, including transportation.

Equipment	= $0.55 × 34.20	= $19	
Mortar mixer at $0.45 per working hour for 40 hr, plus $8 for transportation.		= $26	
Scaffolds are horse scaffolds. Allow 15 cents per 100 block plus $10 for transportation.			
Scaffolds	= $0.15 × 34.20 + $10	= $15	
Total equipment costs			= $ 60
Overhead at 11% of labor costs	= $1,760 × 11%		= $ 194
Profit at 10% of all other costs	= $2,937 × 10%		= $ 294
Total estimated cost			= $3,231

Summary:

Item	Cost per 100 block	Total cost
Materials	$25	$ 923
Labor	48	1,760
Equipment	2	60
Overhead	5	194
Profit	8	294
Total	$88	$3,231

Total number of block is taken as 3,700, considering the lintels to be the equivalent of about 280 block.

Hollow Clay Tile (Terra Cotta) Masonry

Tile Materials Clay or terra cotta hollow tile are made of burnt clay. The tile may be glazed or unglazed and may be classed as wall, partition, furring, or floor tile according to use and as load- or non-load-bearing tile depending upon the load carried. Load-bearing tile are usually 8 in. or more in thickness.

Clay wall and partition tile are usually about 12 by 12 in. face size and 3, 4, 5, 6, 8, 10, or 12 in. thick. Other face sizes are 6 by 12, 8 by 12, and 8 by 8 in. Floor tile are usually 12 by 12 in. in size and vary from 2 to 12 in. or more in thickness. Load-bearing tile are heavier than non-load-bearing tile. Many special tile are also made.

Table 6-12 gives approximate weights and costs of partition and wall clay tile. Prices vary greatly at different times and in different parts of the country. Load bearing tile cost 15 to 25 percent more than partition tile of the same dimensions. Tile with 8- by 8- or 8- by 12-in. faces cost less.

Table 6-12 *Approximate Weights and Costs of Hollow Clay Tile (12- by 12-in. face)*

	Weight per tile, lb		
Thickness, in.	Non-load-bearing	Load-bearing	Cost per 100 tile
2	12 or 14	.....	\$12–\$26
3	14 or 15	.....	\$14–\$28
4	16	.....	\$15–\$30
5	19	.....	\$18–\$37
6	22	30	\$21–\$43
7	25	32	\$24–\$50
8	30	34	\$27–\$56
9	33	36	\$28–\$68
10	35 or 36	40–42	\$30–\$80
12	40	46–52	\$32–\$90

Floor tile average about the same as to weight and costs as wall tile of equal size. Floor tile may be flat or arched. Forms and centering are often required. See Chap. 5 on Concrete for methods of estimating forms.

Roof tile or book tile are used on roofs subjected to comparatively light loads. These tile are often 12 by 24 in. in size. They are usually laid on a structural-steel framework.

Furring tile often come in a split form. They are like partition block which are grooved or scored on the sides so that each partition block can be readily split into two halves, each of which is a furring tile.

Fireproofing tile is tile placed around columns and beams for fireproofing purposes. Regular partition tile or special tile can be used.

The number of 12- by 12-in. tile is usually taken equal to the number of square feet of net wall surface to allow for a slight waste and breakage. Net wall surface equals total area of one side of wall less area of openings.

Mortar Materials The mortar required will vary with the thickness of the tile, the thickness of the joints, and the manner in which the tile is laid (horizontal or vertical). Table 6-13 gives the approximate amounts of mortar required per 100 sq ft of tile surface for tile of different thicknesses. Thickness of joints may vary from 1/4 to 3/4 in., 3/8- or 1/2-in. thickness being common.

Table 6-13 *Approximate Quantities of Mortar for* 100 *Sq Ft of Surface or for* 100 *Tile with* 12- *by* 12-*in. Face*

Thickness of tile, in.	Cubic yards of mortar per 100 sq ft of surface			
	3/8-in. joints		1/2-in. joints	
	Non-load-bearing	Load-bearing	Non-load-bearing	Load-bearing
2	0.05–0.08	....	0.06–0.10	
3	0.06–0.10	....	0.08–0.12	
4	0.08–0.11	....	0.10–0.14	
5	0.10–0.12	....	0.13–0.15	
6	0.12–0.14	0.18–0.24	0.15–0.17	0.23–0.30
7	0.14–0.16	0.20–0.25	0.17–0.20	0.25–0.31
8	0.16–0.19	0.21–0.27	0.20–0.24	0.27–0.34
9	0.18–0.22	0.23–0.30	0.22–0.27	0.30–0.38
10	0.20–0.24	0.26–0.34	0.25–0.30	0.34–0.43
12	0.24–0.29	0.30–0.40	0.30–0.36	0.40–0.50

The mortar proportions may be 1 to 2, 1 to 2½, 1 to 3, 1 to 3½, or 1 to 4 by weight or by volume of portland cement and sand. Hydrated lime should be added, say about 10 percent of the weight of the cement. See Table 6-10 for quantities of cement, lime, and sand per cubic yard of mortar. From 100 to 200 gal of water should be allowed per cubic yard of mortar. The mortar mix usually used is a 1 to 3.

The cost of mortar materials will be about as follows: $1.20 to $2 per sack of portland cement, $25 to $45 per ton of hydrated lime, 25 cents to $1 per 1,000 gal of water, and $1 to $4 per ton, or $1.25 to $5 per cubic yard of sand.

Diagrams 6-2, 6-3, 6-4, and 6-11 can be used for estimating costs of mortar materials and mortar.

Labor The labor required for laying 100 clay tile will vary somewhat with the size and weight of the tile, the number of openings and corners, etc., in the wall, the skill and inclination of the men, etc. One mason can lift and place in the wall tile weighing up to 50 or 60 lb. For heavier tile, two men are often needed. The rate of work (tile laid per hour) does not vary much with the weight of the tile for tile weighing about 40 lb each or less. The number of helpers required per mason may vary from less than one to about two (usually one) depending upon the working conditions at the particular job, the equipment used for mixing mortar and for supplying masons with material, and kind of scaffolding and moves required. On a small job, the crew may consist of 1 to 3

Table 6-14 *Mason Labor Required for Laying 100 Clay Tile*

Kind of work	Thickness, in.	Hr per 100 tile	Tile, per hr
Furring tile	2	2.50–4.00	25–40
	3	2.85–5.00	20–35
Wall and partition tile	2	2.85–5.00	20–35
	3	3.00–5.00	20–33
	4	3.35–5.55	18–30
	6	4.00–6.25	16–25
	8	5.00–7.15	14–20
	10	5.55–8.35	12–18
	12	6.25–10.00	10–16
Floor tile	6	2.25–4.00	25–45
	8	2.50–4.75	21–40
	10	2.85–5.55	18–35
	12	3.35–6.65	15–30
	14	4.00–8.35	12–25
	16	5.00–10.00	10–20
Fireproofing tile	2, 3, or 4	2.50–6.65	15–40
Thin glazed floor tile	$\frac{3}{8}$–$\frac{1}{2}$	5–20	5–20

masons with 1 to 3 helpers. On larger jobs where several masons can work without interference, the crew may consist of 1 foreman, 3 to 12 masons, 1 or 2 hoisting engineers, and 3 to 10 helpers. The use of power hoists and power mortar mixers reduces the number of helpers required.

Table 6-14 gives approximate rates of work. Enough helpers and equipment must be provided to keep the masons busy. Unless otherwise mentioned, all rates are for 12- by 12-in. tile. Times required should be increased 20 to 25 percent for 8- by 12-in. tile and reduced 15 to 20 percent for 12- by 24-in. tile. Load-bearing tile are heavier and require a little more time than non-load-bearing tile.

Labor wages are the same as for brick and stone masons and helpers and may vary from $4.50 to $7 per hour for masons and from $3.50 to $6 per hour for helpers. Hoisting engineers may cost $4.50 to $6 per hour. Payroll taxes and insurance may vary from 8 to 18 percent of the labor wage.

Diagram 6-12 can be used for estimating labor costs when the hourly cost per mason (including helpers, etc.) is known and the hourly output may be reasonably assumed.

Equipment The equipment required and its cost are practically the same as that needed for brickwork (see article in this chapter on Equipment). Power hoists and mortar mixers should be used when their use will save money and tend to speed up the work. Roughly, the equipment expense will be as follows:

Hand equipment.............	\$0.15–\$1.00 per 100 sq ft (or 100 tile)
Hand hoists.................	\$0.05–\$0.20 per hour
Small power mortar mixer.....	\$0.20–\$0.50 per hour
Power hoist or derrick........	\$0.40–\$2.00 per hour
Horse scaffolding.............	\$0.05–\$0.20 per 100 sq ft (or 100 tile)
Pole or post scaffolds.........	\$0.50–\$1.00 per 100 sq ft (or 100 tile)
Scaffold machines............	\$0.20–\$0.50 per hour

The cost of a single tower, 25 to 100 ft in height, may vary from 20 to 60 cents per working hour. The cost of a double tower, 25 to 100 ft in height, may vary from 35 to 85 cents per working hour.

Costs of transportation to and from the job, and erection and dismantling, if required, are in addition to the costs mentioned.

Clay-tile-masonry Estimates The cost estimates for hollow clay tile masonry can be summarized for each kind and size of tile and later totaled as desired. Unit costs may be expressed in costs per 100 sq ft of surface or per 100 tile. Overhead should be included and may range from 8 to 20 percent based on labor costs or from 5 to 15 percent based on the sum of material, labor, and equipment costs. Profit may range from 8 to 15 percent of the sum of all other costs.

Item	Cost per 100 sq ft of surface or per 100 tile	Total cost
Materials (tile and mortar)........	\$	\$
Labor..........................		
Equipment......................		
Overhead.......................		
Profit..........................		
Total..........................	\$	\$

Illustrative estimate. Prepare an estimate of the cost of constructing partitions of non-load-bearing hollow clay tile, 6 by 12 by 12 in. in size. Partitions are four each 9.5 by 24.5 ft with two openings 4 by 7.5 ft, four each 9.5 by 26 ft with two openings 4 by 7.5 ft, and eight each 9.5 by 14 ft with one opening 3.5 by 7 ft.

Cost of materials at job: tile \$28 per 100, portland cement \$1.40 per sack, hydrated lime \$1.50 per 100 lb, sand \$2.75 per ton.

Labor wages are masons \$5.50 per hour, helpers \$4.50 per hour. Payroll taxes and insurance, 10 percent of the labor wage.

Materials:

Total wall area	= 4(9.5 × 24.5) + 4(9.5 × 26) + 8(9.5 × 14)	= 2,983 sq ft	
Openings	= 8(4 × 7.5) + 8(4 × 7.5) + 8(3.5 × 7)	= 676 sq ft	
Net wall area		= 2,307 sq ft	
Number of tile	= 2,307 + 43 (say, for breakage)	= 2,350	
Cost of tile = \$28 × 23.50			= \$ 658

Assuming ½-in. joint, mortar required (see Table 6-13) will be about 0.16 cu yd per 100 tile. For job 0.16 × 23.07 = 3.70 cu yd. Use a 1-to-3 mortar with 5 percent lime.

Sand (assume 2,600 lb = 1 cu yd) = $3.70 \times \frac{2,600}{2,000} \times \$2.75$	= \$13	
Cement 9 sk per cubic yard = 9 × 3.70 × \$1.40	= 47	
Hydrated lime = 42.3 × 3.7 = 156 lb at \$1.50 per 100 lb, say	= 2	
Water, say	= 2	
Cost of mortar materials		= \$ 64
Total cost of materials		= \$ 722

The cost of the mortar materials can be checked by Diagrams 6-2, 6-3, and 6-4 if desired.

Labor: For this type of work, about 5 hr of masons' time per 100 sq ft of wall (or per 100 tile) will be required (Table 6-14). Helpers' time will be about the same.

Masons' cost = \$6.05 × 5 × 23.07	= \$698	
Helpers' cost = \$4.95 × 5 × 23.07	= 571	
Labor cost		= \$1,269

Checking by Diagram 6-12. \$0.55 × 2,307 = \$1,269.

Equipment: For this size of job, mortar will probably be mixed by hand.

Equipment, except scaffolds, say \$0.40 × 23.07 + \$5 transportation	= \$14	
Horse scaffolds, say \$0.13 × 23.07 + \$5 transportation	= 8	
Equipment cost		= \$ 22
Overhead at about 14% of labor costs, \$1,269 × 14%		= \$ 178
Profit at 10% of \$2,191		= \$ 219
Total cost		= \$2,410

Summary:

Item	Cost per 100 tile or 100 sq ft	Total cost
Materials	$31	$ 722
Labor	55	1,269
Equipment	1	22
Overhead	8	178
Profit	10	219
Total	$105	$2,410

Gypsum Block and Tile Masonry

Block and Tile Materials Precast gypsum tile are used for furring, solid or hollow partitions, fireproofing, and other purposes. The face of a gypsum block or tile is usually 12 by 30 in., and the thickness may vary from 2 to 8 in. or more.

Table 6-15 gives approximate weights and costs of 100 gypsum block and tile of different thicknesses.

Table 6-15 *Approximate Weights and Prices of Gypsum Block and Tile**

Thickness, in.	Weight, lb		Cost	
	100 sq ft	1 tile	100 sq ft	100 tile
2 solid	900	2.25	$18–$28	$45–$70
3 solid	1,300	3.25	$20–$30	$50–$75
2 hollow	700	1.75	$18–$28	$45–$70
3 hollow	1,000	2.50	$20–$30	$50–$75
4 hollow	1,300	3.25	$22–$34	$55–$85
5 hollow	1,550	3.88	$24–$38	$60–$95
6 hollow	1,800	4.50	$26–$42	$65–$105
8 hollow	2,300	5.75	$30–$50	$75–$125

* Face area = 12 by 30 in. = 2.5 sq ft.

In the takeoff, the wall areas for each kind and thickness of tile are listed and the number of tile of each kind and thickness computed.

Mortar Materials The mortar materials are gypsum cement and sand. The gypsum cement is sometimes called "gypsum plaster," "patent mortar," "patent plaster," or "prepared mortar." Portland cement can be used in place of gypsum cement, but a portland-cement mortar does not bond well with gypsum tile.

The proportions of the mortar is usually 1 to 3 by volume or weight, with weight preferred. Other mixes used are 1 to 2 and 1 to 2.5.

Table 6-16 gives approximate amounts of mortar required for tile of varying thickness. The thickness of the joints is usually ⅜ or ½ in.

Table 6-16 *Approximate Quantities of Mortar*

Thickness of tile, in.	Cu yd of mortar for 100 sq ft	Cu yd of mortar for 100 tile
2	0.04–0.05	0.10–0.13
3	0.06–0.075	0.15–0.19
4	0.08–0.10	0.20–0.25
5	0.10–0.125	0.25–0.31
6	0.12–0.15	0.30–0.38
8	0.16–0.20	0.40–0.50

Table 6-17 gives the quantities required per cubic yard of mortar.

Gypsum cement costs $25 to $45 per ton.

Sand costs $1 to $4 per ton and $1.25 to $5 per cubic yard.

Diagrams 6-2, 6-3, and 6-4 can be used for estimating the costs of mortar materials, and Diagram 6-11 for the cost of mortar per 100 tile.

Water costs 25 cents to $1 per 1,000 gal, and 100 to 200 gal can be assumed per cubic yard of mortar.

Table 6-17 *Quantities of Gypsum Cement and Sand for 1 Cu Yd of Mortar*

Mix	Mix by volume		Mix by weight	
	Cement, lb	Sand, cu yd	Cement, lb	Sand, lb
1–2	1,250	0.93	1,250	2,500
1–2.5	1,040	0.97	1,040	2,600
1–3	900	1.00	900	2,700

Labor As gypsum block and tile are fairly light and large compared with hollow clay tile, a mason will lay more square feet of wall per hour with gypsum tile than with hollow clay tile. Consequently, more helpers per mason may be required to keep the masons busy. Usually one to two helpers are required per mason. Table 6-18 gives the labor output per mason for laying gypsum block and tile. Face dimensions are 12 by 30 in.

Table 6-18 *Approximate Labor Output when Laying Gypsum Block and Tile*

Thickness of tile, in.	Mason-hours for laying 100 sq ft	Sq ft per hr per mason	Mason-hours for laying 100 tile	Tile per hr per mason
2	2.0–4.0	25–50	5.0–10.0	10.0–20.0
3	2.1–4.2	24–47	5.2–10.5	9.5–19.0
4	2.3–4.4	23–44	5.7–11.0	9.0–17.5
5	2.5–4.6	22–40	6.2–11.5	8.5–16.0
6	2.8–4.9	20–36	7.0–12.5	8.0–14.5
8	3.3–5.3	18–30	8.2–13.5	7.5–13.5

Labor wages may vary from $4.50 to $7 per hour for masons and from $3.50 to $6 per hour for helpers. Payroll taxes and insurance may vary from 8 to 18 percent of the labor wage.

Diagram 6-12 can be used for estimating labor costs per 100 tile.

Equipment The equipment needed will be practically the same as that required for brick masonry and for hollow clay tile masonry. The costs will be practically the same as those given per 100 sq ft or per hour in the article on Equipment (for brickwork) of this chapter. Note that a gypsum tile has a face area of 2.5 sq ft compared with 1 sq ft for hollow clay tile.

Gypsum Block and Tile Estimates The cost estimates for this kind of masonry can be summarized for each size of block as directed for other kinds of masonry. About the same percentages for overhead and profit should be allowed as for brick and hollow clay tile masonry.

Item	Cost per 100 sq ft or per 100 block	Total cost
Materials (block and mortar)..........	$	$
Labor..............................		
Equipment..........................		
Overhead......................		
Profit.............................		
Total..............................	$	$

Architectural Terra Cotta

Terra-cotta Materials Architectural terra cotta or tile may be hollow or solid. Nearly all sizes are hollow, though the smaller and thinner sizes are solid. This material is used for ornamental molding courses, copings, etc., instead of carved-stone trim. The clay can be molded into almost any desired form or shape. The weight is much less than stone and concrete, being about 70 to 75 pcf.

The usual unit is the cubic foot, though the square foot of surface or the lineal foot can be used if desired. The cost of plain architectural terra cotta will be about the same or a little more than that for hollow clay tile of equal size and weight. Specially molded terra cotta will cost more. The price of architectural terra cotta is usually quoted at so much per cubic foot and varies with the finish, ornamentation, coloring, difficulty of molding, number of times a piece is duplicated, cost of materials, labor, transportation, etc.

Mortar Materials The mortar used with architectural terra cotta is usually a 1-to-3 mix by volume or by weight of portland cement and sand. About 10 lb of hydrated lime per sack of cement is usually added for workability. The addition of coloring matter is necessary if the mortar colors are to agree with colors of the tile. Richer mixes, such as 1 to 2 or 1 to 2½, can be used if desired. Iron or steel dowels and anchors (say one to a block) may be required.

Mortar joints are usually ⅜ or ½ in. in thickness.

The amount of mortar required per 100 cu ft of terra-cotta block will vary greatly depending upon the thickness of joints, the size of blocks, and whether or not backing is required. About 0.15 to 0.50 cu yd of mortar may be required per 100 cu ft of architectural terra cotta. Average values will range from about 0.20 to 0.35 cu yd of mortar.

The prices of mortar materials will be the same as those for mortar for clay tile, brick, and stone masonry.

Table 6-8 or 6-10 and Diagrams 6-2, 6-3, and 6-4 can be used for estimating quantities and costs of mortar materials, and Diagram 6-11 for the cost of mortar per 100 cu ft of terra-cotta masonry.

Table 6-19 *Labor for 100 Cu Ft of Architectural Terra Cotta*

Kind of work	Hr per 100 cu ft		Cu ft per hr	
	Mason	Helper	Mason	Helper
Setting	6–10	9–20	10–17	5–11
Backing	3–5	3–5	20–33	20–33
	Hr per 100 sq ft		Sq ft per hr	
Pointing	1–3	1–2	30–100	50–100
Cleaning	1–3	1–2	30–100	50–100

Labor The labor required for laying architectural terra cotta is about the same as for ashlar and cut-stone masonry and for trim-stone masonry. The labor needed will vary considerably with different jobs. One to two helpers will be needed for each mason. Table 6-19 gives approximate labor outputs.

Labor wages for masons and helpers will be practically the same as for other kinds of masonry, namely, about $4.50 to $7 per hour per mason and $3.50 to $6 per hour of helpers. Diagram 6-12 can be used for estimating labor costs. Payroll taxes and insurance may vary from 8 to 18 percent of the labor wage.

Equipment The equipment required will be about the same as that needed for brick and stonemasonry. A hoist should be provided when the size of the job, number and weight of block, and vertical distance to be lifted warrants its use. Scaffolding will usually be needed and may be horse, pole or post, or machine scaffolds.

The cost of the kinds of equipment used will be about the same as that for the equipment used for brick and stonemasonry.

Architectural-terra-cotta Estimates Cost estimates for architectural terra cotta can be itemized and summarized as for other types of masonry. Overhead costs may range from 8 to 25 percent of the labor costs and profit from 5 to 15 percent of the sum of all other costs.

Unit costs may be per 100 cu ft, per 100 sq ft, per 100 lin ft, per 100 block, or per cubic yard as desired.

Item	Unit cost per 100 cu ft	Total cost
Materials (block and mortar)..........	$	$
Labor..............................		
Equipment..........................		
Overhead...........................		
Profit.............................		
Total..............................	$	$

CHAPTER SEVEN

Dampproofing and Waterproofing

Dampproofing and Waterproofing The cost of dampproofing or waterproofing a surface will depend on the kind of surface to be treated or covered, the method of treatment used, the degree of dampproofing or waterproofing desired, the physical conditions of the particular job, and the wages, working conditions, skill and inclination of the workers.

In general, dampproofing is intended to resist dampness and not to resist water pressure. Waterproofing is intended to resist water pressure. The materials should preferably be applied to the exterior or water side of a wall or floor.

The four methods commonly used for dampproofing or waterproofing surfaces are the following:

1. Painting with a water-resisting or -repellant paint or compound. One or more coats may be used.

2. Plastering with water-resisting mortar or compound, usually ½ in. or more in thickness.

3. Placing membranes or layers of waterproofing materials on the surface or between two thicknesses of the masonry or wall material.

4. Integral waterproofing by adding powders or compounds to the concrete or mortar to make it more dense and more water resisting.

The unit of measurement is usually the square yard or the square of 100 sq ft when referring to a surface or the 100 lin ft when referring to the filling of cracks or the dampproofing or waterproofing of narrow strips.

The takeoff should list the type of treatment, kind of surface to which the treatment is to be applied, and other necessary data.

Materials Among the paints or compounds applied in the same manner as a paint for the purpose of dampproofing are alum and soap mixtures (Sylvester process); water glass; alum, lye, and cement washes; cement grout with or without a water repellant; paraffins, waxes, or other mineral bases applied in a melted condition or cold in solution; various varnishes and paints; coatings of common oils; special bituminous (tar and asphalt) coatings; plastic films (such as polyethylene, etc.); silicon products; and various materials provided by different manufacturers.

One gallon of dampproofing paint or compound will cover 60 to 150 sq ft of masonry surface with one coat, depending on the smoothness of the surface. From 125 to 250 sq ft of wood surface can be covered with 1 gal. Paint and mixture prices may range from a few cents to about $4 or more per gallon.

The material used for plastering may be a rich portland-cement mortar with or without the addition of a water-repellant or -resisting powder or compound or may be specially prepared compounds marketed by the manufacturers. Detailed information in regard to price, coverage, and application should be obtained from the manufacturer before making estimates.

Integral waterproofing materials are added either to the cement or to the mixing water used in making the concrete or the mortar. Usually, a certain amount in pounds is added for each sack of cement or a certain amount is added to each gallon of mixing water. The manufacturers of integral waterproofing compounds should be consulted for prices, quantities used, and methods.

Membrane waterproofing usually consists of one or more (usually three to five) layers or membranes applied to the surface. The membrane is usually felt or burlap applied to the surface by means of some coating material, often pitch or asphalt. The specifications should state the number of membranes, the kind and weight of the material, the number of coatings, and the number of pounds of coating material per 100 sq ft of surface. The number of coatings will be one more than the number of membranes. The membranes are usually applied to the exterior surface of a wall or to the underside of a floor.

When the quantity of burlap or felt is estimated, allowance must be made for lapping and waste. From 10 to 15 percent will usually be enough. The burlap or felt may weigh 15 to 45 lb per 100 sq ft, 20 to 30 lb being common. The burlap or felt usually comes in rolls containing about 200 to 500 sq ft. Prices of roofing felt, tar felt, and asphalt felt are often quoted per 100 lb weight. Present prices (1967) range from about $2 to $6 per 100 lb of roofing felt delivered at the job.

The quantity of tar or asphalt required per coat per 100 sq ft of surface may range from about 25 to 45 lb, 30 or 40 lb being common. Tar or asphalt can be purchased by the ton or gallon. A gallon of asphalt compound may weigh about 9 to 10 lb. Asphalt is frequently priced per gallon, and tar pitch per ton. Asphalt prices may range from about $5 to $11 per 100 lb, or from about 40 cents to $1.10 per gallon. Tar pitch may cost $40 to $70 per ton.

Table 7-1 *Labor for Applying Dampproofing and Waterproofing*

Kind of work	Sq ft per hr	Hr per 100 sq ft
Cleaning	35–200	0.5–3.0
Drying (if needed)	25–200	0.5–4.0
Painting, per coat (with brushes)	70–200	0.5–1.5
Plastering, per coat	50–100	1.0–2.0
Membrane:		
Tar or asphalt, per coat	80–170	0.6–1.2
Felt, per layer	80–170	0.6–1.2

Diagrams 7-1, 7-2, and 7-3, (Appendix F) can be used for estimating the materials cost per coat, per layer, or per application when the cost of the material per gallon or per pound is known and the amount per square, including waste, is specified or can be reasonably assumed.

Labor The labor required for applying the dampproofing or waterproofing materials to a surface will include that of cleaning and drying the surface, mixing and otherwise preparing the material, and applying the material to the surface. In the case of integral waterproofing when the material is added to the cement or mixing water used in making the concrete or mortar, but little extra labor is needed.

The labor required for cleaning and drying the wall may vary from less than an hour per 100 sq ft when the wall is dry and requires but little brushing to several hours per 100 sq ft when the wall is dirty or wet and must be dried as with a blow torch.

The labor required for applying dampproofing and waterproofing is given in Table 7-1.

For example, the labor required per square for applying a three-ply membrane and asphalt waterproofing might be (0.8 hr per coat × 4 coats asphalt) + (0.8 hr per layer of membrane × 3 layers), or a total of 3.2 + 2.4 or 5.6 labor-hours per square.

Diagram 7-4 (Appendix F) can be used for estimating labor costs per square when the hourly wage is known and the hourly output can be reasonably assumed.

Equipment The equipment needed will depend upon the kind of waterproofing or dampproofing material used and the method of application.

For cleaning, good stiff brushes are usually sufficient unless the surface must be washed also.

For drying, some form of artificial heat, say a blowtorch, can be used.

For painting, mixing cans, brushes, spraying equipment, ladders, planks, sawhorses, and painters' small tools may be needed.

For plastering, plasterers' small tools and ladders, sawhorses, and plank for light scaffolds will be required.

For applying membranes, about the same equipment will be needed as for painting.

For integral waterproofing, usually no extra equipment is required for adding the waterproofing material either to the cement or to the mixing water.

An allowance of a dollar or two up to $25 or $30, depending on the particular job, may be sufficient for equipment costs.

Overhead and Profit Overhead costs may vary from about 8 to 20 percent of the labor cost or from 5 to 15 percent of the sum of labor and material costs.

Profit may range from about 8 to 20 percent of the sum of all other costs.

Dampproofing and Waterproofing Estimates Estimates for these kinds of work should include costs of all materials at the job, of all labor, of equipment required, and of overhead and profit. Unit costs per square should be computed, and total costs should be computed and tabulated.

Illustrative estimate. Prepare an estimate of the cost of waterproofing a basement wall of concrete for a basement 24 by 40 ft in size and 8 ft high. Specifications require two layers of felt and three coats of asphalt to be used. Each layer of felt must weigh 25 lb per square, and each coating of asphalt must average 35 lb per square. Price of felt is $3.20 per 100 lb and that of asphalt is 60 cents per gallon at the job. Labor wages are $5.00 per hour. Overhead is 14 percent of the labor cost, and profit is assumed at 9 percent of all other costs. Wages include 30 cents per hour for payroll taxes and insurance.

Materials:

Wall surface (24 + 40)2 × 8 = 1,024 sq ft = 10.24 sq	
Felt, 2 layers, allow 12% for laps	
2 × 25 × 1.12 × 10.24 = 575 lb	
Cost at $3.20 per 100 lb	= $ 19
Check with Diagram 7-3. Diagram gives $0.90 per layer per square, and for $0.90 × 2 layers × 10.24 sq = $18.45	
Asphalt, 3 coats 35 lb per coat = 105 lb per square 105 × 10.24 = 1.075 lb	
1 gal = 9.5 lb, gallons needed = 113.5	
Cost at $0.60 per gallon	= $ 68
Check with Diagram 7-2. 1 gal = 9.5 lb	
Cost per pound = $0.60 ÷ 9.5 = $0.063, or $6.30 per 100 lb	
Diagram gives $2.20 per coat per square $2.20 × 3 coats × 10.24 squares = $68	
Total materials	= $ 87

Labor:

Assume 0.9 hr per coat or per layer per square	
Labor-hours = 0.9 × 5 coats and layers × 10.24 sq = 46 hr	
Labor cost = $5.00 × 46	= $230
Check with Diagram 7-4. Cost per square = $22.50 × 10.24 squares = $230	
Equipment, say about $15 for this job	= 15
Overhead, 14% of $230	= 32
Profit, 9% of $364	= 33
Total estimated cost	= $397
Or a cost per square of $397 ÷ 10.24 = $39.	

CHAPTER EIGHT

Wood Construction

Classes and Estimates

Classification Wood construction may be classified according to type of structure or to kind of work. Different types of structures would be mill buildings, bridges, towers, barns, stores, residences, forms, scaffolding, etc. Different kinds of work would be heavy framing, light framing, flooring, planking, sheathing, exterior finishing, and interior finishing. Perhaps a classification as to kinds of work is more suitable for estimating, though an estimate is usually prepared for one certain structure.

In this text, wood construction will be classed as follows:

1. Heavy construction, such as mill buildings and trusses
   a. Heavy framing
   b. Planking
2. Light construction, such as ordinary wooden buildings
   a. Rough carpenter work, including framing, sheathing, rough flooring and roofing, furring, door and window frames, and insulation

b. Finish carpenter work
   (1) Exterior finish, including siding, cornice, porches, doors, windows, etc.
   (2) Interior finish, including finish flooring, doors, windows, stairs, cupboards, trim, panels, etc.

Wood-lathing estimates are discussed in Chap. 11 on Lathing and Plastering, wooden shingles in Chap. 10 on Roofing and Flashing, and forms for concrete in Chap. 5 on Concrete.

Estimates Estimates for wood construction can be divided into materials, labor, plant or equipment, overhead, and profit. The estimate for any particular structure can be prepared as the estimator desires. For example, in estimating the cost of wood construction for a residence, one estimator might list all materials and all labor separately; a second estimator might divide the materials and labor into rough work and finish work and then estimate the costs (materials, labor, and perhaps equipment and overhead) for each of these two kinds of work separately. A third estimator might divide the work further, say into (1) framing; (2) sheathing, rough flooring, and roofing; and (3) other rough carpenter work; (4) exterior finish; (5) interior finish except cabinetwork; and (6) cabinetwork. He might then estimate the materials, labor, and perhaps equipment and overhead for each of these divisions.

Materials The materials included in wood construction are the various kinds of lumber and the necessary nails, screws, bolts, and other hardware. Most lumber is measured by the board foot, which is equivalent to a piece of wood 1 ft square and 1 in. thick. The number of board feet in a piece of lumber is equal to one-twelfth of the product of the width in inches by the thickness in inches by the length in feet. For example, the number of board feet in a 2- by 10-in. joist 12 ft long is $2 \times 10 \times 12/12$, or 20 board feet, or as usually stated, 20 fbm. Molding, trim, and many special shapes are measured by the lineal foot.

All material sawed from logs is called lumber. The larger sizes are called timbers, and the timbers can be resawed to obtain smaller sizes. Rough edge is lumber sawed on two sides. Planed resawed lumber is called dressed lumber. Dressed planks and boards free from all defects are called clear. Lumber can also be classed as No. 1 or No. 2 common, depending on the comparative number, kinds, and sizes of defects.

Rough sawing to standard size means that the timbers will not be over $\frac{1}{4}$ in. scant from the actual size specified; i.e., a 12 by 12 timber shall not measure less than $11\frac{3}{4}$ by $11\frac{3}{4}$ in.

Standard dressing shall mean that not more than $\frac{1}{4}$ in. shall be allowed for dressing each surface; i.e., a 12 by 12 timber after dressing on all four sides will not measure less than $11\frac{1}{2}$ by $11\frac{1}{2}$ in.

Standard lengths are multiples of 2 ft and usually range from 10 to 20 ft. Longer and shorter lengths are special. Fractional lengths are counted as of the next higher standard length.

Standard widths are in multiples of 1 in.

Sizes usually quoted are 1 by 6, 1 by 8, 2 by 4, 2 by 6, 2 by 8, 2 by 10, 2 by 12, 3 by 12, 6 by 12, and 12 by 12 in.

There are several classifications of lumber such as kind of wood (pine, oak, fir), grading (select, common), manufacture or sawing, size, and use. The size classification is as follows:

Yard lumber and structural timber:

1. Strips. Less than 2 in. thick and under 8 in. wide
2. Boards. Less than 2 in. thick and 8 in. or over in width
3. Dimension. Any width, and 2 in. and over, but under 7 in., in thickness
   a. Planks, 2 in. and under 4 in. thick, and 8 in. or more wide
   b. Scantlings, 2 in. and under 6 in. thick, and less than 8 in. wide
   c. Heavy joists, 4 in. and under 6 in. thick, and 8 in. or more wide
4. Timbers (structural). Six inches or larger in least dimensions

Joists and plank are usually 2 to 4 in. thick and 4 in. or more in width.

Beams and stringers are usually 5 in. or more thick and 8 in. or more in width.

Posts, sills, and timbers are usually 6 by 6 in. or larger in size.

The grading classification is briefly as follows:

1. Select lumber that is generally clear and suitable for finishing.
   First class is suitable for natural finishes and includes grades A and B.
   Second class is suitable for paint finishes, and includes grades C and D.
2. Common lumber containing defects making it unsuitable for finishes.
   First class is suitable for use without waste. Grades are Nos. 1 and 2 common.
   Second class is suitable for use with waste. Grades are Nos. 3, 4, and 5 common.
   Common native lumber is sometimes called N. C.

The *construction* and *standard* grades of lumber are about the same as the No. 1 and No. 2 common grades, respectively.

Table 8-1 gives the number of feet board measure, for the more common sizes of lumber for lengths from 8 to 24 ft.

Diagrams 8-1 and 8-2 (Appendix F) show the relations among size, length, and board feet of lumber.

The approximate weights of lumber are given in Table 8-2. These

Table 8-1 *Feet Board Measure*

Size	Length, ft								
	8	10	12	14	16	18	20	22	24
	Feet board measure								
1 by 4	2⅔	3⅓	4	4⅔	5⅓	6	6⅔	7⅓	8
1 by 6	4	5	6	7	8	9	10	11	12
1 by 8	5⅓	6⅔	8	9⅓	10⅔	12	13⅓	14⅔	16
1 by 10	6⅔	8⅓	10	11⅔	13⅓	15	16⅔	18⅓	20
1 by 12	8	10	12	14	16	18	20	22	24
2 by 4	5⅓	6⅔	8	9⅓	10⅔	12	13⅓	14⅔	16
2 by 6	8	10	12	14	16	18	20	22	24
2 by 8	10⅔	13⅓	16	18⅔	21⅓	24	26⅔	29⅓	32
2 by 10	13⅓	16⅔	20	23⅓	26⅔	30	33⅓	36⅔	40
2 by 12	16	20	24	28	32	36	40	44	48
2 by 14	18⅔	23⅓	28	32⅔	37⅓	42	46⅔	51⅓	56
2 by 16	21⅓	26⅔	32	37⅓	42⅔	48	53⅓	58⅔	64
3 by 6	12	15	18	21	24	27	30	33	36
3 by 8	16	20	24	28	32	36	40	44	48
3 by 10	20	25	30	35	40	45	50	55	60
3 by 12	24	30	36	42	48	54	60	66	72
3 by 14	28	35	42	49	56	63	70	77	84
3 by 16	32	40	48	56	64	72	80	88	96
4 by 4	10⅔	13⅓	16	18⅔	21⅓	24	26⅔	29⅓	32
4 by 6	16	20	24	28	32	36	40	44	48
4 by 8	21⅓	26⅔	32	37⅓	42⅔	48	53⅓	58⅔	64
4 by 10	26⅔	33⅓	40	46⅔	53⅓	60	66⅔	73⅓	80
4 by 12	32	40	48	56	64	72	80	88	96
4 by 14	37⅓	46⅔	56	65⅓	74⅔	84	93⅓	102⅔	112
4 by 16	32⅔	53⅓	64	74⅔	85⅓	96	106⅔	117⅓	128
6 by 6	24	30	36	42	48	54	60	66	72
6 by 8	32	40	48	56	64	72	80	88	96
6 by 10	40	50	60	70	80	90	100	110	120
6 by 12	48	60	72	84	96	108	120	132	144
6 by 14	56	70	84	98	112	126	140	154	168
6 by 16	64	80	96	112	128	144	160	176	192
8 by 8	42⅔	53⅓	64	74⅔	85⅓	96	106⅔	117⅓	128
8 by 10	53⅓	66⅔	80	93⅓	106⅔	120	133⅓	146⅔	160
8 by 12	64	80	96	112	128	144	160	176	192
8 by 14	74⅔	93⅓	112	130⅔	149⅓	168	186⅔	205⅓	224
8 by 16	85⅓	106⅔	128	149⅓	170⅔	192	213⅓	234⅔	256
10 by 10	66⅔	83⅓	100	116⅔	133⅓	150	166⅔	183⅓	200
10 by 12	80	100	120	140	160	180	200	220	240
10 by 14	93⅓	116⅔	140	163⅓	186⅔	210	233⅓	256⅔	280
10 by 16	106⅔	133⅓	160	186⅔	213⅓	240	266⅔	293⅓	320
12 by 12	96	120	144	168	192	216	240	264	288
12 by 14	112	140	168	196	224	252	280	308	336
12 by 16	128	160	192	224	256	288	320	352	384

weights are for seasoned lumber with moisture content 15 to 20 percent or less. Weights of green lumber may be from 20 to 35 percent more. When weights per 1,000 fbm were computed, some allowance was made for sawing and dressing.

Table 8-2 *Approximate Weights of Lumber*

Kind of lumber	Lb per cu ft	Lb per 1,000 fbm	
		Rough sawed	Dressed
Cedar	23	1,800	1,650
Cypress	30	2,300	2,150
Fig, Douglas	32	2,500	2,300
Fir, eastern	26	2,000	1,850
Gum	36	2,800	2,550
Hemlock	28	2,150	2,000
Locust	45	3,450	3,200
Mahogany	36	2,800	2,550
Maple, hard	43	3,300	3,050
Oak, red	40	3,100	2,850
Oak, white	45	3,450	3,200
Pine, red	30	2,300	2,150
Pine, white	25	1,950	1,800
Pine, yellow longleaf	42	3,250	3,000
Pine, yellow shortleaf	36	2,800	2,550
Poplar	30	2,300	2,150
Redwood, California	25	1,950	1,800
Spruce	26	2,000	1,850
Walnut	36	2,800	2,550

When quantities of lumber required for any particular purpose are estimated, allowance must be made for waste in cutting, matching, sawing, planing, manufacture, lapping, etc. This allowance may vary from about 5 percent for end cutting up to 40 percent or more for lapping siding. Approximate allowances for waste will be given for different kinds of lumber in the following articles.

The prices asked for lumber will vary greatly as to locality, kind of wood, size of lumber, manufacture, finish, etc. Prices are usually quoted in dollars per 1,000 fbm, though trim and special types may be priced per lineal foot. At the present time (1968), prices may vary from $90 to $180 per 1,000 fbm for the smaller sizes of No. 2 common grade up to about

\$250 per 1,000 fbm for No. 1 common timbers. Hardwood flooring may vary from about \$250 to \$600 per 1,000 fbm according to locality and kind and grade of lumber. When preparing estimates, the estimator must be sure that his prices are correct.

Plywood consists of thin layers of wood (usually five) well glued and firmly pressed together. The wood is usually fir or pine. Plywood may be for exterior or interior use. The finish (often designated by letters, such as AA, AD, BB, CD) may be on one or both sides and may be rough, smooth, sanded, oiled, etc. Thicknesses are $\frac{1}{4}$, $\frac{5}{16}$, $\frac{3}{8}$, $\frac{1}{2}$, $\frac{5}{8}$, $\frac{3}{4}$, and 1 in. Plywood is usually manufactured in strips 4 ft wide and 6, 8, 10, or 12 ft long.

Prices may range from about \$80 per 1,000 fbm to about \$350 per 1,000 fbm, or more, depending on the thickness, grade, finish, and other special requirements.

Allowance must be made in the materials estimate for all construction and builders ironwork and hardware and for the nails, screws, and bolts required. The ironwork and hardware are usually listed from the plans and specifications. The weights of the various items of structural ironwork can be found from any good structural-steel handbook, or the weights can be computed from the dimensions and the weight per cubic foot (iron weighs 450 pcf, and steel 490 pcf). The price is usually given per pound (or per 100 lb) and may vary from about 15 to 40 cents per pound.

Builders hardware such as hinges, locks, handles, and stops should be listed separately, and prices obtained from dealers or from catalogues.

The quantities of nails are usually estimated at so many pounds per 1,000 fbm or at so many pounds per square of 100 sq ft, consideration being given to the size of the lumber and kind of construction.

The prices of wire nails may vary from about 15 to 25 cents per pound. Extra prices are charged for the smaller sizes. Wire nails usually come in kegs weighing 100 lb. Table 8-3 gives the sizes of wire nails, their lengths, and numbers per pound. In general, the length of nail should be from $1\frac{1}{2}$ to 2 times the thickness of the lumber. The quantity required will vary considerably with the thickness of lumber and kind of work. Further information in regard to quantities of nails required will be given in later articles.

When the takeoff is prepared, the materials can be classified as suggested in the article on Classification. Then all the kinds of wood, sizes, and dimensions can be listed under the appropriate headings. Hardware and structural ironwork can be listed separately. Nails, screws, and bolts are usually estimated after the lumber has been listed, consideration being given to the kind of construction. If available, a "reminder" list, such as described in Chap. 20 on Complete Estimates and in Appendix A, should be used so that no important items will be omitted

Labor The classes of labor used in wood construction will include superintendents, carpenter foremen, finish carpenters, rough carpenters, handy men, apprentices, and helpers or laborers. The class of labor and number in each class will depend on the size and kind of job, men available, and local and union regulations. In general, both skilled and

Table 8-3 *Wire Nails. Length and Approximate Number per Pound*

Size	Length, in.	Common	Finish
2d	1	875	
3d	1.25	575	
4d	1.50	320	575
5d	1.75	255	
6d	2	200	300
7d	2.25	155	
8d	2.50	110	190
9d	2.75	90	
10d	3	72	135
12d	3.25	60	
16d	3.50	47	
20d	4	30	
30d	4.5	23	
40d	5	18	
50d	5.5	14	
60d	6	11	

unskilled labor should be used. The laborers should be grouped in gangs in such a manner that the work will go forward efficiently in regard to cost, quality, and quantity. A gang may consist of a foreman, carpenters, handy men, helpers, and apprentices.

The number of labor-hours required for a job will depend on the amounts and kinds of work, the working conditions, and the skill and inclination of the workmen and foreman.

The labor cost will depend upon the hourly wages and the labor-hours required, plus payroll taxes and insurance.

Labor wages vary considerably in different localities and at different times. Wages are usually higher in the larger cities. Approximate ranges of hourly wages in 1967 are given in Table 8-4:

Table 8-4

Kind of labor	*Hourly wage*
Foremen	\$4.00–\$8.00
Carpenters	\$3.50–\$7.00
Handy men	\$3.00–\$5.50
Helpers	\$2.50–\$5.00
Apprentices	\$2.50–\$5.00
Operators for hoists, derricks, etc	\$3.50–\$7.00

Equipment The number and kinds of tools will vary considerably, depending on the size and kind of the job. Practically all jobs will require hand tools such as a carpenter has in his tool chest. Such hand tools include saws, hammers, levels, chisels, squares, and planes. If hand tools are to be provided by the contractor, an allowance of a few cents (say from 3 to 10 cents) per labor-hour should be sufficient. Most carpenters have their own hand tools. Practically every job will require some sawhorses and benches. An allowance of a few dollars per job for transportation and depreciation is usually enough for sawhorses and benches.

On larger jobs, power saws (both portable and stationary), power drills, planers, and other power tools may be efficiently used. Such tools may cost from a few cents up to a dollar or more per hour for ownership costs and power (if power operated). Costs of transportation to and from the job should be added when necessary. Small derricks, cranes, or hoists may be needed on larger jobs. These machines may cost from a few cents up to several dollars per hour. Such costs should include ownership, erection, moving and dismantling, transportation to and from job, and operation. If electric power is used, the costs of running power lines and for providing the necessary switches, fuses, meters, etc., must be included. The figures in Table 8-5 are approximate. Costs of transportation,

Table 8-5

Tool	*Hourly cost*
Portable electric saw	\$0.10–\$0.35
Power drills	\$0.10–\$0.40
Stationary saws (bench)	\$0.20–\$0.60
Stationary planers (bench)	\$0.25–\$0.75
Hand hoists and derricks	\$0.05–\$0.30
Small power hoists	\$0.15–\$1.00
Larger power hoists	\$1.00–\$3.00
Power derrick and cranes	\$3.00–\$7.00
Elevating tower only	\$0.20–\$1.00

erection, moving, and take-down costs should be added. If special operators are used, their wages should be included in the labor costs.

Overhead and Profit Overhead costs should include all cost items such as were listed in the article on Estimating Overhead of Chap. 1. The overhead may be based as a percentage on the sum of the costs of materials, labor, and equipment or may be based on labor costs alone. When based on the material, labor, and equipment costs, overhead costs may range from 5 to 15 percent, and when based on labor costs alone, overhead may vary from 10 to 30 percent. Payroll taxes and insurance should be included in the labor costs.

The percentage allowed for profit on wood construction may range from 5 to 15 percent of the sum of all other costs.

Summary The costs of wood construction can be estimated in as much detail as desired, such as heavy and light construction, rough and finish carpenter work, framing, flooring, roofing, and interior finishing. Some estimators work out their costs per 1,000 fbm. The total and unit costs can be summarized as follows:

Item	Unit cost (usually per 1,000 fbm)	Total cost
Materials..............	$	$
Labor.................		
Equipment............		
Overhead..............		
Profit.................		
Total...............	$	$

Heavy Wood Construction

Materials The materials used in heavy wood construction are wooden timbers and planks, construction steel and ironwork, and spikes, screws, bolts, etc., for fastening the materials together.

The estimating of the quantities of materials in this type of construction is a fairly simple process. However, if, in the takeoff, the length and other dimensions of each piece of framing are listed, together with the number and type of cuts and holes, the estimating of the labor will be aided. In the listing of the planking, it is often advisable to state the spacing of the supporting beams so that the pounds of spikes required can be more accurately estimated.

The takeoff can be divided into general divisions such as framing timbers, planking, steel- and ironwork, and spikes, screws, and bolts. The framing timbers will include such items as sills, columns and posts, girders, and beams in mill buildings and chords, web members, and cross bracing in bridges and trusses. The planking will include plank for flooring, which may or may not be tongued and grooved. The construction steel- and ironwork will include such items as iron and steel column caps and bases, beam hangers, fishplates, tie rods, and beam boxes.

Allowance must be made for waste in the manufacture, cutting, and use of lumber. For heavy framing, if the sticks are listed to the nearest commercial size, an allowance of about 5 to 8 percent is usually sufficient for end cutting. For heavy planking, both end and side waste must be allowed for. End waste will average about 5 or 6 percent and may be more if supports are spaced so that standard lengths cannot be used. Side waste will vary with the width of the plank and may be about 10 percent for plank 8 in. wide, about 13 percent for plank 6 in. wide, and about 19 or 20 percent for plank 4 in. wide. Hence, for plank 6 in. wide, the total waste would be about 5 plus 13, or 18 percent.

Labor The labor for heavy wood construction can be divided into the work of preparing the timbers and the work of erecting or placing the prepared timbers in position. On some jobs, as in the construction of wood mill buildings, the work of erection can be divided into the work of hoisting and that of placing and securing or fastening the timbers. Power tools, such as saws, drills, planers, and hoists, should be used to reduce labor time whenever practical.

The labor for heavy wooden framing can be estimated in hours per 1,000 fbm, due consideration being given to the sizes of the pieces and the sawing, cutting, hoisting, and fastening to be done. A better and more accurate way is to estimate the labor in hours per piece or stick (sill, column, or beam), due consideration being given to the size of the stick and the number of operations required. After the labor-hours, using the piece as a unit, have been estimated, the labor-hours per 1,000 fbm can be computed.

The labor for planking can be estimated in hours per square or per 1,000 fbm, consideration being given to the thickness and width of the plank, the spacing of the supports, and the amount of end cutting required. A portable power saw should be used to reduce the time required for end sawing.

The labor for structural ironwork can be estimated in hours per piece. Frequently this labor is included in the labor of erection. Base plates are usually set by a mason and helper. From 0.50 to 1.50 labor-hours will be required per base plate.

When preparing mill timbers, each carpenter may work alone if the timbers are small, in gangs of two carpenters or of one carpenter and one helper if the timbers are of medium or average size, and in larger gangs for the larger timbers if more than two men are required for handling the timbers.

Table 8-6 gives approximate labor-hours for different operations. For each operation, some of the time will be required for handling the timber, as well as for completing the operation. Sometimes the handling when power tools are used requires more time than when hand tools are used. Small timbers are less than about 100 sq in. in cross section, medium timbers 100 to about 180 sq in. and large timbers over 180 sq in. in cross section. This division as to size is approximate.

Table 8-6 *Approximate Labor-hours for Preparing Timbers*

Kind of work	Labor-hours per operation		
	Small timbers	Medium timbers	Large timbers
End cutting, hand tools	0.3–0.5	0.5–1.0	0.8–1.2
End cutting, power tools	0.2–0.3	0.3–0.6	0.4–0.7
Ripping, lineal foot, hand tools	0.1–0.3	0.2–0.4	0.3–0.5
Ripping, linear foot, power tools	0.1–0.2	0.1–0.3	0.2–0.4
Chamfering, lineal foot	0.01–0.03	0.01–0.03	0.02–0.03
Drilling 1-in. or smaller holes, hand drill	0.02–0.05	0.03–0.06	0.04–0.09
Drilling 1-in. or smaller holes, power drill	0.02–0.03	0.02–0.04	0.03–0.05
Drilling 1.5-in. holes, hand drill	0.03–0.07	0.04–0.09	0.06–0.12
Drilling 1.5-in. holes, power drill	0.02–0.04	0.03–0.06	0.04–0.08
Cutting V notches	0.2–0.3	0.3–0.4	0.4–0.5
Cutting square notches	0.2–0.4	0.3–0.5	0.4–0.6
Cutting mortises	0.1–0.2	0.1–0.3	0.2–0.4

The gang required for hoisting timbers will depend on whether a hand or power hoist or derrick is used and also on the size of the timber. A hand derrick or hoist can be used for structures up to 30 or 40 ft in height. A power hoist or derrick can be used for any height. The use of a crawler crane with a long boom is often economical.

A typical gang for a hand hoist or derrick may consist of

- 1 foreman
- 4 laborers for operating derrick
- 1 laborer for guy line
- 2 carpenters for placing and fastening timber

A typical gang for a power hoist, derrick or crane may be

- 1 foreman
- 1 operator or engineer
- 2 laborers for loading (one may also tend guy)
- 2 carpenters for placing and fastening timber

Sometimes, as in the construction of a multistoried building, a power hoist or derrick is used for hoisting timbers to the floor level and one or more gangs with hand derricks are used for the erection. The personnel of the gangs can be varied for most efficient working. The load for a hand derrick is usually one timber, though two smaller timbers can be hoisted at times. The load for a power derrick is usually larger and may consist of one or two large, two to four medium, or two to six or eight small timbers.

The erection time required per timber will be about the same for timbers of less than about 400 fbm. Larger timbers may require a little more time, a larger gang, or both. The total time required per timber will consist of loading time, hoisting, unloading or placing, and moving derrick (Table 8-7).

Table 8-7

Loading	1–3 min
Unloading and placing timber, hand hoist	3–5 min
Unloading power hoist or derrick	About 1 min per timber
Placing timber	2–4 min
Hoisting, hand	2–5 fpm
Hoisting, power	5–20 fpm
Moving small derrick	10–30 min
Moving large derrick	30–90 min

The work should be planned so that the derrick or hoist moves will be a minimum. The average moving time per timber should not exceed 5 min and should be kept to about 2 or 3 min when practical. The load of a hand derrick will usually be one timber, and the load of a power derrick may be one, two, three, or more timbers, depending on timber size and derrick capacity.

A hand derrick will handle up to about 400 fbm per load, and a power derrick will usually handle larger loads, say up to 800 or 1,200 fbm, depending on the capacity of the derrick.

Plank will ordinarily require no preparation but can be hoisted in suitable loads and then spiked in place. The expense of hoisting plank will depend on the size of load, speed, vertical distance, times for loading and unloading derrick, and times for moving derrick.

The labor of placing and spiking plank will vary somewhat with the thickness and width of plank, spacing of supports, and working conditions. In general, the labor-hours required per 100 sq ft of floor will be about the same. For example, as the distance between supports increases the amount of spiking will decrease and the difficulties of placing the plank and of working will increase.

Table 8-8 gives approximate labor-hours and pounds of spikes required per 100 sq ft of floor area. Lengths of spikes are 3.5, 5.5, and 7 in. for 2-, 3-, and 4-in. plank, respectively.

Table 8-8 *Approximate Pounds of Spikes Required for Planking*

Thickness of plank, in.	Width of plank, in.	Spacing of supports, ft					Labor-hr per 100 sq ft		
		4	6	8	10	12	Tongue and grooved	Splined	Laminated
		Spikes, lb per 100 sq ft							
2	4	4.5	3.3	2.2	1.8	...	3.5–5.0		
	6	3.0	2.2	1.5	1.2	...	2.5–3.5		
	8	2.3	1.7	1.1	0.9	...	2.0–3.0		
3	4	15.5	11.5	7.5	6.5	...	4.0–5.0	5.0–6.0	
	6	10.5	8.0	5.5	4.5	...	3.0–4.0	4.0–5.0	
	8	8.5	6.0	4.2	3.5	...	2.5–3.5	3.5–4.5	
4	6	....	11.5	8.0	6.5	5.7	3.5–5.0	4.5–6.0	6.0–8.0
	8	....	8.5	6.0	5.0	4.3	3.0–4.5	4.0–5.5	5.0–7.0

Table 8-9 gives approximate times in labor-hours for hoisting and erecting timber and plank. The times given include a reasonable allowance for moving the derricks. The height of one story may be taken as varying from 10 to 16 ft. Labor-hours are given per timber and per 100 sq ft of floor area for plank. Timber hoisting includes loading and hoisting with hand derrick and loading, hoisting, and unloading with power derrick. When a power derrick is used, timbers are raised to floor level by power derrick and then raised in place by hand derrick and crew. Planks are raised to floor level and unloaded and then placed and spiked by carpenters.

Whenever possible, the labor should be analyzed in detail and per timber rather than per 1,000 fbm.

For example, an 8- by 8-in. column 12 ft long containing 64 fbm might require 0.4 labor-hour for end cuts, 1.0 labor-hour for chamfering, and 1.1 labor-hours for hoisting and erecting, giving a total of 2.5 labor-hours per timber, or about 39 labor-hours per 1,000 fbm. However, a 16- by 16-in. column 12 ft long containing 256 fbm (four times as much as the

Table 8-9 *Approximate Labor-hours for Hoisting and Erecting Timbers and Plank*

Timber	Load	Hoisting: One story	Hoisting: Two stories	Hoisting: Additional story	Erection or placing and fastening
		Hand derrick, 400-fbm capacity. Labor-hours per timber			
Small	One	1.00	1.50	0.50	0.50
	Two	0.60	0.90	0.30	0.50
Medium	One	1.10	1.65	0.55	0.55
Large	One	1.20	1.80	0.60	0.60
		Labor-hours per 100 sq ft floor area			
2-in. plank	400 fbm	1.20	1.60	0.40	3.00
3-in. plank	400 fbm	1.50	2.10	0.60	3.50
4-in. plank	400 fbm	1.80	2.60	0.80	4.00
		Labor-hours per 1,000 fbm			
2-in. plank	400 fbm	5.00	6.60	1.60	12.5
3-in. plank	400 fbm	4.20	5.80	1.60	9.8
4-in. plank	400 fbm	3.75	5.35	1.60	8.4
		Power derrick, 1,000-fbm capacity. Labor-hours per timber			
Small	One	0.70	0.90	0.20	0.80 + 0.50
	Two	0.40	0.50	0.10	
	Three	0.30	0.37	0.07	
	Four	0.25	0.30	0.05	
	Five	0.21	0.25	0.04	
	Six	0.18	0.21	0.03	
Medium	One	0.80	1.00	0.20	0.90 + 0.55
	Two	0.45	0.55	0.10	
	Three	0.35	0.42	0.07	
Large	One	0.90	1.10	0.20	1.00 + 0.60
	Two	0.50	0.60	0.10	
		Labor-hours per 100 sq ft of floor area			
2-in. plank	1,000 fbm	0.95	1.10	0.15	3.00
3-in. plank	1,000 fbm	1.15	1.40	0.25	3.50
4-in. plank	1,000 fbm	1.30	1.60	0.30	4.00
		Labor-hours per 1,000 fbm			
2-in. plank	1,000 fbm	3.50	3.70	0.20	12.5
3-in. plank	1,000 fbm	2.80	3.00	0.20	9.8
4-in. plank	1,000 fbm	2.20	2.40	0.20	8.4

8 by 8) might require 1.2 labor-hours for end cuts, 1.0 labor-hour for chamfering, and 1.8 labor-hours for hoisting and erecting, giving a total of 4 labor-hours per timber, or about 15.5 labor-hours per 1,000 fbm.

For another example, compare the labor-hours required for a large girder for the second floor and a small crossbeam for the fifth. The larger girder is a 12- by 16-in. timber 20 ft long containing 320 fbm. Labor-hours might be 2.0 for end cuts, 0.8 for chamfering, and 1.7 per hoisting and erecting, giving a total of 4.5 labor-hours per timber, or about 14

Table 8-10 *Approximate Labor-hours per* 1,000 *fbm in Heavy Wood Construction*

Member	Timber size	Labor-hours per 1,000 fbm
Mill building	Small	25–45
	Medium	20–35
	Large	15–25
Roof trusses	Medium and small	30–45
Trestles	Medium and small	15–30
Waling	Large	15–30
	Medium	20–40
Waling	Remove	3–7
Hoisting tower	Small	25–50
Material storage bin	Small and plank	20–40
Post scaffold	Small	12–25
	Remove	5–10
Plank	2 in.	14–22
	3 in.	12–18
	4 in.	10–15

labor-hours per 1,000 fbm. The smaller beam is 8 by 12 in., 10 ft long and contains 80 fbm. Labor-hours might be 1.0 for end cuts, 0.4 for chamfering, and 1.4 for hoisting and erecting, giving a total of 2.8 labor-hours per timber, or about 35 labor-hours per 1,000 fbm.

Hence, from these two examples alone, the labor-hours for 1,000 fbm may vary from 14 to 40, which indicates that labor estimates based on 1,000 fbm will probably be unreliable. In general, the larger the timber, the fewer the labor-hours required per 1,000 fbm. For a rough estimate, the labor-hours per 1,000 fbm may be assumed to vary from 40 to 15 for timbers containing 60 to 400 fbm, with an average value of about 20 labor-hours per 1,000 fbm.

The data given in Table 8-10 have been gleaned from various sources and are approximate only. The labor-hours given are for preparing or framing and hoisting and erecting. It is assumed that most of the work of preparing the timbers will be done on the ground or floor where some power tools are available. If the work of preparing the timbers must be done by hand after the timbers are hoisted, the values given in the table should be increased from 50 to 100 percent. The plank are assumed to be hoisted to the floor and then cut, placed, and spiked.

A statement concerning labor wages is given in the article on Labor in the section on Classes and Estimates in this chapter.

Labor Diagrams. Diagrams 8-3 and 8-4 (Appendix F) can be used for estimating labor costs of preparing, hoisting, and erecting timbers.

Diagrams 8-5 and 8-6 (Appendix F) can be used for estimating the labor cost per 100 sq ft as for planking.

Diagrams 8-7 and 8-8 (Appendix F) can be used for approximate estimates of labor costs per 1,000 fbm for heavy wood construction.

Equipment The equipment or plant required for heavy wood construction will include most of the ordinary carpenter's hand tools and several power tools, such as saws, drills, and planers. Hand or power hoists and derricks will be required on nearly all jobs.

The capacity of a hand derrick is usually about 400 fbm, and that of a power derrick is usually larger, say 600 to 1,200 fbm. The hoisting rates may vary from 2 to 5 fpm for hand derricks and from 5 to 20 fpm for power derricks. An operator is usually required for a power derrick or crane.

See the article on Equipment in this chapter for approximate costs of equipment. When computing equipment costs per hour or per 1,000 fbm, the estimator must include ownership, transportation to and from the job, erection and removal, and operation costs. If the equipment is rented, the estimator must know just what costs are included in the rental prices. When electric power is to be used, the cost of providing this power (power lines, switches, fuses, meter) must be included.

Heavy-wood-construction Estimates The cost estimates for heavy wood construction should be computed and then summarized as directed in the summary in this chapter. As mentioned in the article on Overhead and Profit, overhead expenses may range from 5 to 15 percent of all costs or from 10 to 30 percent of the labor costs. Profit may range from 5 to 15 percent.

Illustrative Estimates The illustrative estimates following show the methods of preparing estimates for this kind of construction. The material prices, labor-hours, wages, equipment costs, etc., assumed in these problems may or may not apply to other jobs because of differences in working conditions, prices, wages, and foremen and of skill and inclination of workmen.

Illustrative estimate. Estimate the cost of constructing a hoisting tower for one hoist, exclusive of the hoist, cable, and pulleys. Platform hoist is 7 ft square. The four corner posts and two guide posts are to be of 4- by 6-in. timbers. Diagonal braces are 1- by 6-in. rough lumber. Horizontal braces are 2 by 6 in. Sills are 6 by 8 in. (two 4- by 6-in. timbers) resting on a bed of 2-in. plank. The top of tower (cat head) is to be covered with 2-in. plank, and four 4 by 8 crossbeams are to be provided. Height of tower is to be 45 ft. Tower is to have four guy wires. Lumber, No. 2 common, costs $130 per 1,000 fmb delivered on the job.

Materials:

Plank for top and base, say 64 × 2 or 128 fbm and 81 × 2 or 162 fbm for base, plus about 10% for waste, say	320 fbm	
Sills, about 2 × 4 × 8, or 64 lin ft of 4 by 6, say	130 fbm	
Posts, about 6 × 44, or 264 lin ft of 4 by 6, say	550 fbm	
Horizontal braces, about 24 pieces each 8 ft long, or 192 lin ft of 2 by 6, say	200 fbm	
Diagonal braces, about 48 pieces, each 19 ft long, or 480 lin ft of 1 by 6, say	250 fbm	
Top crossbeams, four 4- by 8-in. beams each 8 ft long, or 32 lin ft, say about	90 fbm	
Total	= 1,540 fbm	
Four guy wires each 80 ft long, say 320 ft of ⅜-in. wire or	120 lb	
Spikes and bolts, with some waste, say	30 lb	
Cost of material = 1.54 × $130 + $20 (spikes and wire) say		= $220

Labor: Labor-hours required may be somewhere from 14 to 26 per 1,000 fbm or from 22 to 40 for this job if power drills and saws are available. If no power tools are used, the labor-hours may vary from about 25 to 50 per 1,000 fbm or from 40 to 80 for the job. This is not very accurate, so a rough estimate of the labor required for each part of the work will be made. It will be assumed that no power tools are available and that the gang consists of 2 carpenters and 2 helpers.

	Hours
Arrange base plank and construct sill	4
End cutting posts, say 48 cuts	8
Cutting horizontal braces, say 30 cuts	3
Erecting posts and braces by hand	40–48
Placing plank and beams at top	4
Attaching guy wires	3
Total labor-hours	62–70

Say two 8-hr days for the 4-man gang, or 64 labor-hours.

Cost of gang with carpenters at $5.50 per hour and helpers at $4.50 per hour is $20 per hour, plus 15 percent for payroll taxes and insurance, equals $23 per hour.

Cost of gang for two 8-hr days = $23 × 16	= $368
Equipment: Carpenters have their own hand tools. Other equipment will be a few sawhorses and a bench.	
For this equipment and its transportation, allow	20
Overhead: Say 12% of labor costs, or	44
Profit: Say 10% of all other costs, or	65
Total cost	= $717

Illustrative estimate. Estimate the cost of the columns, girders, and planking for a two-story mill building 40 by 48 ft in size. Columns are spaced 10 ft on centers crosswise of the building and 8 ft on centers lengthwise of the building, giving 5 × 7, or 35, columns per floor. Girders run crosswise of the building, and are approximately 10 ft long, with 4 × 7, or 28, girders required per floor. Planking is 3 in. thick and is laid lengthwise of the building. First floor is concrete. Girders for the second floor are 10 by 14 in. in size. First-story columns are 10 by 10 in. in size and 12 ft long. Second-story columns are 8 by 8 in. in size with outer columns 12 ft long, intermediate columns 13 ft long, and center columns 14 ft long. Roof girders are 8 by 12 in. in size and are covered with plank 2.5 in. thick. Steel base plates for the columns have been set by the concrete contractor. Column caps are of steel costing $4.50 each. Lumber for framing costs $150 per 1,000 fbm and plank $130 per 1,000 fbm delivered at the job. Labor wages are foreman $6, carpenters $5, crane operator $5, and laborers $4 per hour. Add 11 percent for payroll taxes and insurance.

Materials:

Columns, first story $= 35 \times 12 \times \dfrac{10 \times 10}{12} = 3{,}500$ fbm

Columns, second story $= (14 \times 12 + 21 \times 14) \times \dfrac{8 \times 8}{12} = 2{,}464$ fbm

NOTE: Fourteen-foot timbers must be purchased for the intermediate columns and then cut.

Girders, second floor $= 28 \times 10 \times \dfrac{10 \times 14}{12} = 3{,}267$ fbm

Girders, roof $= 28 \times 10 \times \dfrac{8 \times 12}{12} = 2{,}240$ fbm

Cost of columns and girders = $150 × 11.471	= $1,721	
Plank, floor and roof, allow 18% for end and side waste, and assume roof area 42 by 50 to allow for slope and overhang (40 × 48 × 3 + 42 × 50 × 2.5)1.18 = 13,000 fbm		
Cost of plank, $130 × 13.000	= $1,690	
Column caps, 2 × 35 × $4.50	= 315	
Spikes, say 25 lb per 1,000 fbm, 25 × 13 × $0.15	= 49	
Bolts, etc., 3 lb per girder, 3 × 56 × $0.15	= 25	
Total materials (24,471 fbm of lumber)		= $3,800

Labor:

Columns: Cut both ends and chamfer 4 edges of each. 1.5 hr each × 70	= 105 hr	
Erection 1st floor = 0.75 hr each × 35	= 26 hr	
Erection 2d floor = 1.15 hr each × 35	= 40 hr	
Column labor-hours		= 171
Girders: Cut both ends, drill bolt holes and chamfer 2 edges of each. 1.5 hr each × 56	= 84 hr	
Erection 2d floor. Hoist, place, and fasten. 1.0 hr each × 28	= 28 hr	
Erection roof. Hoist, place and fasten. 1.3 hr each × 28	= 37 hr	
Girder labor-hours		= 149
Plank, 2d floor. Hoist, place and spike Area = 40 × 48 = 1,920 sq ft = 19.2 sq 4.50 hr per square × 19.2	= 87 hr	
Roof. Hoist, place and spike Area = 42 × 50 = 2,100 sq ft = 21 sq 4.50 hr per square × 21	= 95 hr	
Plank labor-hours		= 182
Total labor-hours		= 502

Gang. Contractor will use a gang of 1 foreman, 1 crane operator, 1 laborer, and 6 carpenters.

Gang-hours	= 502 ÷ 9 = 55.8, say 56	
Gang days	= 56 ÷ 8 = 7 working days	
Labor cost per hour	= ($6 + $5 + $4 + $5 × 6)1.11 = $50.	
Labor cost for job	= $50 × 56 hr	= $2,800

Equipment: The equipment provided by the contractor will be one 4-ton gasoline crawler crane with a long boom, one electric bench saw, one portable electric drill, three small electric portable saws, workbench and sawhorses. Carpenters have their own hand tools.

Ownership and operating cost of crane, computed as explained in Chap. 18 on Construction Plant and Equipment and in Appendix B, is $4.10 per working hour.

Crane cost	= $4.10 × 56 hr	= $230	
Cost of bench saw	= $0.25 per hour × 56 hr	= 14	
Cost of electric drill and saws	= $0.05 × 4 × 56 hr	= 11	
Cost of miscellaneous equipment estimated at		= 10	
Cost of equipment transportation		= 26	
Total equipment cost			= $ 291
Overhead: Assumed at 15% of labor cost of $2,800			= 420
Profit: Assumed at 10% of all other costs ($7,311)			= 731
Total estimated cost			= $8,042

Summary: The unit costs are based on 24,500 fbm of lumber.

Item	Cost per 1,000 fbm	Total cost
Materials	$155	$3,800
Labor	114	2,800
Equipment	12	291
Overhead	17	420
Profit	30	731
Total	$328	$8,042

Light Wood Construction—Rough Work

Light Wood Construction As stated in the article on Classification in this chapter, light wood construction refers to the construction of such structures as ordinary wooden frame stores, barns, residences, and apartments and of buildings about three stories or less in height in which the timbers used are generally less than about 40 sq in. in cross section and are rarely over 2 or 3 in. thick, with the exception of a few sills, posts, or beams.

Light wood construction may be divided into rough and finish carpenter work. This section discusses rough work, and the following section, finish work.

Materials The materials for the rough carpenter work of light wood construction can be classified as follows:

Framing. Sills, studs, joists, rafters, plates, beams, posts, ribbons
Boarding, sheathing, flooring, roofing
Furring and grounds
Door and window frames (if installed by rough carpenters)
Insulation

In the materials takeoff, the sizes and lengths of all lumber should be listed and then computed and totaled in board feet or lineal feet. Almost all lumber for rough carpenter work is priced per 1,000 fbm. Any lumber less than 1 in. thick is classed as 1 in. thick. The use of a good reminder list, such as is described in Appendix A, is helpful in avoiding errors caused by omissions.

Framing. Sills can be scaled from the plans. These may be single sticks or may be built up from 2-in. material. Sills may be 6 by 6, 6 by 8, 6 by 10, etc., in size. When estimating sills, about 6 in. should be allowed for lapping single sticks and about 5 percent allowed for end waste. Sill plates can be used when the walls extend all around the building and offer a firm support. Sill plates may be 2 by 6, 2 by 8, or 2 by 10 in size.

The number, size, and length of posts supporting floor beams or joists can be taken from the plans. Common sizes of wooden posts are 6 by 6, 6 by 8, and 8 by 8. About 5 percent can be allowed for waste. In some instances, cast-iron or structural-steel posts are used.

In some cases, heavy floor beams can be used for supporting the joists. These beams may be single piece, though they are usually built up of 2-in. material. Common sizes are 6 by 8, 6 by 10, 6 by 12, 8 by 8, 8 by 10, 8 by 12, 10 by 10, and 10 by 12.

Ordinary plates and caps, placed below and on top of studs, are usually 2-in. material of the same width as the studs.

Wall and partition studs should be listed as to size and length. These studs are usually 2 by 4 in. in size and spaced 16 in. on centers, though 2 by 3 and 2 by 6 are sometimes used, and spacings may be 12, 20, or 24 in. One extra stud is usually provided at each corner and at each opening. About 5 percent should be allowed for end waste.

A rule for finding the number of wall studs required, spaced 16 in. on centers, is to:

1. Take three-fourths of total length of walls in feet.
2. Add one stud for each length of wall.
3. Add one stud for each intersection with partition.
4. Add two studs for each wall opening.
5. Add, for plates and caps, twice the total length of wall divided by height of studs.
6. Add 5 percent for waste.

For finding the number of partition studs spaced 16 in. on centers:

1. Take three-fourths of total lengths of partitions in feet.
2. Add one stud for each length of partition.
3. Add one stud for each partition intersection.
4. Add two studs for each partition opening.
5. Add, for plates and caps, twice the total length of partition divided by the height of studs.
6. Add 5 percent for waste.

An approximate rule is to allow 1.25 studs for each foot of length of wall and partition.

Joists should be listed according to size and length. They are usually spaced 16 in. on centers. Ordinary sizes of joists are 2 by 6, 2 by 8, 2 by 10, and 2 by 12, though 3-in. material is sometimes used. To find the number of joists, divide the length of floor perpendicular to the joist by $1\frac{1}{3}$ ft (16 in.) and add one end joist. Also add one joist for each partition and header, as joists are usually placed in pairs under partitions and when headers or short crossbeams frame into them. Allow 5 percent for end waste.

Bridging is usually required for all joists. At least one pair of bridging is used between each two joists. Bridging is usually made of 1 by 3 or 1 by 4 strips. The length of each piece will vary according to the spacing and depth of the joists. About 18 in. may be allowed for each piece when joists are spaced 16 in. on centers.

Rafters are usually 2 by 6, 2 by 8, or 2 by 10, though 2 by 4 and 2 by 12 have been used, depending on inclination of roof, span, spacing, and load. The spacing is usually 16 in. or 2 ft. The size, length, and number of rafters should be listed in the takeoff. Allow 5 percent for end waste on ordinary roof work. If there is much special framing as for gables, the waste may be more, say as much as 10 percent in some instances.

Ribbons are usually 1 by 4, 1 by 5, or 1 by 6. The length should be scaled from the plans. Usually one line of ribbons around the walls is required for each story. Ribbons are sometimes included with furring and grounds.

Rough Boarding and Sheathing. Rough boarding or sheathing for floors, walls, and roof may be of unmatched boards 1 by 6, 1 by 8, or 1 by 10 in size or of matched boards (tongue and grooved or shiplap) 1 by 6, 1 by 8, 1 by 10, or 1 by 12 in size. Rough flooring may be laid across the joists or diagonally. Wall sheathing may be placed horizontally or diagonally. Rough roofing is usually placed across the main rafters. No allowance is usually made for small openings in floors or roofs. Small wall openings are not allowed for, and large openings (doors and windows) may or may not be deducted. Some estimators consider the waste in wall sheathing as being equal to the openings, and others allow for waste and deduct for the larger openings.

Suitable grades of plywood (see the article on Materials in this chapter) are often used for rough boarding and sheathing. Plywood usually costs less, has less waste and requires less labor.

Table 8-11 gives approximate allowance for end and side waste for

Table 8-11 *Allowance for Waste in Rough Boarding*

Size, inches	Unmatched		Matched	
	Laid crosswise	Laid diagonal	Laid crosswise	Laid diagonal
1 by 6	15	20–25	20	25–30
1 by 8	13	18–22	18	22–28
1 by 10	11	15–20	16	20–25
1 by 12	10	15–20	15	20–25

matched and unmatched boards laid on floors, walls and flat roofs. For pitched roofs and for gables, add 5 percent more. Wall openings for windows and doors are deducted.

Diagram 8-9 (Appendix F) can be used for estimating the cost of rough boarding and sheathing.

Furring and Grounds. Furring and grounds are usually about 1 by 2 or 1 by 3 in size. The lengths are scaled from the plans. Furring and grounds are usually listed in lineal feet instead of board feet. Approximately 2 lin ft of grounds are required for each linear foot of base or chair rail, and 1 lin ft of grounds for each linear foot of picture mold or casing.

Window and Door Frames. The window and door frames are often set by the rough carpenters as soon as the framing and sheathing permit and, consequently, are included with the rough carpenter work. In some cases, these frames are not set until the finish carpenter work is started, and then this work is included in the finish carpenter work.

In the takeoff, the window and door frames are listed as separate units according to size.

Insulation. Insulating material is of two kinds, that which is placed between the studs or joists and that which is placed outside the studs either on the outside to take the place of the rough sheathing or on the inside to take the place of the lath.

Insulating material that is placed between the studs or joists usually comes in strips of suitable widths for studs or joists spaced 16 in. on centers and in various lengths. Some of the softer material comes in rolls. The total length should be given, with an allowance of about 5 percent for laps and waste. Some insulating material comes in blocks or bats about 15 in. wide and of varying lengths and should be listed as to number of units of a given length or as to total length. Other material comes loose and is sold by the pound or cubic foot, 1 cu ft being enough for an area of 3 or 5 sq ft between joists.

Insulating material that is to be placed on the outside of the studs usually is in widths of 32 or 48 in. and in lengths from 6 to 12 ft. The thickness is usually ⅝ or ¾ in. This material is listed and priced per 1,000 sq ft of surface. There are many varieties on the market. When this material is used for outer sheathing, it is usually placed by the rough carpenters and classed as rough carpenter work. When it is used instead of lath, it can be placed by rough carpenters or by lathers and classed accordingly. When it is used as interior finish, it is placed by finish carpenters and classed as interior finish. Waste, when classed as rough carpenter work, should be small, say 5 percent. No deductions should be made for small openings, but large openings (ordinary windows and doors) should be deducted. There are many varieties of plywood or wallboard or plaster board on the market. Prices for this type of material vary from about $45 and up per 1,000 sq ft; average prices for plaster and wallboard material ordinarily used vary from about $50 to $150 per 1,000 sq ft, depending on kind, quality, and thickness.

Miscellaneous material may be listed according to article, weight, length, area, or volume as the case may be.

Nails. The nails used should be from about 1.5 to 2 times the thickness

Table 8-12 *Units of Measurement, Approximate Waste, and Nails Required*

Material	Unit of measurement	Percentage of waste	Pounds of nails
Framing:			
Sills, single piece	1,000 fbm	5	5–10
Sills, built up	1,000 fbm	5–8	10–20
Posts	1,000 fbm	5	None
Beams, built up	1,000 fbm	5–8	10–25
Studs	1,000 fbm	5	10–15
Plates and caps	1,000 fbm	5–8	10–20
Joists	1,000 fbm	5	10–25
Bridging	1,000 fbm	5–10	20–25
	1,000 lin ft	2–5	5–8
Rafters, simple roofs	1,000 fbm	5–8	8–15
Rafters, gables and dormers	1,000 fbm	10–20	10–20
Ribbons	1,000 fbm	5–8	20–25
	100 lin ft	2–4	1–2
Boarding, flooring, and sheathing:			
Rough flooring, unmatched			
Laid crosswise	1,000 fbm	15	20–30
Laid diagonal	1,000 fbm	20–25	20–30
Rough flooring, matched			
Laid crosswise	1,000 fbm	20	20–30
Laid diagonal	1,000 fbm	25–30	20–30
Roofing, unmatched			
Flat and plain	1,000 fbm	15–20	20–30
Pitched, gables	1,000 fbm	20–30	20–30
Roofing, matched			
Flat and plain	1,000 fbm	20–25	20–30
Pitched, gables	1,000 fbm	25–30	20–30
Wallboard, plasterboard, etc	1,000 sq ft	5–10	12–20
Sheathing, unmatched			
Laid crosswise	1,000 fbm	15	20–30
Laid diagonal	1,000 fbm	20–25	20–30
Sheathing, matched			
Laid crosswise	1,000 fbm	20	20–30
Laid diagonal	1,000 fbm	25–30	20–30
Miscellaneous:			
Furring and grounds	100 lin ft	8–12	1–2
Window and door frames	Unit	.....	1–2

Table 8-13 *Approximate Time in Man-hours per Timber for Light Framing*

Timber and work	Size of timber, in.	Hours per timber		
		Preparation	Erection	Total
Sills, cutting, and splicing	4 by 6 6 by 6	0.3–0.6	0.2–0.4	0.5–1.0
	6 by 8 8 by 10	0.4–0.9	0.3–0.6	0.7–1.5
Joists, fire cutting	2 by 8 2 by 10 2 by 12	0.2–0.4	0.2–0.3	0.4–0.7
	3 by 10 3 by 12	0.3–0.5	0.2–0.4	0.5–0.9
Joists, square cutting	2 by 8 2 by 10 2 by 12	0.1–0.3	0.2–0.3	0.3–0.6
	3 by 10 3 by 12	0.2–0.4	0.2–0.4	0.4–0.8
Wall studs and plates. Cutting ends and ribbon notches	2 by 4 2 by 6	0.1–0.3	0.1–0.3	0.2–0.5
Partition studs and plates and caps. Cutting both ends	2 by 3 2 by 4 2 by 6	0.1–0.2	0.1–0.2	0.2–0.4
Rafters, main, jack, and hip and valley	2 by 4 2 by 6	0.2–0.3	0.2–0.3	0.4–0.6
	2 by 8 2 by 10	0.2–0.4	0.2–0.4	0.5–0.8

of the board. At least two nails should be allowed for each board at each support. Boards over 8 in. wide will require three or more nails at each support. The waste in nails is comparatively large. From 10 to 25 percent may be allowed for this purpose.

The information given for materials for rough carpenter work is summarized in Table 8-12.

Labor

Framing. The labor time for framing light wood construction will vary with the times required for preparing and erecting each piece. These times will vary according to the number and kind of cuts required per piece, size of piece, and difficulties of erection. Consequently, some estimators prefer to estimate light wood framing using a piece (or 100 pieces) as a unit and others use 1,000 fbm as a unit.

Table 8-13 gives the carpenter-hours required for light framing, using the piece as a unit. Cutting will require about two-thirds of the time required for preparation, and measuring and laying out about one-third. The values given in the table are for softer woods such as hemlock, spruce, and soft pine. From 15 to 25 percent more time must be allowed for

Table 8-14 *Approximate Time in Man-hours per* 1,000 *Fbm for Light Framing*

Kind of timber	Hr per 1,000 fbm		
	Preparation	Erection	Total
Sills, single piece	12–18	8–12	20–30
Built up	15–25	8–12	25–35
Posts, single piece	8–12	8–12	16–24
Girders, single piece	12–18	10–15	24–35
Built up	15–25	10–15	27–40
Joists, 2 by 6, 2 by 8, 2 by 10	12–18	9–15	22–33
2 by 12, and larger	10–15	8–12	18–27
Ceiling beams, 2 by 4, 2 by 6, 2 by 8	15–20	10–16	25–35
Bridging, per 1,000 pieces	10–15	10–15	20–30
Per 1,000 fbm	30–40	30–40	60–80
Studs, wall, 2 by 4, 2 by 6	12–25	8–12	18–37
Partition, 2 by 3, 2 by 4, 2 by 6	12–25	8–15	20–40
Plates and caps	.....	.....	20–40
Ribbons, 1 by 4, 1 by 5	.....	.....	30–50
Rafters, main	10–20	10–15	20–35
Jack	15–25	12–20	25–40
Hip and valley	20–30	15–30	30–45
Small roof trusses	25–30	15–20	40–50

preparing harder woods such as hard pine, chestnut, and oak. The time required for preparing the timbers depends on the cuts required and not on the length of the piece. If templates can be used considerable time can be saved in laying out and measuring.

For bridging between joists, allow 2 to 4 labor-hours per 100 pieces.

The times given are carpenter-hours if no helpers are used. On work where helpers may be used, the times given are man-hours.

Diagrams 8-3 and 8-4 may be used for estimating labor costs per timber for light framing.

Table 8-14 gives approximate times in man-hours required for light framing, using 1,000 fbm as a unit. Usually one-half to two-thirds of the time will be preparation (measuring, laying out, and cutting) and the remainder for erection. Special and complex framing will require more time.

Diagrams 8-7 and 8-8 can be used for estimating labor costs per 1,000 fbm for lighting framing.

Rough Boarding. The labor for rough boarding such as rough flooring, roofing, and sheathing depends on the width of the boards, number of openings, manner of laying (crosswise or diagonal), openings, and jogs in walls. The unit of measurement may be the square of 100 sq ft or 1,000 fbm. A square will require more than 100 fbm on account of waste.

Table 8-15 *Approximate Man-hours for Rough Boarding and Sheathing*

Work	Hr per square of 100 sq ft	Hr per 1,000 fbm
Rough flooring, unmatched, crosswise	1.6–2.9	14–25
Diagonal	2.1–3.5	17–29
Matched, crosswise	1.9–3.3	16–27
Diagonal	2.4–4.0	19–31
Roofing, unmatched, plain and flat	2.0–3.0	17–25
Gables and dormers	2.7–4.0	22–32
Matched, plain and flat	2.3–3.5	19–28
Gables and dormers	3.0–4.5	24–35
Sheathing, unmatched, crosswise	1.8–3.0	16–26
Diagonal	2.3–3.7	19–30
Matched, crosswise	2.0–3.5	17–29
Diagonal	2.5–4.1	20–32
Wallboard and plasterboard	1.5–2.8	14–26

When wall or plaster board is used for rough sheathing, the time required is usually less than for rough boarding. Approximate time in man-hours for rough boarding and sheathing is given in Table 8-15.

Some estimators add 1 hr of labor for every 8 to 12 lin ft of hip, ridge, and valley in the roof and 1 to 2 fbm of lumber for each lineal foot of roof hip, ridge, and valley.

Diagrams 8-5, 8-6, 8-7, and 8-8 can be used for estimating labor costs of rough boarding and sheathing.

Miscellaneous. There is always more or less miscellaneous labor connected with the rough carpenter work which cannot be readily classified with other labor and, consequently, should be listed separately. Table 8-16 gives approximate man-hours required for several items of work.

Diagrams 8-5 and 8-6 can be used for estimating furring and grounds, etc., using rate of work per 100 lin ft instead of 100 sq ft.

Equipment The equipment used for rough carpenter work in light wood construction often consists of hand tools only, of which each carpenter should have a set. Sometimes workbenches, power bench saws, and portable electric saws can be used to advantage. If electric power is to be used, the expense of providing for this power must not be omitted. See the article on Equipment in this chapter for allowances for equipment.

Scaffolding will be usually required for most rough carpenter work. On large jobs, the scaffolding should be estimated as a separate item. The costs will include materials (less salvage), transportation, and labor of erection and removal. On small jobs, the scaffolding may be estimated at so much per square (100 sq ft) of wall surface. These costs may range from about $1.50 to $4 per square. When the same scaffolding can be used for other purposes, such as exterior finish, the cost per square for rough carpenter work will be reduced proportionately.

Table 8-16 *Approximate Man-hours for Miscellaneous Work*

Kind of work	*Man-hours*
Furring on wood	1–2 per 100 lin ft
Furring on brick and tile	1.5–3 per 100 lin ft
Grounds on wood	2–5 per 100 lin ft
Grounds on brick and tile	5–6 per 100 lin ft
Insulating material, between studs	2–8 per 100 lin ft
Insulating material, between joists	2–6 per 100 lin ft
Setting single outside door frames	1.5–2.5 each
Setting single inside door frames	1–3 each
Setting double inside door frames	2–3 each
Setting single window frames	1.0–1.5 each
Setting double window frames	1.5–2.0 each
Setting triple window frames	2.0–3.0 each
Cutting holes and fitting plugs in masonry walls	2.5–5.0 for 10
Bridging	2–4 per 100 pieces
Building paper	0.8–1.5 per 100 sq ft

Estimates for Rough Carpenter Work Estimates for this kind of work are usually prepared by computing the materials, labor, equipment, overhead, and profit separately for all the rough carpenter work and then totaling. However, some estimators prefer to separate the rough carpenter work into framing, rough boarding, and miscellaneous and then to estimate the materials, labor, equipment, and overhead separately for each of these three items. Such a method is satisfactory on the larger jobs. If unit estimates are desired, they are usually computed per 1,000 fbm of the lumber used, though in some instances they may be computed per 100 sq ft, or square, or per wall panel (16 in. wide and about 8, 9, or 10 ft high).

Overhead costs may range from 10 to 30 percent of the labor costs or from 5 to 15 percent of the costs of materials and labor. Profit may vary from about 5 to 15 percent of the sum of all other costs.

Illustrative estimate. Prepare an estimate of the total cost of the rough carpenter work (materials, labor, equipment, overhead, and profit) for a simple two-story residence.

Materials prices for materials delivered at the job are as follows:

Rough lumber for framing and sheathing $120 per 1,000 fbm
Nails = 22 cents per pound. Insulating exterior sheathing board $90 per 1,000 sq ft
Outside door frames $10. Inside door jamb sets $4. Window frames $7
Labor wages are $5.50 per hour for carpenter and $4 per hour for helper
Payroll taxes and insurance, 10 percent of the labor wage

The materials takeoff gave the following information:

Size of basement, 24 × 30. Basement wall in place
Plate or sill on top of wall, 2 × 6 in.
Posts in basement, 8 × 8 by 7 ft 0 in. high
Built-up girder, 8 × 12 by 29 ft long
First-floor joists, 2 × 12 by 12 ft long, 16 in. on centers
Second-floor joists, 2 × 10 by 12 ft long, 16 in. on centers
Second-floor ceiling beams, 2 × 8 by 12 ft long, 16 in. on centers
Rafters, 2 × 6 by 16 ft long, 16 in. on centers
Outside wall studs, 2 × 4 by 17 ft 6 in. long, 16 in. on centers
Attic wall joists at ends, 2 × 4 up to 9 ft long, 16 in. on centers
Basement partitions, 22 lin ft less 1 single door, 8 ft high

First-floor partitions:

29 lin ft less 1 single door 32 in. wide and 1 opening 5 ft 6 in. wide
23 lin ft less 2 single doors, each 32 in. wide
Height of partitions is 8 ft 6 in.

Second-floor partitions:

29 lin ft, less 2 single doors, 32 in. wide
29 lin ft, less 2 single doors, 32 in. wide
9 lin ft, less 1 single door, 32 in. wide
9 lin ft
9 lin ft, less 1 single door, 28 in. wide
10 lin ft, less 1 single door, 28 in. wide
10 lin ft, less 1 single door, 28 in. wide
3 lin ft
3 lin ft
Height of partitions is 8 ft

Outside doors, two, 36 in. wide, 6 ft 9 in. high
Attic doors are 6 ft 9 in. high
Windows, first floor, two double, 4 ft 8 in. by 5 ft 6 in.; four single, 2 ft 4 in. by 5 ft 6 in.; one single, 2 ft 0 in. by 4 ft 2 in.

Windows, second floor, two double, 4 ft 8 in. by 5 ft 2 in.; five single, 2 ft 4 in. by 5 ft 2 in.

Windows, attic, two single, 2 ft 0 in. by 3 ft 10 in.

Materials: Framing:

Basement. 2 posts 8 × 8 by 7 ft 0 in.	= 75 fbm
1 beam 8 × 12 by 29 ft 0 in. (say 30 ft)	= 240
1 sill plate 2 × 6 by 112 ft	= 112
Joists. First-floor joists, each joist 2 × 12 by 12 ft = 24 fbm	
Number of joists = ($\frac{3}{4}$ × 30) × 2 = 23 × 2 = 46 plus 2 end joists and 2 partition joists = 50	= 1,200
Second-floor joists, each 2 × 10 by 12 ft = 20 fbm	
Number of joists = ($\frac{3}{4}$ × 30) × 2 = 23 × 2 = 46 plus 2 end joists and 4 partition joists = 52	= 1,040
Second-floor ceiling joists, each 2 × 8 by 12 ft = 16 fbm	
Number of joists = ($\frac{3}{4}$ × 30) × 2 = 23 × 2 = 46	= 736
Bridging 132 pairs of 1 × 3 by 18 in. long = 396 lin ft	= 100
Studs:	
Outside wall studs, each 2 × 4 by 17 ft 6 in. (say 18 ft) = 12 fbm	
$\frac{3}{4}$(30 + 30 + 24 + 24) + 4 corners + 4 partition intersections + 18 (openings) = 108 at 12 fbm	= 1,296
Attic wall studs $\frac{3}{4}$(24 + 24) + 2 ends = 38	
Average height 4 ft 6 in., or 3 fbm	= 114
Plates and caps 2(30 + 30 + 24 + 24) = 216 lin ft of 2 × 4	= 144
Ribbons 108 ft of 1 × 5, say	45
Fire stops at second floor, say 100 lin ft 2 × 4	= 67
Basement partition studs, each 2 × 4 by 8 ft = 5$\frac{1}{3}$ fbm	
$\frac{3}{4}$(22) + 1 end + 2 openings = 20	= 107
Plates and caps 2 × 22 = 44 lin ft	= 30
First-floor partitions, 2 × 4 by 8 ft 6 in. = 5$\frac{2}{3}$ fbm	
$\frac{3}{4}$(29) + 1 end + 4 openings = 27	
$\frac{3}{4}$(23) + 1 end + 4 openings = 23	
Total of 50	= 284
Caps and plates 2(29 + 23) = 104 lin ft	70
Second-floor partition, 2 × 4 by 8 ft = 5$\frac{1}{3}$ fbm	
$\frac{3}{4}$(29) + $\frac{3}{4}$(29) + $\frac{3}{4}$(9) + $\frac{3}{4}$(9) + $\frac{3}{4}$(10) + $\frac{3}{4}$(10) + $\frac{3}{4}$(3) + $\frac{3}{4}$(3) + 9 (ends) + 16 (openings) = 22 + 22 + 7 + 7 + 7 + 8 + 8 + 3 + 3 + 9 + 16 = 112	= 598
Plates and caps 2(29 + 29 + 27 + 20 + 6)	= 158
Rafters, 2($\frac{3}{4}$ × 30) + 2 ends = 48	
Each 2 × 6 by 16 ft. long = 16 fbm	= 768
Total framing net	= 7,174
Allow 5 percent waste	= 358
Total	= 7,532
Say	= 7,500 fbm

Rough boarding:

First floor, 1 × 8 laid diagonal, gross area, 24 × 30 = 720 sq ft, allow 20 percent waste	=	865
Second floor, same	=	865
Attic floor, shiplap laid crosswise 24 × 30 = 720 sq ft. Add 20 percent for waste	=	865
Total rough flooring	=	2,595
Say	=	2,600 fbm

Exterior sheathing board for outside walls.

$$17.5(30 + 30 + 24 + 24) + 2\left(\frac{24 \times 9}{2}\right) \text{ (attic ends)}$$

$$= 1{,}890 + 216 = 2{,}106 \text{ sq ft}$$

Add about 5% for waste, say 2,200 sq ft

Note that as no deductions have been made for openings, no allowance for waste may be needed.

Exterior sheathing board for roof, say 32 × (2 × 16) = 1,024 sq ft

Add 5 percent for waste	=	1,100
Total sheathing board	=	3,300 sq ft

Grounds, about 432 lin ft on first floor, plus about 600 lin ft on second floor for base, plus about 516 lin ft for molds gives 1,548 lin ft. Allowing 5 percent for waste gives a total of 1,625 lin ft, or fbm, say about 550 fbm

Door and window frames:

Door frames, outside	=	2
inside	=	12
Window frames, double	=	4
single	=	12

Nails, framing, say 10 lb per 1,000 fbm	=	75 lb
Rough boarding and grounds, say 20 lb per 1,000 fbm	=	65 lb
Exterior sheathing board, say 12 lb per 1,000 sq ft	=	40 lb
Total	=	180 lb

Materials, summary:

	Cost
Framing, 7,500 fbm at $120	$ 900
Rough boarding, 2,600 fbm at $120	312
Sheathing board, 3,300 sq ft at $90	297
Grounds, 550 fbm at $120	66
Door frames, 2 at $10 and 12 at $4	68
Window frames, 4 at $14 and 12 at $7	140
Nails, 180 lb at $0.22	40
Total materials for framing	$1,823

Labor: Framing. This labor may be estimated per piece or per 1,000 fbm. Some estimators estimate per piece and check per 1,000 fbm.

	Labor-hours
Basement, 2 posts, square cut and erect	3
1 beam, prepare and erect	8
1 sill plate	4
Joists:	
First, floor, 50 pieces, 2 × 12 by 12 ft at 0.50	25
Second floor, 52 pieces, 2 × 10 by 12 ft at 0.50	26
Ceiling, 46 pieces, 2 × 8 by 12 ft at 0.45	21
Bridging, 132 pieces or 396 lin ft	6
Studs:	
Outside wall, 108 pieces, 2 × 4 by 17 ft 6 in. at 0.30	32
Attic wall, 38 pieces, 2 × 4 by 4 ft 6 in. average at 0.25	10
Plates and caps, 2 × 4, 216 lin ft	6
Ribbons, 1 × 5, 108 lin ft	4
Fire stops, 2 × 4, 100 lin ft	4
Studs, partitions, basement, 20 pieces 2 × 4 by 8 ft at 0.20	4
Caps and plates, 2 × 4, 44 lin ft	1
First floor, 50 pieces, 2 × 4 by 8 ft 6 in. at 0.25	13
Caps and plates, 2 × 4, 104 lin ft	3
Second floor, 112 pieces, 2 × 4, by 8 ft at 0.25	28
Plates and caps, 2 × 4, 111 lin ft	3
Rafters, 48 pieces, 2 × 6 by 16 ft at 0.50	24
Total	225

Check by 1,000 fbm method.

Item	Fbm	Hours per 1,000 fbm	Labor-hours
Basement, posts	75	20	2
Beam	240	35	9
Sill plate	112	25	3
Joists, all	2,976	26	78
Bridging	100	60	6
Studs, all	2,400	30	72
Plates and caps	402	30	12
Fire stops	67	60	4
Ribbons	45	40	2
Rafters	768	30	23
Total	.....	...	211

Check is reasonably close.

	Labor-hours
Rough boarding and sheathing:	
Flooring, laid diagonal, first and second floors with no deductions for openings.	
1,440 sq ft, or 14.4 sq at 2.8	40
Laid crosswise, attic, shiplap, 720 sq ft, or 7.2 sq at 2.5	18
Sheathing, walls, 2,106 sq ft, or 21.06 sq, at 2.3	48
Roof, 1,024 sq ft, or 10.24 sq, at 2.0	20
Grounds, 1,548 lin ft at 3 per 100	46
Total	172

Door and window frames:	
Door frames, outside, 2 at 2	4
Inside, 12 at 1.5	18
Window frames, double, 4 at 1.75	7
Single, 12 at 1.25	15
Total	44

Check by 1,000 fbm method.

Item	Fbm	Hours per 1,000 fbm	Labor-hours
Flooring, plain boards laid diagonal	1,730	24	42
Attic, shiplap laid crosswise	865	22	19
Sheathing, walls	2,200	23	50
Roof	1,100	20	22
Grounds	550	60	33
Total	.....	...	166
Door and window frames	30	...	45

Labor-hours summary:

Item	Labor-hours	
	Piece method	1,000 fbm method
Framing	225	211
Rough boarding and sheathing	172	166
Door and window frames	44	45
Total	441	422

Labor cost: The labor used may be all carpenters or part carpenters and part helpers and apprentices. By assuming a gang of 3 carpenters (one acting as straw boss) and 2 helpers and apprentices, average hourly wage will be $(3 \times \$5.50 + 2 \times \$4)1.10 \div 5$, or \$5.39 per labor-hour. Assume 440 labor-hours.

$$\text{Total labor cost} = \$5.39 \times 440 = \$2,372$$

Equipment: The equipment required will be ordinary carpenter's hand tools of which each carpenter will have his own set. Two electric handsaws, some sawhorses, and workbenches will be required. Possibly an allowance of about \$50 would be sufficient, including transportation and power outlet.

For scaffolding, an allowance of about \$1 per 100 sq ft of wall surface will be ample, especially if the same scaffolding can be used later for the exterior finish. About \$20 should be sufficient.

Total equipment costs will thus be \$70.

Overhead and profit: For overhead costs, 15 percent of labor costs can be assumed. For profit, 10 percent of all other costs will be assumed.

This estimate may be summarized as follows:

Item	Framing	Rough boarding	Doors and windows	Cost per 1,000 fbm	Total cost
Material	$ 923	$ 692	$208	$130	$1,823
Labor	1,210	926	236	170	2,372
Equipment	23	33	14	5	70
Overhead	182	139	35	25	356
Profit	234	179	49	33	462
Total	$2,572	$1,969	$542	$363	$5,083

The costs per 1,000 fbm are based on a total of approximately 14,000 fbm.

When commenting on this estimate, it should be noted that different estimators would make different allowances for waste and would assume different rates of work depending upon their experience and judgment.

Some estimators would summarize the costs for framing, for rough boarding and sheathing, and for door and window frames separately and then obtain totals.

Light Wood Construction—Finish Work

Finish Work As stated previously, finish work can be divided into exterior and interior finish and these subdivided as desired. As indicated by the terms, exterior finish refers to the finish carpenter work required on the outside or exterior of the building, and interior finish to the finish carpenter work required on the inside or interior of the building. The following classification will give an idea of the items included.

Exterior finish:

- Siding and shingling on walls, including building paper
- Trim, including molding, cornice, corner boards, water tables, etc.
- Doors, windows, blinds, and shutters
- Roofing and flashing. Discussed in Chap. 10
- Steps and stoops
- Porches, with various details listed such as floor and roof framing, sheathing, flooring, ceiling, roofing, columns, and various finish details
- Exterior hardware

Interior finish:

- Flooring, including deadening and sanding
- Trim such as base, chair rail, plate rail, picture mold, beam casings, wainscot, panels

Stairs, stringers, treads and risers, balusters, rail, newels, coves
Windows and doors
Cabinet- and millwork, including cupboards, ironing boards, cabinets, china closets, linen closets, medicine cabinets, bookcases, columns, mantels, breakfast sets, drawers
Wooden ceiling
Miscellaneous, such as shelving and other items not previously listed
Interior hardware

Some estimators consider all doors and windows as interior trim. However, in this chapter, door and window frames have been included

Table 8-17 *Units of Measurement for Materials for Finish Work*

Exterior finish	Unit	Interior finish	Unit
Siding	1,000 fbm, 100 sq ft	Flooring	1,000 fbm, 100 sq ft
Shingles, (siding)	100 sq ft, bundles, 1,000 pieces	Deadening felt	100 sq ft, roll
Building paper	Roll, 100 sq ft	Trim, base, chair rail, base rail, moldings, etc.	100 lin ft
Doors and windows	Piece	Panels	100 sq ft
Trim	Side, or set	Ceiling	1,000 fbm, 100 sq ft
Blinds and shutters	Piece, or pair	Windows and doors	Piece
Trim, various	100 lin ft	Trim	Side, or set
Porches, framing	1,000 fbm		
Boarding, sheathing	1,000 fbm	Cabinet and mill work, various items	Piece
Flooring, ceiling, roofing	100 sq ft	Miscellaneous	Various
Trim	100 lin ft	Interior hardware	Piece, or pair, dozen, set pound, etc.
Posts	Piece		
Hardware	Piece, or pair, dozen, set, pound, etc.	Nails and bolts	Pound
Nails and bolts	Pound		

with the framing, outside trim of doors and windows with exterior finish, outside doors and windows and inside trim with interior finish, and inside doors with interior finish. Some estimators place outside door and window trim with exterior finish and the doors and windows and interior trim with interior finish.

Materials There are several units of measurement for the materials used for finish work. Table 8-17 gives many of the materials and the units by which they are estimated.

Much of the molding is measured and priced by the *molding inch*, which is the equivalent of a piece of lumber 1 in. square and 1 ft long. The cross-sectional area is figured on the full size of stock from which the molding is made. For example, a $\frac{3}{4}$- by $1\frac{3}{4}$-in. molding would be made from 1- by 2-in. material. and a strip 1 ft long would contain 2 molding inches. Prices are often quoted per 100 lin ft per molding inch and vary with different kinds of woods and moldings. Waste in moldings will be end waste and may vary from 5 to 10 percent.

Wood siding may be drop or novelty siding or lap or bevel siding. Table 8-18 gives approximate materials required.

Table 8-18 *Amount of Siding per* 100 *Sq Ft of Wall Surface*

Width and type	Exposure, in.	Lap and waste, percent	Fbm per 100 sq ft	Nails, lb per 100 sq ft
6-in. lap or bevel........	4.75	25	125	2–3
6-in. lap or bevel........	4.50	33	133	2–3
5-in. lap or bevel........	3.75	40	140	2–3
5-in. lap or bevel........	3.50	50	150	2–3
4-in. lap or bevel........	2.75	50	150	3–4
4-in. lap or bevel........	2.50	60	160	3–4
6-in. drop..............	5.00	25	125	3–4

Wood siding may cost (1967) from about $150 to $300 or more per 1,000 fbm depending on the kind of wood, quality or grade, and locality. Diagram 8-9 may be used for estimating the cost of wood siding.

There are many kinds of special siding such as aluminum, asphalt, asbestos, galvanized steel, etc. These special sidings vary considerably in width and exposure. Special sidings usually require accessories such as backing paper or felt, starter strips, moldings, outside and inside corners, window and door trim, special nails, etc. Waste may range from about 5 to 10 percent for ordinary walls. Special sidings are usually applied by specially trained workmen. A special siding may cost about $40 or more per square.

There are many kinds of shingle sidings available. These shingles may come in rolls or strips, or in cartons or bundles. Usually 3 or 4 bundles are required per square. Shingle sidings usually require accessories. Waste may vary from about 3 to 10 percent for ordinary walls. Shingle sidings may cost from about $15 to $40 or more per square, including costs of accessories.

Ceiling materials vary a little in thickness and width. Ordinary ceiling is about $\frac{5}{8}$ in. thick and $3\frac{1}{2}$ in. wide and is matched and beaded. The waste is about 25 percent, and 20 to 25 lb of nails are required per 1,000 fbm, or 2 to 3 lb per square of 100 sq ft.

For most trim, such as is measured per 100 lin ft, 5 to 10 percent should be allowed for waste. From 2 to 3 lb of nails per 100 lin ft will usually be sufficient.

The materials for flooring are usually estimated per 1,000 fbm, or per 100 sq ft with allowance for waste. Finish flooring is usually about 5⁄16 to 13⁄16 in. thick and of various widths from 1.5 to 3.5 in. or more.

Board feet required for 100 sq ft of floor area for flooring 11⁄16 in. thick is based on the assumption that this thickness is considered as 1 in. when measuring board feet (Table 8-19). If considered as 1.5 or 2 in., then

Table 8-19 *Amount of Finish Flooring per* 100 *Sq Ft of Floor Surface*

Flooring		Waste, percent	Fbm per 100 sq ft floor area	Nails per 100 sq ft floor area
Width, in.	Thickness, in.			
1.5	3⁄8	40	140	4–7
1.5	13⁄16	40	140	6–8
2.25	13⁄16	35	135	5–7
3.25	13⁄16	30	130	5–7
4.25	13⁄16	25	125	4–6
5.25	13⁄16	20	120	4–6
2.25	11⁄16	35	135	6–8
3.25	11⁄16	30	130	6–8
4.25	11⁄16	25	125	7–10
5.25	11⁄16	20	120	7–10

values given must be multiplied by 1.5 or 2, respectively.

Diagram 8-9 can be used for estimating cost of finish flooring.

Takeoff. Perhaps the simplest and best method of taking off materials and labor for finish work is to list the items under headings and subheadings such as

Exterior finish. Siding, shingling, trim, doors, windows, porches (framing, flooring, posts, ceiling, roofing, trim, rail, etc.), miscellaneous, etc.

Interior finish:

Basement. Doors, windows, stairs, shelves, ceiling, etc.
First floor. Flooring, trim, ceiling, paneling, doors, windows, stairs, cabinetwork (mantels, cabinets, cupboards, etc.), hardware, etc.,
Second floor and other floors. Same as for first floor

A good reminder list suitable for the type of structure considered should be checked so as to avoid omissions. See Appendix A for a typical reminder list. The plans and specifications must be scrutinized carefully so that no details are omitted.

Whenever possible, stock sizes of materials should be used to save time and expense of making specials.

Dimensions, style, kind, catalogue number, quality, and other specifications should be noted.

For example, doors should be described as to kind of wood, style, width, length, thickness, panels, etc. Window sash should be described as to size, number of lights, kind of glass, etc. Door and window frames should have style, quality, and dimensions stated. Door and window trim should be described as to wood, dimension, style, etc. Stairs should be described in detail as to treads, risers, balusters, rails, newels, etc. Cabinetwork should be described in detail with sketches. Interior and exterior trim should be described according to wood, style, size, etc.

If desired, interior finish may be given in considerable detail. For example, an exterior door with frame and trim may be divided into side jambs, head jamb, side casings, head casing, sill, drip cap, inside casing, buck band, plinth, door, lock, hinges, etc. However, such detail is rarely necessary unless the door, for example, is to be fabricated almost entirely on the job. Usually a listing such as exterior door complete will be satisfactory, or the listing may be divided into three or four items as frame, exterior trim and interior trim, door, and hardware.

For preliminary and approximate estimates, the various items can be priced from catalogues. For exact estimates, all items should be listed and prices obtained from the mill or dealer. No attempt will be made here to give approximate prices for all the various items of finished woodwork.

Labor The labor required for finish work will vary considerably according to skill, ability, and inclination of carpenter and working conditions, kind of wood (hard or soft), etc. Finish work with a hard wood may require 20 to 35 percent more time than like work with a soft wood. Consequently, the values given in the Tables 8-20 and 8-21 are approximate and should be varied according to the conditions of the particular job. Diagrams 8-5, 8-6, 8-7, and 8-8 can be used for estimating labor costs of finish carpenter work.

Many estimators consider a door, a window, or a flight of stairs as one unit and estimate labor required for a complete unit. Others go into more detail. For example, they may divide the labor for a door into frame, trim, hanging, and hardware and for a flight of stairs into stringers, treads and risers, nosings, coves, newels, railings, balusters, etc. Such detail is usually not necessary except for special work.

Equipment The equipment required for finish work will invariably include a fairly complete set of hand tools for each carpenter and may include a few sawhorses and benches, a small portable electric saw, and electric sander. The costs will usually be the transportation charges for the carpenters' kits and other equipment, plus a small charge for horses and bench, and necessary charges for portable electric saws and electric sanders, as the case may be. The costs of providing the power (lines, switches, meters) should be included.

Table 8-20 *Approximate Labor Required for Exterior Finish*

Item	Unit	Labor-hours	Item	Unit	Labor-hours
Ceiling, matched	1,000 fbm	25–40	Door, exterior, double	Complete	7–12
	100 sq ft	3–5	Frame	One	2–3
Siding, lap, narrow (lap or bevel).	1,000 fbm	20–50	Trim	Set	1–3
	100 sq ft	3–8	Hanging	Pair	2–3
Lap, wide (lap or bevel).	1,000 fbm	15–45	Hardware	Set	2–3
	100 sq ft	2–6	Window, single	Complete	4–7
Drop	1,000 fbm	12–40	Frame	One	1–2
	100 sq ft	2–5	Trim	Set	1–2
Shingles (siding) on walls	100 sq ft	3–8	Hanging	2 sash	1–2
Building paper on walls	100 sq ft	1–2	Hardware	Set	1–2
Cornice	100 lin ft	4–20	Window, double	Complete	6–9
String course	100 lin ft	4–8	Frame	One	1–2
Mold	100 lin ft	4–8	Trim	Set	1–2
Water table	100 lin ft	3–6	Hanging	4 sash	2–3
Corner board	100 lin ft	3–5	Hardware	Set	1–3
Blinds and shutters	Pair	1–4	Window, triple	Complete	8–12
Porches:*			Frame	One	2–3
Columns	Each	1–2	Trim	Set	1–3
Rail, complete	100 lin ft	40–70	Hanging	6 sash	3–4
Steps, set of 4	Set	8–16	Hardware	Set	2–3
Door, exterior, single	Complete	5–8	Screens, doors	Each	3–5
Frame	One	1–2	Windows	Each	1
Trim	Set	1–2			
Hanging door	One	1–2			
Hardware	Set	1–2			

* For other porch details, see tables on framing, flooring, roofing, etc.

Exterior finish work will usually require the use of some scaffolds and ladders. Frequently, the same scaffolds that were erected for the rough carpenter work can be left in place and used for the finish work. Then the cost of scaffolds will be part of the lumber cost plus the labor of removal. An allowance of 35 to 75 cents per 100 sq ft of wall surface is usually sufficient. However, if separate scaffolding must be provided for exterior finish, the costs will be greater and may range from about 80 cents to $2.50 per 100 sq ft of wall surface. Scaffolding should be estimated separately and in detail for large jobs.

Estimates for Finish Work Estimates for finish work can be prepared by considering all finish items together and estimating the materials, labor, and equipment. Or the estimates can be prepared by considering the items separately and estimating costs of materials, labor, and equipment for each item and then totaling. Either way is satisfactory, provided that no costs are omitted.

Some estimators divide the carpenter work into rough work and finish work without subdividing the finish work into exterior and interior finish. This procedure is frequently followed on small jobs.

Table 8-21 *Approximate Labor Required for Interior Finish*

Item	Unit	Labor-hours	Item	Unit	Labor-hours
Wood ceiling	1,000 fbm	25–40	Door, double	Complete	4–8
	100 sq ft	3–5	Frame	One	1–2
Paneling and plywood	100 sq ft	3–15	Trim	Set	1–3
Plaster board	1,000 fbm	10–20	Hanging	Two	1–2
	100 sq ft	1–3	Hardware	Set	1–2
Floors, finished hard-wood, matched narrow	1,000 fbm	25–40	Opening, framed	Complete	2–4
	100 sq ft	4–6	Frame	One	1–2
Wide, 3¼ in. and over	1,000 fbm	20–35	Trim	Set	1–2
	100 sq ft	3–5	Stairs, cellar, complete	10 ft or one story	4–8
Floors, finished, soft-wood, matched	1,000 fbm	16–30			
	100 sq ft	2–4	Stairs, back, or enclosed	12 ft or one story	8–20
Building paper between floors	100 sq ft	1–2	Stairs, front, open	12 ft or one story	20–60
Deadening felt	100 sq ft	1–3			
Sanding floors, hand	100 sq ft	2–4	Stairs, front, open and ornate	12 ft or one story	40–100
With sander	100 sq ft	1–2			
Wainscot	100 sq ft	3–5	Mantels	One	3–10
Baseboard	100 lin ft	4–7	Installing ready-made medicine cabinets, closets, ironing boards, broom closets, cupboards, china closets, bookcases, linen closets, etc.	One small	1–4
Chair rail	100 lin ft	4–6		One medium	2–6
Plate rail	100 lin ft	4–10		One large	3–8
Picture mold	100 lin ft	2–6			
Mold, various	100 lin ft	1–6			
Beam casings	100 lin ft	4–8			
Door, single	Complete	3–6			
Frame	One	1			
Trim	Set	1–2			
Hanging	One	1			
Hardware	Set	1			

Note that costs of door and window frames were included in the previous estimate for rough carpenter work. Frequently, the rough carpenter sets all door and window frames, the finish carpenter installs outside doors and windows and outside trim with other exterior finish, and the finish carpenter does not install the inside trim until after the lathing and plastering has been completed. The cost of door and window frames were included in the estimate for the rough work, and the cost of doors and windows, less frames, will be included with the finish work.

Overhead costs may vary from about 10 to 30 percent of the labor costs, or from 5 to 15 percent of the costs of both material and labor. Profit may range from about 5 to 15 percent of all other costs.

Illustrative estimate. Prepare an approximate estimate for the finish work for the simple two-story residence for which an estimate for the rough carpenter work was prepared in the Illustrative Estimate of the section on Light Wood Construction—Rough Work. Carpenter's wages are $5.50 per hour, plus 10 percent for payroll taxes and insurance. The takeoff gave the following information. The prices given were quoted by a local dealer.

Exterior finish. Bevel or lap siding, 6 in. wide, 4.5-in. exposure. Total area 2,106 sq ft

- No deductions for openings
- Building paper, area 2,106 sq ft. Price $2 per roll
- 1 front door, special, complete with trim and hardware less frame. Price $36
- 1 rear door, complete with trim and hardware. Price $27
- 4 double windows, complete with trim and hardware, less frame
- 12 single windows, complete with trim and hardware, less frame
- Screens, 2 doors, 4 double windows, 12 single windows
- Cornice on ends, allow 70 lin ft
- Cornice on sides, allow 70 lin ft
- Water table, 108 lin ft

Interior finish:

- Basement: Basement stairs
  - 1 door with trim and hardware, less frame
- First floor. 4 inside doors, with trim and hardware, less frames
  - Building paper between rough and finish flooring, area 700 sq ft
  - Flooring, hardwood $^{13}/_{16}$ by $2\frac{1}{4}$ plain, first grade, area 700 sq ft
  - Base, base mold, and quarter round, length 220 ft
  - Picture mold (crown and picture), length 220 ft
  - Mantel, living room
  - China closet, dining room
  - Kitchen, ironing-board closet
    - Broom closet
    - 2 cupboards or dressers
  - Stairs, open, first to second floor
  - 1 coat closet
- Second floor. 8 inside doors with trim and hardware, less frames
  - Deadening felt between rough and finish floors, area 700 sq ft
  - Flooring, hardwood, $^{13}/_{16}$ by $2\frac{1}{4}$ plain, first grade, area 700 sq ft
  - Base, base mold, and quarter round, length 300 ft
  - Picture mold, single, length 300 ft
  - Bathroom, medicine cabinet, and towel cabinet
  - Linen closet, complete with door, drawers, and shelves
  - 3 clothes closets, poles and shelves
  - Stairs, second floor to attic, enclosed

The finish materials and labor can be estimated and tabulated as follows. Nails are included with material estimates. Labor required for cabinetwork, doors, windows, and screens is for fitting and installing.

Labor wage per hour = $\$5.50 \times 1.10 = \$6.05$.

Material and Labor Estimates

Item and description	Material cost	Labor hours	Labor cost
Siding:			
Building paper, 2,100 sq ft at $2 per roll of 216 sq ft	$ 20	25	$ 151
Bevel siding 6 in., 4.5 in. exposed, 2,100 sq ft at $240 per 1,000 fbm. Allow 33% for lap and waste, plus nails	680	105	635
Other exterior trim:			
Cornice on sides, 70 lin ft at $42 per 100, 10% waste	34	6	36
Cornice on ends, 70 lin ft at $35 per 100, 10% waste	28	6	36
Water table, 108 lin ft at $20 per 100, 10% waste	24	6	36
Doors and windows with trim and hardware, less frame:			
Outside doors, front	36	5	30
Rear	27	5	30
Inside doors, basement (1)	10	3	18
First floor (4)	60	20	121
Second floor (8)	96	34	206
Double windows (4)	80	30	182
Single windows (12)	180	60	363
Screens:			
Door screens (2) complete at $10	20	7	42
Double window screens (4) at $7	28	6	36
Single window screens (12) at $5	60	12	73
Flooring:			
Building paper, first floor, 700 sq ft at $2 per roll of 300 sq ft	5	8	48
Deadening felt, second floor, 700 sq ft at $3 per roll of 200 sq ft	11	12	73
Hardwood flooring, 1,400 sq ft at $350 per 1,000 fbm plus 35% for waste and nails	690	70	424
Interior trim:			
Base, first floor 220 ft at $20 per 100 plus nails	45	16	97
Second floor 300 ft at $20 per 100 plus nails	62	16	97
Picture mold, first floor 220 ft at $8 per 100 plus nails	18	8	48
Second floor 300 ft at $8 per 100 plus nails	25	8	48
Stairs:			
Basement, complete	35	8	48
First floor, complete	300	32	194
Second floor, complete	80	16	97
Cabinetwork:			
Mantel	150	6	36
China closet	105	4	24
Ironing-board set	20	3	18
Subtotal, carried forward	$2,929	537	$3,249

Material and Labor Estimates (CONTINUED)

Item and description	Material cost	Labor-hours	Labor cost
Brought forward	$2,929	537	$3,249
Broom closet	26	2	12
Kitchen cupboards (2) at $95	190	8	48
Medicine cabinet	20	2	12
Towel cabinet	20	2	12
Linen closet	60	6	36
Shelving, etc., in closets:			
Basement shelves	20	4	24
First-floor closet	20	4	24
Second-floor closets (3) at $12	36	8	48
Total	$3,321	573	$3,467

Recapitulation, Material and Labor Estimates

Siding	$ 700	130	$ 787
Other exterior trim	86	18	109
Doors and windows	489	157	950
Screens	108	25	151
Flooring	706	90	545
Interior trim	150	48	290
Stairs	415	56	339
Cabinetwork and shelving	667	49	296
Total	$3,321	573	$3,467

The equipment needed will be the ordinary hand tools of which each carpenter has a set, plus a few sawhorses and a bench. Some scaffolding will probably be needed for exterior siding and trim. An allowance of about $40 should be sufficient.

For overhead, 12 percent of labor costs will be assumed.

For profit, 10 percent will be used.

Summary. The costs for finish work can be summarized as follows:

Item	*Cost*
Materials	$3,321
Labor	3,467
Equipment	40
Overhead (12% of labor)	416
Profit (10% all other costs)	724
Total	$7,968

CHAPTER NINE

Structural Steel

Estimating of Structural Steel The correct estimating of structural steel is a comparatively complex and detailed problem. The form and manner in which the estimate is made will depend on the work that is to be done by the estimator's company. For example, fabricating estimates will vary considerably in detail from construction or erection estimates. As this text is intended more for the use of the general contractor, this chapter will consider only the work done by the general contractor or builder, i.e., by assuming that the contractor buys the steel already fabricated and erects the steel with his own construction gang. However, in some instances, the providing of the steel, the fabrication, and also the erection are sublet to a steel company that specializes in this type of work. On other jobs, the general contractor may purchase the steel ready fabricated and sublet the erection to another company. The estimate should be divided into the five general headings of material, labor, equipment, overhead, and profit.

The materials estimate should include all quantities required and all costs of the materials delivered either at the job or at the nearest and most convenient railway siding or dock. The cost of material delivered at the job is recommended, even if the contractor provides the handling and trucking between railway siding and job. The materials costs may include:

Mill or warehouse costs, including base prices and various extras.

Transportation costs from mill or warehouse to fabrication shops.

Fabrication costs, including shop drawings and details, handling, fabrication, painting, inspection, and overhead. For some jobs, the fabrication costs may equal the mill or warehouse costs of the materials.

Transportation costs from fabrication shop to most convenient railway siding or dock, or trucking costs to the job.

Transportation costs from railway siding or dock to job, including handling from railway car to truck, trucking to job, and unloading from truck. Some estimators prefer to include these costs with the erection costs, especially if the general contractor furnishes the labor, trucks, and handling equipment.

The labor estimate should include all labor costs of erecting the steel at the site. Items included may be labor required for:

Hoisting the steel in place and temporarily fastening with bolts. (The labor required for erecting, operating, moving or resetting, and taking down of hoists and derricks is usually included here.)

Plumbing the structure
Riveting, welding, or bolting as specified
Painting after erection

The equipment estimate should include all equipment costs such as ownership, rentals, operation, erection, moving, dismantling, and transportation to and from the job. If not included with the labor estimate, the labor for erection, moving or resetting, takedown, and operation should be included also.

The overhead expenses may be based on labor alone, on the sum of material and labor costs, or part may be based on labor and part on materials if the overhead costs can be readily divided between these two items. On structural-steel jobs, the overhead expenses are high and may vary considerably. Inspection and general supervision, and preparation of drawings (if needed) are important items. Overhead expenses may range from 20 to 50 percent of the labor costs.

The profit is based on the sum of all other costs and may range from 5 to 25 percent, depending upon the hazards and uncertainties related to the particular job, the competition expected, what the contractor thinks is fair, and also what the contractor thinks he can get.

The Takeoff, or List of Materials When a list of materials is prepared, the plans and specifications should be studied so that the estimator will have a general understanding of the requirements of the job. Then the takeoff is carefully prepared, all the materials being listed in a logical order or sequence. This order may vary with different types of structures and with different estimators. Usually it is not necessary to make an erection estimate as detailed as a fabrication estimate. All main items should be listed separately, but smaller items such as beam fastenings, bolts, rivets, and minor details can be approximated. The takeoff should show the grade of steel, kind of member and structural-steel shape, size, length, and weight in pounds or tons. As structural steel is purchased by the pound, all weights must be listed.

The use of a good reminder list, suitable for the type of structure being estimated, is advisable, especially as a check on all important members and also as a check on many of the minor details, some of which may not be shown on the plans or mentioned in the specifications but which must be included with the job. Some of these minor items are bearing plates for beams, various field connections, gusset plates, field bolts, ties, beam separators, anchors, and turnbuckles.

The order or sequence of the takeoff may be about as follows for:

- Truss bridges
  - Each truss. Lower and upper chords
    - Web members, vertical and inclined
    - End posts
    - End bearing plates, rockers, etc.
  - Portals
  - Cross, sway, and wind bracing
  - Floor systems, girders, beams, floor covering, connections, and fastenings
  - Miscellaneous items and details
- Girder bridges
  - Each girder. Various structural shapes and plates
    - End bearing details
  - Floor systems, small beams and girders, and connections and flooring
  - Miscellaneous items
- Mill buildings
  - Walls. Columns, including plates, brackets, etc.
    - Crane runway girders
    - Wall girts and purlins
    - Wall beams and tie rods
    - Wall wind or sway bracing
    - Miscellaneous details
  - Roof trusses. Lower and upper chords
    - Web members
    - Cross bracing
    - Purlins
    - Miscellaneous details
  - Side covering. Corrugated steel or other wall covering
    - Steel sash for windows
    - Doors
    - Miscellaneous

Roof covering. Corrugated steel or other roofing
Skylights including sash and glazing
Ventilators
Miscellaneous
Steel frame office (multistoried) buildings. (These are usually estimated by the floor)
Basement. Column bases and caps
Columns and details
Girders and beams and details
Metal lumber, if included in contract
Bracing, if any
Steel flooring, if any
Miscellaneous
First and other floors
Columns and column details
Girders and beams and details
Lintels
Metal lumber
Bracing
Miscellaneous
Roof. Columns and column details
Girders and beams and details
Purlins
Metal lumber
Bracing
Roof covering, if of steel
Miscellaneous
Exterior finish. All items of steel such as cornice and cove
Steel sash and glazing
Siding
Miscellaneous
Large girders. Each girder estimated in same manner as for a bridge girder
Building trusses. Each truss estimated in same manner as for a bridge truss

Some estimators when estimating steel-frame buildings prefer to take off the material in the following order:

Columns including main items, plates, caps, and connections
Floor beams and girders including details
Bracing, such as wind bracing
Girders, as for a bridge girder
Trusses, as for a bridge truss
Metal lumber, joists, and studs
Roofing materials
Steel sash and glazing
Miscellaneous items

After completing the takeoff in a manner similar to that described, many estimators group all the materials for convenience in pricing somewhat as follows:

Structural Shapes. Grouped according to grade (A36, A242, A440,

A441, etc.), shape (I beams, H beams, channels, angles, T's, Z's), with subdivisions according to standard or special sizes and subgroups according to large, medium, and small sizes or weight. The grouping into large, medium, and small sizes will vary with the different estimators. For example, I beams 6 in. or less in depth may be called small, those 6 to 15 in. called medium, and those over 15 in. in depth called large. The "base" size or weight may be called the median size.

Bars and Rods. According to size and shape.

Flats. According to size and thickness.

Plates. According to size and thickness, specials being noted.

Connections. According to kind, etc.

Rivets, bolts, small details, etc., often as a percentage of the weights of other materials.

Corrugated steel. Finish, gage, and length noted.

Steel sash. Details noted.

Metal lumber. Kind, dimensions, weight and other details.

Miscellaneous items.

For an illustration of a takeoff, see the illustrative estimate at the end of this chapter.

Materials

Structural Steel. After the takeoff has been completed, the unit weights of the various items should be noted opposite the corresponding items. Then the total weights of each of the items should be computed. Sometimes the weight computations are made at the same time the materials are listed. Structural-steel handbooks should be consulted for dimensions and unit weights of structural-steel shapes and of other steel materials. The total estimated weight, as computed by the estimator, should agree within 3 percent of the actual shipping weight of the fabricated steel. Hence, it is advisable for the estimator slightly to overestimate the weights in the takeoff so as to allow for small overruns and omissions.

As structural steel is priced per 100 lb (or per ton), the weights of the various items must be found. These weights should be based on

Rectangular dimensions for all plates
Overall lengths of all structural shapes
No allowances for clipping, milling, punching, or boring

Allowances for overrun calculated according to the percentage of overrun permitted by the ASTM standards

Shop rivets can be estimated as in Table 9-1

Field rivets should be estimated at actual weights, with 10 percent allowed for waste. About 20 to 30 field rivets will be needed for each ton of steel, and more may be required in special cases.

Table 9-1

Diameter, in.	*Weight per* 100 *rivets, lb*
½	20
⅝	30
¾	50
⅞	100
1	125

Field bolts should be estimated at actual weights, with 5 percent allowed for waste. For temporary bolting, about 3 to 10 bolts will be required for each ton of steel. For permanent bolted connections, about 15 to 30 high-strength bolts will be needed for each ton of steel.

Paint will increase the weight of structural steel about 8 or 10 lb per ton per coat.

Details can be approximately estimated as in Table 9-2.

Table 9-2 *Approximate Weights of Details*

Shape	Rivets, or bolts, percent	Details, percent
Columns	3–4	10–15
Beams	1–2	5–20
Built-up girders	5–6	10–12
Roof trusses	3–4	15–20

Beam details vary considerably. The weights of standard connections can be found in structural-steel handbooks. Some estimators allow about 4 percent of the weights of all members for rivets and about 20 percent of the weights of main members for details.

The weight of steel for the steel frame only of a mill building, one story high (15 to 25 ft to bottom chord of truss), with no crane runways, may vary from about 10 to 15 lb of steel per square foot of floor area. If crane runways are provided, the weight of steel frame may vary from about 12 to 20 lb of steel per square foot of floor area. The weight of crane runways will depend upon the size or capacity of the crane and the crane span. The weight of a crane runway may vary from about 150 lb per lineal foot for a 10-ton crane with a 40-ft span to about 400 lb per lineal foot for a 30-ton crane with an 80-ft span. Note that 2 lin ft of crane runways is required for each lineal foot of crane travel. Weights of steel frames for buildings several stories high may vary from about 2 to 4 pcf of building.

Structural steel is usually priced and sold by the hundredweight (100 lb). The price quoted is usually that f.o.b. the mill or a main warehouse. In addition to the base price per 100 lb, there are several extras which may be added. The base price usually applies to certain standard sizes. The base price (1967) is about $6 per 100 lb f.o.b. the mill and about $10 per 100 lb at a main warehouse.

Extras are charged for such items as the following:

Size extras. These extras may vary from about 35 cents to $2 per 100 lb, the smaller sizes being more expensive.

Quantity extras. For shipments less than 2,000 lb, add from $1 to $9 per 100 lb depending upon the weight of the shipment. For shipments 2,000 to 4,999 lb, base price. For shipments 5,000 lb and over, deduct from 10 to 70 cents per 100 lb, depending upon the weight of the shipment. Quantity extras are determined by the total weight of each individual size in the same order, regardless of the total weight of all individual sizes in the same order. Each grade, weight, gage, and thickness of a structural section is considered as an individual size.

Grade extras. These extras may vary from about 15 cents to $1.25 per 100 lb.

Extras are charged for all sections other than standard.

Extras are also charged for special cutting and milling, cambering, testing, and special qualities and specifications. These extras may range from 25 cents to $3 or more per 100 lb.

The reader is referred to up-to-date price lists of the steel companies for correct charges for extras.

The extras may increase the price so that the price paid (including extras) may range from base price to base plus 8 to 30 percent.

Overrun or overweight caused by permissible variations in dimensions and weight may increase the costs up to about 5 percent.

For general estimates on structural steel, the total weight can be computed and the price for each kind of material taken at about base price (mill or warehouse) plus 10 to 20 percent.

Whenever possible, the material list (or plans and specifications) should be referred to a steel company and the correct prices obtained.

Shop drawings are required for practically all fabrication of structural steel. Shop drawings are usually prepared on 24- by 36-in. sheets. The cost of preparing one sheet may range from about $40 to $175. The number of sheets required for a job will vary with the complexity of the job and the number of details and to some extent on the size of the job. A small and simple job may not require more than one or two sheets, while a large and complex job may require more than 100 sheets. Almost all fabricating shops maintain drafting departments for preparing their shop drawings.

The cost of shop drawings is about the same for welded as for riveted or bolted structures. The following are approximate costs for shop drawings for various steel structures:

Steel mill and factory buildings, per ton	\$ 40–\$ 80
Steel-skeleton office buildings, hotels, etc., per ton	\$ 30–\$ 50
Auditoriums, per ton	\$ 50–\$100
Plate and rolled girders, per ton	\$ 15–\$ 40
Roof trusses, per 24- by 36-in. sheet	\$100–\$175
Highway bridge trusses, per lin ft	\$ 15–\$ 25

Fabricating costs will depend upon the amount of shop work done upon the steel. These costs are usually quoted per ton and vary considerably; hence the estimator should secure the fabricator's figures whenever possible.

For riveted or bolted connections, the fabricating costs include handling, cutting, punching, milling, planing, and shop riveting.

For welded connections, the fabricating costs include handling, cutting, punching for temporary bolts, milling, planing, beveling, and shop welding. Welded connections may cost up to \$10 a ton more than riveted connections.

Table 9-3 gives approximate 1968 costs of steel fabrication.

Table 9-3 *Approximate Costs* (1968) *of Steel Fabrication*

Description	Cost per ton	Cost per 100 lb
Foundations, anchor bolts, grillage beams	\$30–\$70	\$1.50–\$3.50
Columns and struts:		
I and H sections	\$30–\$65	\$1.50–\$3.25
Built-up sections	\$40–\$90	\$2.00–\$4.50
Beams, I, H, and channel sections, punched and framed	\$15–\$45	\$0.75–\$2.25
Built-up girders	\$40–\$85	\$2.00–\$4.25
Misc. framing, angles and girts	\$20–\$60	\$1.00–\$3.00
Skylight and dormer frames	\$50–\$120	\$2.50–\$6.00
Roof trusses:		
Under 1,500 lb	\$80–\$110	\$4.00–\$5.50
1,500–3,000 lb	\$65–\$90	\$3.25–\$4.50
Over 3,000 lb	\$50–\$65	\$2.50–\$3.25

Shop painting may cost from \$5 to \$9 per coat per ton, depending upon the size of the sections, cost of paint (\$4 to \$6 per gallon), wage of the painter (\$3 to \$7 per hour), and whether spray guns can be used. One gallon of paint should cover about 325 to 400 sq ft of surface, and a painter should paint from 1 to 2 tons per hour.

Transportation costs from a mill or a main warehouse to the fabricator and to the job will depend upon the rates in force between the places considered. These rates will vary according to classification and also from time to time. The estimator should consult with transportation agents and the mill or fabricator so as to obtain the best freight classification and rate. Carload lots should be shipped whenever practical.

In summing up, the cost of structural steel f.o.b. at a convenient siding or at the job may be roughly as in Table 9-4.

Table 9-4

Item	*Cost per* 100 *lb*
Base plus extras	\$ 8.00–\$13.00
Shop drawings	\$ 1.50–\$ 4.50
Fabrication	\$ 1.50–\$ 7.00
Shop painting	\$ 0.40–\$ 0.75
Transportation	\$ 0.50–\$ 2.00
Giving a total of about	\$11.50–\$27.00

Corrugated Steel. Corrugated sheet steel can be used for exterior surfaces as siding and roofing. This material is available in regular open-hearth or rust-resisting (copper alloy) steels and in black (unpainted mill), painted, galvanized, ceramic coated, and asbestos covered.

Corrugated steel sheets can be obtained in many different patterns of different thicknesses, widths, lengths, and corrugations. Manufacturers' catalogs should be consulted for information on corrugated sheets available. Data for two of the more commonly used sizes is given in Table 9-6. More information about corrugated steel roofing is given in the article on Metal Roofing in Chap. 10.

Common lengths are 5 to 12 ft with 1-ft variations.

Widths may vary from about 22 to 36 in.

Corrugations vary in pitch and depth. Common values are 2⅔ (2½) in. pitch with ½-in. depth, and 3-in. pitch with ¾-in. depth.

Side laps are usually 2 in. or one corrugation for siding, and 3.5 in. or 1.5 corrugations for roofing.

Siding sheets usually have both edges turned the same way and roofing sheets usually have one edge turned up and the other edge turned down.

End laps are usually 4 in. for siding, 6 in. for roof pitches of 4 to 12 or more, and 8 in. for roof pitches less than 4 to 12.

U.S. gages are 12, 14, 16, 18, 20, 22, 24, 26, 28, and 29.

Information about gages, thicknesses, and weights of black (mill) and galvanized flat and corrugated steel sheets is given in Table 9-5. Painted black (mill) sheets will weigh about 1.5 to 2 lb more per square (100 sq ft) than the unpainted sheets.

Table 9-5 *Gage, Thickness and Weight of Black and Galvanized Flat and Corrugated Sheets*

U.S. gage	Black				Galvanized			
	Thickness, in.	Weight, lb per sq ft			Thickness, in.	Weight, lb per sq ft		
		Flat	Corrugated			Flat	Corrugated	
12	0.1046	4.38	4.77	5.05	0.1084	4.53	4.94	5.23
14	0.0747	3.13	3.41	3.61	0.0785	3.28	3.58	3.79
16	0.0598	2.50	2.73	2.89	0.0635	2.66	2.90	3.07
18	0.0478	2.00	2.18	2.31	0.0516	2.16	2.35	2.49
20	0.0359	1.50	1.64	1.73	0.0396	1.66	1.81	1.91
22	0.0299	1.25	1.36	1.44	0.0336	1.41	1.53	1.62
24	0.0239	1.00	1.09	1.15	0.0276	1.16	1.26	1.33
26	0.0179	0.75	0.82	0.87	0.0217	0.91	0.99	1.05
28	0.0149	0.63	0.68	0.72	0.0187	0.78	0.85	0.90
29	0.0135	0.56	0.60	0.65	0.0172	0.72	0.78	0.83
Corrugations, in.			$2\frac{2}{3} \times \frac{1}{2}$	$3 \times \frac{3}{4}$	Corrugations, in.		$2\frac{2}{3} \times \frac{1}{2}$	$3 \times \frac{3}{4}$

Corrugated steel sheets may be fastened by special clips, self-tapping screws, bolts, closing rivets, and other devices. Fasteners should be spaced at 18 in. maximum for siding and 12 in. maximum for roofing. Fasteners are also required at about every third corrugation at each girt or purlin. About 2 to 4 lb of fastening devices will be required per square (100 sq ft).

Sheet steel flashing should be provided at doors, windows and other openings, roof ridges, and eaves, and wherever needed for water tightness.

The number of square feet of corrugated sheets required to cover a square (100 sq ft) for any width, length, and lap is given by the formula:

$$N = \frac{100\,WL}{(W - S)(L - E)}$$

where N = square feet of corrugated sheets required per square
W = width of corrugated sheet, in.
L = length of corrugated sheet, in.
S = side lap, in.
E = end lap, in.

The weight of corrugated steel sheets in lb per square equals $N \times$ weight per square foot (from Table 9-5).

Table 9-6 *Data for "Standard" and "Wide" Corrugated Steel Siding and Roofing Sheets*

Items	Siding				Roofing			
	Standard		Wide		Standard		Wide	
Width W, in.	26	26	31.25	32	27.50	27.50	32.75	33.50
Side lap S, in.	2	2	2.25	2	3.50	3.50	3.75	3.50
Cover, in.	24	24	29	30	24	24	29	30
End lap E, in.	4	4	4	4	6–8	6–8	6–8	6–8
Corrugations:								
Pitch, in.	2.67*	3	2.67*	3	2.67*	3	2.67*	3
Depth, in.	0.50	0.75	0.50	0.75	0.50	0.75	0.50	0.75
Number in cover	9	8	11	10	9	8	11	10

* Nominal pitch is 2.50 in.

The number of corrugated sheets required per square equals

$$\frac{14{,}400}{(W - S)(L - E)}$$

Approximate prices (1968) per 100 lb for corrugated steel sheets delivered at the job are about $16 to $22 for black (mill) sheets and about $18 to $25 for galvanized finish. Special finishes cost more. These prices include base price f.o.b. warehouse plus size and quantity extras plus delivery to the job.

When an estimate is prepared, the local dealer or manufacturer should be consulted so that the correct prices of the materials may be had.

Sometimes an anticondensation lining is applied to the corrugated-steel surfaces. This lining may be asbestos fastened to the corrugated sheets or may consist of asbestos paper over wire mesh installed separately from the corrugated steel. On roofs, a layer of about 19-gage galvanized wire with 2-in. mesh can be laid over the purlins and tied with brass wire. A layer of asbestos paper (about 30 lb per square) is placed over the wire netting. Then the corrugated roofing is laid over the top of the asbestos.

Metal Lumber. Metal lumber of various types manufactured by several companies is available. The three main types made from steel are joists (open type, plate girder, I beam, and channel), studs, roof decking, floor plates, furring, lath, corrugated siding and roofing, pressed-steel siding and ceiling, and ornamental pressed steel. The term *metal lumber* is sometimes used to include joists and studs only.

Metal lath and furring will be discussed in Chap. 11 on Lathing and Plastering. Corrugated steel roofing will be discussed in Chap. 10 on Roofing and Flashing, and also in this chapter.

Open-type joists are usually made in depths of 8, 10, 12, 14, and 16 in.

and in spans varying from 4 to 31 ft. Weights per foot of length may vary from about 2 to 6 lb.

Plate-girder and I-beam types of joists are made in depths of about 4 to 14 in. and of varying lengths up to about 30 ft. Weights per foot of length may vary from about 3 to 12 lb.

Channel joists are made in depths of 4 to 12 in. and of varying lengths up to about 30 ft. Weights per foot of length may vary from about 2 to 6 lb.

Steel studs may vary in depth from 1 to 8 in., though ordinary depths are 2, 3, or 4 in. The weights may vary from about 0.30 to 1.00 lb per lineal foot for the lighter sections and from about 2 to 10 lb per lineal foot for the heavier sections used in supporting partitions.

Special fittings such as cap and sill plates, braces, and connections are required for most special lumber. About 5 to 15 percent should be allowed for details.

Metal lumber is often sold by the pound or by the lineal foot with other details given. The price may vary from 1.5 to 2.5 times that charged for standard structural shapes. The grade of steel is usually A36.

The metal-lumber joists and studs are used in place of corresponding wooden joists and studs.

Metal lumber is usually not included in a structural job.

Steel Sash. Steel sash are available in standard units to take glass of various sizes. Most steel sash are of the pivot type. Steel sash vary in size from 1 light high and 3 lights wide to 7 lights high and 6 lights wide. Almost all wall sash use lights 14 by 20, 12 by 18, or 10 by 16 in. in size. When the steel sash has ventilators, the width of glass in the ventilator sections is usually 1 in. less on the outer edges.

The catalogues of the sash manufacturers give full information regarding this sash.

Steel-sash prices vary constantly with the steel market and other causes. Consequently, the estimator, when preparing an estimate for a job, should list the steel sash required and secure prices from the manufacturer. The steel sash can be installed by the steel gang, by carpenters or metalworkers, or by a special crew provided by the sash company.

Miscellaneous and Ornamental Steel. Miscellaneous and ornamental structural-steel and iron work and various types of pressed steel should be listed separately with adequate descriptions. Prices should be secured from the manufacturers or dealers.

Paint. Paint required for structural steel can be estimated at ½ to 1 gal per coat per ton of steel. One gallon will cover about 400 to 500 sq ft of steel surface with one coat. Small sizes of structural steel may average from 350 to 450 sq ft per ton of steel, medium sizes from 250 to 350 sq ft, and large sizes from 175 to 250 sq ft, including some allowance for connections and framing details. Small trusses and built-up frames will average from 350 to 450 sq ft per ton. Suitable paint may cost $4 to $7 per gallon. See Chap. 15 on Painting, Papering, and Glazing for information on painting structural steel.

Labor The labor required on a structural-steel job will depend upon the wages and skill of the workers, upon the ability of the foreman, and upon the labor-saving equipment used, as well as upon the kind of job, exactness of fabrication of the materials, weather, and working conditions.

The labor for erecting structural-steel framing may include two or more of the following:

1. Labor of handling steel at railway siding and at site
2. Labor of erecting, moving, and taking down of hoists and derricks
3. Labor of erection, including plumbing and temporary bolting
4. Labor of riveting, bolting, or welding
5. Labor of painting
6. Labor of installing corrugated siding and roofing, metal sash, and miscellaneous items

The labor for handling structural steel (unloading railway cars or trucks, loading trucks, etc.) may be done by a gang of two or more men with hand and power equipment as needed.

Steel erection is usually done by a gang of skilled men working under a foreman. Such a gang may consist of a foreman, one or two crane or derrick operators, and from about three to six ironworkers as needed.

Riveting may be done by one or more small gangs of about four men each (a heater, a catcher, and two riveters). One helper may be needed for every three or four riveting gangs to care for compressor and hose.

Bolting may be done by one or more small gangs of two or four bolters each. About one helper may be needed for every four to six bolters to care for compressor and hose, etc.

Welding may be done by one or more small gangs of one or two welders each. One helper may be needed for every one or two welders.

Painting may be done by a gang of painters starting at the top of the structure and working down. One or more ground men or helpers may be needed to assist the painters.

Corrugated siding and roofing, steel sash, and miscellaneous items may be installed by men working in one or more small gangs of two to four men each. If the material must be hoisted or raised more than about two stories, extra men and equipment may be needed.

Steel lumber is not usually installed by structural-steel workers and may be installed by other metalworkers or carpenters.

Steel sash is usually not installed by the structural-steel gang but may be installed by other steelworkers, such as a special crew provided by the sash manufacturer or dealer.

Ornamental steel of various kinds is usually not installed by the structural-steel gang but may be installed by other steel- and ironworkers, tinsmiths, or special workers provided by the manufacturers or dealers.

At the railroad siding, the material may be unloaded on the ground and then on trucks or it may be unloaded directly from the cars to the trucks. Small pieces may be handled by men without a hoist or derrick. Medium-weight and heavy pieces may be handled by a hand or power derrick, or a crawler or wheel-mounted power crane. After the material is trucked to the site, it is unloaded by hand, pole derrick (hand operated), or power crane or derrick as the case may be. A crawler or wheel-mounted power crane is suitable here also.

The contractor must estimate the sizes and weights of the structural members in advance and have suitable gangs and equipment to handle it. Columns usually come fabricated in two-story lengths (20 to 30 ft long). Built-up beams and girders are entirely assembled before shipping. Light trusses of short spans may be entirely assembled. Longer trusses may be assembled in two or three parts so that their erection can be simplified on the job. Heavy trusses often have to be almost entirely assembled on the job.

Before approximate estimates are prepared of the labor required for erection in labor-hours per ton of steel, the general method or plan of erection for the particular structure should be considered, as well as the various sizes and weights of the various structural members and the equipment to be used.

If the structure is a mill building of a story in height with roof trusses, it can be erected with the aid of a gin pole or a small auto or crawler crane. The columns are erected first and plumbed and braced. The trusses are assembled on the ground and then raised in position and secured. Trusses may be shipped in one, two, or three pieces according to size. After the trusses are in place, the purlins and horizontal bracing are installed. After all parts of the structure are plumbed and bolted, the required riveting, bolting, or welding can be done. Painting comes next. After painting, the corrugated siding and roofing, steel sash, skylights, ventilators, etc., can be installed, glazing done, and other painting done as required.

If the structure is several stories high, such as an office building, the columns are erected and plumbed and then the beams and girders are bolted in place. As the columns are usually two stories in length, the framing for two stories is erected and then the derrick or hoist is placed on the last floor and the framing for the next two stories erected. This process is repeated until the framing is completed. If a structural steel tower with a truss boom is used, the tower is erected (and the boom raised) one or more floors at a time as needed. After the frame has been erected, it is plumbed by means of cross bracing of wire rope with turnbuckles and

Table 9-7 *Approximate Labor-hours Required for Equipment*

Kind of work	*Labor-hours*
Gin pole, large, unload and erect	24–44
Move horizontally	12–24
Take down and load	8–20
Gin pole, small, unload and erect	12–24
Move horizontally	6–12
Take down and load	6–10
Guy derrick, large, unload and erect	36–80
Move horizontally	10–20
Move vertically (2 floors)	16–36
Take down and load	16–36
Stiffleg derrick, large (25-ton), unload and erect	120–240
Take down and load	60–120
Stiffleg derrick, medium (10-ton), unload and erect	50–100
Take down and load	24–42
Derrick, small (5-ton), unload and erect	12–24
Take down and load	6–16
Breast derrick, small, unload and erect	6–10
Move horizontally	2–4
Move vertically (2 floors)	3–5
Take down and load	3–5
Air compressor, large, install	20–36
Remove	20–36
Air compressor, portable, diesel, electric, or gas power	
Install	1–2
Remove	1–2

additional bolts provided to hold the frame in place. Some erectors are careful to keep the frame plumb as it is erected so that but little final plumbing is required. Other erectors plumb the frame every few stories, so that riveting, bolting, or welding can be started before the entire frame is erected. Special equipment and extra men may be required for erecting extra-heavy or large pieces like girders or trusses. After erection, all connections are riveted, bolted, or welded as the case may be. The steel frame is usually painted after the riveting, bolting, or welding is completed, the painters working from the top down. The derrick can be removed after the frame is erected, after the riveting is completed, or after painting. If the structure is not too high, the use of a long-boom auto or crawler crane may be economical.

The trusses for small truss bridges may come assembled or may be assembled on the ground and then raised in position by a crane or derrick. Then the cross bracing, portals, and floor system are installed and riveted or otherwise secured in place. Painting follows.

The trusses for medium-size truss bridges may be assembled on the ground and then raised in position by one or two cranes; they may be

Table 9-8 *Approximate Labor-hours Required for Handling and Erecting Structural Steel*

Kind of work	*Labor-hours per ton of steel*
Handling, such as unloading from railway car to truck and from truck to ground, with aid of derrick or crane when needed	1–2
(average)	(1.3–1.5)
Erection labor, erect, bolt, and plumb only:	
Foundation work	3–6
Columns and struts	4–8
Beams and channels, ordinary	3–6
Beams and channels, special	4–8
Plate girders	3–6
Crane rails	3–6
Knee braces	6–10
Floor plates	4–8
Fittings, bolts, rods, anchor plates	2–4
Girts, angles, angle braces, purlins	4–8
Skylight frames and curbs	6–12
Monitor frames	6–14
Dormers	6–12
Door frames	8–16
Roof trusses	5–12
Transmission towers	16–30
Light steel trestles	12–24
Steel framing, as for power or heating plant	10–16
Steel mill building (with roof trusses, roofing, and siding)	4–12
Steel-frame multistoried building (as office buildings)	3–10

partly assembled on the ground, the parts raised in position and then bolted; or the trusses may be assembled in position by means of falsework and small derricks and cranes. After the trusses are erected, the floor system, bracing, portals, etc., are installed. Painting follows.

Large trussed bridges may be erected piece by piece on falsework by the aid of traveling or movable cranes or derricks, or they may be erected by the cantilever method, the frame being designed to support its own weight and that of the derrick, as erection progresses. The main bracing is usually erected with the trusses. Floor systems and other details are placed later.

The approximate amounts of labor required for various items of work connected with the erection of structural-steel work are given in Tables 9-7, 9-8, and 9-9. As there are many variables that may affect the labor-hours required, the tabular values are to be considered as approximate only. It is assumed that the laborers are skilled in their work and are organized in suitable gangs under the direction of competent foremen.

Equipment labor will vary with type of power units (boiler and steam engine, gas or oil engine, electric motor, etc.).

Labor wages per hour (1968) are approximately as follows: foremen, $5 to $9; structural-steel and iron workers and other skilled labor, $4 to $7; and unskilled labor, $3 to $5. Payroll taxes and insurance are high and may vary from about 12 to 35 percent of the labor wage.

The following diagrams can be used for estimating labor costs: Diagrams 9-1, 9-2, 9-3, 9-4, and 9-5 for handling, erecting, riveting, bolting, and welding structural steel; Diagram 15-3 for painting structural steel; and Diagram 9-6 for applying corrugated steel.

Equipment The equipment used on a structural-steel job will vary with the kind of structure, size of job, and the equipment the contractor has available and thinks is suitable for the particular work. The equipment may include:

For unloading and transporting steel from railway siding to site:
- Gin poles, derricks or cranes, hand or power operated
- Dollies and other small tools
- Autotrucks and auto trailers
- Special trucks and trailers as needed

For erection of steel:
- Derrick, crane, or steel tower with truss boom, preferably power operated, of any kind and capacity suitable for the job.
- Power unit for derrick
- Fuel and oil for power unit (coal, water, gasoline, oil, electricity)
- Winches and block and tackle.
- Long-boom auto or crawler crane
- Wire and manila rope
- Bolts, turnbuckles, drift pins, and various small tools
- Small portable building for office and tool house

For riveting:
- Rivets
- Riveting tools, air hammers, etc.
- Air piping and hose
- Hand lines and hand hoist
- Forge and tools
- Scaffolds
- Air compressor with power unit
- Fuel for forges and for compressor operation

For welding:
- Welding metal
- Welding machines
- Welding tools

Table 9-9 *Approximate Labor-hours Required for Bolting, Riveting, and Other Work*

Kind of work	*Labor-hours*
Temporary bolting (3 to 10 bolts per ton)	5–7 per 100 bolts
Riveting, air driven (20 to 40 rivets per ton):	
On ground, easy work	6–10 per 100 rivets
Trusses	7–12 per 100 rivets
Steel office buildings	10–15 per 100 rivets
Steel mill buildings	10–13 per 100 rivets
Light trestles and towers	14–20 per 100 rivets
Riveting, hand driven:	
Easy work	12–16 per 100 rivets
Difficult work	16–25 per 100 rivets
Bolting (15–30 high-strength bolts per ton)	3–7 per 100 bolts
Welding (5–10 ft of ¼-in. weld per ton)	14–30 per 100 ft
Painting structural steel, one coat:	
Heavy-weight members	0.5–0.9 per ton
Medium-weight members	0.7–1.4 per ton
Light-weight members	1.0–2.0 per ton
Corrugated-steel siding and roofing, on wood:	
26 U.S. gage and lighter	0.5–1.5 per square
Heavier than 26 gage	1.0–2.0 per square
Corrugated-steel siding and roofing on metal frame:	
26 U.S. gage and lighter	1.0–3.0 per square
Heavier than 26 gage	1.5–4.0 per square
Asbestos covered	3.0–6.0 per square
Anticondensation lining for corrugated roofing	2–4 per square
Steel roofing trim, ridge, valley, cornice, etc	2–6 per 100 lin ft
Steel lumber:	
Joists, 4–8 in. depth	0.15–0.30 per each 1.5–2.5 per 100 lin ft
Joists, 8–12 in. depth	0.25–0.50 per each 2–4 per 100 lin ft
Studs, light, 1 lb per ft or less	0.06–0.15 per each 1–2 per 100 lin ft
Studs, heavy, 2 lb per ft or more	0.12–0.25 per each 1.2–2.5 per 100 lin ft
Furring, and grounds	3–6 per 100 lin ft
Steel sash, installing only	3–12 per 100 sq ft of opening area
Painting or calking	2–5 per 100 lin ft of joints
Installing and calking	5–16 per 100 sq ft

Scaffolds
Hand lines and hand hoists
Gas for welding
Electricity for welding

For bolting:
Bolts (high-strength), washers, and nuts

Bolting tools, air wrenches, etc.
Air compressor with power unit
Air piping, hose, and connections
Hand lines and hand hoist
Scaffolds
Fuel for compressor operation

For painting:

Paint
Brushes, pails, spray guns, and other painters' tools
Ladders, rope slings, small scaffolds, boatswain's chairs
Hand lines

Approximately 1 ton of coal or 200 gal of fuel oil or gasoline will be required for erecting 30 to 50 tons of steel.

About 1 ton of coal or 200 gal of fuel oil or gasoline will be required for the compressor for every 400 to 600 field rivets.

The various types of cranes and derricks that can be used for structural-steel work include:

Locomotive cranes on tracks, 40- to 105-ft booms
Tractor or crawler cranes, 20- to 75-ft booms
Stiffleg derricks, varying length of booms and capacities
Guy derricks, guyed at top of mast
Breast derricks
Jinniwink, A frame, one stiffleg, and boom
Gin pole, single pole guyed at top
Hoists and block and tackle

The cranes and derricks of the larger capacities are invariably power operated. The ones of smaller capacities (say 5 tons or less) may be either hand or power operated. Each type has its own advantages and disadvantages in construction use.

On many jobs, an estimate of the cost of equipment per ton of steel is satisfactory. On other jobs, the estimator may need to estimate the costs of each large or important piece of equipment separately. To find the equipment costs per ton of steel, the total cost of all equipment used on the job is first estimated and then this total cost is divided by the number of tons of steel in the job.

For approximate estimates of the cost of equipment used in structural-steel work, the following values can be used. It is thought that these ranges of values will be fairly representative of the equipment costs per ton for most of the ordinary structural-steel jobs. The lower values may be expected on simple and easy jobs where the tonnage is large and the individual prices are easy to handle. The higher values may be expected on the more difficult jobs, when the tonnage is small, and when the individual pieces are not easy to handle.

Handling and transporting from railways siding to site
  Each crane or derrick, hand operated = \$0.10 to \$0.20 per ton of steel handled
    Power operated = \$0.15 to \$0.30 per ton of steel handled
  (Power-operated cranes and derricks are faster and usually save labor costs)
  Transportation in autotrucks, per ton of steel hauled = \$3 + \$0.25 × miles hauled (i.e., the truck cost is \$3 per ton for time of loading and unloading plus \$0.25 for each ton mile)
    Or for the average job = \$3 to \$6 per ton of steel handled

Erection:
  Crane or derrick including power = \$2.00 to \$10.00 per ton of steel
  All other tools and equipment = \$0.30 to \$1.00 per ton of steel

Riveting:
  Compressor, \$6 to \$10 per 100 field rivets, or \$1.00 to \$2.00 per ton of steel
  Other tools, \$1 to \$4 per 100 field rivets, or \$0.30 to \$1.10 per ton of steel

Bolting:
  Compressor, \$4 to \$7 per 100 bolts, or \$1 to \$2 per ton of steel
  Other tools, \$1 to \$3 per 100 bolts, or \$0.25 to \$1 per ton of steel

Welding:
  Electric welders = \$1.00 to \$2.00 per ton of steel
  Other tools = \$0.20 to \$1.00 per ton of steel

Painting of steel frame = \$0.20 to \$0.70 per ton of steel
Corrugated steel, erection of = \$0.10 to \$0.40 per square

For more detailed estimates of equipment costs, the cost for the job of each important or large piece of equipment is found separately. Tools and minor equipment can be considered together and the cost for the job estimated. The cost of equipment should include costs of transportation to and from the job, rentals, depreciation, repairs, maintenance, fuel, oil, etc., as the case may be. The labor costs of erection, operation, moving, and taking down are usually included with the labor estimate. When not included in the labor estimate, these costs must be included with the equipment costs.

Further information is given in Chap. 18 on Construction Plant and Equipment and in Appendixes B and C. The "Contractors' Equipment Ownership Expense" prepared and published by The Associated General Contractors of America, Inc., Washington, D.C., contains about the best information on owner's equipment costs. Other worthwhile information can be found in schedules prepared and distributed by different government agencies and by various manufacturers. Information in regard to equipment cost is sometimes published in engineering periodicals.

Overhead and Profit Overhead costs on structural-steel work will be comparatively large because of the hazards and uncertainties of the work and the larger amount of office work and supervision required. The overhead costs are usually based on labor. The percentage of the labor costs may range from 20 to 35 percent.

Sometimes the overhead costs are estimated separately for materials, labor, and equipment, using different rates for each item. These rates are determined by the past experiences and the cost records of the company plus the judgment of the estimator. These rates may be based on materials costs or quantities per ton, on labor wages or hours, and on equipment costs or time.

The percentage of profit figured on structural-steel work is usually comparatively high and may range from 10 to 25 percent of the sum of all other costs or from 15 to 30 percent of the sum of labor, equipment, and overhead costs. Sometimes the percentages allowed for profit vary with the items, being lower for materials and higher for labor costs.

Structural-steel Estimates Structural-steel estimates will include the following:

Materials. Cost at nearest railway siding or cost at job
Labor. All labor costs involved
Equipment. Costs of all equipment required for the particular job
Overhead. Usually based on labor costs
Profit. Based on sum of all other costs
Total costs. Sum of all other costs

Unit costs per ton or per hundredweight can also be computed.

Estimates can be summarized and tabulated to show unit costs per ton or per hundredweight of steel as well as total costs.

Some estimators subdivide their estimates as follows:

Materials. Cost f.o.b. nearest siding, or delivered at the job
Handling and transportation. From railway siding to job site
Labor. Costs of all labor used for this purpose
Equipment. Costs of all equipment used
Erecting, riveting, bolting, or welding, and painting
Labor. Costs of labor for erection and riveting, bolting, or welding, and painting
Equipment. Costs of erection and riveting, bolting, or welding equipment
Overhead. If not subdivided and included with other costs
Profit.

This form of estimate has the advantage of giving the estimated costs of the materials f.o.b. siding, the costs of moving the steel from siding to job site, and the costs of erecting, riveting, bolting, or welding, and painting the steel. A few estimators go into further detail and figure costs of erecting, riveting, bolting, or welding, and painting separately, as different gangs are used for these three kinds of work.

Illustrative estimate. Prepare a cost estimate for the structural-steel frame for a small steel mill building 60 by 105 ft in size, 20 ft from ground level to bottom of roof trusses. The building has seven 15-ft bays, thus requiring eight trusses of 60-ft span. Diagonal cross bracing is provided in the second and sixth bays for walls and for the roof trusses in planes of both the upper and lower chords. All steel is structural steel grade A36.

Work Plan: There is a railway siding at the site so that the material can be unloaded at the site. The unloading, erection, temporary bolting, and plumbing will be done by a crew of 1 foreman, 1 crane operator, and 4 structural-steel workers with a diesel crawler crane of 6 tons' capacity and a 40-ft boom and the necessary wire rope, small tools, etc. The riveting will be done by a crew of 4 (1 heater, 1 catcher, and 2 riveters) with a forge, a portable air compressor, 100 ft of air hose, and the necessary tools, etc. The painting will be done by a crew of 3 painters with hand spray guns, ladders, etc.

The total steel tonnage (see Materials Takeoff, Table 9-10) is a little less than 62 tons.

For unloading at site, erection, temporary bolting, and plumbing, allow 7 labor-hours per ton for the 14.5 tons of roof trusses or 102 labor-hours and allow 6 labor-hours per ton for the 47 tons of other steel, or 282 labor-hours. Total labor-hours equal 384, or 64 crew-hours.

For riveting, assume 25 field rivets per ton, or about 1,540 rivets. At 12 labor-hours per 100 rivets, or 3 crew-hours, about 46 crew-hours will be required.

Painting this frame with hand spray guns should not be difficult. Allow 1.15 labor-hours per ton. Required labor-hours equal 71, or 24 crew-hours.

Materials: See Materials Takeoff (Table 9-10).

Cost delivered at fabricating shop, including base price f.o.b. mill, extras, and freight, 123,110 lb at $7.60 per 100 lb		= $ 9,356
Fabrication, including drawings, shop riveting, and painting:		
Roof trusses, 29,080 lb at $143 per ton		= 2,079
Other steel, 94,030 lb at $85 per ton		= 3,996
Freight to job site, 123,100 lb at $0.75 per 100 lb		= 923
Cost of structural steel, f.o.b. railway siding		= $16,354
Miscellaneous materials:		
Field rivets (25 per ton) about 1,540. Say 800 lb at $20 per 100 lb	= $160	
Field bolts, about 250 at $0.30 each	= 75	
Paint, about 41 gal for one coat at $5.60	= 230	
Total miscellaneous materials		= $ 465
Total materials cost		= $16,819

Labor: Assume payroll taxes and insurance at 20 percent of the labor wage. Required crew hours are given in the Work Plan. Erection crew consists of 1 foreman at $6.50 per hr, 1 crane operator at $5.25 per hr, and 4 structural-steel workers at $5.50 per hr each.

Crew cost per hour = ($6.50 + $5.25 + $5.50 × 4) × 1.20 = $40.50	
$40.50 × 64 crew-hour	= $2,592
Riveting crew, cost per hour = $5.50 × 4 men × 1.20 = $26.40	
$26.40 × 46 crew-hour	= 1,215
Painting crew, cost per hour = $5.50 × 3 men × 1.20 = $19.80	
$19.80 × 24 crew-hour	= 475
Total labor cost	= $4,282

Table 9-10 *Materials Takeoff*

One-story Steel Mill Building 60 by 105 ft. Job No.______

No. of pieces	Size, etc.	Length, each, ft	Weight per ft-lb	WF's, I's, [s, lb	∠s, lb	Plates, lb	Rivets, lb	Total
22	Wall columns, 14 × 8WF	20.0	43.0	18,920				
	Details, 15 %.........	....	....	......	......	2,840		
	Rivets, 4 %..........	....	....	......	......	......	760	22,520
22	Wall beams, 10 × 4B....	15.0	17.0	5,610				
	Details, 10 %.........	....	....	......	......	560		
	Rivets, 3 %..........	....	....	......	......	......	170	6,340
	Wall bracing:							
8	∠s 2 × 2 × ¼.........	25.0	3.2	......	640			
16	∠s 2 × 2 × ¼.........	12.5	3.2	......	640			
	Details, 10 %.........	....	....	......	......	130		
	Rivets, 3 %..........	....	....	......	......	......	40	1,450
120	Wall girts [s, 6 in........	15.0	8.2	14,760				
	Rivets, 4 %..........	....	....	......	......	......	590	15,350
	One truss:							
4	∠s 5 × 3½ × ⅜.......	35.0	10.4	......	1,455			
4	∠s 3 × 3 × ¼.........	19.0	4.9	......	375			
2	∠s 3 × 3 × ¼.........	23.0	4.9	......	225			
4	∠s 3 × 2½ × ¼.......	19.0	4.5	......	340			
4	∠s 3 × 2½ × ¼.......	9.0	4.5	......	160			
4	∠2 3 × 2½ × ¼.......	8.0	4.5	......	145			
4	∠s 2 × 2 × ¼.........	9.5	3.2	......	120			
4	∠s 2½ × 2 × ¼.......	4.0	3.6	......	60			
1	∠ 2 × 2 × ¼..........	15.0	3.2	......	50			
	Total...............	....	....	......	2,930			
	Details, 20 %.......	....	....	......	......	585		
	Rivets, 4 %........	....	....	......	......	......	120	(3,635)
(8)	Trusses...............	....	....	......	23,440	4,680	960	29,080
(16)	Knee braces:							
32	∠s 3½ × 3 × ⅜.......	11.0	7.9	......	2,745			
	Rivets, 4 %..........	....	....	......	......	......	110	2,855
	Roof purlins:							
182	I's 7 in................	15.0	12.5	34,125				
	Details 10 %..........	....	....	......	......	3,415		
	Rivets, 3 %..........	....	....	......	......	......	1,025	38,565
	Upper chord bracing:							
16	∠s 2 × 2 × ¼.........	22.0	3.2	......	1,130			
	Details, 10 %.........	....	....	......	......	115		
	Rivets, 3 %..........	....	....	......	......	......	35	1,280
	Lower chord bracing:							
4	∠s 4 × 3 × 5⁄16........	25.0	7.2	......	720			
8	∠s 4 × 3 × 5⁄16........	20.0	7.2	......	1,155			
20	∠s 4 × 3 × 5⁄16........	15.0	7.2	......	2,160			
4	∠s 2 × 2 × ¼	27.0	3.2	......	345			
8	∠s 2 × 2 × ¼.........	25.0	3.2	......	640			
	Total...............	....	....	......	5,020			
	Details, 10 %.......	....	....	......	......	500		
	Rivets, 3 %........	....	....	......	......	......	150	5,670
	Summary:							
	Trusses...............	....	....	......	23,440	4,680	960	29,080
	All other...............	....	....	73,415	10,175	7,560	2,880	94,030
	Total...............	....	....	73,415	33,615	12,240	3,840	123,110

Equipment: See Chap. 18, Construction Plant and Equipment, and Appendix B for method of computing equipment costs.

Diesel crawler crane, 6-ton, 40-ft boom, with wire rope and tackle. Contractor's books show an ownership cost of $5.10 per working hour and an operating cost of 50 cents per working hour. Allow $30 for transportation to and from the job.

Crane cost = ($5.10 + $0.50) × 64 hr + $30	= $388		
Small tools. Allow for the job	= 30		
Total for erection equipment		= $418	
Riveting forge and fuel, $0.60 per hour	= $ 28		
Two riveting hammers, at $0.20 per hour each	= 19		
Portable air compressor and fuel at $2.75 per hour	= 127		
Air hose and couplings, two 50-ft sections	= 8		
Small tools for riveting crew, allow	= 10		
Transportation of riveting equipment	= 20		
Total for riveting equipment		= $212	
Painting equipment. Portable 1-hp, 45-psi industrial sprayer with gasoline engine, 150-ft hose and connections three spray guns, ladders and miscellaneous equipment, plus transportation, allow for job		= $ 65	
Total cost of equipment			= $ 695
Overhead at 25% of labor costs			= 1,071
Profit at 10% of all other costs			= 2,287
Total estimated cost			= $25,154

This estimate can be summarized as follows, with unit costs based on 123,110 lb of steel.

Item	Unit cost per 100 lb	Unit cost per ton	Total cost
Materials	$13.65	$273	$16,819
Labor	3.50	70	4,282
Equipment	0.55	11	695
Overhead	0.85	17	1,071
Profit	1.85	37	2,287
Total	$20.40	$408	$25,154

In this estimate it was assumed that the contractor bought the steel from the mill and had the fabricating shop make the shop drawings and fabricate the members.

For a job of this size, the usual procedure at the present time (1968) would be for the contractor to give copies of the plans and specifications to one or more steel companies and to secure their bids for the materials, prefabricated and delivered at the job site ready for erection.

CHAPTER TEN

Roofing and Flashing

Roofing and Flashing Roofing or roof covering is usually considered to include the actual covering material. Roofing often does not include the rafters, purlins, or roof sheathing.

The paper or felt that may be placed over the roof boards and under the covering material is usually included with the covering material as it is placed at about the same time.

In some instances, purlins are included with the roofing. If the purlins are placed by the same gang that erects the roof trusses, the purlins are included with the trusses. If the purlins are placed by the roofing gang, they can be included with the roofing material.

Flashing is the waterproofing material placed around joints in the roof to keep the water from penetrating the roof at those places or from getting under the roof covering. Examples are the pieces of tin placed around chimneys when the roof covering butts up against the chimney, roof valleys, etc.

Roof trim includes such items as the metal, asphalt, or wooden strips at the ridges or hips, gutters, and downspouts.

The cost of roofing will depend upon the kind of area to be covered (simple as a plain roof or complex when the roof is cut up by hips and valleys), kind of roofing material used, cost of materials, and wages, skill, and inclination of the workmen.

When estimating roofing areas, the estimator must be careful to get the actual area overall from eaves to eaves and to be sure that the true lengths of all inclinations are used. The takeoff should include all details of the roofing areas, materials, flashings, and trim required. The units of measurement are usually as follows:

Roof areas. By the square of 100 sq ft. In shingling, sometimes by the bundle of or by the 1,000 shingles

Flashing. By the piece, square foot, or lineal foot

Trim. By the lineal foot, with other details noted

Roofing Materials The materials used in roofing and flashing may be classified as to kind of material (wood, asphalt, composition, metal, canvas, etc.) or as to type of material (shingles, rolls, tile, strips, pieces, etc.).

In the absence of other information, the weights per square of various roofing materials can be taken from Table 10-1.

Roofing Paper and Felt Roofing paper usually comes in rolls of varying lengths and weights. A roll may contain 108, 216, or 432 sq ft or enough for one, two, or four squares with a 2-in. lap. Weights of roofing

Table 10-1 *Approximate Weights per Square of Various Roofing Materials*

Roofing material	*Weight per square, lb*
Sheathing, wood	225–350
Felt, per layer	15–30
Tar or asphalt, per coat	20–30
Tar and slag roof, 4- ply	400
Asphalt roll roofing	55–105
Canvas roofing	15–25
Copper, No. 22 B and S gauge	125
Corrugated galvanized steel 22 U.S. gage	175
Corrugated galvanized steel 24 U.S. gage	145
Corrugated galvanized steel 26 U.S. gage	115
Corrugated galvanized steel 28 U.S. gage	100
Lead, 1/8 in. thick	750
Glass, 1/8 in. thick	175
Shingles, asbestos	200
Shingles, asphalt	150–400
Shingles, wood	200
Shingles, slate	450–700
Tile, clay, and cement	800–1,200

paper may vary from 4 or 5 lb up to 30 lb or so per square. Prices may range from about 60 cents to $2 per square. Roofing paper is often fastened with large-headed tacks about $\frac{9}{16}$ in. long and costing 15 to 25 cents per $\frac{1}{2}$-lb package (enough for one square).

Roofing felt usually comes in rolls of varying areas and of weights from about 15 to 40 lb per square. The felt may be priced per roll with area and weight given or per 100 lb with weight per square given. Cost per square may vary from 50 cents to $2.25, and cost per 100 lb from $2 to $6. The felt is impregnated with tar or asphalt, with the asphalt usually being considered as a better material and costing a little more. About 0.50 lb of $\frac{5}{8}$-in. to 1.25 lb of 1-in. large-head galvanized nails may be required per square when the felt is nailed to a wood surface.

Diagrams 10-1 and 10-3 (Appendix F) can be used for estimating the costs per square of paper and felt.

Wood Shingles Wood shingles usually come in random widths (say from 4 to 12 in.) packed in bundles. Each bundle contains the equivalent of two-hundred and fifty 4-in. shingles. The length of the shingles is usually 16 or 18 in. In general, the length should equal three exposures plus 1.5 to 3 in. in addition for head or end lap so as to provide a roof covering three layers thick. The thickness of the shingles at the butt is usually 0.40 (five make 2 in.) or 0.45 in. (five make 2.25 in.). The exposure is usually 4, 4.5, 5, 5.5, or 6 in. Some manufacturers pack their shingles in special bundles or cartons, each carton containing enough shingles to cover 25, 33$\frac{1}{3}$, or 50 sq ft or so, with a stated exposure of 5 or 5.5 in. The wood used is usually red cedar of No. 1, No. 2, or No. 3 grade. Prices of shingles may range from about $2 to $6 per standard bundle of 250 shingles or from $8 to $25 per square.

Wood shingles are usually nailed to the shingle strips or wood roof sheathing with two nails per shingle. Usually no building paper is placed between the roof shingles and the roof sheathing, though building paper is frequently placed between siding shingles and the wall sheathing. The number of nails required per square may vary from 3 to 5 lb, depending upon the exposure and consequent number of shingles, or about 4 lb of 4-penny shingle nails per 1,000 shingles. Nails may cost 15 to 25 cents per pound for plain-wire nails and 20 to 35 cents per lb for galvanized-wire nails.

The amount of waste with wood shingles may vary from about 8 to 25 per cent, depending on how much the roof is cut up by ridges, valleys, dormers, etc. A double row of shingles is usually laid along all eaves.

Diagrams 10-1, 10-2, and 10-5 (Appendix F) can be used for estimating the costs per square for wooden shingles.

Table 10-2 gives theoretical quantities of wood shingles and nails required per 100 sq ft of roof area with no waste.

Table 10-2 *Wooden Shingles and Nails Required per Square of* 100 *Sq Ft*

Length, in.	Exposure, in.	Number of shingles	Nails, lb
16	4	900	4.0
	4.5	800	3.6
	5	720	3.3
18	5.5	655	3.0
	6	600	2.8
24	8	450	2.3
	10	360	2.0

Wood shakes are usually packed in bundles, each bundle containing enough shakes for 20 sq ft. The butt thickness may vary from ½ to 1¼ in. Lengths are 16, 18, and 24 in. From about 2 to 4 lb of 4d, 5d, or 6d nails may be required per square, depending on the butt thickness and exposure. Shakes may cost from about $15 to $40 or more per square, depending on the kind and quality of the wood and the butt thickness and length.

Asphalt Shingles Asphalt shingles are usually packed in bundles or cartons. Each bundle may contain enough shingles for 25, 33⅓, or 50 sq ft of roof. Exposures are usually 4, 4.5, or 5 in. Single shingles vary from about 12 in. in length for 4-in. exposure to 16 in. in length for 5-in. exposure. Strip shingles are usually 36 by 10, 11, or 12 in. in size, and each strip is usually cut or marked to indicate three or four shingles, i.e., three or four tabs. Asphalt shingles are usually laid over a covering of building or sheathing paper.

Asphalt shingles are made in varying weights and finishes. Weights per bundle may vary from 40 to 125 lb. Weights per square when laid may vary from about 150 to 400 lb. Prices may range from about $5 to $16 per square or from $1.50 to $6 per bundle.

Waste in laying asphalt shingles may vary from about 4 up to 15 percent, depending on how much the roof is cut up by ridges, valleys, dormers, etc.

Asphalt starting strips and ridge and valley strips come in rolls 9 and 18 in. wide by 36 ft long, weighing about 22 and 44 lb per roll and costing about $1 and $2 per roll, respectively.

Large-headed galvanized roofing nails, 1 to 1.75 in. long, are used with asphalt shingles. These nails cost 20 to 35 cents per pound. From 4 to 5 lb per square is required for individual 9- by 12-in. shingles, from 2.5 to

4.0 lb per square for individual 12- by 16-in. shingles, and from 2.5 to 4.0 lb per square for three- and four-tab strip shingles.

Diagrams 10-1 and 10-2 can be used for estimating the costs per square for asphalt shingles.

Asbestos Shingles Asbestos shingles are available in a number of sizes and styles. Hexagonal shingles about 16 by 16 in. or square styles about 8 by 16 in., 12 by 24 in., and 14 by 30 in. are common. These shingles come in bundles containing enough shingles for 25, 33⅓, or 50 sq ft of surface. The lap is about 2 in., so the exposure is equal to one-half (length minus end lap). The cost per square may range from about $25 to $45. The average weight is about 250 lb per square. The waste may vary from about 8 to 25 percent. From 3 to 4 lb of 1.5- or 1.75-in. galvanized roofing nails are needed per square.

Asbestos shingles are usually laid over a layer of roofing felt, which may weigh 15 to 35 lb per square.

Diagrams 10-2 and 10-5 can be used for estimating the cost per square of asbestos shingles.

Slate Shingles Slate shingles are usually rectangular in shape and vary in width from 6 to 16 in. and in length from 12 to 24 in. The end lap is usually about 3 in., and the exposure is one-half (length − end lap). These shingles are often priced per square of 100 sq ft (lap and other dimensions given). The price per square may range from $25 to $100. The thickness of the individual shingle may vary from about 0.20 to 2.00 in. Average weight of slate shingles is about 500 to 600 lb per square.

Waste may vary from 5 to 25 percent, depending on the form and irregularities of the roof.

Slate shingles are laid over a layer of roofing felt, as are asbestos shingles.

From 2 to 6 lb of nails may be required per square, depending on size and thickness of shingle. Copper nails are recommended. From 3 to 4 lb is usually enough per square. These nails may cost 40 to 80 cents per pound.

Diagrams 10-2 and 10-5 may be used for estimating the cost per square for slate shingles.

Metal Shingles Stamped metal shingles of copper, tin (steel with tin surfaces), galvanized steel, etc., are available in many shapes and sizes. Quantities required per square will depend upon size, overlap, and exposure. The shingles are often quoted per square and sometimes per 100. Prices vary considerably, the cost per square ranging from about $6 per square and up.

Metal flashing shingles are usually about 5 by 7 in. in size and are made of comparatively lightweight tinned or galvanized steel material of about No. 28 or 29 gage. They may cost about $2 to $3 for a bundle of 100. Copper flashing shingles are also used and may cost from four to six times as much as the galvanized-steel shingles, depending on size of shingle and gage.

Diagrams 10-2 and 10-5 can be used for estimating the costs per square of metal shingles.

Tile Cement tile, clay tile, and steel tile (imitation of clay tile) are available in many sizes and shapes. These tile may cost $20 to $200 per square, depending upon material and style. Hips and valleys may cost 50 cents to $2 per lineal foot.

The tile are usually laid over a layer of roofing felt weighing about 30 to 40 lb per square. Furring strips and lath are required with some styles.

As most tile are furnished and laid by the manufacturer or dealers, the estimator should give the plans and specifications to them so as to secure their estimates.

Asphalt-roll Roofing Asphalt-roll roofing usually comes in rolls 32 or 36 in. wide and of a length sufficient to cover 100 sq ft of surface with an allowance for 2-in. laps. The actual area per roll is usually about 108 sq ft. Sometimes a 3-in. lap is provided for. The finish may be a mica or slate surface. The weights per roll or per square may vary from 35 to 110 lb per square by 5-lb intervals. The 45- to 105-lb weights are common. The price per square (or per roll) may vary from about $2 to $7, depending upon weight, finish, and style.

Nails and roofing cement for laps may or may not be included with the rolls. From 2 to 4 lb of 1- to 1.75-in. galvanized large-headed nails are required per square or roll. Roofing cement is needed at the laps and around flashings. This costs about 12 to 22 cents per pound, depending on composition, color, and quantity. From 2 to 5 lb should be enough per square.

Asphalt-roll roofing is usually laid on wood roof sheathing or over old roofs. Building or sheathing paper is rarely used under asphalt-roll roofing.

Diagram 10-1 can be used for estimating the cost per square of asphalt-roll roofing.

Composition or Built-up Roofing Composition or built-up roofing may consist of two to five plies of roofing felt or similar material with alternate layers or coats of tar or asphalt and sometimes a surfacing of gravel or slag. The roofing felt used may vary from 15 to 40 lb per ply per square, the tar or asphalt from about 20 to 30 lb per coat per square, (40 to 60 lb per square for the top coat under gravel or rock surfacing), gravel or rock surfacing from 250 to 450 lb per square, and slag surfacing from 200 to 350 lb per square.

Two layers of asphalt impregnated felt (about 15 lb) with coatings of hot asphalt well mopped in, and a top layer of 90 or 105 lb mineral surfaced asphalt rolled roofing makes a good roof covering.

Roofing felt usually comes in rolls of various weights, each roll containing 108 sq ft or enough material for one square with 2-in. laps. Roofing felt may be priced per roll or per 100 lb as previously stated in the article on Roofing Paper and Felt.

Tar pitch usually comes in 350-lb barrels and may cost $40 to $70 per ton. Asphalt is usually priced per gallon (1 gal weighing about 9.5 lb) and may cost 45 to 95 cents per gallon or $4.50 to $10 per 100 lb.

Slag and gravel suitable for roofing may cost $5 to $25 per ton.

A little roofing cement may be needed for patching and also for flashings.

The composition roofing is used on comparatively flat surfaces only. If the surface is of concrete, the concrete is first given a priming coat of asphalt or tar roof paint or a coat of the regular tar or asphalt thinned a little before the first layer of felt is placed. Hence, on a concrete roof, there will be one more coating of tar or asphalt than there are plies of felt. The same is true for metal surfaces when a composition roof is placed over an old metal roof. For wooden surfaces, a layer of felt is placed first, and this layer may or may not be counted as one ply. Sometimes the composition roofing is extended for 6 in. to 1 ft up the sides of brick or other walls to serve as flashing.

The materials estimate should note:

Squares. Number of squares, including any extras as for flashing

Felt. Number of plies and weight per ply per square. Tar or asphalt impregnated

Tar or asphalt. Number of coats and pounds per coat per square

Slag or gravel. Pounds per square

Miscellaneous materials. Such as nails, roofing cement, asphalt priming paint, and roofing paper per square

In addition to the types of composition roofing mentioned, there are many special types prepared and applied by various manufacturers and dealers.

Diagrams 10-1 and 10-3 can be used for estimating the costs per square of composition roofing materials.

Canvas Roofing Canvas roofing of two- or three-ply construction may come in rolls up to 36 in. in width and 50 yd in length. The weight may vary from about 12 to 24 oz per square yard. The canvas roofing should be laid in paint on a smooth surface, and the finished roof given one or two coats of paint. The canvas should be fastened by copper or galvanized-steel or iron tacks about $\frac{5}{8}$ to $\frac{3}{4}$ in. long and spaced about 1 in. or less apart. From 8 to 12 percent should be allowed for laps and waste in cutting and fitting. Flashings may be formed by extending the canvas 6 to 8 in. up the sides of the walls. This extra material must be allowed for. The canvas may cost $30 to $80 per square, according to weight and number of plies. The cost of the paint is additional.

The materials estimate for a canvas roof should include canvas, tacks, and paint.

Metal Roofing Metal roofing is usually rolled open-hearth steel roofing of about 22 to 29 U.S. gauge and may be plain (black), painted, tinned, or galvanized. Rust-resisting steel alloys (usually copper bearing) are also available. Other metals, especially copper and aluminum, are also used. Sheathing or roofing paper may or may not be placed under metal roofing.

Flat or Sheet Roofing. This roofing comes in sheets. Material may be sheet steel, tin plated, with the underside painted, or copper. The sizes of tinned steel sheets are 14 by 20 or 20 by 28 in., the larger sizes being preferred. Common thicknesses or weights are U.S. gage 26, 28, and 29. Weights per square foot for these gauges are a little more than the corresponding weights of black finish flat sheets given in Table 10-5. This material may come in boxes of 112 sheets of the 14- by 20-in. size or 56 sheets of the 20- by 28-in. size or in boxes containing 100 lb of material. This material can be laid with standing seams or laid flat and soldered.

For a standing-seam roof, 32 sheets of the 20- by 28-in. size or 68 sheets of the 14- by 20-in. size are required per square for covering and laps for seams.

The materials needed for a flat soldered-seam roof are given in Table 10-3. The table gives the number of sheets required, including laps.

Table 10-3 *Flat Soldered-seam Roof. Materials per Square*

Material	Using 14- by 20-in. sheets	Using 20- by 28-in. sheets
Sheets, number	62	29
Solder, lb	7–9	4–6
Nails, lb	3–5	3–4
Rosin, lb	1	1
Charcoal, lb	2	2

From 4 to 8 percent should be added to tabular values for waste of cutting and fitting.

This roofing material may cost $18 to $45 per square for standing seams and $22 to $50 per square for flat soldered seams, including solder and other materials. The material may be priced per sheet, per box, per pound, or per 100 lb, with gauge and other details being given. At the present time, tin-plated steel roofing may cost $12 to $16 per 100 lb at the job.

Copper roofing sheets come in varying sizes and weights. Common weights are 14 to 20 oz per square foot. The roof can be laid flat or with standing seams. Copper roofing can be laid over roofing paper or lightweight roofing felt. Copper or bronze nails should be used if nailing is required. About 5 percent should be allowed for waste in addition to laps. Copper roofing may cost $40 to $90 per square.

Corrugated Steel Roofing. Sheets of corrugated steel roofing are available in rust-resisting alloy steel (copper bearing) and open-hearth steel, black (unpainted mill), painted, and galvanized finishes. Common gages are 12, 14, 16, 18, 20, 22, 24, 26, 28, and 29 U.S. Lengths are 5 to 12 ft with 1-ft variations. Widths range from about 22 to 36 in. Common widths are the standard 27.5 in. and the wide 33.5 in. Corrugations vary in pitch and depth, common values are 2.67 (2.5) in. pitch with 0.50-in. depth and 3-in. pitch with 0.75-in. depth.

Corrugated steel roofing sheets are usually priced per 100 lb with extras for quantities, sizes, finishes, transportation, etc. Prices may be f.o.b. mill or warehouse or delivered at the job. When preparing an estimate, the estimator should consult the local material dealer for the correct price.

At the present time (1968) approximate price per 100 lb. delivered at the job may range from $16 to $22 for black (mill) finish and from $18 to $25 for galvanized finish.

Many dealers price corrugated steel roofing by the sheet with gauge, width, length, finish, and other details given, or by the lineal foot with gauge, width, finish and other details given. This method of pricing is popular in many localities.

Steel flashing is required on ridges, valleys, eaves, roof openings, etc., whenever needed for watertightness.

Corrugated steel roofing sheets may be fastened by special clips, self-tapping screws, bolts, closing rivets, and other devices spaced at 12 in. maximum. Fasteners are also required at about every third corrugation at each purlin. About 2 to 4 lb of fasteners are required per square (100 sq ft).

Side laps for corrugated steel roofing sheets are usually 3.5 in. or 1.50 corrugations.

End laps are usually 6 in. for a roof with a pitch of 4 to 12 or more, and 8 in. for a roof pitch of 4 to 12 or less.

The relations among gages, thicknesses, and weights of black (mill) and galvanized corrugated steel roofing are given in Table 9-5.

Some data for standard 27.5-in. and wide 32.75-in. and 33.50-in. corrugated steel roofing is given in Table 9-6.

Painted sheets will weigh 1.5 to 2 lb more per square than unpainted black (mill) sheets.

The number of square feet of corrugated roofing sheets required to cover one square (100 sq ft) of roof area for any width, length, and laps is given by the formula:

$$N = \frac{100\,WL}{(W - S)(L - E)}$$

where N = square feet of corrugated sheets required per square
W = width of corrugated sheet, in.
L = length of corrugated sheet in.
S = side lap (usually 3.5 in.)
E = end lap (usually 6 or 8 in.)

The weight of corrugated steel sheets in pounds per square equals $N \times$ weight per square foot (from Table 9-5).

The number of corrugated sheets required per square equals

$$\frac{14{,}400}{(W - S)(L - E)}$$

with W, L, S, and E in inches.

Tables 10-4, 10-5, and 10-6 have been prepared for corrugated steel roofing sheets 27.5 and 33.5 in. wide with a side lap of 3.5 in. and end laps of 6 and 8 in. If desired, like tables can be prepared for corrugated steel roofing sheets of other widths and side and end laps.

Diagram 10-4 (Appendix F) can be used for estimating the cost per square of metal roofing.

Flashing Materials used for flashing may be tinned, painted, or galvanized steel; copper; or asphalt. Metal sheets or shingles about

Table 10-4 *Square Feet of Corrugated Steel Roofing and Number of Sheets Required per Square. Standard Width W = 27.5 in., Side Lap S = 3.5 in.*

	Sq ft per square		Sheets per square	
Length, ft	6-in. lap	8-in. lap	6-in. lap	8-in. lap
5	127	132	11.11	11.55
6	125	129	9.09	9.38
7	124	127	7.69	7.90
8	123	125	6.67	6.82
9	122	124	5.88	6.00
10	121	123	5.26	5.36
11	120	122	4.76	4.84
12	120	122	4.35	4.41

Table 10-5 *Square Feet of Corrugated Steel Roofing and Number of Sheets Required per Square. Wide Width W = 33.5 in., Side Lap S = 3.5 in.*

Length, ft	Sq ft per square		Sheets per square	
	6-in. lap	8-in. lap	6-in. lap	8-in. lap
5	111	116	8.89	9.23
6	109	113	7.27	7.50
7	108	111	6.15	6.32
8	107	109	5.33	5.46
9	106	108	4.71	4.80
10	106	107	4.21	4.29
11	105	107	3.81	3.87
12	105	106	3.48	3.53

Table 10-6 *Square Feet of Roof Area Covered by One Sheet of Corrugated Steel Roofing, of Standard 27.5-in. Width and Wide 33.5-in. Width. Side Lap = 3.5 in. End Laps = 6 and 8 in.*

Length, ft	Standard width, 27.5 in.		Wide width, 33.5 in.	
	6-in. lap	8-in. lap	6-in. lap	8-in. lap
5	9	8.67	11.25	10.83
6	11	10.67	13.75	13.33
7	13	12.67	16.25	15.83
8	15	14.67	18.75	18.33
9	17	16.67	21.25	20.83
10	19	18.67	23.75	23.33
11	21	20.67	26.25	25.83
12	23	22.67	28.75	28.33

5 by 7 in. in size are generally used around chimneys and small openings. For valleys (and sometimes for ridges), rolls of metal or asphalt roofing can be used. The cost of metal flashing will depend upon the material, gage, width, and length. Valley strips are usually 12 to 18 in. wide. Ridge strips may be 9 or 12 in. wide. Cost of tinned-steel flashing shingles 7 by 9 in. may vary from about \$3 to \$6 per 100. Cost of copper flashing shingles or strips may vary from 60 cents to \$1 per lb. Steel

valley, 14 to 18 in. wide, 28-gage galvanized, comes in rolls about 25 or 50 ft long. The 25-ft rolls may cost $3 to $6 per roll, depending on width and gage. Asphalt valley strips come in rolls about 18 in. wide and 36 ft long and cost about $3 per roll. Asphalt ridge strips come in rolls about 9 in. wide and 36 ft long and cost about $1.50 per roll.

Roofing Trim Roofing trim such as ridge strips, gutters, and downspouts are usually estimated per lineal foot. Ridge ends (finials), end caps, eaves trough corners, elbows, shoes, drop outlets, cutoffs, etc., are usually estimated by the piece, other details being given. Hooks, hangers, etc., are listed by the dozen. Approximate prices (1967) are as follows:

Ridge strips, 26- to 28-gage galvanized steel	$1.50–$2.60 per 10 lin ft
Finials	$0.50–$1.50 each
Gutters, 26- to 28-gage, galvanized steel, 3.5–6 in	$1.50–$3.00 per 10 lin ft
Rain pipe, 26- to 28-gage, galvanized steel, 3–4 in	$1.50–$3.00 per 10 lin ft
Drop outlets, elbows, shoes, corners, etc	$0.50–$2.00 each
Rain-pipe cutoffs	$2.00–$4.00 each

These prices are approximate; hence, dealers' price lists should be consulted when preparing estimates. Special work should be estimated separately by a competent tinsmith.

Sheet-metal Work Frequently most of the items listed under roofing trim are classed as sheet-metal work. These items are purchased ready made from the dealers, and the contractor erects them. On some jobs, the sheet-metal work (providing of the materials and their installation) is let as a subcontract. For estimates on special work such as cornices, skylights, ventilators, and windows, the estimator should consult the manufacturer or dealer and secure his prices.

Labor The cost of labor for roofing and flashing will depend upon kind of labor (roofers, carpenters, tinsmiths, painters, handy men, helpers, etc.) used, their hourly output, and hourly wages. The hourly output will depend on the particular job (cutting and fitting required, materials to be used, etc.) and upon the skill and inclination of the workers. Labor wages may vary from $3 to $5 per hour for unskilled labor and from $4 to $7 per hour for skilled labor. Payroll taxes and insurance may vary from about 10 to 18 percent of the labor wage.

For roofing work, the men should be organized in gangs consisting of both skilled laborers and helpers (if local rules permit the use of helpers). If a gang consists of three or more men, one should be the straw boss, or foreman. The number of men in the gang will vary with the amount of roofing and kind of material. For ordinary shingling or the laying of asphalt roofing, a gang may consist of two to five men. For built-up or composition roofing, one or two men may be needed on the ground for about every two men on the roof. Enough helpers should be available to do the work on the ground and to keep the roof men supplied with materials. The proportion may vary from about one helper for four roofers to one helper for one roofer. If the gang is large (say five or more men) or if two or more gangs are used on the job, a foreman should be provided to supervise the work and keep the men busy.

Tables 10-7 and 10-8 give approximate labor required for different kinds of roofing and flashing work. The labor-hours given refer to work on buildings of about two or three stories or less in height. For multistoried buildings, additional labor time must be allowed for hoisting materials to the roof. From 0.2 to 0.5 additional labor-hour per square per

Table 10-7 *Approximate Labor-hours Required per Square for Applying Roofing Materials*

Roofing materials	Hours per square	
	Simple roofs	Complex roofs
Shingles, wood, single	2–6	5–10
Asphalt, single	2–6	5–10
Asphalt, 3 and 4 strip	1–4	3–6
Asbestos, single	3–8	5–12
Slate, single	3–8	5–12
Metal, single	3–6	5–10
Tile, clay	4–10	6–16
Metal	3–9	6–15
Asphalt roll roofing	0.5–2	1–3
Canvas, laying canvas	1.5–2.5	
Painting, per coat	0.3–1.0	
Composition felt and tar or asphalt:		
2-ply and 2-coat	0.8–1.2	
2-ply and 3-coat	1.0–1.5	
3-ply and 3-coat	1.2–1.8	
3-ply and 4-coat	1.4–2.1	
4-ply and 4-coat	1.6–2.4	
4-ply and 5-coat	1.8–2.7	
5-ply and 5-coat	2.0–3.0	
1 coat gravel or slag	0.2–0.4	
1 coat tar or asphalt mopped on	0.2–0.3	
Metal, tinned, soldered:		
14 × 20	6–9	8–14
20 × 28	4–7	6–10
Tinned, seamed:		
14 × 20	5–7	6–12
20 × 28	3–6	5–9
Crimped and corrugated on wood sheathing	0.5–1.5	1–3
On metal frame	1.0–2.5	2–5
Roofing paper and felt	0.3–1.0	1–2

Table 10-8 *Approximate Labor-hours Required for Roof Flashing and Trim*

Kind of work	Unit of measurement	Labor hours
Ridges	100 lin ft	1–3
Valleys	100 lin ft	1–3
Flashing	100 lin ft	3–10
Flashing	100 shingles or pieces	3–8
Gutters	100 lin ft	2–5
Downspouts	100 lin ft	2–5
Miscellaneous items	Each	0.1–1.0

story may be required, depending on the materials used and hoisting equipment available. If materials have to be carried by the men, 0.4 to 1.0 additional labor-hour per square per story may be needed.

When the sheet-metal work is let as a subcontract, the estimator can use the subcontractor's figures and will not need to make his own estimate of materials and labor for this work.

Diagram 10-6 (Appendix F) can be used for estimating labor costs of roofing per square or per 100 lin ft.

Equipment The equipment required will vary with the kind of roofing and the conditions on the particular job.

Shingling. Hand tools, strips for footholds, ladders, and sometimes hoists and scaffolds.

Tile. About the same as for shingling job.

Asphalt roll. Hand tools, including brushes, ladders, sometimes special shoes with rubber or nonskid soles, and sometimes hoists and scaffolds.

Canvas. Hand tools, ladders, painting tools.

Composition. Hand tools, brushes, heaters, pails, ladders, and sometimes hoists.

Metal. Hand tools including metal shears, special tools for bending seams, soldering outfits, ladders, and sometimes scaffolds and hoists. Corrugated and crimped roofing will not require seaming or soldering tools.

Roofing paper and felt. Hand tools, ladders, and sometimes scaffolds and hoists.

The cost of equipment can be estimated by the job or by the square. Transportation, depreciation, and replacement of hand tools must be considered. If heaters are used, fuel for them is often included with the equipment costs. If hoists are required, their costs must be included. A simple beam hoist (beam, pulleys, blocks, tackle, etc.) may not cost over a dollar or so per job. However, if a power hoist is required, the cost may vary from a few dollars for the job up to a cost of $1 or more per hour, plus transportation and operator. Scaffolding costs may vary from a few dollars and up according to the job. Sometimes scaffolding erected for the framing is available for the roofing. Each job must be considered by itself when the equipment requirements and costs are estimated.

Overhead and Profit Overhead costs on roofing jobs will usually be comparatively high because of the variables present. The overhead costs may vary from about 8 to 25 percent of the labor costs or from about 5 to 15 percent of the sum of material and labor costs.

Profit may range from about 8 to 20 percent of all other costs.

Roofing and Flashing Estimates The estimate for a complete roofing job should include the following:

Materials. All roofing materials including roofing, fastening, paper, felt, paint, flashing, and trim that are required for the particular job

Labor. All labor required for installing all the roofing materials required for the job

Equipment. All equipment needed for the job, including any special tools or equipment that may be required

Overhead. Based on labor costs or on sum of material and labor costs

Profit

The summary or tabulation should show unit costs per square as well as total costs.

	Cost per square	Total cost
Materials.	$	$
Labor.		
Equipment.		
Overhead.		
Profit.		
Total.	$	$

Six illustrative estimates. These six illustrative estimates show how estimates may be prepared for wood-shingle, asphalt-strip shingle, asphalt-roll, corrugated-steel, three-ply asphalt composition, and four-ply tar composition roofs.

Wood-shingle Roof. Prepare a cost estimate of covering a residence roof with wooden shingles. Ground size of residence is 26 by 36 ft. Roof ridge runs lengthwise of roof. Slope of roof is 3 vertical to 4 horizontal. Eaves project 1 ft. There is a chimney at one end of roof, requiring about 7 lin ft of flashing. Galvanized eaves troughs are to be provided along eaves at sides. There are to be four downspouts. Cedar shingles (No. 1 red cedar) are to be used laid with 4.5-in. exposure on wooden roof sheathing.

All materials are to be purchased at local dealers at dealers' prices.

Labor wage is \$5 per hour for shinglers, plus 12 percent for payroll taxes and insurance.

Roof area. Length 36 + 2 (eaves) = 38 ft. Width (allowing for incline) is $[26 + 2 \text{ (eaves)}] \dfrac{(3^2 + 4^2)^{1/2}}{4} = 35$ ft. Area = 1,330 sq ft, or 13 3 sq.

Materials: From Table 10-2, 800 shingles are needed per square. Allowing 10 percent for waste, double layer at eaves, and fitting around chimney, gives 880 shingles per square.

Number of bundles of 250 each = 880 × 13.3 ÷ 250 = 46.8 bundles, buy 47 bundles at dealer's price of \$6.75 delivered = \$317

Nails, say 4 lb per square × 13.3 sq × \$0.22 per lb = 12

Flashing around chimney, about 25 flashing shingles at about \$0.03 each, less than 1

Metal ridge, about 38 ft at \$2 per 10 ft plus \$2 for finials, and nails, about 10

No valleys are needed

Roofing materials = \$340

Eaves troughs, 2 × 38 = 76 lin ft at \$2.60 per 10 ft = 20

Allow \$2 for end caps, \$4 for 4 drop outlets, and \$3 for hangers = 9

Downspouts, 4 (one at each corner) × 20 = 80 ft at \$2.40 per 10 ft = 27

Allow \$10 for 1 doz. elbows and necessary hangers = 10

Eaves, troughs, and downspouts = \$ 66

Total materials = \$406

Labor: For a simple roof of this type, about 4 hr of labor per square should be sufficient to do careful work including time for flashing around chimney and placing metal ridge.

Cost of labor = \$5 × 1.12 × 4 hr per square × 13.3 squares = \$300

For this type of work, about 4.3 hr per 100 lin ft of eaves troughs and downspouts should be sufficient.

$$\$5 \times 1.12 \times 4.3 \text{ hr.} \times \left(\frac{76 + 80}{100}\right) = 38$$

Total labor cost = \$338

Overhead costs are assumed at 12% of all labor costs = \$ 41

Equipment costs will be low for a job of this size.

About \$7 will be assumed for roofing and \$3 for eaves, troughs, and downspouts = \$ 10

Profit: Taken as 10% of all other costs = 10% of \$795 = \$ 80

Total estimate = \$875

This estimate can be tabulated as in Table 10-9, using 13.3 squares of roofing and 232 lin ft of trim.

Usually, this estimate would be divided into one for the roofing and another for the eaves troughs and downspouts, as indicated in the tabulation.

Diagrams 10-5 and 10-6 can be used for checking material costs for shingles and labor costs per square if desired.

Table 10-9

Item	Eaves, troughs, and downspouts 156 lin ft		All roofing 13.3 squares		
	Per 100 lin ft	Total	Per square	Total	Total
Material	$42	$ 66	$26	$340	$406
Labor	24	38	22	300	338
Equipment	2	3	1	7	10
Overhead	3	5	3	36	41
Profit	8	12	5	68	80
Total	$79	$124	$57	$751	$875

Asphalt-strip Shingle Roof. Prepare a cost estimate of covering the residence roof in the preceding illustrative estimate, using a good grade of asphalt-strip shingles (three tab) and placing a layer of roofing paper between the wooden sheathing and the shingles. Cost of eaves troughs and downspouts will be the same as before and will not be repeated here.

Materials: Area = 1,330 sq ft, or 13.3 squares. Price of shingles of grade selected by owner is $4 per bundle covering 33⅓ sq ft, or $12 per square. Bundles required = 13.3 × 3 = 39.9, or 40.

Allowing 7% for waste, fitting, and 2 layers at eaves, bundles required will be 43. $4 × 43	= $172
Nails at about 4 lb per square = 4 × 13.3 × $0.25	= 13
Flashing around chimney, say	= 1
Metal ridge, as before	= 10
Roofing felt, 13.3 squares will take 4 rolls of 432 sq ft each at $3 per roll for the 15-lb weight	= 12
Tacks for the felt, at $0.30 per square	= 4
Roofing materials	= $212

Labor: About 2 hr of labor per square should be sufficient for laying roofing paper, strip shingles, flashing and metal ridge.

Labor cost = $5 × 1.12 × 2 × 13.3	= $150
Overhead at 12 percent of labor cost	= 18
Equipment, about	= 7
Profit at 10% of all other cost	= 39
Total cost	= $426

This can be tabulated as follows, using 13.3 squares of roof:

Item	Cost per square	Total roofing cost
Materials	$16	$212
Labor	11	150
Equipment	1	7
Overhead	1	18
Profit	3	39
Total	$32	$426

If desired, Diagram 10-2 can be used for checking shingle costs per square and Diagram 10-6 for labor costs per square.

Asphalt-roll Roof. Prepare a cost estimate for covering a barn roof measuring 44 by 96 ft over all with 65-lb mica-finished asphalt roofing laid directly on wooden sheathing. Cost of roofing per roll with nails and roofing cement included is $3.65 per roll of 108 sq ft delivered at the job.

Materials: Roof area = 44 × 96 = 4,224 sq ft, or 42.24 squares. Allow about 4 percent for end laps, extra lap at ridge, and for edges and waste. This gives 42.24 × 1.04, or about 44 sq.

Cost of materials = $3.65 × 44 = $161

Labor: This material may be laid quite rapidly on a simple roof. About 1 hr per square should be sufficient, or 42.25 labor-hours. With roofers at $4 per hour, plus 12 percent for payroll taxes and insurance, the labor cost = $4 × 1.12 × 42.25 = $189

Equipment cost will be small, say = 4

Overhead, at 12% of labor cost = 23

Profit, say 10% of all other costs = 38

Total cost = $415

This can be summarized as follows for 42.25 squares:

Item	Cost per square	Total cost
Materials	$3.80	$161
Labor	4.50	189
Equipment	0.10	4
Overhead	0.55	23
Profit	0.90	38
Total	$9.85	$415

If desired, Diagram 10-1 can be used for checking the materials cost per square and Diagram 10-6 the labor costs.

Corrugated-steel Roof. Prepare a cost estimate for covering the barn roof of the preceding illustrative estimate with corrugated galvanized-steel roofing of

No. 28 gauge and 27.5 in. wide. The roof is 96 ft long and has a central ridge running lengthwise. No. 28 gage corrugated galvanized roofing 27.5 in. wide costs 58 cents per lineal foot in standard lengths, delivered at the job. Lead-headed nails 1.75 in. long cost 40 cents per pound. Galvanized ridge costs $1 per 10 lin ft. Finials or ends cost $1 each.

Materials: The length of roof from eaves to ridge is 44 ÷ 2, or 22 ft. This length will take one 12-ft length and one 11-ft length, will allow for more than enough end lap, and will permit the ends to extend an inch or so beyond the sheathing at the eaves.

Number of 12-ft strips = (96 ÷ 2 ft net width) × 2 sides = 96	
Number of 11-ft strips = (96 ÷ 2 ft net width) × 2 sides = 96	
Total lineal feet of corrugated steel = 96 × 12 + 96 × 11 = 2,208 ft	
Cost of corrugated-steel roofing = $0.58 × 2,208	= $1,280
Length of ridge = 96 ft. Use 10 strips 10 ft long, allowing for laps. Use 2 ends or finials. Cost = $2 × 10 + $1 × 2	$ 22
Nails, say 50 lb at $0.40	= 20
Total materials	= $1,322

Labor: Labor may be estimated per sheet, per square, or per 100 lin ft. About 1 hr per square or about 2 hr per 100 lin ft should be sufficient, as no cutting or fitting is required.

2,208 lin ft at 2 hr per hundred gives 44 hr
Add 1 hr for ridge. Total labor = 45 hr

This job could be handled by 2 metalworkers. Wage is $4 per hour plus 12 percent for payroll taxes and insurance.

Labor cost = $4 × 1.12 × 45 hr	= $ 202
Equipment costs for this job should be low, say	= 6
Overhead at 12% of labor costs	= 24
Profit at 10% of other costs	= 156
Total cost	= $1,710

This estimate can be itemized as follows for 42.25 squares:

Item	Cost per square	Total cost
Materials	$31.30	$1,322
Labor	4.80	202
Equipment	0.15	6
Overhead	0.55	24
Profit	3.70	156
Total	$40.50	$1,710

If desired, this estimate could have been prepared by using materials and labor required per square.

Three-ply Asphalt Composition Roof. Prepare an estimate for a three-ply felt and asphalt roof over a concrete roof deck measuring 40 by 101 ft. The concrete deck has a slope of ½ in. per foot with no obstacles except a brick parapet wall along the two sides and one end. There are to be two layers of 15 lb per square asphalt-saturated roofing felt and a top layer of 90 lb per square mineral-surfaced top sheet. There will be two coatings of hot asphalt, well mopped on, of about 25 lb per square each. The concrete deck will be given a coat of asphalt primer paint.

The prices of materials delivered at the job are:

Asphalt 15-lb roofing felt........................	\$2.60 per roll of 324 sq ft
Asphalt 90-lb mineral surfaced roofing............	\$4.25 per roll of 108 sq ft
Asphalt (West Coast price)......................	\$2.25 per 100 lb
Asphalt primer paint............................	\$1.00 per gal

Labor, including payroll taxes and insurance is \$5.50 per hour.
Equipment cost for the job, estimated at \$14.
Overhead is 14 percent of the labor cost.
Profit is 8 percent of all other costs.

Materials: Roof area, allowing 1 ft for flashing along one end and each side = (40 + 2) × (101 + 1) = 4,284 sq ft, say 43 squares.

Asphalt primer, ½ gal per square = \$1 × 21.5 gal.................. =	\$ 22
Asphalt, 2 coats at 25 lb per square each = 2 × 25 × 43 × \$2.25.... =	48
Asphalt 15-lb felt, 2 layers × 43 squares at ⅓ roll per square = 28.67 rolls, say 29. \$2.60 × 29.................................... =	76
Asphalt 90 lb roofing, \$4.25 × 43 squares.......................... =	183
Total materials.. =	\$329

Labor: At about 1.1 hr per square, say 48 hr. \$5.50 × 48............. = \$264

Equipment:.. =	\$ 14
Overhead: 14% of the labor cost................................ =	37
Profit: 8% of all other costs.................................. =	52
Total estimated cost.. =	\$696

The summarized costs for 43 squares are as follows:

Item	Cost per square	Total cost
Materials..............	\$ 7.65	\$329
Labor..................	6.10	264
Equipment............	0.35	14
Overhead..............	0.85	37
Profit..................	1.20	52
Total...............	\$16.15	\$696

Four-ply Tar Composition Roof. Prepare a cost estimate for a four-ply tar and felt roof for the roof of the preceding illustrative estimate, using 25 lb per square of tar per coat, except 50 lb of tar and 250 lb of gravel per square for the top coat, and felt weighing about 15 lb per square. Price of tar delivered at the job is \$50 per ton and \$5 per ton for gravel. Other prices are the same.

Materials:

Roof area with flashing = 43 squares		
Roofing felt, 4 plies × 43 squares × 1/3 roll per square = 57.3 rolls, say 58. \$2.60 × 58	= \$151	
Asphalt primer, 1/2 gal per square at \$1 per gallon	= 22	
Tar, 4 coats at 25 lb per coat and one 50-lb coat = 150 lb per square × 43 squares = 6,450 lb at \$50 per ton	= 161	
Gravel, 5.40 tons at \$5 per ton	= 27	
Total materials		= \$361
Labor: Assume 1.85 hr per square for 43 squares. Or 80 hr at \$5.50 per hour		= \$440
Equipment for the job		= 14
Overhead at 14% of the labor cost		= 62
Profit at 8% of all other costs		= 70
Total estimated cost		= \$947

Summarized costs for 27.25 squares are:

Item	Cost per square	Total cost
Materials	\$ 8.40	\$361
Labor	10.20	440
Equipment	0.30	14
Overhead	1.45	62
Profit	1.65	70
Total	\$22.00	\$947

Diagrams 10-3, 10-4, and 10-6 can be used for checking the material and labor costs per square if desired.

CHAPTER ELEVEN

Lathing, Plastering, and Drywall Construction

Lathing

Lathing Lathing estimates include materials, labor, equipment, overhead, and profit. The unit of measurement for lathing work is usually the square yard (9 sq ft), though in some instances such units as the square foot, 100, or 1,000 sq ft of surface area; the square yard or 100 sq yd of surface area; 100 or 1,000 wood lath; 1 or 100 sq ft, one piece, or 100 lb of wire lath; one piece or 100 sq ft of surface area of plaster board have been used. Furring, corner bead, etc., are usually estimated by the lineal foot, 100 lin ft being common. When preparing lathing estimates, the estimator should be familiar with the local trade practices.

Materials

Wood Lath. Wood lath vary in price according to time, locality, kind of wood, and width and length of lath. The common size is 3/8 in. thick, 1½ in. wide (actual width is 1 3/8 in.), and 48 in. long. Lath are usually placed 3/8 in. apart so that a 1 3/8-in. lath covers 1¾ in. As the lathing area is usually "taken off" in square yards, the number of lath required per square yard must be known. Wood lath are usually sold by the thousand or in bundles of 50 or 100 lath. Prices vary from about $25 to $90 per 1,000 lath, depending upon the kind of wood, size, and quality. When estimates are prepared, prices should be obtained from the dealers from whom the lath is to be purchased. If the net yardage (all openings deducted) is given in the takeoff, from 5 to 10 percent should be added for waste. If the small openings are not deducted, no allowance is needed for waste. Wood lath are rarely used.

The nails are usually 3d (3-penny) weight, and the blued finish is preferred. These nails will average about 600 to the pound and may cost from $16 to $25 per 100 lb.

Plasterboard. Plasterboard or gypsum board lath comes in various sizes and thicknesses. Gypsum board lath may be perforated or plain, and may or may not have a foil back. Common thicknesses are ¼, ⅜, and ½ in., widths are 18, 24, 32, 36, and 48 in., and lengths are from 2 to 12 ft. The width or length should be such that the boards will fit 12- and 16-in. nailing centers without cutting and waste. The size should be such that one board can be readily applied by one man. Weights may vary from 6 to 25 lb per square yard, depending on material and thickness. If net areas are given in the takeoff, 5 to 15 percent should be allowed for waste. The waste will increase with the number of openings and jogs to be framed.

Plasterboard may be sold by the board (thickness, length, and width specified) or by 1,000 sq ft with thickness given. Prices per 1,000 sq ft vary from about $35 to $95, depending on thickness and quality. Finishing nails about 3d (3-penny) weight are often used. From 0.15 to 0.18 lb of nails will be required per square yard for nailing centers 16 in. apart, and from 0.20 to 0.24 lb of nails per square yard for 12-in. nailing centers, with allowance for waste.

Estimates for plasterboard are usually based on the square yard, though the square foot, square of 100 sq ft, or 1,000 sq ft can be used.

Metal Lath. The general function of metal lath is to form a support, background, or base on which plaster or stucco may be applied. Metal lath can also be used for other purposes, such as lockers, register faces, and concrete reinforcement.

All metal lath is furnished painted, asphaltum dipped, or galvanized. The galvanizing may precede or follow the manufacturing process.

There are many kinds of metal lath. The two kinds in common use for lathing and plastering are the ordinary expanded metal lath (often diamond mesh), which is applied on wood or steel furring, beams, or studs, and the ribbed or self-furring lath of expanded metal, which can be used on flat surfaces without furring. The painted copper alloy and the galvanized finishes are preferred.

Estimating the cost of the metal lath is not difficult, since it is only necessary to take off the actual areas of surfaces to be covered and to make allowances for lapping and waste. Metal lath should be lapped about 1 in.

Allowances for waste and lapping are as follows:

Plain surfaces, add 8 to 10 percent
Beams, pilasters, etc., add 12 to 16 percent
Cornices, add 20 to 25 percent
Domed or groin work, add 25 to 30 percent

Price quotations can be readily obtained on all kinds of metal lath, and it is only necessary to multiply the quantity by the price (delivered at the job) to obtain the material costs.

The price of metal lath is usually quoted per 100 sq yd with the weight per square yard mentioned. Ordinary weight for interior work is about 2.5 lb per square yard. Heavier lath, weighing up to 3.4 lb per square yard, can be used for exterior work. Average range of prices may be from $45 to $75 per 100 sq yd for the 2.5 lb per square yard weight and from $55 to $90 per 100 sq yd for the 3.4 lb per square yard weight, depending upon whether the lath is painted or galvanized and upon the locality.

The metal lath can be applied to wooden studs and beams, to wood furring strips, or to metal furring strips. If furring is required, this is usually estimated separately.

For ordinary metal lath applied to wood, about 1 lb of nails or staples will be required for each 15 to 20 sq yd of lath. The nails should be about 1 in. long and should have large flat heads, say about ½ in. in diameter. Nails should be spaced about every 6 in. Some constructors tie the sheets of metal lath about every 6 or 8 in. along the laps, using a galvanized wire of about No. 18 gage.

Galvanized metal lath of the heavier weight should be used for exterior plastering and stuccowork.

When the expanded metal lath is carefully bent around corners and all joints (especially at corners) are lapped and wire tied, no corner reinforcement is required.

In the manufacture of expanded metal lath, the sheets are usually cut to some exact size, though sometimes the sheets are made or cut a little large to allow for laps of about 1 in. The length of the sheets is 8 ft, and the widths vary.

Galvanized woven wire (stucco mesh or stucco netting) is often used for exterior stuccowork. This woven wire comes in rolls 3 ft wide and 150 ft long, of 1- and 1.5-in. mesh, and of 17, 18, or 20 gage. Approximate prices (1967) are:

17 gage, 1½-in. mesh, 3- by 150-ft roll, per roll..........	\$12–\$18
18 gage, 1-in. mesh, 3- by 150-ft roll, per roll............	\$18–\$25
20 gage, 1-in. mesh, 3- by 150-ft roll, per roll............	\$10–\$14

Galvanized or blued finished nails with felt washers are usually used for fastening the stucco netting to the studs or sheathing. These nails may cost from 20 to 35 cents per pound.

For exterior stuccowork, a layer of heavy building paper, 15- or 30-lb felt, should be placed between the stucco netting and the sheathing or studs. A roll (324 sq ft) of 15-lb felt may cost about \$2 to \$3.50, and a roll (216 sq ft) of 30-lb felt may cost about \$3 to \$4.50.

When there is no sheathing on the exterior walls, the walls should be first covered with galvanized wire (No. 18 gage) spaced not over 6 in. on centers horizontally and stretched taut. This wire forms a base or backing for the building paper.

Furring. Wooden or metal furring may be required when lath is to be placed on brick or concrete walls. Wooden furring usually consists of 1- by 2-in. strips of pine or spruce spaced 12 or 16 in. on centers. Both wood and metal lath can be applied to wood furring. The wooden furring is usually estimated per 100 lin ft, about 11 lin ft with 12-in. spacing, or 8 lin ft with 16-in. spacing, being required per square yard of lath, with allowance for waste.

Metal furring may consist of ¾-in. steel channels fastened to the masonry or other surface by hangers, bars, or plugs, as the case may be. From 5 to 15 lb of metal furring may be required for each square yard of metal lath, depending on the amount and complexity of the framing. About 10 percent should be allowed for waste.

Grounds. Grounds are wooden strips, usually about 1, 2, or 3 in. in width and of varying thicknesses, depending on the thickness desired for the lath and plaster. Common thicknesses of grounds are about as shown in Table 11-1.

Table 11-1 *Thickness of Grounds*

Item	Plaster, coats	Grounds, in.
Wood lath...........	2	⅝, ¾, ⅞
	3	¾, ⅞, 1
Metal lath...........	3	⅝, ¾
Brick and tile........	2	⅝
	3	⅝, ¾
Plaster board........	2	⅝ to 1
	3	¾ to 1¼

Grounds are carefully placed around all openings and along the bottom of walls near the floor so as to serve as a guide for the surface of the plaster. Grounds are placed directly on the framing in wood construction, and the lath or plaster board butts up against the grounds. Sometimes the grounds are placed over the plaster board. Grounds are usually placed over the brick or tile in this type of construction.

Corner Beads. Corner beads are frequently used to protect all corners and edges. Metal corner bead is usually priced per 100 lin ft. Standard corner bead comes in 6, 8, or 10 ft lengths and may cost $5 to $10 per 100 lin ft. Expansion corner bead may cost $8 to $13 per 100 lin ft.

Corner Reinforcement. Corner reinforcement, made of strips of expanded metal lath may cost $5 to $9 per 100 lin ft. Corner reinforcement is usually required when wood lath or plaster board is used. It is not required with expanded metal lath when this lath is bent or lapped and wire tied at the corners.

Diagrams. Diagrams 11-1 and 11-2, (Appendix F) can be used for estimating lathing materials. The diagrams for plaster board include a reasonable allowance for nails.

Labor Labor costs of lathing will depend upon the time in hours required per square yard and the hourly labor wage. Hourly wages for lathers may vary from about $4 to $7. Frequently, the labor cost of lathing is based on the square yard, allowing a certain price per square yard for ordinary work with extras for special work. Payroll taxes and insurance may vary from 8 to 15 percent of the labor wage.

All lath are usually applied by skilled lathers. The amount of lath applied per hour will depend on the skill of the lathers and on the particular job.

When the labor cost of lathing is based on the area or on the amount of lath, it is necessary to know how the yardage or amount of lath is computed. Trade practices vary in different localities. The measurement of yardage may be based upon the following:

1. *Gross area.* No deductions for openings.
2. *Net area.* All openings deducted.
3. *Gross area less certain deductions.* For example, in a certain locality no opening of less than 2 sq yd is deducted. In another locality, openings less than 2 ft wide are not deducted. In another location, half of the openings larger than 2 sq yd are deducted.
4. *Flat and curved work.* Curved work is often estimated to cost twice as much as flat work.

It is necessary for an estimator to study the practices in the place in which the job is located and to govern himself accordingly when the work is to be sublet or paid for on a unit-price basis.

If the contractor intends to employ the lathers himself and wishes to estimate accordingly, he should consider the actual net areas and make allowances for corners and other items.

There are so many kinds of metal lath that it is almost impossible to give accurate information in regard to the labor time required for applying every kind of metal lath under all the different conditions. The labor of applying metal lath will vary with the difficulty of handling the materials, depending on the kind of lath and the gauge or weight.

Wood furring and grounds are usually placed by carpenters. Metal furring and grounds can be placed by carpenters or lathers. Lathers usually place the corner bead and corner reinforcement.

Tables 11-2 to 11-4 give approximate labor-hours required for applying different kinds of lath and furring.

Table 11-2 *Labor Required for Applying Plasterboard*

Nailing centers, in.	Hr per 1,000 sq ft	Sq ft per hr	Sq yd per hr	Hr per 100 sq yd
12	6–11	90–160	10–18	6–10
16	5–9	110–180	12–20	5–8

Diagrams. Diagram 11-3 (Appendix F) can be used for estimating labor costs of lathing, and Diagram 11-4 (Appendix F) for labor costs of placing, furring, corner beads, and reinforcement when reasonable rates of work can be assumed.

Equipment Wooden horses, platforms, plank, and scaffolding are required on all lathing jobs. The proportionate cost (depreciation and repairs) of this equipment should be charged to the job. Transportation costs to and from the job must be included. If scaffolding is to be erected and taken down by other labor, this labor cost must be included. For a small lathing job, like that for a residence, probably an allowance of $5 for the material and $10 for trucking would be enough. For metal lath and furring, metal punching shears and perhaps bending and other special tools will be needed. The costs of such tools should be included in the estimate.

Overhead and Profit Overhead on lathing jobs is usually based on labor costs and may vary from about 10 to 25 percent of labor costs or from 5 to 12 percent of the cost of materials and labor.

Profit may range from about 5 to 15 percent of all other costs.

Lathing Estimates The estimate for the total cost of lathing should include the following:

Materials. Lath, nails, and staples. Plaster board and nails. Corner beads and corner reinforcement. Furring, if this must be provided. Proper allowances must be made for waste.

Table 11-3 *Labor Required for Applying Metal Lath*

Kind of work	Sq yd per hr		Hr per 100 sq yd	
	12-in. nailing centers	16-in. nailing centers	12-in. nailing centers	16-in. nailing centers
On wood furring, studs, or sheathing:				
Walls and partitions	7–10	9–13	10–15	8–11
Columns	3–4	4–5	24–32	20–25
Flat ceilings	5–9	8–12	10–18	8–12
Paneled, arched, or groined ceilings	5–7	6–10	14–20	10–16
Beams and girders	3–4	4–5	24–32	18–26
Simple coves and cornices	3–5	4–7	21–30	15–24
Complex cornices	3–4	3–5	25–35	20–30
Stucco netting, exterior walls	11–17	14–20	6–9	5–7
On metal furring:				
Walls and partitions	5–10	8–11	10–18	9–13
Columns	3–4	3–5	25–35	20–30
Flat ceilings	5–8	7–10	12–20	10–15
Paneled, arched, or groined ceilings	4–7	5–8	15–24	12–18
Beams and girders	3–4	4–5	24–32	20–25
Coves and cornices	3–5	4–7	22–30	15–24
Complex cornices	3–4	3–5	25–35	20–30

Table 11-4 *Labor Required for Applying Furring and Grounds*

Kind of work	Hr per 100 lin ft	Hours per 100 lb of metal	Hr per 100 sq yd	
			12-in. nailing centers	16-in. nailing centers
Wood furring on wood	1.0–1.6	....	9–18	7–14
Wood furring on brick or tile	1.5–3.0	....	16–35	12–25
Metal furring, alone (¾-in. channel)	1.5–3.0	3–6	16–35	12–25
Metal furring, with ties and bars	3.0–5.0	8–12	35–55	25–40
Wood grounds	1.5–4.0			
Metal corner beads	3.0–5.0			
Metal corner reinforcement	2.0–3.5			

Labor. All labor including helpers and foreman (if provided).

Equipment. All equipment costs.

Overhead. Based on labor costs or on the sum of material and labor costs.

Profit.

Unit costs per square yard should also be computed based on the actual net yardage. Unit costs per square of 100 sq ft can be computed if desired.

Illustrative estimate. Prepare a lathing estimate (1) with plaster board, and (2) with wire lath for the downstairs rooms of a residence. The dimensions of the rooms and openings are as follows:

Living room:

12 by 20 by 8.5 ft high
2 triple windows, 5 by 8 ft each
2 windows, 3 by 5 ft each
1 double door, 5 by 7 ft

Dining room:

12 by 14 by 8.5 ft high
1 triple window, 5 by 8 ft
1 double door, 5 by 7 ft
1 single door, 3 by 7 ft

Kitchen:

10 by 12 by 8.5 ft high
1 double window, 5 by 5 ft
2 doors, 3 by 7 ft each

Hall:

8 by 10 by 8.5 ft high
2 double doors, 5 by 7 ft each
1 outside door, 3.5 by 7 ft
1 closet door, 2.5 by 7 ft

Closet:

4 by 4 by 8.5 ft high
1 door, 2.5 by 7 ft

The following data are assumed:

Wage of lathers is $5 per hour, plus 10% for payroll taxes and insurance
An allowance of $15 for equipment
An allowance of 12% of labor costs for overhead
An allowance of 10% of all other costs for profit
Grounds were placed by carpenters
Nailing centers are of wood and are spaced 16 in. on centers
No furring is required. No corner bead required
Corner reinforcement required for all corners for plasterboard

Prices of materials are as follows:

Gypsum board lath, ½ in., $40 per 1,000 sq ft

Wire lath, 2.5 lb per square yard, black painted, \$50 per 100 sq yd
Corner reinforcement expanded metal lath, \$5 per 100 lin ft
Nails \$0.25 per pound

Results wanted are the costs of materials, labor, equipment, overhead, profit, and totals per square yard and for the job.

Solution: Yardage equals

Room	Walls		Ceil-ing		Open-ings		Net
Living room..........	60.5	+	26.7	−	16.1	=	71.1
Dining room..........	49.1	+	18.7	−	10.7	=	57.1
Kitchen..............	41.6	+	13.3	−	7.5	=	47.4
Hall.................	34.0	+	8.9	−	12.4	=	30.5
Closet...............	15.1	+	1.8	−	2.0	=	14.9
Total................						=	221.0

1. Plasterboard, 221 sq yd = 1,989 sq ft

Adding 10% for waste gives about 2,200 sq ft	
Cost of plaster board at \$40 per 1,000 sq ft	= \$ 88
Nails, about 37 lb at \$0.25	= 9
Corner reinforcement, about 400 lin ft	= 20
Total materials	= \$117
Labor, say 6 hr per 100 sq yd will require 13 hr	
Labor applying corner reinforcement, say 7 hr	
Total labor = 20 hr at \$5 × 1.10	= \$110
Equipment, as stated	= 15
Overhead, 12% of \$110	= 13
Profit, 10% of \$255	= 26
Total	= \$281

This is a unit cost of about \$1.28 per square yard

2. Metal lath, 221 sq yd

Add 12% for waste and laps, giving a total of 247 sq yd	
247 sq yd at \$50 per 100 sq yd	= \$124
Nails, 1 lb for every 15 sq yd, say 15 lb at \$0.25	= 4
Corner reinforcement is not needed	
Total materials	= \$128
Labor, about 8 hr per 100 sq yd of surface, will require 18 hr	
This allowance is a little high to allow for bends and laps at corners.	
Total labor 18 hr at \$5.50	= \$ 99
Equipment, as stated	= 15
Overhead, 12% of \$99	= 12
Profit, 10% of \$254	= 25
Total	= \$279

This is a unit cost of about $1.26 per square yard.

These two estimates may be summarized as follows:

Item	Cost per square yard		Total cost	
	Plaster board	Metal lath	Plaster board	Metal lath
Materials	$0.53	$0.58	$117	$128
Labor	0.50	0.45	110	99
Equipment	0.07	0.07	15	15
Overhead	0.06	0.05	13	12
Profit	0.12	0.11	26	25
Total	$1.28	$1.26	$281	$279

If desired, the unit costs for materials and labor may be found from Diagrams 11-1 to 11-4.

Plastering

Plastering Plastering estimates include materials, labor, equipment, overhead, and profit. The unit of measurement for plastering is usually the square yard (9 sq ft), though sometimes the square foot or square of 100 sq ft is used. The yardage may be based on net area, on gross area, or on gross area less large openings, with allowances for special and difficult work. The unit of measurement for special molding is usually the lineal foot, with width given. When the molding is over 1 ft wide, or when it extends over a considerable area, the square foot or square yard can be selected as a unit. When preparing plastering estimates, the estimator should be familiar with the local trade practice.

Plastering work can be classified as to interior or exterior work, as to the number of coats applied (one, two, or three), as to the materials used, as to the surface finish, or as to the type of surface on which the plaster may be applied.

The cementing materials for interior plaster are usually lime, gypsum, and plaster of paris. For exterior plaster, or "stucco," portland cement or magnesite can be used.

When plaster is applied to lath, three coats (scratch, brown, and finish) are usually required, though in some instances only two coats are used. When it is applied to brick, tile, or plaster board, two coats (brown and finish) are usually sufficient, though in some instances one coat or three coats may be required. Sometimes state or local codes specify the number of coats to be used.

Materials There are a number of plasters and plastering materials on the market, each of varying cost, different covering capacity, and different method of application. There are also a large number of plaster finishes. As it is obviously impractical to attempt to discuss all these in this text, only the more common types will be considered.

Before preparing material estimates for plastering, the estimator should familiarize himself with the various kinds of plaster materials available in his market, their cost, covering capacity, workability, etc.

Interior Plaster. In three-coat work, applied to wood or metal lath, the scratch coat must be sufficient to fill the space between the lath, form clinches back of the lath, and cover the front of the lath. The thickness of the front covering may vary from about $\frac{1}{16}$ to $\frac{1}{8}$ in. for wire lath to $\frac{1}{8}$ to $\frac{1}{4}$ in. for wood lath. When the coat is applied to plasterboard or to brick or tile, the thickness is about $\frac{1}{4}$ in. The brown coat in three-coat work is usually $\frac{1}{4}$, $\frac{5}{16}$, or $\frac{3}{8}$ in. thick, the $\frac{1}{4}$-in. thickness being common. The finish coat may be $\frac{1}{16}$ or $\frac{1}{8}$ in. thick, the $\frac{1}{8}$-in. being preferred.

In two-coat work, the base or scratch coat may vary in thickness from $\frac{1}{4}$ to $\frac{3}{8}$ in. Enough material must be provided to give this thickness and to fill spaces and form clinches when placed on lath. The finish coat may be $\frac{1}{16}$ to $\frac{1}{8}$ in. thick.

The sum of the thicknesses of the plaster coats plus the thickness of the lath or plasterboard must be equal to the thickness of the grounds.

When quantities of plaster are computed for the scratch coat, no allowance can be made for the space occupied by metal lath, as this is offset by the plaster required for clinching, an allowance equal to half the thickness of the wood lath can be made and have enough plaster to fill spaces between the lath and to form clinchers, and an allowance equal to the thickness of the plain plasterboard can be made when this plasterboard butts against the grounds. When perforated plasterboard is used, there must be enough plaster to fill the holes and to form clinchers.

The mix used for the scratch coat is usually a 1-to-2 or a 1-to-3 mix by volume. When patent hard wall, gypsum, and other prepared plasters weighing about 100 pcf are used, a 1-to-2 or a 1-to-3 mix by weight is equally satisfactory. The richer mix (1 to 2) is often preferred on account of its greater strength.

The mix for the brown coat (in three-coat work) may be a 1-to-2 or a 1-to-3 mix by volume or a like mix by weight when the cementing material weighs about 100 pcf. The 1-to-3 mix is preferred by many, though other plasterers prefer to keep the mixes the same for both the scratch and brown coats.

The mixes for the finishing coats vary considerably. For a white lime finish, a volume mix of 2 parts lime to 1 part plaster of paris can be used or a weight mix of about 2 lb hydrated lime to 1 lb plaster of paris. When a gypsum or other prepared plaster and sand finish is used, the proportions are about 1 to 2 either by volume or by weight.

Tables 11-5 and 11-6 give information as to the quantities of material required for various plaster coats of varying thickness for 100 sq yd of plaster surface. If about 6.7 percent is allowed for waste and unevenness of surface, about 5 cu ft of material will be required per 100 sq yd of surface for each 1/16-in. thickness of plaster.

Table 11-5 *Approximate Quantity of Plaster Required per* 100 *Sq Yd of Surface*

Thickness of plaster, in.	1/16	1/8	3/16	1/4	5/16	3/8	1/2	5/8	3/4	7/8	1
Cu ft of plaster	5	10	15	20	25	30	40	50	60	70	80
Cu yd of plaster	0.19	0.37	0.56	0.74	0.93	1.10	1.50	1.85	2.25	2.60	2.95

Table 11-6 *Approximate Quantity of Plaster Required for Filling Spaces in Lath and Providing Clinchers for* 100 *Sq Yd of Surface*

Lath	Size	Cubic feet of plaster	Cubic yards of plaster
Plaster board lath perforated	1/4-in. thick	3.3	0.12
	3/8-in. thick	3.8	0.14
	1/2-in. thick	4.3	0.16
Metal lath	3/16 in. thick	15	0.56
	1/4 in. thick	20	0.74
	5/16 in. thick	25	0.93
	3/8 in. thick	30	1.10

Table 11-7 gives information as to quantities of materials required for plaster mixes. As measurement by weight is more accurate than by volume, the weights are given. Hydrated lime may come in bulk (ton lots), in 50-lb sacks, or in barrels. Gypsum and other plaster cement usually comes in 100-lb sacks. Plastering sand may weigh 2,400 to 2,700 lb per cubic yard. Tabulated values are based on sand weighing about 100 pcf. Hydrated lime, gypsum, and prepared plaster may weigh 50 to 80 pcf.

Plaster cement may be sold by the sack, barrel, cubic foot, or ton. Plaster aggregates may be sold by the sack, cubic foot, ton, or cubic yard.

When materials for interior plaster are estimated, consideration must be given to the number of coats, the thickness of lath or plasterboard, the thickness of the grounds, and the mixes to be used. Then the thickness of each coat can be assumed, the number of cubic feet (or cubic yards) per 100 sq yd of surface for each coat found from Tables 11-5 and 11-6, and the quantities of materials from Table 11-7. Then the various quantities

Table 11-7 *Approximate Quantities of Materials Required per Cubic Foot and per Cubic Yard for Plaster*

Kind of plaster	Mix by wt	Materials	Quantities per cu ft	Quantities per cu yd
Ordinary lime for scratch and brown coats	1–2	Hydrated lime	40–45 lb	1,100–1,200 lb
		Hair and fiber	0.07–0.10 bu	2.0–2.5 bu
		Sand	80–90 lb	2,200–2,400 lb
	1–3	Hydrated lime	30–33 lb	800–900 lb
		Hair and fiber	0.07–0.10 bu	2.0–2.5 bu
		Sand	90–100 lb	2,400–2,700 lb
Gypsum, hard, patent, or prepared for scratch and brown coats	1–2	Plaster cement	40–45 lb	1,100–1,200 lb
		Sand	80–90 lb	2,200–2,400 lb
	1–3	Plaster cement	30–33 lb	800–900 lb
		Sand	90–100 lb	2,400–2,700 lb
White finish	2–1	Finish hydrated lime	40–50 lb	1,100–1,350 lb
		Plaster of paris	20–25 lb	550–700 lb
Gypsum, sanded	1–2	Gypsum	40–45 lb	1,100–1,200 lb
		Sand	80–90 lb	2,200–2,400 lb
Keene's	...	Keene's cement	40–50 lb	1,100–1,350 lb
		Hydrated lime	20–25 lb	550–700 lb
Prepared white finish	...	....	60–75 lb	1,600–2,000 lb

(and their costs) required per 1 or 100 sq yd and the totals for the job can be tabulated.

There are so many kinds, varieties, and designs of ornamental plastering on the market that it is impractical to prepare any tables or diagrams for this kind of work. The quantities of materials can best be estimated after the plans and specifications have been carefully studied. The quantity of plaster can be first computed, and then, the mix and the materials being known, the quantities of the various materials can be calculated.

Diagrams 11-5 and 11-6 (Appendix F) show the costs of materials per 100 sq yd of surface when the prices of the materials are known and the quantities required per cubic foot can be reasonably assumed.

Exterior Plaster or Stucco. There are three kinds of exterior plaster or stucco in common use: lime, portland cement, and magnesite. Lime stucco may not prove satisfactory in cold (freezing) climates, and magnesite in wet climates.

Lime stucco is a mixture of slaked or hydrated lime, sand, and sometimes other ingredients.

Portland cement stucco is a mixture of portland cement and a small amount of lime, sand, and perhaps other ingredients such as grit, crystals, or pebbles.

Magnesite stucco is a mixture of a calcined and ground magnesite rock (magnesium oxide), sand, and other ingredients such as asbestos, grit, mineral colors, crystals, or pebbles. The liquid used is magnesium chloride. Magnesite stucco is sold under various trade names such as Kellastone and Rockbound.

Stucco can be applied directly to masonry such as rock, concrete, brick, or tile walls, or it can be applied to metal lath furred out a short distance from the walls. Wood lath is rarely used, except in dry climates, because of the effect of moisture on the volume of the wood.

Lime stucco is usually applied in two or three coats, no coat being less than ¼ in. thick and the total thickness of the coats being ¾ in. to 1 in. The mixes for the scratch and brown coats may be 1 to 2 or 1 to 3 by weight, and that of the finish coat, 1 to 2 by weight.

Portland cement stucco is usually applied in two or three coats, no coat being less than ¼ in. thick, and the total thickness being about ¾ in. or a little more. The mix for the scratch and brown coats is usually 1 to 2 or 1 to 3 by weight or volume, and 1 to 2 for the finish coat.

Magnesite stucco is usually applied in two coats, each ¼ in. thick. This stucco usually comes with the dry materials ready mixed and in 100- or 200-lb bags.

Special finishes, such as spatter or slap dash, stippled, broomed, pebble dash, or crystal, can be applied to the outer surface of the stucco.

The quantities of materials for lime stucco can be readily estimated from Tables 11-5, 11-6, and 11-7. It should be noted if the thickness specified for the scratch coat includes or excludes the lath.

The quantities of materials for portland cement and magnesite stucco are given in Table 11-8.

Labor Care must be exercised in estimating labor time and costs for plastering because of variations in materials, finishes, specifications, local customs, kind of jobs, and the abilities of different workmen.

Specifications must be carefully studied. Some specifications state that the work must be done in accordance with all state and local regulations. In some localities, there must be at least two coats of plaster applied, and in some places three coats are required.

When the labor cost of plastering is based on the number of square yards, it is necessary to know how the yardage is computed. Trade practices vary in different localities. Yardage may be taken equal to

Table 11-8 *Approximate Quantities of Materials Required per* 100 *Sq Yd for a Stucco Coat ¼ In. Thick*

Kind of stucco	Mix by weight	Materials	Quantity per 100 sq yd, ¼ in. thick
Hydrated lime.............	1–2	Lime	800–900 lb
		Sand	18 cu ft, or 1,600–1,800 lb
	1–3	Lime	600–670 lb
		Sand	20 cu ft, or 1,800–2,000 lb
Portland cement...........	1–2	Cement	9–10 sacks, or 900–1,000 lb
		Lime	80–100 lb
		Sand	20 cu ft, or 1,800–2,000 lb
	1–3	Cement	6–7 sacks, or 600–670 lb
		Lime	60–90 lb
		Sand	20 cu ft, or 1,800–2,000 lb
Magnesite stucco..........	...	Dry mixture	20 cu ft, or 1,500–1,800 lb

(1) the gross area, (2) the net area (gross area less openings), or (3) the gross area less openings over a certain size. In some places, curved work is charged for at one and a half or two times the rate for flatwork.

In general, in work such as scratch and brown coats, two plasterers will require one helper to keep the plasterers supplied with materials. In special work, such as finish work, paneling, and molding, the number of helpers needed will vary from one-third to two-thirds the number of plasterers. The labor estimates include the necessary time for handling simple horse and plank scaffolds. If special scaffolds are required (as for very high ceilings), an allowance should be made for the work of erection and removal.

Table 11-9 gives the approximate labor required for various kinds of interior plastering. The labor-hours required per 100 sq yd will vary somewhat according to kind of surface, (lath, tile, plasterboard, plaster), surface condition (plain, curved, smooth, rough), thickness of plaster coat, trueness of finish required, etc.

Table 11-9 also gives the labor required for exterior plaster or stuccowork. The labor-hours given are based on the assumption that simple horse and plank scaffolding can be used or that the scaffolding used by carpenters and lathers is available. If special scaffolding must be provided, an allowance must be made for this.

Labor wages per hour (1967) may vary from about $4 to $7 for plasterers, and from about $3 to $6 for helpers.

Diagram 11-7 (Appendix F) can be used for determining labor costs per square yard when the hourly wage is known and the labor-hours per 100 sq yd can be reasonably assumed.

Table 11-9 *Approximate Labor Required for Plastering*

Kind of work	Sq yd per hr per plasterer	Plasterer labor-hr per 100 sq yd*
Interior plastering		
Scratch or brown coats, per coat:		
Plain surface	14–25	4–7
Curved or irregular surface	8–16	6–12
Finish coats, per coat:		
Plain finish on plain surface	14–25	4–7
Plain finish on curved or irregular surface	8–16	6–12
Paneling	4–8	12–25
On beams	5–10	10–20
Ornamental	3–7	14–33
Ruling to imitate tiling (lin ft per hr and hr per 100 lin ft	15–33	3–7
Exterior plastering and stucco on plain surfaces		
Portland cement or lime stucco:		
Scratch coat	14–25	4–7
Brown coat	14–25	4–7
Finish coat	12–20	5–8
Magnetiste stucco:		
Scratch or base coat	12–20	5–8
Finish coat	10–20	5–10
Special finishes such as pebble dash, slate, splatter dash, etc	6–9	11–17

* Helper's time can be estimated at 40 to 75 percent of plasterer's time.

Equipment The equipment needed for plastering will include sawhorses, plank, platforms, scaffolds, mortar and mixing boxes, mixing tools, hods, mortarboards, and various hand tools needed by the plasterers and helpers. On most jobs, a small power mixer may be used. Hoisting equipment will be needed for work on multistoried buildings. The allowance for equipment should include transportation to and from the job, depreciation and maintenance on ordinary scaffolding and tools, and costs of special scaffolding, mixers, and hoists if these are used.

On most building jobs, an allowance of $20 to $50 will usually be sufficient for each gang of about four or five plasterers (or less) and their helpers.

Overhead and Profit Overhead on plastering jobs is usually based on labor costs and may vary from about 10 to 25 percent of labor costs or from 5 to 12 percent of the cost of labor and materials.

Profit may range from about 5 to 15 percent of all other costs.

Plastering Estimates The estimate for the total cost of plastering should include the following:

Materials. Lime, plaster, cement, hair, fiber, sand, etc.
Labor. Plasterers and helpers. Foreman also if the number of workmen warrants
Equipment. All equipment costs
Overhead. Based on labor costs or on sum of material and labor costs
Profit.

The summary or tabulation should show unit costs per square yard of surface (or per 100 sq yd) as well as total costs.

Illustrative estimate. Prepare estimates of the cost of plastering the rooms of the illustrative estimate on lathing estimates (page 288), assuming that either metal lath or plasterboard can be used. The plaster will be mixed by a power mixer.

If on metal lath, the grounds are ¾ in. and three coats of plastering are to be used.

If on plasterboard, ½ in. thick, the grounds are 1 in. and two coats of plaster are to be used.

Plaster mix is 1 to 2 by weight of gypsum plaster and sand for the scratch and brown coats, and for the gypsum sanded finish coat.

Prices of materials delivered at the job are as follows:

Gypsum plaster.............	\$33 per short ton in 100-lb sacks
Sand......................	\$ 2 per short ton
Water.....................	\$ 1 proportional charge for the job

Wages are plasterer \$5 per hour, helper \$4 per hour. Payroll taxes and insurance are 10 percent of the labor wage.

Equipment: Allow \$35.

Overhead: Allow 10 percent of labor costs.

Profit: Allow 8 percent of all other costs.

Solution: From the illustrative estimate on lathing, the net area is 221 sq yd.

The mixes for all coats are the same. One cubic foot will require 45 lb of gypsum plaster and 90 lb of sand.

1. *Plaster applied on ½-in. perforated gypsum plasterboard*

Materials: Amount of plaster required, Tables 11-5 and 11-6.

Scratch coat, ⅜ in. = 30 + 4 = 34 cu ft per 100 sq yd
Finish coat, ⅛ in. = 10 cu ft per 100 sq yd
For 221 sq yd, $(34 + 10) \times 221 \div 100 = 97.24$ cu ft, say 98

Gypsum plaster $= \dfrac{98 \times 45}{2{,}000} \times \$33$	= \$ 73
Sand $= \dfrac{98 \times 90}{2{,}000} \times \$3$	= 13
Water, say	= 2
Total materials	\$ 88

Labor: Assume plasterer's hours per 100 sq yd at 6 for scratch coat and 8 for finish coat.

$$\text{Plasterer's} = \frac{(6 + 8) \times 221}{100} \times \$5 \times 1.10 \qquad = \$170$$

Assume helper's hours at one-half of plasterer's hours

$$\text{Helper's} = \frac{(3 + 4) \times 221}{100} \times \$4 \times 1.10 \qquad = \quad 68$$

Total labor	= $238	
Equipment	= 35	
Overhead, 10% of labor cost	= 24	
Profit, 8% of all other costs	= 31	
Total estimated cost		= $416

Cost per square yard = $1.88

2. *Plaster applied on metal lath*

Materials: Amount of plaster required, Tables 11-5 and 11-6.

Scratch coat, ⅜ in. = 30 + 10 = 40 cu ft per 100 sq yd
Brown coat, ¼ in. = 20 cu ft per 100 sq yd
Finish coat, ⅛ in. = 10 cu ft per 100 sq yd
For 221 sq yd, = (40 + 20 + 10) × 221 ÷ 100 = 154.7 cu ft, say 155

$$\text{Gypsum plaster} = \frac{155 \times 45}{2{,}000} \times \$33 \qquad = \$115$$

$$\text{Sand} = \frac{155 \times 90}{2{,}000} \times \$3 \qquad = \quad 21$$

Water, say	= 2
Total materials	= $138

Labor: Assume plasterer's hours per 100 sq yd at 6 for scratch coat, 6 for brown coat and 8 for finish coat.

$$\text{Plasterer's} = \frac{(6 + 6 + 8) \times 221}{100} \times \$5 \times 1.10 \qquad = \$243$$

Assume helper's hours at half of plasterer's hours

$$\text{Helper's} = \frac{(3 + 3 + 4) \times 221}{10} \times \$4 \times 1.10 \qquad = \quad 97$$

Total labor	= $340	
Equipment	= 35	
Overhead, 10% of labor cost	= 34	
Profit, 8% of all other costs	= 44	
Total estimated cost		= $591

Cost per square yard = $2.67

If desired, the estimated costs of lathing and plastering may be summarized as follows (based on 221 sq yd):

	Cost per sq yd		Total cost	
	On plasterboard	On wire lath	On plasterboard	On wire lath
Lathing	$1.28	$1.26	$281	$279
Plastering	1.88	2.67	416	591
Total	$3.16	$3.93	$697	$870

Diagrams 11-5, 11-6, and 11-7 may be used in preparing or checking estimates of materials and labor for plastering jobs.

Illustrative estimate on stuccowork. Prepare an estimate of the cost of lathing and plastering the exterior walls of a one-story residence. Building is 25 by 72 ft in size and 11 ft high from ground to eaves. There is no exterior sheathing. The eaves extend 2 ft horizontally all around the building and are furred out so that the underside can be wired and plastered. Specifications require the covering of the exterior walls and the underside of the eaves with No. 18 gauge galvanized wire spaced not over 6 in. apart and stretched taut. The wire is to be covered with a layer of asphalt building paper, 15-lb felt. This paper is to be covered with stucco netting, No. 17 gage with 1½-in. mesh. The stucco is to be applied in three coats. The thickness of each coat is to be at least ¼ in. The mix for the first two coats is to be 1 part of portland cement to 3 parts of sand by weight. The mix for the third, or finish, coat is to be 1 part of portland cement to 2 parts of sand by weight. Lime, equal to 10 percent by weight of the cement, is to be added for workability.

Solution: The costs of materials delivered at the job are

No. 18 gauge galvanized wire, per 1,000-ft roll = $1.05
15-lb felt, 324 sq ft per roll, per roll = $2.60
Nails per pound, average price = $0.25
No. 17 gauge, 1½-in. mesh, stucco netting, 3-by 150-ft roll, per roll = $12.50
Portland cement, per sack of 94 lb = $1.35
Hydrated lime, per 100-lb sack = $1.60
Sand, per ton = $2.75
Coloring for finish coat = $15.00
Hourly labor wages: lather $5, plasterer $5, helper $4
Payroll taxes and insurance = 8% of labor wage

Equipment needed: scaffolding for lathers and plasterers, mortar mixer for plaster, and small tools. An allowance of $10 for scaffolding and small tools and $16 for the mortar mixer, including transportation, should be sufficient.

Overhead = 10% of labor costs
Profit = 8% of all other costs

An examination of the plans shows that there are 2 doors and 11 windows with a total area of 262 sq ft.

Exterior wall area = 2(25 + 72) × 11 = 2,134 sq ft
Area under eaves = 2(25 + 72) × 2 + 2 × 2 × 4(corners) = 404 sq ft
Net area to be covered = 2,134 + 404 − 262 = 2,276 sq ft = 253 sq yd

Lathing Estimate:

Materials:

Wire, 25 ft per square yard, including turns and waste, etc., 25 × 253 × $1.05 per 1,000 ft	=	$ 7	
Nails for wire, about 6 lb at $0.25	=	2	
Building paper, 15-lb felt. Rolls = 2,276 ÷ 300 = 7.6 rolls at $2.60	=	20	
Nails for paper, say 7 lb at $0.25	=	2	
Stucco mesh, 2,276 ÷ 450 sq ft per roll = 5.1 rolls at $12.50	=	64	
Nails for mesh, say 8 lb at $0.25	=	2	
Total materials	=	$ 97	

Labor:

Fastening wire, 253 sq yd at 2 hr per 100 sq yd = 5 hr			
Applying building paper, 253 sq yd at 4 hr per 100 sq yd = 10 hr			
Applying stucco mesh, 253 sq yd at 5 hr per 100 sq yd = 13 hr			
Labor cost, 28 hr at ($5 × 1.08) per hour	=	$151	
Equipment: Half of cost of scaffolding and small tools	=	5	
Overhead at 10% of labor cost	=	15	
Profit at 8% of all other costs	=	21	
Total lathing cost ($1.14 per sq yd, net area)			= $289

Plastering Estimate:

Materials: Each coat is ¼ in. thick. See Table 11-8.

First and second coats. Sand, 2,000 lb per 100 sq yd per coat. 253 sq yd require 2.53 × 2,000 × 2 = 10,120 lb = 5.1 tons		
Cement, ⅓ of sand = 3,340 lb = 36 sk		
Lime, 10% of cement = 334 lb = 3.4 sk		
Third coat. Sand at 2,000 lb per 100 sq yd, 253 sq yd require 5,060 lb = 2.6 tons		
Cement, ½ of sand = 2,530 lb = 27 sk		
Lime, 10% of cement = 253 lb = 2.6 sk		
Cost of sand = $2.75 × (5.1 + 2.6)	=	$ 21
Cost of cement = $1.35 × (36 + 27)	=	85
Cost of lime = $1.60 × (3.4 + 2.6)	=	10
Cost of coloring for finish coat	=	15
Cost of water, say	=	3
Total materials	=	$134

Labor: Assume a gang of 2 plasterers and 1 helper. Each plasterer should apply about 18 sq yd per hour per coat (5.55 hr per 100 sq yd) for the first and second coats and about 14 sq yd per hour (7.15 hr per 100 sq yd) for the finish coat.

Plasterer's hours, $2.53 \times (5.55 + 5.55 + 7.15) = 46$		
Helper's hours, half of plasterer's hours = 23		
Gang time for 2 plasterers and 1 helper = 23 hr		
Gang cost per hour, $\$5 \times 1.08 \times 2 + \$4 \times 1.08 = \$15.12$		
Total labor cost = $\$15.12 \times 23$	= $348	
Equipment: Half of scaffolding and small tools ($5) plus mortar mixer (1$16)	= 21	
Overhead at 10% of labor cost	= 35	
Profit at 8% of all other costs	= 43	
Total plastering cost ($2.30 per square yard, net area)		= $581
This includes an allowance of about 6 percent waste of materials.		
Total cost of job, $289 (lathing) + $581 (plastering)		= $870
This is a cost of $3.44 per square yard.		

Drywall Construction

Drywall Construction Estimates for drywall construction include materials, labor, equipment, overhead, and profit.

The unit of measurement is usually the square yard of surface area, though the square foot or the square of 100 sq ft may be used. The number of square yards may be based on the net area, the gross area, or on the gross area less large openings. Allowance should be made for special or difficult work. Nailing strips may be estimated by the lineal foot, or by feet board measure; adhesive by the gallon or pound; tape by the foot; nails, staples, and clips by the pound or by the box or number.

Drywall construction may be classified as to exterior or interior work, kind of material, kind of surface finish, manner of placing or fastening, etc.

Materials There are many kinds of drywall materials on the market such as gypsum wallboard, wood panels, tiles, etc., of varying costs and different methods of application. As it is impractical to describe all of these in this text, only a few of the more common kinds will be considered.

Gypsum Wallboard. This material may have different surface finishes, edges (tapered, V-grooved, tongue and grooved), thicknesses (¼ to ¾ in.), and sizes (usually 4 ft wide and 6, 8, 10, or 12 ft long). This wallboard may be placed over old plaster or fastened to nailing strips, studs, or joists by nails, staples, clips, or glue. The wall boards are usually laid horizontally. The ¼- and ⅜-in. thicknesses are suitable when the wallboard is applied over old plaster. Tape and paste are needed for the joints.

The price of gypsum wallboard varies with the time, locality, thickness, and finish. Approximate prices (1967) may vary from about $50 to $100 per 1,000 sq ft. The price may or may not include nails, staples, clips, joint filler, etc.

Wood Panels. There are many kinds of wood panels on the market, made of different woods and with different finishes. Edges are usually V-grooved or tongue and grooved. Panels are usually 4 ft wide and from 6 to 16 ft long. Thicknesses may vary from ⅛ to ¾ in. Wall panels are usually applied vertically on nailing strips, and may be fastened by nails, clips, or glue.

Prices (1967) range from about $100 to $1,000 or more per 1,000 sq ft depending on the kind of wood, thickness, and finish. Prices are sometimes given per square yard, per square foot or per 100 sq ft.

Tiles. Drywall tiles may be classified in many ways such as acoustical or nonacoustical, perforated or plain, wall or ceiling, edges, finish, material, thickness, and size.

The usual thickness is ½ in. Common sizes are: 12 × 12 in., 16 × 16 in., 12 × 24 in., and sometimes 16 × 32 in., 24 × 24 in., and 32 × 32 in.

The tile may be applied over old plaster, or to nailing strips, studs, and joists by means of nails, staples, clips, cement, glue, etc.

Present prices (1967) range from about $100 to $300 or more per 1,000 sq ft. Prices are sometimes quoted per piece, per square foot, or per square yard.

Accessories. Some of the accessories (and their approximate prices) which may be required in drywall construction are:

Accessory	*Approximate price*
Nailing strips, ¾ × 3 or 4 in.	$125 per 1,000 fbm
Nails, special, 5 lb per 1,000 sq ft	$0.75 per lb
Staples	$0.35 per lb
Clips, 1,000 per box	$5 per box
Cement paste or glue, 1 gal for 175 sq ft	$5 per gal
Adhesive, 1 gal for 65 sq ft	$3 per gal
Joint filler for gypsum wallboard:	
Joint filler cement	$4 per 25-lb box
Vinyl ready mix	$5 per 5-gal box
18 lb cement and 250 ft tape	$5 per set

Diagram 11-8 may be used for estimating drywall materials such as gypsum wallboard, panels, and wall tile.

Labor Drywall materials are usually applied by skilled workers especially trained for this work. The number of labor hours required per 100 sq ft or per 100 sq yd of surface area will depend on the material, the particular job, how the surface area is computed, and on the skill and inclination of the workmen. Hourly wages (1967) for skilled drywall workers may range from about $4 to $7.

Table 11-10 gives the approximate labor hours required for applying drywall materials.

Diagram 11-9 may be used for estimating the labor costs of applying gypsum wallboard, paneling, and wall and ceiling tile.

Diagram 11-4 may be used for estimating the labor cost of applying nailing strips.

Table 11-10 *Approximate Labor-hours for Applying Drywall Materials*

Item	Hr per 100 sq ft	Hr per 100 sq yd
Gypsum wallboard		
Walls	1.3–2.3	12–21
Ceilings	1.7–3.0	15–27
Joints	0.7–1.2	6–11
Paneling, varies greatly	1.8–6.0	16–55
Tile walls		
Small pieces*	2.0–3.5	18–32
Large pieces†	1.5–3.0	13–27
Tile ceilings		
Small pieces	2.5–4.0	22–36
Large pieces	2.0–3.5	18–31
Nailing strips		
3 or 4 in. wide per 100 lin ft	1.0–2.2	

* Small tile pieces are 16 × 16 in. or smaller.

† Large tile pieces are larger than 16 × 16 in.

Equipment Wooden horses, plank, platforms, and some scaffolding will be required on almost all drywall jobs. Transportation costs for the equipment to and from the job should be included. If the scaffolding is to be erected and taken down by other labor, this labor cost should also be included.

The workmen will usually have all needed small tools in their tool chests. The contractor may provide special tools such as small electric saws, staplers, etc. The costs of these extra tools should be included.

For a small drywall job, such as for a residence, an allowance of $10 to $25 for equipment will usually be enough.

Overhead and Profit Overhead costs are usually based on the labor costs and may vary from about 10 to 25 percent of the labor costs, or from about 5 to 15 percent of the costs of materials and labor.

Profit may vary from about 7 to 20 percent of all other costs.

Drywall Estimates The estimate for the total cost of a drywall construction job should include all of the costs of materials, labor, equipment, and overhead charged to the job plus a reasonable profit.

Unit costs per square yard or per square foot should be computed, based on the net surface area in square yards or square feet.

Illustrative estimate. Prepare a cost estimate of drywall construction for the rooms of the residence described in the illustrative estimate on page 288, using ½-in. gypsum wallboard.

The price of the ½-in. gypsum wallboard delivered at the job is $8.15 per 100 sq ft. This price includes nails, joint filler, and tape.

Wages are $5.50 per hr, including payroll taxes and insurance. Allow $15 for the cost of equipment for this job.

Overhead costs are 10% of the labor costs.

Profit is 8% of all other costs.

Materials: From the illustrative estimate on page 288, the net areas are:

Walls	= 151.6 sq yd	= 1,364.4 sq ft
Ceilings	= 69.4 sq yd	= 624.6 sq ft
Total	= 221.0 sq yd	= 1,989.0 sq ft

For waste, allow about 12% for walls (several openings) and about 5% for ceilings. Then the square feet of wallboard required will be:

Walls	= 1,364.4 × 1.12	= 1,528 sq ft
Ceilings	= 624.6 × 1.05	= 656 sq ft
Total		= 2,184 sq ft. Order 2,200 sq ft.

$$\text{Cost of gypsum wallboard} = \frac{\$8.15 \times 2{,}200}{100} = \$179$$

Labor: Estimate labor at 1.9 hr per 100 sq ft for walls (several openings), and at 2.2 hr per 100 sq ft for ceilings.

$$\text{Labor hours} = \frac{1.9 \times 1{,}365 + 2.2 \times 625}{100} = 26 + 14 = 40$$

Labor cost = $5.50 × 40	$220
Equipment allowance	15
Overhead, 10% of labor cost	22
Profit, 8% of all other costs	35
Total estimated cost	$471

This estimate may be summarized as follows (based on 1,989 sq ft or 221 sq yd):

Item	Cost per sq ft	Cost per sq yd	Total cost
Materials	$0.09	$0.81	$179
Labor	0.11	0.99	220
Equipment	0.01	0.07	15
Overhead	0.01	0.10	22
Profit	0.02	0.16	35
Totals	$0.24	$2.13	$471

CHAPTER TWELVE

Heating, Cooling, and Air Conditioning

Heating, Cooling, and Air Conditioning Systems

Estimating of Heating, Cooling, and Air Conditioning Detailed estimates of heating, cooling, and air conditioning are almost always prepared by subcontractors who specialize in this kind of work. The general contractor (or his estimator) rarely has the specialized knowledge required for making accurate and detailed estimates, but he should have a general knowledge of the work so that he can take off the required materials with reasonable accuracy and prepare approximate estimates of the quantities and costs of materials and labor and of the costs of equipment, overhead, and profit. These approximate estimates can be used for the purpose of checking and determining the reasonableness of the bids submitted by the subcontractors or for making approximate estimates of the costs when subcontractors' bids are not available.

Cost estimates of heating, cooling, and air conditioning systems will include the five main items of materials, labor, equipment, overhead, and profit. Unit costs can be computed as desired. Some units are cost per outlet in warm-air heating, cost per radiator or the cost per square foot (or 100 sq ft) of radiation in steam and hot-water heating, the cost per cubic foot (or 1,000 cu ft) in air conditioning, the cost per square foot (or 100 sq ft) of floor area, and the cost per cubic foot (or 1,000 cu ft) of volume of space heated or cooled. Other information should be included, such as size of structure, kind of structure (wood, brick, steel, etc.), occupancy or use (residence, office, school, factory, store, etc.), locality and climate, temperature ranges, and kind of plant. Such information is necessary if the costs are to be used for comparative purposes.

In the preparation of an approximate estimate by the contractor (or his estimator), it is assumed that the plans and specifications furnished him are complete enough so that no designing is necessary and also that enough information is given so that all materials required can be readily taken off. On many of the larger jobs, special plans and drawings are provided. Many of the heating and cooling details are shown on the plans by symbols. Hence, the contractor must have a knowledge of the meanings of these various symbols.

There are so many types and combinations of heating, cooling, and air conditioning systems that it is not practical to try to include all of them in one chapter. The estimating of only a few of the more common types will be discussed.

Heating Systems The main purpose of a heating system is to provide and maintain temperatures. There are many types of heating plants in use. Some are simple; others are complex. Some use but one method of heating; others combine two or more methods. Some are combined with air conditioning systems; others are not.

There are several ways of classifying heating systems such as by kind (central or unit), type (steam, hot water, forced air, etc.), heat flow (radiation, conduction, convection), heat source or fuel (coal, oil, gas, electricity), location (floor, basement, ceiling, attic, wall, closet); etc. Perhaps the classification as to type is the best for general estimating purposes. This classification is as follows:

1. Steam heating systems. Central systems
   a. Pressure. One- or two-pipe systems
   b. Vapor. Very low pressure two-pipe systems
   c. Vacuum. Two-pipe system with vacuum pump
   d. Zone control. Pressure system with different pipelines independently controlled
2. Hot-water heating systems. Central two-pipe systems
   a. Gravity. Open and closed types. Rarely used
   b. Pressure. Closed type. Rarely used
   c. Forced. Water forced through system by a pump or pumps

   d. Flow control. Boiler maintained at a definite temperature. One pump and controls for each independent piping circuit
3. Warm-air systems
   a. Forced-air systems. Usually central systems. Often combined with an air conditioning system. Fuel may be coal, oil, or gas. Location may be basement, floor, ceiling, attic, wall, or closet
   b. Gravity systems. Usually unit systems, as the old type gravity central system is rarely used. Fuel is usually gas. Location is usually floor or wall
4. Other systems, such as electric, radiant gas, etc.

The design of a heating system usually includes:

1. Computation of heat losses, usually for each room as well as for the entire building.

2. Selection of system. Consideration is given to the kind and size of structure, occupancy, cost, maintenance, climate, desires of architect and owner, etc.

3. Design of heat-distribution units, such as radiators, etc.

4. Design of heat-conveying system, such as piping.

5. Selection of furnace or boiler and fuel-firing unit.

6. Selection of accessories.

7. Selection of heat-control system.

Air Cooling Systems These may be classified as follows:

1. Unit coolers.
   a. Water cooled. Usually small and portable with manual controls. Suitable for one room.
   b. Refrigeration cooled. Various capacities. Usually placed in wall or window. Manual controls. Suitable for one to five rooms.

2. Small central systems. Capacity up to about 100,000 Btu requires ducts and fans. Usually thermostat controls. Suitable for a home or an apartment.
   a. Water cooled. Should be connected to a public water system.
   b. Refrigeration cooled. Requires an outside vent. Ducts and vents should be rustproof. Cooling unit may be placed in closet, below floor, or in attic. Often combined with a forced warm-air heating system

3. Large central systems. Capacity up to 300,000 Btu (25 tons) or more. Suitable for large buildings. Thermostat controls. May have several independent circuits. May be water cooled or refrigeration cooled. Often combined with an air conditioning system. Designed and installed by specialists.

Air Conditioning Systems Complete air conditioning includes:

Temperature. Including humidity and air motion
Ventilation. Air supply and distribution
Purification. Dust, bacteria, odors, gases, etc.

There are many types of air conditioning systems. They may be central or unit systems. They may care for one, two, or all three of the items of temperature, ventilation, and purification. For example, a complete central system in a large office building may care for temperature (heating, cooling, and humidity), ventilation, and purification (filtering and washing).

Steam and Hot-water Heating Systems

Materials Takeoff Before the takeoff of the materials can be made, the estimator should have available a fairly complete set of heating plans showing the location of the boiler, the arrangement of the horizontal piping in the basement, the arrangement of all vertical piping, the location of all radiators, and all necessary details in regard to makes, kinds, sizes, etc., of all items required. Detailed specifications should accompany the plans, giving all additional information needed.

Sometimes no detailed plans of the heating system are provided. When such is the case, the general contractor may ask competent subcontractors for bids accompanied by sketches showing the layout of the heating system. Or the general contractor may have a competent heating engineer design the heating system, prepare sets of plans and specifications, and give a set to each subcontractor who is to bid on the job. Then the general contractor can use one of these sets of plans for checking the bids received.

Perhaps the best way of taking off the required materials is to go over the plans in detail and list all materials under the five main headings, using as many subheadings and remarks as necessary. Some estimators prefer to go over the plans floor by floor and list all materials on each floor before proceeding to another floor. Other estimators prefer to go over the plans, list all heat-distributing units, and then repeat the process for pipes, etc. A good idea is to check each item on the plans as it is listed, using a red or other colored pencil. Many estimators prefer to use ruled sheets, sometimes with the various items printed on them, for the takeoff. A reminder list is helpful. Such a list is given in Appendix A. If the job is fairly large, a summary sheet should be used.

The takeoff should include the following:

1. *Heat-distributing units.* Radiators, valves, air valves, etc., with notes as to sizes, heating surface, height, etc. These may be taken off floor by floor and then summarized.

2. *Heat-conveying system or piping.* All pipes, valves, joints, tees, elbows, unions, traps, tanks, etc., should be listed with details as to sizes, etc., starting at the furnace or boiler and working around each circuit in order.

3. *Heat-generating units.* Boiler, stoker, burner, etc., with smoke pipe, dampers, valves, gauges, etc., such as usually accompany and are attached to these units.

4. *Accessories.* Various items not usually included with radiators, pipes, boilers, and heat controls. Some of these items may be extra tanks, traps, pumps, valves, etc. An estimator must use his own judgment in regard to classifying certain items as accessories or as parts of the boiler or piping system.

5. *Heat-control system.* All thermostats, wiring, transformers, piping, pressure-stats, rheostats, cutouts, and various other items included in the particular system selected.

Materials. Heat-distributing Units The materials usually included are radiators, air valves, valves, unions, reducers, covers, hangers, and fans as required. Some estimators include radiator valves, unions, reducers, etc., with the piping system. Radiators are usually made of cast iron but may be of other materials. They are of the tube type, may be mounted on floor or wall, may be concealed, may have covers and enclosures, and may have fans.

Radiators are usually listed as to type, height, number of tubes, number of sections, heating surface in square feet per section and per radiator. Hot-water radiators are invariably larger than steam radiators. An average cast-iron radiator may be assumed to have a heat emission of about 240 Btu per square foot per hour with steam and about 160 Btu per square foot per hour for hot water. Hence, the heating surface of a hot-water radiator should be approximately 1.5 times that of a corresponding steam radiator. Cast-iron radiators are often priced per section with height, number of tubes, and square feet of heating surface given. However, the price per square foot of heating surface is practically the same for all sections. This price (1967) may range from about 75 cents to $1.35 per square foot of heating surface. Hence, the estimator can find the total square feet of heating surface for all radiators and use a suitable price per square foot. Dealers' catalogues should be consulted for information concerning sizes and square feet of heating surface.

Table 12-1 gives some information regarding cast-iron radiators. Average length of section is 2.5 in.

Air valves come in many varieties. For average room radiators, simple air valves may range from $1 and up. About $2.20 is a fair average value. Air valves suitable for use on a vacuum system may cost $2 and up. About $3.30 is a fair average value.

Table 12-1 *Square Feet of Heating Surface per Section of Cast-iron Tube Floor Radiators*

Height from floor, in.	Number of tubes				
	3	4	5	6	7
14	.....	.....	....	.....	2.50
17	.....	.....	....	.....	3.00
20	1.75	2.25	2.67	3.00	3.67
23	2.00	2.50	3.00	3.50	4.25
26	2.33	2.75	3.50	4.00	5.00
32	3.00	3.50	4.33	5.00	
38	3.50	4.25	5.00	6.00	
Width, in.	4.625	6.3125	8.00	9.6875	11.375

Table 12-2 gives approximate prices for various radiator fittings.
Radiator wall hangers may cost $2 to $5 per radiator.
Humidifiers to place on radiators may each cost about $2 and up.
Radiator covers may each cost $2 and up.
Combined radiator covers and humidifiers may each cost $5 and up.

Table 12-2 *Approximate* 1967 *Prices of Radiator Fittings*

Item	Size, in.				
	½	¾	1	1¼	1½
Valves	$2.75	$3.50	$4.25	$6.25	$7.50
Floor and ceiling plates	$0.25	$0.30	$0.35	$0.40	$0.45
Brass unions and elbows	$1.75	$2.15	$3.00	$4.00	$5.00
Unions	$0.75	$1.00	$1.25	$1.60	$2.00
Reducers and bushings	$0.25	$0.30	$0.35	$0.40	$0.50

Materials. Piping Systems The pipe generally used for steam and hot-water piping systems may be of black or galvanized finish and made of wrought iron or open-hearth or copper-bearing steel. The weight may be nominal or standard, extra strong, or double extra strong. Copper-bearing steel pipe is preferred because of its ability to resist rust.

The base price for pipe is $200 per net ton f.o.b. mill with various extras and discounts, depending upon kind of finish (black or galvanized), kind of pipe (wrought iron, open-hearth or copper-bearing steel), and nominal diameter of pipe. Freight, trucking, and handling costs must be added to the prices f.o.b. mill.

Table 12-3 gives information concerning pipes of different sizes. The values given are for black finish. Galvanized pipes will be a little heavier.

Table 12-3 *Sizes, Diameters, and Weights of Steel and Wrought-iron Pipes*

Size, in.	External diameter, in.	Internal diameter, in.			Weights per foot, plain ends, lb		
		Standard	Extra strong	Double extra	Standard	Extra strong	Double extra
0.125	0.405	0.269	0.215	.....	0.244	0.314	
0.250	0.540	0.364	0.302	.....	0.424	0.535	
0.375	0.675	0.493	0.423	.....	0.567	0.738	
0.500	0.840	0.622	0.546	0.252	0.850	1.087	1.714
0.75	1.050	0.824	0.742	0.434	1.130	1.473	2.440
1.00	1.315	1.049	0.957	0.599	1.678	2.171	3.659
1.25	1.660	1.380	1.278	0.896	2.272	2.996	5.214
1.50	1.900	1.610	1.500	1.100	2.717	3.631	6.408
2.0	2.375	2.067	1.939	1.503	3.652	5.022	9.029
2.5	2.875	2.469	2.323	1.771	5.793	7.661	13.695
3.0	3.500	3.068	2.900	2.300	7.575	10.252	18.583
3.5	4.000	3.548	3.364	2.728	9.109	12.505	22.850
4	4.500	4.026	3.826	3.152	10.790	14.983	27.541
5	5.563	5.047	4.813	4.063	14.617	20.778	38.552
6	6.625	6.065	5.761	4.897	18.974	28.573	53.160
8*	8.625	8.071	......	.....	24.696		
8	8.625	7.981	7.625	6.875	28.554	43.388	72.424
10*	10.75	10.192	......	.....	31.201		
10*	10.75	10.136	......	.....	34.240		
10	10.75	10.020	9.750	.....	40.483	54.735	
12*	12.75	12.090	......	.....	43.773		
12	12.75	12.000	11.750	.....	49.562	65.415	

* Special sizes furnished when specified with weights given.

Weights are for black finish. Galvanized finish will weigh slightly more. Permissible variation in weight is 5 percent, more or less.

Pipe is often purchased per 100 ft of length. Lengths of individual pieces will vary somewhat, with an average length of about 20 ft. Pipe may come with plain ends or with threaded ends. Frequently, one coupling is included with each length of pipe. Table 12-4 gives approximate 1967 local prices of standard steel pipe.

Table 12-4 *Approximate* 1967 *Prices for Standard-weight Steel and Wrought-iron Pipe per* 100 *Lin Ft*

Size, in.	Ordinary steel		Copper-bearing steel		Wrought iron		Extra for threads, each end
	Black	Galvanized	Black	Galvanized	Black	Galvanized	
¼	$ 11	$ 12	$ 13	$ 14	$ 16	$ 18	$0.10
½	13	15	16	18	24	28	0.10
¾	18	20	20	22	30	35	0.12
1	25	29	29	33	45	50	0.14
1¼	34	38	37	41	56	62	0.16
1½	39	46	45	52	70	77	0.18
2	54	63	62	72	87	100	0.23
2½	88	105	103	115	130	160	0.35
3	115	135	137	148	195	220	0.45
4	177	210	200	225	....	....	0.55

Table 12-5 gives approximate 1967 local prices for pipe fittings of standard-weight black pipe.

In addition to the information given in Tables 12-2 to 12-5, inclusive, further information in regard to pipes and pipe fittings, is given in Tables 13-1 to 13-11, inclusive, of Chap. 13 on Plumbing.

Pipe larger than 3 or 4 in. is usually designed with flanged fittings.

Table 12-5 *Approximate* 1967 *Prices per* 100 *Fittings for Standard Steel Pipe**

Size of pipe, in.	Caps, plugs, nipples, and bushings	Couplings	Reducers	Malleable		
				Elbows	T's and Y's	Unions
½ in. or less	$ 7–$ 9	$ 9–$ 11	$ 13–$ 17	$ 14–$ 18	$ 27–$ 32	$ 45–$ 50
¾	$ 9–$12	$ 13–$ 17	$ 21–$ 25	$ 18–$ 32	$ 56–$ 63	$ 50–$ 55
1	$10–$14	$ 17–$ 23	$ 27–$ 37	$ 37–$ 42	$ 90–$100	$ 65–$ 75
1¼	$17–$20	$ 24–$ 30	$ 35–$ 50	$ 58–$ 65	$115–$125	$ 90–$100
1½	$20–$27	$ 35–$ 45	$ 45–$ 60	$ 80–$ 90	$160–$180	$110–$125
2	$26–$35	$ 60–$ 70	$ 70–$ 90	$110–$125	$300–$325	$135–$180
2½	$35–$50	$ 85–$100	$110–$150	$200–$235	$330–$360	$225–$300
3	$50–$65	$115–$140	$150–$200	$310–$370	$370–$400	$400–$500

* The cost of fittings with galvanized finish may be about 20% more than the cost with black finish.

The cost of fittings for copper-bearing steel pipe may be about 20% more than for ordinary steel pipe.

Flanges may be screwed or welded to the pipe. The joint is made by inserting a gasket between the flanges and bolting them together.

Fusion-welded joints are sometimes used for pipe larger than 3 or 4 in.

Brass pipe and fittings can be used. The pipe should be of the standard commercial iron-pipe-gauge thickness for the various diameters. Fittings should have thicknesses equal to those of brass pipes of the same diameter. Brass tubing should have a thickness equal to or more than No. 18 gauge.

Copper tubing, either in straight lengths or in 60-ft coils, and fittings are available in various diameters up to about 1.5 in. in size. Copper tubing of the smaller diameters can be readily bent around joists, corners, etc., thus saving many fittings and considerable labor.

Weights and 1967 prices for copper tubing and 1967 prices for fittings are approximately as shown in Tables 13-9, 13-10, and 13-11.

Pipe hangers are required for supporting horizontal pipes. The cost of the hanger will vary with the type and size. Prices may range from about 35 cents to $2 for each.

Pipe covering or pipe insulation may be composed of such materials as magnesia, asbestos, mineral wool, and combinations. A suitable, though not always long-lasting, covering is made of asbestos in combination with a cellular material. Prices vary with material, thickness, and size of pipe.

Shutoff valves should preferably be provided for each circuit so that the entire plant need not be shut down in order to repair one line. Prices of valves will depend upon kind, quality, and make. Prices may range from about $2 or more for a 1-in. valve to about $6 or more for a 2-in. valve, $10 or so for a 3-in. valve, $20 or more for a 4-in. valve, and more for the larger sizes.

Brass radiator valves may cost from about $7 to $12 each, depending on size (3/8 in. up to 1 in.) and quality.

Reducing valves, steam traps, steam separators, and other items may be required on some systems.

The estimator should refer to dealers' catalogues and price lists when preparing estimates of piping and pipe fittings and accessories. Prices vary according to time and place. Prices given in this chapter are approximate only and should not be used except for very approximate estimates.

Materials. Boilers and Accessories The boiler selected should be of the proper kind for the fuel used and of sufficient size to do the work. There are many types of boilers used for heating work, the more common being the cast-iron boiler, the steel fire-tube boiler, and the steel water-tube boiler.

Heating boilers are rated according to the number of square feet of radiation they can handle.

The design of the boiler will also vary according to the fuel and method of firing. With coal, boilers may be hand or stoker fired. Oil burners are used for oil fuel, and gas burners for gas.

Round cast-iron boilers, consisting of three to five main castings, can be obtained in capacities up to about 1,600 sq ft of radiation.

Sectional cast-iron boilers, consisting of 5 to 10 sections, can be obtained in capacities up to about 18,000 sq ft of radiation.

Steel boilers may be riveted or welded. The fire-tube type can be used in capacities up to about 15,000 sq ft of radiation, and water-tube boilers for larger capacities.

The boiler selected must be large enough to provide the heat required:

1. By all radiators with normal room temperature and minimum outside temperature
2. For all heat losses by the piping system
3. By all attached water heaters
4. For bringing all room temperatures from a lower temperature up to the desired temperature during the warming-up time
5. For raising the temperature of the metal in the system to the requisite working temperature

Items 1, 2, and 3 make up the design load, and items 4 and 5, the starting load.

Heat requirements are approximately as follows:

240 Btu per hour per square foot of steam radiation
160 Btu per hour per square foot of hot-water radiation
2 Btu per hour per square foot of uncovered pipes
0.5 to 1 Btu per hour per square foot, of insulated (covered) pipes

The maximum load that a boiler will be expected to carry will be the sum of the design and starting loads. The starting load may vary from about 65 to 50 percent of the design load for design loads varying from almost nothing up to 1,200,000 Btu per hour.

Prices of boilers will vary according to kind, make, and capacity. Approximate prices for 1967 are about as given in Table 12-6 for boilers without accessories except pet cocks, water glass, gauge, safety valve, and damper for steam boilers and pet cocks, thermometer, and damper for hot-water boilers.

Coal stokers for boilers with capacities up to 3,000 sq ft of radiation may range in price from about $250 to $1,000, depending on make and size.

An oil burner with tank and accessories may cost $250 to $1,000.

Combined boilers and oil burners will cost less than when the boiler and burners are considered separately. Special designs may be more expensive.

Table 12-6 *Approximate Boiler Costs per* 100 *Sq Ft of Radiation*

Radiation sq ft	Cost per 100 sq ft of radiation	
	Hot water	Steam
300	$55	$90
600	47	75
1,000	40	62
1,500	36	55
2,000	33	51
3,000	31	48

Combined boilers and gas burners will cost about the same as boilers with oil burners or stokers. Special designs will cost more.

Other accessories with hot-water boilers may be expansion tanks, valves, circulating pumps, hot-water heaters, boiler feed pumps, etc.

Other accessories with steam boilers may be valves, air or vent valves, special returns, safety valves, hot-water heaters, traps, low-water cutouts, vacuum pumps, boiler feed pumps, etc.

Prices of various accessories can be obtained from local dealers, out-of-town dealers, and various catalogues.

Materials. Temperature Controls There are various types of temperature-control systems in use. They can be classified as follows:

Central Thermostat. A central thermostat, supplemented by pressure control for steam heat and by temperature control for hot-water and warm-air heat, is almost the standard type for residences and other small buildings.

Individual Room Thermostats. Individual room thermostats may be of the compressed-air, electrical, or self-contained radiator-valve type. The compressed-air and electrical types are often used in public buildings. These types are comparatively expensive to install and maintain.

Central Control. Central control with radiator orifices, variation in steam pressure, variation in hot-water temperatures, or with a varying supply are sometimes used in large buildings to give an approximate temperature control and to save fuel.

Zone Control. Approximate temperature control by a zone or circuit system results in some economy in fuel consumption. This system can be used advantageously in combination with other types of control.

The cost of a temperature-control system will vary with the type of system and make. A central-thermostat system may cost $100 to $300 plus installation. An individual room-thermostat system may cost $20 or $30 and up per room. The cost of a central-control system will vary greatly with the type and make selected and the number of radiators, etc., to be controlled. With the zone system used in combination with a thermostat system, the cost for each zone may vary from about $60 to $150 plus the cost of extra pumps, valves, checks, etc., as required.

Labor The installing of steam and hot-water heating systems is usually done by teamwork. For ordinary work, the team may consist of one steam fitter and one helper. On heavy work, the team may consist of one steam fitter and two or more helpers or two or more two-men teams may be used. On large jobs, there may be several teams working at the same time under the direction of a foreman or superintendent.

The time required for doing a certain item of steam- or hot-water-heating work will vary considerably, depending upon the particular work, the working conditions, and the skill and inclination of the workmen.

The time required for installing radiators can be estimated per radiator.

Labor for pipe work can be more accurately estimated per joint instead of per 100 lin ft, with kind of pipe and diameter given. The labor per joint will include all threading, hanging, and connecting needed. It is assumed that an adequate supply of nipples, couplings, unions, and other fittings is available. The larger the pipe, the more labor-hours required. When threaded joints are estimated, each coupling, valve, or union counts as two joints and a T as three joints, etc.

For a pipe up to 3 or 4 in. in size, a team of one pipe fitter and one helper is satisfactory; for pipe sizes of 4 to 6 in., the team should be composed of one pipe fitter and two helpers. For pipe sizes from 6 to 12 in., a team of two pipe fitters and three helpers may be required. For pipe sizes of 14 to 24 in., a team of two pipe fitters and three to five helpers may be needed. The labor-hours given for pipework in Table 12-7 are based on the assumption that one pipe thread must be cut for each joint. When it is not necessary to cut a thread, the time given can be reduced 35 to 50 percent.

Furnaces, fittings, and accessories are usually installed by gangs of two or more men, depending upon the weights to be handled. An electrician will be needed if any electrical work is required.

Temperature-control systems are usually installed by a man or men skilled in this work, assisted by helpers. The skilled workmen may be pipe fitters or electricians or both, depending upon the kind of system to be installed.

Table 12-7 gives approximate labor-hours required for various items of work in connection with the installation of steam and hot-water systems.

The wages of pipe fitters, electricians, and other skilled labor may vary from about $4 to $7 per hour. Wages of helpers may vary from about $3 to $5 per hour. Payroll taxes and insurance may vary from about 8 to 14 percent of the labor wage.

Diagrams 12-1 and 12-2 (Appendix F) can be used for estimating the

Table 12-7 *Approximate Labor-hours Required for Installing Steam and Hot-water Heating Systems*

Item	Labor-hours	Item	Labor-hours
Radiators, one-pipe system, including steam and air valves:		Boiler, small, 600 sq ft steam radiation or less	8–16
Floor	1.4–2.0	Medium, 600 to 1,000 sq ft steam radiation	12–24
Wall	1.7–2.5	Large, over 1,000 sq ft steam radiation per 1,000 sq ft radiation	10–20
Radiators, two-pipe system, including steam and air valves:		Coal stoker, small	4–8
Floor	2.0–3.0	Large	6–20
Wall	2.4–3.5	Oil burner and tanks, small	4–10
Pipe work, per joint:		Large	6–20
Threaded, ½ in. and smaller	0.3–0.5	Valves, each	0.5–1
Threaded, ¾ and 1 in	0.4–0.6	Traps, each	5–15
Threaded, 1¼ and 1½ in	0.5–0.8	Tanks, each	2–10
Threaded, 2 and 2½ in	0.7–1.1	Pumps, each	5–20
Threaded, 3 and 3½ in	0.9–1.6	Temperature control:	
Threaded, 4 in	1.3–2.0	Central thermostat	3–6
Threaded, 5 in	2.0–3.0	Room thermostats, each room	2–4
Threaded, 6 in	2.5–4.0	Central control, variable	3–12
Flanged, 4 in	1.4–2.2	Zone system, each zone	3–6
Flanged, 5 in	2.0–3.0		
Flanged, 6 in	2.5–4.0		
Flanged, 8 in	3.5–5.0		
Flanged, 10 in	5.0–6.5		
Flanged, 12 in	6.5–8.5		
Flanged, 14 in	8.0–10.0		
Flanged, 16 in	10.0–12.0		
Flanged, 20 in	13.0–17.0		
Flanged, 24 in	15.0–20.0		
Asbestos pipe covering for pipe in 3-ft lengths:			
¾ in. and less, per 100 ft	2.0–3.0		
1, 1¼, and 1½ in. per 100 ft	2.5–3.5		
2 and 2½ in. per 100 ft	3.0–4.0		
3 and 3½ in. per 100 ft	3.5–4.5		
4 and 5 in. per 100 ft	4.0–5.0		
6 in. per 100 ft	4.5–5.5		
8 in. per 100 ft	5.0–6.0		
10 in. per 100 ft	6.0–7.5		
12 in. per 100 ft	7.0–9.0		

labor cost of installing pipework and other items when the labor wage is known and the time required can be reasonably assumed.

Equipment The equipment needed will vary somewhat with the character and size of the job. Each team of pipe fitter and helpers may need pipe cutters, pipe threaders, vise and vise frame, wrenches, hammers, chisels, and other small tools. If an electrician is required, he will need his set of tools. Sometimes ladders, sawhorses, plank, etc., are also needed.

The cost of a set of tools for one team will be comparatively little for any one job. Transportation costs to and from the job should be included. The equipment cost will vary with the size of the job and the length of time required. Equipment costs per team per job may range from about \$5 or \$10 to \$25 or \$30.

Overhead and Profit Overhead may be based on labor costs alone or on the sum of materials, labor, and equipment. Overhead costs should include all general overhead items plus costs of all fees and permits. General overhead costs may range from 8 to 25 percent of the labor costs.

If no profit has been included in the material prices, the percentage allowed for profit may range from about 8 to 20 percent of the sum of all other costs. If profit on materials has been included in the material prices, job profit should be figured on labor, equipment, and overhead costs.

Steam and Hot-water Heating Estimates Estimates of the costs of hot-water and steam heating systems should be summarized to show total costs of materials, labor, equipment, overhead, and profit for the entire job. Unit costs can be computed per radiator, per square foot of radiation surface, per square foot of heated floor area, or per cubic foot of heated space as desired. The summary is as follows:

Item	Unit cost as desired	Total cost
Materials............	\$	\$
Labor...............		
Equipment..........		
Overhead...........		
Profit...............		
Total.............	\$	\$

Illustrative estimate. The general contractor building a medium-sized brick-veneer residence has requested you (as his estimator) to prepare an approximate estimate of a single-pipe heating system for this residence as a rough check on the bids submitted by the heating contractors.

The plans and specifications gave the following information:

1 steam boiler, capacity 550 sq ft of radiation, complete with smoke pipe and damper, jacket, and trim

1 stoker for coal fuel
15 cast-iron radiators having a total of 169 sections and 507 sq ft radiation
15 air valves, one for each radiator
2 large air valves, one for each return pipe to furnace
15 radiator valves
15 unions for radiators
15 bushings for radiators
15 floor plates for radiators
84 ft 2-in. copper-bearing black-finish steel pipe
92 ft 1.5-in. copper-bearing black-finish steel pipe
145 ft 1.25-in. copper-bearing black-finish steel pipe
14 T's, 2 reducers, 7 elbows for 2-in. pipe
9 elbows, 2 reducers for 1.5-in. pipe
37 elbows and couplings for 1.25-in. pipe
84-ft asbestos pipe insulation for 2-in. pipe
92-ft asbestos pipe insulation for 1.5-in. pipe
1 thermostat complete (to be selected by owner)

Material price estimates can be obtained from local wholesaler or from catalogues with allowance for freight and trucking.

A gang of 2 pipe fitters and 2 helpers (two 2-men teams) will be used. Wages are \$6 per hour for pipe fitters and \$4 for helpers, including payroll taxes and insurance.

Assume a reasonable cost of equipment.

Contractor's overhead costs to be assumed as 22 percent of labor cost, including all fees and permits.

Assume profit at 10 percent of all other costs.

Materials cost (to nearest dollar). Prices are costs (without profit) of materials delivered at the job.

		Local prices
1 steam boiler, capacity 550 sq ft of radiation, with jacket and trim	\$350	
1 stoker	295	
15 radiators, 169 sections, 507 sq ft at \$0.75	380	
15 air valves for radiators at \$2.00	30	
2 large air valves at \$11.00	22	
15 radiator valves at \$5.35	80	
15 radiator unions at \$1.35	20	
15 floor plates at \$0.30	5	
15 bushings at \$0.25	4	
84 ft 2-in. copper-bearing steel pipe at \$0.70	59	
14 T's, 2 reducers, 7 elbows at \$0.80	18	
92-ft 1½-in. copper-bearing steel pipe at \$0.60	55	
9 elbows, 2 reducers at \$0.70	8	
145-ft 1¼-in. copper-bearing steel pipe at \$0.45	65	
37 elbows and couplings at \$0.70	26	
176-ft asbestos covering for 2-in. and 1½-in. pipe at \$0.30	53	
1 thermostat complete (as selected by owner)	90	
Total		\$1,560

Labor cost (2 pipe fitters at $6 per hour and 2 helpers at $4 per hour, average wage of $5 per hour):

	Labor-hours
1 steam boiler and trim	12
1 stoker	6
15 radiators	25
2-in. pipe equivalent of (42 + 4 + 12) or 58 joints	47
1.5-in. pipe equivalent of (18 + 4) or 22 joints	16
1.25-in. pipe equivalent of 74 joints	45
176 ft of asbestos covering	6
1 thermostat	5
Total (81-team-hours)	162

Labor cost, 162 hr at $5	= $ 810
Equipment cost for 2 teams of 1 pipe fitter and 1 helper each	= 20
Overhead cost: asume 22% of labor cost	= 178
Profit: assume 10% of all other costs	= 257
Total estimated cost	= $2,825

This estimate may be summarized as follows:

Item	Unit cost per sq ft of radiation	Unit cost per radiator	Total cost
Materials	$2.84	$104	$1,560
Labor	1.47	54	810
Equipment	0.04	1	20
Overhead	0.32	12	178
Profit	0.47	17	257
Total	$5.14	$188	$2,825

Warm-air Heating Systems

Warm-air Heating Systems Warm-air heating systems are briefly classified in the article on Heating Systems in this chapter.

The cost estimate of a warm-air heating system will include the five main items of materials, labor, equipment, overhead, and profit. Unit costs can be computed, if desired, per warm-air outlet, per outlet (sum of warm- and cold-air outlets), per square foot of heated floor area, and per cubic foot of heated space.

Materials Takeoff It is assumed that the estimator has a fairly complete set of plans and specifications at his disposal so that he can readily take off and list all the required materials. The use of a good reminder list is advisable.

The takeoff for a central system may be as follows:

1. *Outlets.* Both warm and cold air. List size, finish, and other details.

2. *Pipes and ducts.* List size, shape, elbows, bends, junctions, outlet boxes, pipe insulation (if required), etc., for all warm- and cold-air ducts. Kind and gauge of metal should be noted.

3. *Furnace.* Complete with jacket, damper, smoke pipe, trim, etc. Size and heating capacity. Note if fan, humidifier, and air conditioning equipment are included with the furnace. Also note kind of fuel (coal, oil, gas) and firing equipment.

4. *Accessories.* Humidifier, fans, blower, washer, filters, and other air conditioning equipment not included with furnace. Coal stoker, oil burner, or gas burner if separate from furnace. Other miscellaneous accessories.

5. *Temperature and other controls.* Thermostats, timers, switches, and other controls with all needed wiring, piping, etc.

The takeoff for a unit system will include the complete unit or units ready to install. The size, capacity, and other details should be noted, including wiring, piping, etc.

Materials

Central Systems. Warm-air and cold-air outlets vary in price according to size, material, finish, etc. Prices may range from about $2 each up to $10 or more.

Pipes and ducts are usually made in a sheet-metal shop and erected on the job. Prices of pipe are often quoted per foot of length or per section (2, 2.5, or 3 ft long) and vary with the kind of metal, gauge, finish, size, and shape of the pipe. Prices of specials, such as register or outlet boxes, junctions, elbows, etc., are usually quoted per each with size, gauge, finish, and kind of metal given. Costs of pipes and pipe specials delivered at the job can be found by consulting a local sheet-metal shop.

Pipes and pipe specials may be made of tin, galvanized steel, copper-bearing steel, copper, or aluminum. A rust-resisting metal should be used for an air conditioning system. Number 26 gauge is about the lightest weight that should be used for piping, except that No. 28 gauge is suitable for the smaller sizes. Weights per 100 sq ft are as follows: No. 28 gauge 78 lb, No. 27 gauge 85 lb, No. 26 gauge 91 lb, and No. 24 gauge 116 lb. Tin plate may cost $15 or more per 100-lb box. Galvanized steel sheets may cost $12 or more per 100 lb. Approximate 1967 prices for galvanized pipe and ducts are given in Table 12-8.

Asbestos paper, suitable for covering pipes, comes in rolls about 18 in. wide and up to 100 yd in length. A 100-yd roll weighs about 50 lb and may cost about \$9. Aluminum foil on one side of 40-lb Kraft paper may cost about \$2.50 per 100 sq ft.

Smoke pipe and vent pipe may cost from 75 cents to \$2 for a 2-ft length, depending upon the diameter and gauge of metal. Bends and elbows will each cost about as much as a 2-ft pipe section. Dampers may cost from \$1 to \$3 each.

A separate coal stoker may cost from \$250 to \$900 plus installation costs. A separate oil burner, with tanks and accessories, may cost from \$200 to \$800 plus installation costs. A good thermostat with accessories may cost from \$50 to \$150.

Table 12-8 *Approximate* 1967 *Prices of Galvanized Steel Pipes and Ducts per* 100 *Lin Ft**

Pipe		Galvanized ordinary steel		Galvanized copper-bearing steel	
Diameter, in.	Circumference, in.	28 gauge	26 gauge	28 gauge	26 gauge
3	10	\$22	\$ 26	\$24	\$ 28
4	12–14	28	33	31	36
5	16	34	39	37	43
6	18–20	40	46	44	51
8	24–26		58		65
10	30–34		70		80
12	36–40		82		105
15	44–50		100		125

* Elbows (45 or 90 degrees) cost as much as the corresponding 2.5-ft section of straight pipe.

Warm-air furnaces, combined with air conditioning units, are available in various combinations. A furnace with a refrigeration cooling unit may cost about twice as much as the furnace alone.

Forced-air furnaces (oil or gas fired) for central systems for residences (4 to 10 heating outlets) are available in "packaged" units complete with a built-in burner and all needed accessories.

Complete forced-air furnaces combined with air conditioning units (i.e., heating and refrigeration cooling) are available in packaged units complete with accessories. The same piping system is used for both heating and cooling. The pipes should be of rust-resisting metal.

Forced-air furnaces for central systems may be priced according to heating capacity in cubic feet of space or in Btu per hour. Approximate 1967 prices and capacities are given in Table 12-9.

Table 12-10 gives the approximate amount of heat in Btu per hour required to heat 1 cu ft of space to 75 or 80°F.

Table 12-9 *Approximate* 1967 *Prices of Warm-air Central-heating Systems**

Furnace with built-in oil burner		Forced-air gas-fired furnace with automatic controls, thermostats, blower, filter, etc.		
Heating capacity, cu ft	Approximate price	Heating capacity Btu per hr	Warm-air outlets	Approximate price
6,000	$300	60,000	4–6	$200–$300
8,000	350	80,000	5–8	$230–$330
10,000	400	100,000	6–10	$280–$380
12,000	475	120,000	8–12	$350–$450
16,000	600	150,000	10–15	$425–$550

* Cost of materials for warm-air outlets and cold-air returns may vary from about $20 to $40 each, depending on size and length.

Table 12-10 *Approximate Btu per Hour Required per Cubic Foot of Space Heated*

Min outside temp, °F	Very good insulation, Btu per hr	Average insulation, Btu per hr	Poor insulation, Btu per hr
0	8	12	15
+10	7	10	13
+20	6	9	11
+30	5	7	9

Unit Systems. The materials required for a warm-air unit system are usually a gas floor or wall furnace with the necessary gas piping, vent pipes, and manual or automatic controls. Some units may have a built-in or a wall thermostat, and some of the larger units may have a fan or blower. Table 12-11 gives approximate 1967 prices for single floor and wall gas-fired furnaces with manual and automatic (thermostat) controls. A dual wall furnace may cost about $10 to $25 more than a single wall furnace of the same capacity. Separate thermostats may cost from $20 to $35 each, with installation costs of about $10.

A 3-in. vent or smoke pipe is needed for gas furnaces of 15,000- to 25,000-Btu capacity; a 4-in. smoke or vent pipe for gas furnaces of 35,000 to 55,000 Btu. Larger furnaces will require 5- or 6-in. vent pipes.

Labor The labor required for a warm-air heating system is for the installation of the pipes, registers, furnaces, accessories, and controls.

Table 12-11 *Approximate* 1967 *Prices for Unit Warm-air Gas-fired Furnaces**

Capacity Btu per hr	Single floor furnaces		Single wall furnaces		
				Automatic control	
	Manual control	Automatic control	Manual control	Built-in	Wall
25,000	$ 90	$115	$ 50	$ 65	$ 75
35,000	100	125	65	80	95
50,000	110	135	75	90	105
70,000	140	165	110	125	135

* A dual wall furnace costs from $10 to $25 more than a single wall furnace of the same capacity.

This labor may be done by skilled mechanics or by skilled mechanics aided by helpers.

The labor required for any particular installation may vary considerably according to the kind of job (large or small, new or old building, long or short leaders and stacks, number of bends and turns in pipes, number and kind of accessories, etc.), the working conditions, and the skill and inclination of the workmen. A team composed of a skilled workman with one or two helpers is a good working combination.

Approximate labor-hours required for different items of work are given in Table 12-12. The metal warm- and cold-air pipes and register boxes are assumed delivered at the job made up in correct sizes of junctions, elbows, bends, outlet boxes, and straight lengths of pipe and ready for installation. Labor-hours given for installing furnace include time for installing jacket, smoke pipe, and damper.

The labor required for installing pipes and ducts larger than 400 sq in. in cross-sectional area will vary with the cross-sectional area rather than with the circumference because of the heavier weights (gauge) of metal used and the lesser ease of handling.

The labor required for installing thermostats and controls will vary almost directly with the number of instruments, 1 to 3 hr being required for each instrument.

The labor required for installing various accessories will vary with the kind and size of accessory and the installation requirements.

Diagram 12-3 can be used for estimating the labor costs of installing various parts of warm-air heating systems.

The labor required for installing larger and heavier furnaces and equipment can be estimated according to the methods suggested in Chap. 16 on Heavy Machinery.

Labor wages per hour (1967) may vary from $4 to $7 for skilled workers and from $3 to $5 for good helpers. Payroll taxes and insurance may vary from 7 to 12 percent of the labor wage.

Equipment The equipment required for installing warm-air heating systems in buildings up to about 50,000 cu ft in volume will include all the hand tools required by the mechanics and perhaps a few special and portable tools such as saws, drills, and small soldering outfits. Sometimes ladders and sawhorses and planks for temporary scaffolding may be needed. Special equipment may be required for handling furnaces and other machinery that are too heavy to be handled by a small crew of men.

Table 12-12 *Approximate Labor-hours for Installing Warm-air Heating and Air Conditioning Systems*

Item, central systems	Labor-hours	Item, central systems	Labor-hours
Furnace, burner and accessories, package deal	8–20	Warm- or cold-air pipes:	
Separate humidifier	2–4	100 sq in. or less, per 10 ft	1–3
Furnace only	6–14	100–400 sq in. per 10 ft	2–5
Auxiliary fan	2–4	Registers with boxes, each (outlets or inlets)	1–3
Separate air conditioner, cooler only	6–14	Covering pipes with asbestos paper or aluminum foil per 100 ft	2–6
Combined forced-air systems, heating and cooling	10–24	Heaters, wall or floor, including pipe and connections	3–6
Coal stokers	4–10	Additional piping per 10 ft	1–2
Oil burners and tanks	6–14	Wall thermostat and controls	1–3
Thermostat and controls	2–8	Air conditioners, coolers only (window, wall, or ceiling)	2–6
Gas piping and connections, per 10 ft	1–2	Electrical outlet	2–4

The equipment cost for a warm-air heating job, including transportation, may vary from about $5 to $15 or $20, depending on the size of the job and other conditions. Equipment costs for buildings larger than 50,000 cu ft will be proportionately higher.

Overhead and Profit Overhead costs on a warm-air heating job are usually based on labor costs alone but may be based on the sum of the costs of materials, labor, and equipment. When based on labor costs, the general overhead may range from 15 to 35 percent.

Profit may range from about 8 to 20 percent on the sum of all other costs. If profit on the materials or on part of the materials has been included in the materials costs, this profit should not be included a second time.

Warm-air Heating Estimates Estimates for warm-air heating systems can be summarized to show the total costs of materials, labor, equipment, overhead, and profit. Unit costs can be computed, if desired, per warm-air outlet, per outlet (based on sum of warm- and cold-air outlets and inlets), per square foot of floor area heated, and per cubic foot of space heated. The summary can be tabulated as follows:

Item	Unit cost (as desired)	Total cost
Materials.	$	$
Labor.		
Equipment.		
Overhead.		
Profit.		
Total.	$	$

Illustrative estimate. As a general contractor, you desire to prepare an approximate estimate of the cost of a warm-air heating system for a small two-story frame residence, 26 by 28 ft in size. Fairly complete plans and specifications are available. The warm-air heating system is a forced-air system with an oil-fuel furnace, blower and air filter.

Materials: An examination of the plans and specifications and of wholesalers' catalogues available gave the following information:

Forced-air furnace with a built-in oil burner and complete with all items and accessories to meet all specification requirements and with 8 heating outlets (a package deal)		= $470
Registers, 8 warm-air and 8 cold-air returns		
4 registers, 6 × 10 in. at $2.50 each	= $ 10	
12 registers, 8 × 10 in. at $3 each	= 36	
16 register boxes at $1.50 each	= 24	
Total, all registers		= 70
Warm-air and cold-air piping (26 gauge)		
48 ft 3½ × 10 in. at $0.70 per ft	= $ 34	
8 elbows at $1.60 each	= 13	
150 ft 3½ × 12 in. at $0.85 per ft	= 128	
24 elbows at $2 each	= 48	
24 ft 10 × 14 in. trunk lines at $1.25 per ft	= 30	
4 elbows at $2.75 each	= 11	
Asbestos paper covering for all warm-air pipes estimated at	12	
Total, all piping		276
Total, materials		$816

Labor: Assume a team of one mechanic at $5.00 per hour and one helper at $3.50 per hour, including payroll taxes and insurance of 10 per cent of the labor wage.

Team cost per hour = \$5.00 plus \$3.50 = \$8.50		
Installing furnace, etc., complete	= 18 hr	
Installing registers and boxes, 16 at 1.40 hr	= 23 hr	
Installing piping system, assuming each elbow as the equivalent of a 2-ft pipe. 288 equivalent feet at 1.5 hr per 10 ft	= 43 hr	
Covering warm-air pipes with asbestos paper	= 4 hr	
Total labor-hours	= 88 hr	
Team-hours = 88 ÷ 2 = 44		
Labor cost = \$8.50 × 44		= \$ 374
Equipment for the job, estimated at		15
Overhead estimated at 20% of labor cost		= 75
Profit at 10% of all other costs		= 128
Total estimated cost		= \$1,408

This gives a unit cost of about \$176 per warm-air outlet or a unit cost of about \$88 for the sum of all warm- and cold-air outlets.

Air Cooling Systems

Air Cooling Systems Air cooling systems are briefly classified in the article on Air Cooling Systems in this chapter.

The cost estimate of an air cooling system will include the five main items of materials, labor, equipment, overhead, and profit. Unit costs may be computed per outlet, per square foot of cooled floor area, or per cubic foot of cooled space.

The amount of energy required for cooling (or for heating) is often stated as the number of Btu (British thermal units) required to cool (or heat) one cubic foot of space for different kinds of structures, varying temperature conditions, different exposures and uses. The amount of cooling energy required will vary considerably with the amount of insulation, window area, exposure to the direct rays of the sun, and use.

Another unit sometimes used is the amount of energy required to freeze one ton of ice in one hour. This is equal to 12,000 Btu per hour.

Table 12-13 gives the approximate Btu per hour required to cool 1 cu ft of space under varying conditions.

Materials Takeoff The estimator should have a fairly complete set of plans and specifications at his disposal so that he can takeoff and list all of the required materials.

Unit Air Cooling Systems. The takeoff should list the number of each kind (water-cooled or refrigeration), make, size or capacity, installation (window or wall), electric wiring and circuits required, outlets, volts, amperes, etc. Note if any thermostats or other controls are needed. Also note if the units are complete (package deal) and ready to install. A summary sheet may be prepared if there are many units.

Small Central Air Cooling Systems. The takeoff for a small central air cooling system should list such items as separate or combined system, water or refrigeration cooled, electric or gas, outside vent, number, gauge, length and size of all pipes and ducts, grills, location of cooler, electric wiring (volts, amperes, circuits, etc.), water piping, thermostats and controls, etc. If the cooling system is combined with a heating system, note how they are connected and if extra pipes, circuits, and controls, etc. are needed. Also note if the cooler comes in a packaged form ready to install.

Table 12-13 *Approximate Btu per Cubic Foot of Space Cooled*

Difference between outside and inside air, °F	Insulation and conditions		
	Very good	Average	Poor
10	1.5–2	2–3	3–5
15	2–2.5	2.5–3.5	3.5–6
20	2.5–3	3–4.5	4–7
25	3–5	4–7	5–9

Large Central Air Cooling Systems. Usually a general contractor prepares an approximate estimate instead of a complete detailed estimate. Hence the materials takeoff is quite general in character. The takeoff lists the kind of structure, size, cubic feet of space to be cooled, amount of cooling, kind (water or refrigeration), separate or combined, electric or gas, kind and amount of controls, and other items and information which will aid the general contractor in preparing a good approximate estimate.

Materials

Small Unit Systems. A small water cooler, suitable for one room, usually comes ready to operate as soon as some water is added and the cooler is plugged in to an electric circuit. The cost of this unit cooler may vary from about $25 to $40.

A small electric refrigeration cooled unit, suitable for about one to five rooms, usually comes ready to install and operate. Such a unit may vary in capacity from about 7,000 to 12,000 Btu per hour for a 110-volt circuit and up to 30,000 Btu per hour for a 220-volt circuit. The cost per unit may vary from about $100 to $500 (retail price) depending on the capacity.

The operating load is about 1,000 watts per 10,000 Btu per hour capacity. The initial or starting load may be momentarily about twice the operating load. The electrical circuit must be able to care for this starting load. All electrical units should be grounded. If extra electric circuits are required, the cost of these circuits must be included in the estimate.

A small electrical refrigeration cooled unit is usually installed in a window or in a wall. The cost of installation in a window is usually not more than $5 or $10. The cost of installation in a wall may range from about $25 to $60 per unit.

Small Central Systems. A small central air cooling system with a refrigeration unit and several cold-air outlets will cost about the same as a corresponding forced warm-air heating system. Refrigeration coolers often come in a packaged form ready for installation. The compressor should be installed at the outside of the building, or have an adequate vent or duct extending to the outside.

Table 12-14 gives approximate 1967 prices of refrigeration cooling units complete with accessories. If special electric circuits are needed, the cost of these circuits is additional.

Table 12-14 *Approximate* 1967 *Prices of Small Refrigeration Air Cooling Units*

Number of cold-air outlets	Cooling capacity Btu per hr	Approximate price including accessories
4–7	40,000–50,000	$200–$300
5–9	50,000–60,000	$225–$335
8–12	65,000–80,000	$260–$400
10–15	80,000–100,000	$300–$475
12–18	100,000–120,000	$375–$550

The cost of cold air outlets and returns will be about the same as the cost of warm-air outlets and returns of corresponding size and length. The cost of materials for cold-air outlets and returns may be estimated in the same way as the cost of materials for warm air outlets and returns as described in the section on Warm-air Heating Systems in this chapter. The cost of the material for a cold-air outlet or return may vary from about $20 to $50 depending on the material, gauge, size, and length.

The total cross-sectional area of the cold-air returns should equal the total cross-sectional area of the cold-air outlets. The material used should be rust resisting.

Large Central Systems. As stated in the article on Materials Takeoff (Air Cooling Systems) in this chapter, the general contractor rarely prepares a complete and detailed list of all of the materials required. He usually prepares an approximate estimate for the complete job, listing only such items as will aid him in selecting a reasonable unit price. Then he multiplies this unit price (in cents per cubic foot of space cooled) by the amount of space cooled (in cubic feet) to obtain the total estimated cost.

Labor The labor required for an air cooling system is for the installation of the cooling unit, accessories, pipes and ducts, vents, registers, controls, etc. The labor required may vary considerably because of the particular job, working conditions, and the skill and inclination of the workers. The comments stated in the article on Labor (Warm-air Heating Systems) in this chapter apply also to the labor for air cooling systems.

A team composed of one skilled workman and a helper is a good working combination. The approximate labor hours required for installing various items of an air cooling system are given in Table 12-12 of this chapter.

Labor wages per hour (1967) may vary from about $4 to $7 for skilled workman and from about $3 to $5 for good helpers. Payroll taxes and insurance may vary from about 8 to 15 percent of the labor wage.

Equipment The equipment required for installing air cooling systems in buildings up to about 100,000 cu ft in volume will include all of the hand tools used by the workmen and perhaps some special tools. Ladders and scaffolding may be needed on some jobs.

The cost of equipment for an air cooling job, including transportation, may vary from about $5 to $25 depending on the size of the job and other conditions.

Overhead and Profit Overhead costs on an air cooling job are usually based on the labor costs, but may be based on the sum of the costs of materials, labor and equipment. When based on the labor costs, the general overhead costs may vary from about 15 to 35 percent.

Profit may range from about 8 to 20 percent of the sum of all other costs.

Air Cooling Estimates Estimates of the cost of an air cooling system may be summarized to show the total costs of materials, labor, equipment, overhead, and profit. If desired, unit costs may be computed per cold-air outlet, per square foot of floor area cooled, or per cubic foot of space cooled. The summary may be tabulated as follows:

Item	Unit cost (as desired)	Total cost
Materials..........	$	$
Labor.............		
Equipment........		
Overhead..........		
Profit.............		
Total...........	$	$

Air Conditioning Systems

Estimating of Air Conditioning Systems As mentioned in the article on Air Conditioning Systems in this chapter, a complete air conditioning system includes temperature (heating and cooling), ventilation, circulation, and purification. Heating and cooling estimates have been discussed in the preceding articles of this chapter.

Perhaps the best way of estimating an air conditioning system is to study the plans and specifications and note the kind of system, whether it comes in a packaged form or in many separate parts, the work it is to do, and if it is a combined or separate system.

In a central system some of the items to be noted are method of ventilation, heating system (steam, hot water, forced air), cooling (water or refrigeration), blowers, humidifiers and dehumidifiers, air filters and washers, ducts or pipes for conveying air, outside-air connections, thermostats and other controls, etc.

In a unit system, the units are often complete in packaged form ready to install. Note whether the system is a combined or separate system. Note the kind of heating, the kind of cooling, piping connections, electrical connections, thermostats and controls, etc.

The amount of energy required for heating or cooling is often stated as the number of Btu per hour required to heat or cool 1 cu ft of space for different kinds of structures, for varying temperature conditions, and for different exposures and uses.

Table 12-10 gives the approximate Btu per hour required to heat 1 cu ft of space. Table 12-13 gives the approximate Btu per hour required to cool 1 cu ft of space. The amount of cooling required will vary considerably with the kind of insulation, amount of window space, exposure to the direct rays of the sun, and use.

A general contractor rarely prepares a complete detailed estimate for a large air conditioning job. He usually prepares an approximate estimate, like an approximate estimate for a large air cooling job as mentioned in the article on Materials (Air Conditioning Systems) in this chapter.

Materials Takeoff When preparing a takeoff for a central system, many estimators follow the procedure described in the articles on Materials Takeoff (in the sections on Steam and Hot-water Heating Systems and on Warm-air Heating Systems) in this chapter, including both heating and cooling items, and listing all items as to kind, size, dimensions, capacity, etc.

When preparing a takeoff for a unit system including a number of units, some estimators like to consider each unit by itself and list it and the items connected with it before proceeding to the next unit. Other estimators prefer to list all units first, then all piping, then all electrical connections, then the controls, etc. If the job is a fairly large one, a summary sheet can be prepared, summarizing the units, piping, electrical work, controls, etc.

After the takeoff has been completed, the various items can be priced and estimates made for the labor, equipment, overhead, and profit.

Materials

Central Systems. The assembly of a blower for a central system includes housing or casing, fan, and motor. The capacity of a fan is usually given in cubic feet of air per minute for a given outlet velocity, static pressure, and barometric reading. Manufacturers' catalogues and fan tables should be consulted for sizes, capacities, motor horsepower required, outlet velocities, weights, and prices. Outlet velocities are often limited to 1,200 or 1,400 fpm.

Fans are usually driven by electric motors. The horsepower required will vary with the fan output and static pressure. The electric motors selected should have a horsepower 10 to 25 percent greater than that listed in the fan tables because of possible variations in static pressures.

Several types of heaters are used in fan systems. There are several designs of each on the market.

The fan, motor, heaters, and casing or housing are usually mounted on brick or concrete foundations. The casing should be designed to reduce friction resistances as much as possible.

The duct or pipe system is usually a trunk system with as many branches as desired. Sometimes two or more main-trunk ducts are used, depending upon the structure and the design. Branches lead from the mains or trunks to the outlets. The ducts are often constructed of galvanized steel or iron of suitable gage and reinforced with angles, straps, or other small structural shapes. Rust-resisting metal should be used for air ducts.

The outlets are usually of grillwork somewhat like warm-air registers in appearance.

Humidifiers, filters, or washers are usually located near the blower and heater. Filters or washers are usually placed between the fresh-air inlet and the blower.

The thermostatic control is usually a little more complicated than that for an ordinary warm-air heating system. Instruments may be located in the fresh-air inlet, near the air washer or filter, on the far side of the fan, and also as desired in the building. If desired, a thermostat and mixing damper may be provided for each room.

Unit Systems. There are several kinds of unit air conditioners on the market with varying capacities. Each unit may consist of a fan, motor, heater and cooler, filter, and sometimes a humidifier or dehumidifier.

These units are usually sold by the manufacturer completely assembled and ready to install. Hence, it is necessary to list the units required by size, dimensions, capacity, and manufacturers' number as desired.

These units often require extra piping and electrical connections.

Labor The labor required for installing central air conditioning systems can be estimated by considering the different parts of the system and estimating the labor required for each part. Most central air conditioning systems are installed by manufacturers, dealers, or contractors who specialize in that line of work. The bids submitted usually include all costs of the plant installed and ready to operate. The general contractor will rarely have to do the work of installation.

The labor for installing heavy blowers, heaters, motors, fans, casings, filters, washers, etc., can be estimated according to the methods suggested in Chap. 16 on Heavy Machinery.

The labor for installing duct systems will vary with size and weights of the ducts and the working conditions.

The labor for installing control systems will vary almost directly with the number of instruments. From 1 to 3 labor-hours per instrument may be allowed.

The labor required for installing various parts of small and medium-sized central air conditioning systems may be estimated by the methods given in the articles on Labor in the sections on Steam and Hot-water Heating Systems, on Warm-air Heating Systems, and on Air Cooling Systems. Table 12-12 gives the approximate labor-hours required for installing various parts of a small or medium-sized system.

A labor gang consisting of 1 or 2 mechanics and 1 to 4 helpers is usually sufficient for installing small and medium-sized systems. Sometimes the services of an electrician, pipe fitter, or a sheet-metal worker may be needed.

Diagram 12-4 (Appendix F) may be used for estimating the labor costs.

The wages of skilled mechanics may vary from about $4 to $7 per hour and of helpers from about $3 to $5 per hour. A labor gang may be composed of a foreman, mechanics, and helpers. Mechanics may include electricians, pipe fitters, sheet-metal workers, masons, and carpenters.

Payroll taxes and insurance may vary from about 7 to 12 percent of the labor wage.

Equipment The equipment required for installing small and medium-sized central systems will include the hand tools of the mechanics, special tools, ladders, and some scaffolding. Equipment costs for a job may range from about $5 or $10 up to $50.

Overhead and Profit Overhead for installing air conditioning systems is often based on labor costs. General overhead may range from about 10 to 30 percent of the labor cost.

Profit on air conditioning jobs may range from about 8 to 20 percent of the sum of all other costs. If, when estimating unit systems, the price of a unit includes profit, this profit should not be included again.

Air Conditioning Estimates Estimates of the cost of an air conditioning system should be summarized to show total costs of materials, labor, equipment, overhead, and profit. Unit costs of central systems can be computed per cubic foot of space or per square foot of floor area. Unit costs of unit systems can be computed per unit, per cubic foot of space, or per square foot of floor area. Unit cost per square foot of floor area appears to be the unit selected by many estimators. Central air conditioning systems may cost from about $1 to $2 per square foot of floor area. In each case, information should be given in regard to kind of building, occupancy, and kind of air conditioning system. The estimate can be summarized as follows:

Item	Unit cost as desired	Total cost
Materials..........	$	$
Labor.............		
Equipment........		
Overhead..........		
Profit.............		
Total...........	$	$

Illustrative estimate. As a general contractor you desire to prepare an approximate estimate of the cost of a central air conditioning system for a new one-story residence that you are building. Fairly complete plans and specifications are available. The house is being built in a warm climate.

An examination of the plans and specifications gave the following information:

Seven rooms plus hall and bathroom. No basement and a low attic.
Floor area is 1,875 sq ft. Volume is 16,200 cu ft.
A combined forced-air central air conditioning system is required with eight floor outlets (two main trunks with three outlets each and two single outlets). The ducts for the outlets are located under the floor. There is one return duct in the attic with a wooden grill in the hall ceiling.
The air conditioning unit is located in an inside closet about 8 ft from the rear of the building. There is a vent for the furnace extending through the roof, and an air duct under the floor extending from the compressor to the outside of the house. There are thermostat controls with two electric switches so that the heater and the cooler can be turned on and off as desired.
All needed electric wiring is included in the electric contract.
All needed gas piping for the heater is included in the plumbing contract.

Materials:

One complete air conditioning unit with accessories		$ 712
One wooden grill for air-return duct	$14	
8 floor registers and boxes @ $4.75	38	
Total, registers, boxes and grill		52
Air-return duct, 26 gauge, 14 × 16 in.		
18 ft at $2.20 per ft	40	
1 elbow at $5.60	6	
Two main trunks, 26 gauge, 10 × 12 in.		
40 ft at $1.35 per ft	54	
2 elbows at $3.35 each	7	
Eight branch ducts, 28 gauge, 4 × 10 in.		
102 ft at $0.85 per ft	87	
8 elbows at $2.10 each	17	
Air duct, compressor to outside of building		
26 gauge, 9 ft at $2.20 per ft	20	
1 outside grill	2	
One roof vent for heater	13	
Asbestos paper covering for pipes and ducts	17	
Total for pipes, ducts and vents		263
Total materials		$1,027

Labor: Assume a team of one mechanic at $5.50 per hour and one helper at $4 per hour, giving a team cost of $9.50 per hour. This labor cost includes payroll taxes and insurance.

Labor-hours	
Installing the air conditioning unit	24
Installing all registers and boxes	16
Installing all trunks, pipes, ducts, etc	60
Covering trunks, pipes, and ducts with asbestos paper	8
Connecting up and testing system	4
Total labor-hours	112
Team hours = 56	
Labor cost = $9.50 × 56	$ 532
Equipment. Estimated	22
Overhead. 15 percent of the labor cost	80
Profit. 10% of all other costs	166
Total estimated cost	$1,827

This gives a unit cost of $228 per outlet, or $203 each for eight outlets and one return duct.

The estimated unit cost per square foot of floor area is about $1.00, and the estimated cost per cubic foot of space is about $0.113.

CHAPTER THIRTEEN

Plumbing

Estimating of Plumbing The plumbing work on a job is almost always sublet to a subcontractor specializing in this kind of work. This subcontractor (or his plumbing estimator) is a man capable of preparing detailed and accurate estimates. Consequently, the general contractor (or his estimator) rarely needs to be able to prepare an accurate and detailed plumbing estimate, but he should have a good general knowledge of estimating so that he can check intelligently and can determine the reasonableness of the detailed bids submitted by the plumbing contractors. It is the purpose of this chapter to help supply the general knowledge that the general contractor or his estimator may need for checking plumbing bids and estimates.

On some jobs, a separate set of plumbing plans, diagrams, and specifications are furnished in detail so that it is an easy matter for the estimator to take off accurately all plumbing materials required. With such plans, a good estimator can prepare an accurate list of everything that will be required.

On other jobs, the plans show merely the locations of the fixtures and the specifications state the types and makes. In such cases, the plumbing estimator must be able to visualize the work in detail and perhaps make drawings or sketches of the piping so that he can estimate reasonably accurately all the required pipings and fittings as well as the fixtures.

Some architects, when preparing the plans and specifications, do not describe all the plumbing work in detail but insert one or more general clauses stating that all plumbing work must be done in accordance with all state and local codes and regulations and also that the work must pass the inspection of all authorities having jurisdiction, usually a municipal plumbing inspector. Such clauses place upon the estimator the need of determining all the requirements relating to kinds and sizes of pipes, stacks, vents, traps, and other details.

Plumbing work is usually divided into rough plumbing and finish plumbing. Rough plumbing includes all sewer pipes, water and gas pipes, vents, drains, etc., required, and finish plumbing includes the fixtures and their accessories.

The plumbing estimate for either rough or finish work may be divided into materials, labor, equipment, overhead, and profit. Rough and finish plumbing are usually estimated separately, though in some instances the materials, labor, and equipment can be estimated separately and overhead and profit figured on the total plumbing work.

Rough Plumbing

Rough-plumbing Takeoff The takeoff for the rough plumbing should include all materials for

Water pipes, both exterior and interior
Gas pipes, both exterior and interior
Sewer pipes
Drains
Soil and vent stacks
Soil, waste, and vent pipes
All roughing work required for fixtures

The takeoff must show the size, length, kind (cast iron, copper, lead, glazed clay, galvanized, black), weight (standard or extra heavy), and other details of all pipe; materials needed for calking cast-iron soil, waste, and vent pipes; and all pipe fittings such as unions, nipples, caps, hangers, elbows, or bents, Y's, and T's. The takeoff should also list kinds and sizes of all valves, traps, plugs, exterior grease traps, cleanouts, etc. If trenching is required, the length, average depth, and kind of soil should be noted.

Explanations of some of the rough plumbing terms follows:

A sewer pipe is the section of drain piping extending from the street sewer to the building. This pipe may be of cast iron or glazed vitrified clay, 4 in. or larger.

The house drain is the drain pipe inside the building (horizontal) extending from the sewer pipe to the stacks. It may be of cast iron, wrought iron, or steel.

Stacks and vertical pipes extend from the house drain to the roof. Stacks are usually made of cast-iron pipe, though wrought-iron or steel pipe with sanitary fittings can be used. Stacks may be soil, waste, or vent.

Between the various fixtures and the stacks and connecting them are three kinds of pipes, namely:

Soil pipes, draining a toilet or closet

Waste pipes, draining all other fixtures

Vent pipes, affording ventilation and preventing siphonage of water in traps

Soil and waste pipes must be located below the fixtures they serve.

Vent pipes are usually connected above the fixture or trap to be vented.

The rough-plumbing takeoff should be accurate in detail. All materials should be listed in an orderly manner so that totals can be readily computed, prices inserted, and costs computed.

The local water company usually runs the water pipe from the water main to the lot line and installs a shutoff there. The plumber runs the water pipe from the shutoff to the building. The water company may furnish and install the water meter, or the plumber may do the installing.

The municipality usually lays the sewer from the nearest main to the lot line, and the plumber lays the sewer from the lot line to the building.

The local gas company usually lays the gas pipe from the mains into the building and furnishes and installs the meter.

Materials for Rough Plumbing Pipe of various kinds are usually priced per linear foot with other details being given. Sometimes the price is per piece, with length and other details being given, or per pound or ton with size, thickness, and other details being given.

Pipe fittings may be priced per piece, per dozen pieces, or per 100 pieces, with kinds, sizes, and other details being given. Sometimes they are priced by the pound.

Valves, traps, and special fittings are usually priced per piece with size, kind, material, and other details noted.

The data given in this chapter in regard to materials and prices are approximate only and are subject to change from time to time, as manufacturers modify their products and prices change. The data are included for the purpose of giving the estimator some information for making approximate estimates and for checking estimates submitted by plumbers.

Lead and oakum are usually sold by the pound. About 3 lb of pig lead and 1/4 lb of oakum are needed for one joint of 4-in. cast-iron pipe. Lead may cost 20 to 35 cents per pound, and oakum 20 to 30 cents per pound.

Vitrified-clay sewer pipe comes in various sizes and lengths (usually 2 to 4 ft). The different diameters are priced per foot of length delivered at the job. Average price ranges are about as follows. The 6, 8 and 12 in. are single-strength, and the 24 in. is double-strength pipe.

6-in. diameter, \$0.50 to \$0.85 per lineal foot
8-in. diameter, \$0.60 to \$1.00 per lineal foot
12-in. diameter, \$1.15 to \$2.00 per linear foot
24-in. diameter, \$5 to \$10 per lineal foot
T's, Y's, and L's will cost about twice as much as straight pipe.

Cast-iron soil pipe comes in various diameters and weights. The length of each section is usually 5 ft, and each section may be single hub or double hub. Double-hub sections are useful when short lengths are required. Each double-hub section, when cut into two parts, will provide two short sections each with a hub at one end. Diameters vary from about 2 to 10 in., the 3-, 4-, 6-, and 8-in. diameters being more common. The pipe may come in standard weight or extra-heavy weight. The extra-heavy weight is preferred because of its greater strength.

Cast-iron soil pipe is usually sold by the manufacturers to wholesalers and large dealers by the ton with size, thickness, and other details given. Prices (1967) may range from about \$145 to \$185 per ton, f.o.b. factory. Prices vary somewhat with the pipe diameters, the smaller diameters costing more. Small dealers, plumbers, and retailers usually price the soil pipe by the 5-ft section, diameter and weight being given. It is easier for the estimator to figure soil pipe by size and length than by weight. Prices delivered at the job may range from about \$200 to \$250 per ton.

Table 13-1 gives approximate weights and prices of 5-ft cast-iron soil pipe. Dealers price lists should be consulted for accurate estimates.

Three-inch-diameter cast-iron soil pipe is about the smallest size that should be used. Four-inch is recommended as a minimum size. Table 13-2 gives the approximate maximum number of fixtures that should be connected to a soil pipe.

Cast-iron soil-pipe fittings are usually estimated by the piece, and the price per piece obtained from the dealer. However, fittings are often sold by the manufacturer by weight, the price per ton averaging about 1.5 to 2.5 times that of the regular soil pipe, depending upon the manufacturing

Table 13-1 *Sizes, Approximate Weights, and Prices of 5-ft Sections of Cast-iron Single-hub Soil Pipe**

Diameter, in.	Weight per 5-ft length, lb		Approximate prices (1967) of 5-ft lengths	
	Standard	Extra heavy	Standard	Extra heavy
2	18	27	$ 2.75	$ 3.80
3	26	47	3.65	6.00
4	35	65	4.50	7.00
5	45	85	6.00	10.00
6	55	100	6.80	11.50
8	...	170	12.00	17.50
10	...	225	20.00	36.00
12	...	275	28.00	46.00

* A double-hub section will weigh (and cost) about 5 percent more than the corresponding single-hub section.

Table 13-2 *Maximum Number of Fixtures Connected to Pipe*

Diameter of pipe, in.	Soil pipe alone, water closets, and toilets	Combined waste and soil-pipe fixtures
2	...	8
3	2	18
4	15	60
5	35	140
6	70	280
7	120	500
8	200	800
10	400	1,500

Table 13-3 *Approximate* 1967 *Prices of Simple Cast-iron Soil-pipe Fittings*

Size of pipe, in.	Quarter and eighth bends		T's and Y's		Cast-iron plain taps and traps	
	Standard	Extra heavy	Standard	Extra heavy	Standard	Extra heavy
2	$0.70	$ 0.90	$ 1.10	$ 1.40	$1.50	$1.85
3	0.90	1.20	1.70	2.20	1.90	2.45
4	1.30	1.60	2.40	2.90	2.50	3.30
5	1.80	2.30	3.40	4.40		
6	2.30	2.90	4.75	6.00		
8	6.75	8.00	11.00	16.00		
10	.....	11.00	......	21.00		
12	.....	15.00	......	25.00		

difficulties. For rough estimating, approximate prices of various fittings can be used without causing appreciable errors in the totals. Dealers' catalogues and price lists should be consulted for accurate estimates.

Steel and wrought-iron pipe may be of ordinary steel, copper-bearing steel, or wrought iron and black or galvanized in finish. Galvanized finish is preferred except for gas pipes. Steel and wrought-iron pipe may be sold by weight or by the foot. Table 13-4 gives the sizes and weights of steel and wrought-iron pipe.

Approximate retail prices are given in Table 13-5. Pipes come in ran-

Table 13-4 *Black-finish Steel and Wrought-iron Pipe Diameters and Weights**

Nominal size, in.	Weights, lb per 100 ft			Nominal size, in.	Weights, lb per 100 ft		
	Standard	Extra strong	Double extra strong		Standard	Extra strong	Double extra strong
¼	43	54	...	2	367	502	902
⅜	57	74	...	2½	581	766	1,369
½	85	108	171	3	761	1,025	1,858
¾	113	147	244	4	1,088	1,498	2,754
1	168	217	365	5	1,481	2,077	3,855
1¼	228	300	521	6	1,918	2,857	5,316
1½	273	363	640	8	2,856	4,339	7,243

* Galvanized-finish pipe will weigh a little more. Permissible variation in weight is 5 percent. Pipes 2.5 to 3 in. in diameter are butt-welded. Pipes 3.5 to 6 in. in diameter are lap-welded. See Table 12-3 for more data on pipe.

Table 13-5 *Approximate* 1967 *Prices of Standard-weight Steel and Wrought-iron Pipe per* 100 *Lin Ft*

Size, in.	Ordinary steel		Copper-bearing steel		Wrought-iron		Extra for threads, each end
	Black	Galvanized	Black	Galvanized	Black	Galvanized	
¼	$ 11	$ 12	$ 13	$ 14	$ 16	$ 18	$0.10
⅜	12	13	14	16	20	23	0.10
½	13	15	16	18	24	28	0.10
¾	18	20	20	22	30	35	0.12
1	25	29	29	33	45	50	0.14
1¼	34	38	37	41	56	62	0.16
1½	39	46	45	52	70	77	0.18
2	54	63	62	72	87	100	0.23
2½	88	105	103	115	130	160	0.35
3	115	135	137	148	195	220	0.45
4	177	210	200	225	....	....	0.55

dom lengths, with an average of 20 ft. Pipes can be purchased with or without threaded ends. If ends are threaded, one coupling may be furnished with each random length.

Approximate prices of various fittings for standard steel pipe are given in Table 13-6.

Lead Pipe. Bends and traps should not weigh less than the values given in Table 13-7. Lead pipe may cost about 25 cents per pound.

Table 13-6 *Approximate* 1967 *Prices per* 100 *Fittings for Standard Steel Pipe*

Fitting	Size of pipe, in.							
	½ or less	¾	1	1¼	1½	2	2½	3
Caps, plugs, nipples, and bushings....	\$ 7–\$ 9	\$ 9–\$12	\$10–\$14	\$ 17–\$ 20	\$ 20–\$ 27	\$ 26–\$ 35	\$ 35–\$ 50	\$ 50–\$ 65
Couplings.....	\$ 9–\$11	\$13–\$17	\$17–\$23	\$ 24–\$ 30	\$ 35–\$ 45	\$ 60–\$ 70	\$ 85–\$100	\$115–\$140
Reducers.....	\$13–\$17	\$21–\$25	\$27–\$37	\$ 35–\$ 50	\$ 45–\$ 60	\$ 70–\$ 90	\$110–\$150	\$150–\$200
Malleable elbows.....	\$14–\$18	\$18–\$22	\$37–\$ 42	\$ 58–\$ 65	\$ 80–\$ 90	\$110–\$125	\$200–\$235	\$310–\$370
T's and Y's.	\$27–\$32	\$56–\$63	\$90–\$100	\$115–\$125	\$160–\$180	\$300–\$325	\$330–\$360	\$370–\$400
Unions.....	\$45–\$50	\$50–\$55	\$65–\$ 75	\$ 90–\$100	\$110–\$125	\$135–\$180	\$225–\$300	\$400–\$500

The cost of fittings with galvanized finish may be about 20 percent more than the cost with black finish. The cost of fittings for copper-bearing steel pipe may be about 20 percent more than for standard steel.

Table 13-7 *Weights of Lead Pipe*

Inside diam, in.	Wt per ft, lb	Inside diam, in.	Wt per ft, lb
1	2.0	2	4
1¼	2.5	3	6
1½	3.5	4	8

Brass Pipe and Fittings. Brass pipe should be of the standard commercial iron-pipe-gauge thickness for the various diameters. Fittings should have thicknesses equal to those of brass pipes of the same diameter. Brass tubing should have a thickness equal to No. 18 gauge or thicker.

Copper Tubing and Fittings. Copper tubing is available in both soft and hard tempers. The soft-tempered tubing comes in coils of 30, 45, or 60 ft in length and in sizes up to 1¼ in. in diameter. Both soft- and hard-tempered tubing are available in straight 20-ft lengths. Copper tubing can be bent around joists, corners, etc., and thus save many fittings and much labor. Solder fittings are required for hard-tempered tubing.

Soft-tempered tubing comes in two weights or types: standard (type L) and heavy (type K). Hard-tempered tubing comes in three weights or types: standard (type L), heavy (type K), and light (type M). The sizes, or diameters, most commonly used are ⅜, ½, and ¾ in., though sizes are available from ¼ up to about 12 in. diameter. Table 13-8 gives approximate weights and 1967 prices for copper tubing of the ⅜, ½, ¾, and 1 in. diameters. Approximate 1967 prices for fittings for copper tubing are given in Tables 13-9 and 13-10.

Table 13-8 *Approximate Weights and* 1967 *Prices for Hard and Soft Tempered Copper Tubing**

Tube diam, in.	Approximate wt per 100 ft, lb†			Approximate price per 100 ft‡		
	Light M	Standard L	Heavy K	Light M	Standard L	Heavy K
⅜	15	20–24	30–28	$13–$25	$16–$ 30	$25–$ 45
½	21	30	40–37	$15–$30	$20–$ 40	$28–$ 55
¾	33	50	70	$25–$50	$35–$ 70	$50–$100
1	50	75	110	$35–$70	$50–$100	$70–$140

* Soft tempered tubing is not available in the light M grade.

† When two weights are given, the first is for the hard-tempered tubing.

‡ Approximate warehouse prices for copper tubing (1967) range from about $0.70 to $1.40 per pound.

Table 13-9 *Approximate* 1967 *Prices for Fittings for Flexible or Soft-tempered Copper Tubing*

Tube size, in.	T's	Union elbows	Elbows	Unions	Adapters	Flaring tool
⅜	$1.50	$3.00	$1.20	$1.00	$0.75	$0.75
½	2.00	3.50	1.50	1.25	1.00	0.85
¾	2.50	4.00	2.00	1.70	1.25	1.00

Brass Valves. Brass valves vary in price according to size, kind, and care used in manufacture. Approximate prices (1967) are given in Table 13-11. These prices are for medium-grade goods. Prices may be cheaper when the valves are bought in quantity lots.

Traps. Cast-iron drain traps, about 4 by 5 or 4 by 8 in. in size, may cost $2 to $4 each. A metal grease trap, about 18 in. in diameter and 24 in. high, may cost $15 to $30. Vitrified clay pipe (two lengths) with concrete base and cap may cost $12 to $25.

Increasers. Cast-iron increasers are needed where the soil stack goes through the roof. Approximate prices are $3 to $4 for 2-in. soil pipe, $4 to $6 for 3-in. soil pipe, and $5 to $8 for 4-in. soil pipe, depending upon size and weight.

Table 13-10 *Approximate* 1967 *Prices for Solder Fittings for All Types of Copper Tubing*

Tube size, in.	T's	Unions and union elbows	Other types of fittings
3/8	$0.20	$0.70	$0.15–$0.30
1/2	0.25	0.60	$0.15–$0.35
3/4	0.45	0.65	$0.20–$0.45
1	1.15	1.50	$0.40–$0.75
1 1/2	2.25	2.50	$0.65–$0.90
2	4.50	6.00	$1.00–$1.60

Table 13-11 *Approximate* 1967 *Prices of Brass Valves*

Size of pipe, in.	Gate valve	Globe valve	Stop and waste cock	Check valve	Angle valve
3/8	$ 2.50	$ 2.50	$2.75	$ 2.75	$ 2.15
1/2	2.75	2.75	3.25	3.00	2.50
3/4	3.25	3.10	4.00	3.75	3.25
1	4.75	4.00	4.75	4.50	4.75
1 1/4	7.00	6.75	6.25	6.40	7.00
1 1/2	9.00	8.00	7.00	8.50	9.00
2	12.50	10.50	8.50	11.00	11.50

Miscellaneous. Pig lead may cost about 30 cents per pound, oakum about 50 cents per pound, and wire solder about $1.35 per pound. The estimator should consult a dealer's supply catalogue for prices of miscellaneous materials.

Labor for Rough Plumbing Plumbing work is almost always done by teamwork, the team usually consisting of two men, one plumber and one helper, for all ordinary work. On very heavy work, the team may consist of one plumber and two or three helpers. On large jobs, there may be several plumbing teams working at the same time under the supervision of a plumbing foreman or master plumber.

In this chapter, a plumbing team will be assumed to consist of one plumber and one helper unless otherwise specified. Hence, the actual labor-hours will be twice the team-hours.

The time required for a certain item of plumbing work will vary considerably, depending on the particular work, the working conditions, and the skill and inclination of the plumber. Roughing work for plumbing fixtures can be estimated in labor-hours or team-hours required per fixture without going into further details. Labor for pipework can be more accurately estimated per joint instead of per 100 lin ft of pipe, with kind of pipe and diameter given. The labor per joint will include all threading, hanging, and connecting needed. The larger the pipe, the more labor-hours required. It is assumed that an adequate supply of nipples, unions, couplings, and other fittings are available.

Approximate labor-hours and team-hours required for various kinds of rough plumbing work are given in Table 13-12.

When threaded joints are estimated, each coupling and valve counts as two joints, a T as three joints, etc. The labor-hours and team-hours given for threaded pipe joints in the table are based on the assumption that one pipe thread must be cut for each joint. When no thread needs to be cut, the time given can be reduced 35 to 50 percent. Each fitting will require two joints on an average.

The wages for plumbers may vary from about $4 to $7 per hour. Wages of plumbers' helpers may vary from about $3 to $5 per hour. Payroll taxes and insurance may vary from 8 to 14 percent of the labor wage.

Diagram 13-1 (Appendix F) can be used for estimating the labor cost of pipework and Diagram 13-2 (Appendix F) for the labor cost of rough-plumbing work when the hourly combined wage of plumber and helper is known and the time required can be reasonably assumed.

Equipment The equipment needed will vary with the character and size of the job. Each team may need the following:

Pipe cutters
Pipe threaders
Vise, and frame for holding vise
Workbench
Wrenches
Calking tools
Lead-melting pot
Lead
Oakum for calking
Blowtorch
Hammers, chisels, and other small tools
Ladders, sawhorses, and planks, etc., for temporary scaffolds

Table 13-12 *Approximate Team- and Labor-hours Required for Plumbing Work*

Kind of work	Team-hours	Labor-hours
Digging a 6-ft trench and backfilling, per lineal foot	0.25–0.60	0.5–1.2
Installing sewer pipe, 4 in. to 12 in., in trench, per ft	0.10–0.30	0.2–0.6
House drains and stacks, cast-iron soil pipe:		
3- and 4-in. diameter, lead and oakum joints, per joint	0.25–0.60	0.5–1.2
5- and 6-in. diameter, lead and oakum joints, per joint	0.45–0.75	0.9–1.5
Pipe work, steel and wrought-iron threaded pipe, including threading, hanging, and connecting, assuming one thread cut per joint:		
½-in. pipe, and smaller, per joint	0.15–0.25	0.3–0.5
¾- and 1-in. pipe, per joint	0.20–0.30	0.4–0.6
1¼- and 1½-in. pipe, per joint	0.25–0.40	0.5–0.8
2- and 2½-in. pipe, per joint	0.35–0.55	0.7–1.1
3- and 3½-in. pipe, per joint	0.45–0.75	0.9–1.5
4-in. pipe, per joint	0.65–1.00	1.3–2.0
Copper tubing:		
⅜- and ½-in. diameter, per joint	0.15–0.35	0.3–0.7
¾-in. diameter, per joint	0.20–0.40	0.4–0.8
(Usually fewer joints, ⅓ to ¼, will be required for copper tubing than for steel pipe of equal length)		
Roughing work:		
Urinal with stall	3–8	6–16
Water closet	6–12	12–24
Lavatory	4–10	8–20
Shower with stall	6–10	12–20
Bathtub	4–8	8–16
Bathtub with shower	6–12	12–24
Kitchen sink	4–7	8–14
Slop sink	3–6	6–12
Laundry tubs, pair	4–7	8–14
Floor drain	2–4	4–8
Grease trap	3–6	6–12
Valves, faucets, etc., installed with rough plumbing:		
1-in., or less, pipe connection, each	0.25–0.50	0.5–1.0
1- to 2-in. pipe connection, each	0.35–0.65	0.7–1.3
2-in. and over pipe connection, each	0.50–0.75	1.0–1.5
Testing plumbing system when required, per fixture	1–2	2–3

The cost of a set of tools required for one team will be comparatively little for one job. The cost of transporting equipment to and from the job must not be neglected. The cost will vary with the magnitude of the job and the length of time required. Equipment costs per team per job may range from $5 or so up to $20 or $25.

Overhead and Profit Overhead on rough-plumbing jobs may be based on labor costs alone or on the sum of labor, materials, and equipment costs. If costs of materials are estimated at retail prices delivered at the job, then no overhead and profit on materials should be included in the estimate because these retail prices include allowances for these items.

General overhead, based on labor, may range from 10 to 25 percent of the labor costs.

Other items that are usually included in the overhead are the costs of all fees and permits to be paid for by the contractor. The water company and sewer department charge for extending the water pipes and sewers to the lot line. The gas utility charges for laying the gas pipe from the main into the building and installing the meter. Sometimes these charges are paid by the owner directly, and sometimes they are paid by the contractor. If paid by the contractor, such charges can suitably be included in the overhead costs. The costs of these items are in addition to the percentages given.

If profit has not been included in the material prices, the percentage allowed for profit may range from 8 to 20 percent of the sum of all other costs. If profit on materials has been included in the material prices, the job profit should be figured on the labor and equipment costs.

Summary The rough-plumbing estimate should be tabulated and summarized to show the total costs of materials, labor, equipment, overhead, and profit for the entire job. Unit costs per plumbing fixture are sometimes computed. The summary may be as follows:

Item	Cost per fixture	Total cost
Materials..........	$	$
Labor.............		
Equipment........		
Overhead..........		
Profit.............		
Total...........	$	$

Illustrative estimate of rough plumbing. Prepare an approximate cost estimate for the rough plumbing for a small two-story six-room house. All pipes enter the basement. An examination of the plans and specifications shows that there is to be a 30-gal automatic gas water heater, twin laundry tubs, and a small two-burner gas laundry stove in the basement. On the first floor, there is to be a 42-in. left-hand-drain apron-front sink in the kitchen. Gas pipe is to be run for a gas stove in the kitchen (stove is not furnished by the contractor). On the second floor over the kitchen, the bathroom has a built-in corner bathtub with shower, a toilet, and a wall lavatory. Laundry tubs, kitchen sink, bathtub and shower, and lavatory are to have hot and cold water. Plumbing fixtures are to be selected by owner from dealer's catalogue. The contractor is to provide the plumbing permit, costing $8. The gas company runs pipe from gas main into basement and installs meter for $30. Water and sewer are available at the lot line.

The rough-plumbing takeoff gave the following information:

Gas line. In the house, 52 ft of ¾-in. black steel pipe with 2 T's, 14 elbows, and 1 valve are required.

Water line. From lot line to house, 35 ft of ¾-in. galvanized pipe are needed. In the house, there are required 165 ft of ¾-in. galvanized-steel pipe with 12 T's, 38 elbows, 2 faucets for outdoor outlets for sprinkling, and 1 valve near meter. Water company provides meter, but plumber installs it. Plumber must dig trench for water pipe. Trench is 35 ft long and 5 ft deep.

Sewer line. From lot line to house, 35 ft of 4-in. cast-iron soil pipe are required. Plumber must dig trench 35 ft long and 7 ft deep. In the house, there are needed 30 ft of drain pipe of 4-in. cast-iron soil pipe, 35 ft of the same pipe for a stack, and 15 ft of 2-in. drain pipe. All 4-in. soil pipe is standard. Also 7 T's, Y's, and elbows are needed for the 4-in. pipe and 4 for the 2-in. pipe, with two 5- by 8-in. cast-iron drain traps and one increaser for top of stack where it passes through the roof. One floor drain and trap in basement.

Materials estimate, local prices (1967) are used.

Sewer line:

100 ft (20 pieces), 4-in. soil pipe at $4 per piece	= $ 80	
15 ft (3 pieces), 2-in. drain pipe at $2.40 per piece	= 7	
7 fittings for 4-in. pipe at $2 each	= 14	
4 fittings for 2-in. pipe at $1 each	= 4	
2 cast-iron traps at $3 each	= 6	
1 increaser for 4-in. pipe at $4 each	= 4	
1 basement-floor drain and trap at $7	= 7	
Total	= $122	

Water line:

200 ft, ¾-in. galvanized-steel pipe at $21 per 100 ft	= $ 42	
50 fittings at $0.30 each	= 15	
2 faucets ¾ in. at $2.50 each	= 5	
1 valve ¾ in. at $3 each	= 3	
Total	= $ 65	

Gas line:

52 ft ¾-in. black steel pipe at $17 per 100 ft	= $ 9	
16 fittings at $0.25 each	= 4	
Total	$ 13	
Materials total		= $ 200

Labor estimate:

Local wages, plumber $6, helper, $4, total for team of $10 per hour, including payroll taxes and insurance.		
Sewer line. Trench, 35 ft long, 7 ft deep, dug and backfilled by laborer, 35 hr at $3	= $105	
15 joints of 4-in. pipe at 0.40 hr = 6.00 team-hours		
10 joints of 2-in. pipe at 0.36 hr = 3.60 team hours (most pipe ends threaded)		
9.60 team-hours at $10	= 96	
Water line. Trench, 35 ft long, 5 ft deep, dug and backfilled by laborer. 25 hr at $3	= 75	
97 joints of $\frac{3}{4}$-in. pipe at 0.12 = 11.67 team-hours at $10 (almost all pipe ends threaded at shop)	= 117	
Gas line. 29 joints of $\frac{3}{4}$-in. pipe at 0.12 hr = 3.52 team-hours at $10 (most pipe ends threaded at shop)	= 35	
Roughing work, water closet 8 hr, lavatory 5 hr, bathtub and shower 10 hr, kitchen sink 6 hr, laundry tubs 5 hr, floor drain installed 3 hr, or a total of 37 team-hours at $10	= 370	
Total labor		= $ 798

Equipment:

All equipment charges, including lead and oakum for calking, and transportation of tools to and from shop, estimated		= $ 40
Overhead: Taken as 14% of labor	= $112	
Gas company charge $30 plus plumbing permit $8	= $ 38	
Total overhead		= $ 150
Profit: Taken as 10% of all other costs		= $ 119
Rough-plumbing total		= $1,307

Finish Plumbing

Finish-plumbing Takeoff The takeoff for the finish plumbing will include a list of all items and fixtures that are to be furnished and installed by the contractor. The listing should include for each fixture a brief note stating the kind, size, and make of fixture and any other information necessary for pricing. For example, a complete description of a bathtub should include the following:

Tub, size, leg or built in, corner or recess (if built in), style, finish, and color

Accessories and fittings, number of each, style, make, finish

Shower if included with tub. Shower accessories, names, number, style, make, finish

Pole and curtain, if included, together with pole fittings, make, finish

Items that may be included in a finish-plumbing takeoff are lavatories, urinals, closets, showers, bathtubs, foot baths, slop sinks, kitchen sinks, laundry tubs, washers, water softeners, and all extra and special items given in the plans and specifications. Accessories are usually included with the fixtures. If not so included, all accessories must be listed separately. Other items that are sometimes furnished and installed by the plumber include gas stoves (kitchen), laundry stoves, and water heaters. It should be noted as to what fixtures are to be piped for hot, cold, and soft water as the case may be. The water meter is usually furnished by the local water company and may or may not be installed by the plumber. The gas meter is usually furnished and installed by the local gas company.

Listing the plumbing fixtures, though a simple process, requires care in noting and counting all fixtures shown on the plans and noting all the specification requirements regarding make of fixture, catalogue number if given, size, finish, and accessories. If the catalogue number is given, the price can be obtained from the catalogue (with notes in regard to discounts, freight, etc.). It should be noted whether the price is for fixtures complete with accessories or for fixtures and accessories priced separately. Accessories for a bathtub include faucets, supply pipes, wash and waste pipes, traps, shower, shower faucets and pipes, pole and curtain, and soap dishes.

Accessories for a kitchen sink would include faucets, supply pipes, waste and drain pipes, screens, traps, legs (if any), cabinet (if any), and drainboard (if separate).

Materials for Finish Plumbing The costs of the fixtures and accessories vary greatly with regard to make, kind, type, and quality. The best and most accurate way of obtaining the costs of the materials is to consult the dealer (or dealers) and enter the price of each fixture in its proper place in the takeoff. Fixtures and accessories are usually delivered packed or crated at the job but may be delivered unpacked and ready to install.

Table 13-13 gives approximate prices of medium-grade fixtures and accessories.

Labor The labor required for finished plumbing is that required for installing and connecting up the fixtures and accessories. Sometimes the fixtures and accessories are delivered packed and crated. Then they must be unpacked and assembled as well as installed.

Installing and connecting fixtures and accessories are usually done by men working in two men teams, each team consisting of one plumber and a helper. Sometimes union rules do not permit the use of helpers.

The team time required for installing fixtures will depend on the type, kind, and size of fixture; the number of accessories; the working conditions; and the skill and inclination of the men.

Table 13-13 *Approximate Prices of Medium-grade Plumbing Fixtures and Accessories*

Item	Fixture	Acces-sories	Total fixture and ac-cessories
Lavatory	\$15–\$ 35	\$ 5–\$15	\$ 20–\$ 50
Water closet, complete	....	....	\$ 30–\$ 80
Urinal, one stall with stall	\$40–\$ 65	\$ 5–\$15	\$ 45–\$ 80
Shower, without stall	\$ 5–\$ 15	\$ 5–\$15	\$ 10–\$ 30
Shower, with stall	\$50–\$100	\$10–\$15	\$ 60–\$115
Bathtub, with legs	\$30–\$ 60	\$10–\$15	\$ 40–\$ 75
Bathtub, built-in	\$70–\$120	\$10–\$25	\$ 80–\$145
Bathtub and shower	\$70–\$120	\$20–\$50	\$ 90–\$170
Foot bath	\$60–\$ 70	\$ 5–\$15	\$ 25–\$ 85
Kitchen sink and dish-washing compartment	\$90–\$150	\$15–\$25	\$105–\$175
Kitchen sink, double drain	\$50–\$110	\$10–\$20	\$ 60–\$130
Kitchen sink, single drain	\$30–\$ 75	\$10–\$15	\$ 40 \$ 90
Kitchen sink, small single drain	\$20–\$ 30	\$ 5–\$15	\$ 25–\$ 45
Slop sink	\$10–\$ 20	\$ 5–\$ 8	\$ 15–\$ 28
Laundry tubs, pair	\$15–\$ 30	\$ 5–\$10	\$ 20–\$ 40
Gas range, cooking	\$70–\$400	....	\$ 70–\$400
Gas laundry stove	\$ 5 \$ 20	....	\$ 5–\$ 20
Water softener:			
Large, 60,000 grain	....	....	\$200–\$300
Medium, 40,000 grain	....	....	\$100–\$200
Water heater, gas, automatic			
20 gal	....	....	\$ 50–\$100
40 gal	....	....	\$ 80 \$150
Garbage disposal	....	....	\$ 45 \$100
Tanks, for water, vary with size up to 100 gal:			
Uninsulated	....	....	\$ 20–\$ 50
Insulated	....	....	\$ 30–\$ 70
Dishwasher	....	....	\$150–\$230

Approximate team- and labor-hours are given in Table 13-14.

Wages for finish-plumbing work are the same as for rough plumbing. Plumbers' wages vary from about \$4 to \$7 per hour, and helpers' wages from about \$3 to \$5 per hour. Payroll taxes and insurance may vary from 8 to 14 percent of the labor wage.

Diagram 13-3 (Appendix F) can be used for estimating the labor cost of finish-plumbing work when the hourly wages of plumber and helper are known and the time required can be reasonably assumed.

Equipment The equipment required for installing fixtures and accessories will usually include only the ordinary plumber's hand tools. Other equipment, such as pipe threaders, scaffolding, and calking outfits, will rarely be needed.

Equipment costs for finished-plumbing work may range from about $2 or $3 to $5 or $10 per team, depending upon the magnitude of the job and the time required.

Overhead and Profit Overhead on finish-plumbing work is usually based on labor costs. The percentage will be practically the same as for rough-plumbing work, say 10 to 25 percent of the labor costs.

Table 13-14 *Approximate Team- and Labor-hours Required for Installing Fixtures and Accessories*

Fixture	Team-hours	Labor-hours
Lavatory	3–5	5–9
Water closet	3–6	5–10
Urinal with stall	6–10	12–20
Shower with stall	12–16	20–30
Shower without stall	3–5	5–10
Bathtub	5–10	10–20
Bathtub with shower	7–15	14–30
Foot bath	4–8	8–15
Laundry tubs, pair	3–5	6–10
Kitchen sink, small	3–6	6–12
Kitchen sink, large	4–8	8–15
Slop sink	3–5	6–10
Gas laundry stove	2–3	3–5
Gas kitchen range	2–4	3–7
Water heater, gas, automatic with tank	3–7	5–12
Dishwasher	3–6	5–10
Water tank	3–5	5–9
Water softener	3–6	5–12
Garbage disposal	2–4	4–8

Permits for installing fixtures are sometimes required and may cost about $1 per fixture. The cost of these permits can be included in the overhead.

Profit is usually based on the sum of all other costs if no profit has been charged on the fixtures. If the prices for the fixtures include profit, then the job profit should be based on the sum of labor, equipment, and overhead costs. Percentages charged for profit may range from about 8 to 20 percent.

Summary Estimates for a finish-plumbing job will include materials, labor, equipment, overhead, profit, and total cost. The cost per fixture is sometimes computed. If desired, a summary can be prepared as in the Summary article in the section on Rough Plumbing.

Illustrative estimate of finish plumbing. Prepare an approximate cost estimate for the finish plumbing for the six-room house of the Illustrative Estimate of Rough Plumbing.

It is assumed that the owner has looked over the plumbing-fixture catalogue with the contractor and has selected the fixtures desired. The following list gives the fixtures and prices. The prices are retail and include plumbers' profit. Prices are for fixtures complete with accessories.

Materials:

30-gal automatic gas water heater	= $ 70	
Twin laundry tubs	= 30	
Two-burner gas laundry stove	= 10	
42-in. left-hand-drain apron-front kitchen sink	= 85	
Built-in corner bathtub with shower and curtains	= 120	
Closet or toilet	= 55	
Wall lavatory	= 20	
Total		= $390

Labor installation: Plumber and helper, $10 per hour including payroll taxes and insurance:

	Team-hours	
Gas water heater	4	
Twin laundry tubs	4	
Gas laundry stove	2	
Kitchen sink	5	
Bathtub, shower, etc	10	
Closet	4	
Lavatory	3	
Total team-hours	32 at $10	= $320

Equipment, estimated as $8 per job	= 8
Overhead, 14% of labor	= 45
Profit, 10% of labor, equipment, and overhead (profit on materials included in materials prices)	= 37
Finish-plumbing total	= $800

Total or Combined Plumbing Estimates

Combined Plumbing Estimates On many jobs, the plumbing contractor estimates both the rough and finish plumbing and submits a bid for the entire job. When such estimates are prepared, the rough- and finish-plumbing costs are usually estimated separately and then the sum of both costs is found. Sometimes the estimator estimates the materials, labor, and equipment separately for the rough and finish plumbing and then estimates overhead and profit for the combined job.

Illustrative estimate. For an example of an estimate for a combined plumbing job, the following tabulation has been prepared, using the costs estimated in the preceding estimates for rough and finish plumbing.

Combined Plumbing Job Estimate

Item	Rough plumbing	Finish plumbing	Total
Materials	$ 200	$390	$ 590
Labor	798	320	1,118
Equipment	40	8	48
Overhead	150	45	195
Profit	119	37	156
Total	$1,307	$800	$2,107

CHAPTER FOURTEEN
Electrical Work

Electrical Work in General

Electrical Work The electrical work on a building is almost always done by a subcontractor experienced in this kind of work. The estimates for this work are usually prepared by the subcontractors, who submit bids. The general contractor or his estimator rarely has to prepare estimates for electrical work except, perhaps, as a check on the estimates and bids submitted by the subcontractors.

The purpose of this chapter is to give enough information regarding estimates for electrical work so that the general contractor or his estimator will be able to prepare approximate estimates for checking and for determining whether or not the bids submitted for electrical work are reasonable.

The electrical work required for a building can be divided into two parts, the wiring, or *rough*, work and the fixtures, or *finish*, work.

The electrical wiring, or rough work, is installed as the building is erected. This rough work includes nearly everything from the place where the supply mains enter the building up to the outlets in the walls or ceilings. Some of the materials to be considered are wire, conduit, switches, fuses, fuse boxes, circuit breakers, outlet boxes, and floor plugs.

Electrical wiring may be *open* or *concealed*. Open wiring is rarely used except on temporary and hasty construction jobs. Many localities now prohibit the installation of open wiring. The methods used in concealed wiring are *conduit* and *cable*. The conduit used is the *flexible* type, *thin-wall* type or *rigid* type. The electric cable usually used is the BX cable or the Romex cable.

The electrical fixtures are usually installed after the building is practically completed, i.e., after the lathing and plastering are done and, in many cases, after the painters are through.

An estimate for the electrical fixtures is rarely called for because the selection of the fixtures is usually left to the owner or architect. Sometimes a lump sum is provided for in the contract for the electrical fixtures with the understanding that if the fixtures selected cost more than this sum, the owner pays the difference, and vice versa.

At the present time, the National Electrical Code is used to a large extent throughout the United States. Some states and localities have their own special codes which differ somewhat from the national code. In many instances, the national code is required with some exceptions and additions. The code regulations governing the particular job should be carefully studied before a detailed electrical estimate is prepared.

A set of standard symbols for electrical wiring work has been adopted and is in use in most parts of this country. Some localities have their own special symbols which they use in place of or in addition to the standard symbols. Consequently, the electrical estimator should be familiar with local symbols as well as with the standard symbols.

Estimating Electrical Work The items included in an estimate of electrical work are materials, labor, equipment, overhead, and profit.

The cost of electrical work will depend upon the amount, kind, quality, and price of the required materials; the kind of work; the skill, inclination, and wages of the workmen; the equipment required; the overhead expense, including shop expense, and general overhead; and the profit charged. Code requirements, local ordinances, and labor regulations also affect the costs.

The accuracy of the estimate will depend upon the time and effort exerted when preparing the estimate, upon the skill and qualifications (experience and judgment) of the estimator, and upon the method followed.

Estimating with Plans In most cases, it is undoubtedly more scientific, businesslike, and accurate to make a detailed analysis of the circuits. Such an analysis is almost necessary when the experience of the estimator is limited or when unusual conditions are met. An analysis is best made from the architect's or electrical engineer's plans and specifications. The plans should be gone over in a systematic manner, and all outlets as well as corresponding paragraphs in the specifications should be checked with a red pencil in order to be certain that everything is included and also to avoid estimating twice on the same items. If complete and detailed data are given in the plans and specifications, the estimator has only to write down the details on an analysis sheet. From this, the cost of material is readily determined and the labor-hours are estimated.

If complete detailed wiring plans are not available, it is often desirable for the electrical contractor to prepare his own detailed layouts or wiring plans and have the plans approved before starting work. The layout should show the location of each circuit on each floor and should show all cutouts, switches, meters, plugs, fuse boxes, etc.

Estimating without Plans When detailed plans and specifications for electrical work are not furnished, it is advisable for the estimator to go over the general plans and specifications very carefully and note the location of all outlets, such as switches, meters, plugs, fuse boxes, and cutouts.

A very useful aid for the estimator in determining the details is for him to sketch the floor plans of the building upon sheets of cross-sectional paper having eight divisions per lineal inch. One or more sheets should be used for each floor. A scale of 1/4 or 1/8 in. to the foot will usually be satisfactory. By jotting down other information as to the kind of building, walls, and especially any unusual conditions, the estimator can readily get the necessary data for making reliable estimates. Not only do sketches prepared in this way aid materially in making estimates of material and labor, but they are also of great value to the workmen who install the wiring. These sketches also help in eliminating a large part of the disputes with customers, since disputes are usually due to some misunderstanding concerning the work to be done.

After the sketches are prepared, the estimator should go over them with the architect or owner to determine if the estimator's ideas are reasonably correct and satisfactory. Then, after the sketches have been approved, the estimator can prepare the estimate.

Approximate Estimates Based on Cost per Outlet An approximate and common method of estimating electrical-wiring costs on small- and medium-sized jobs is to base the estimate on the average cost per outlet. When the estimator has complete and up-to-date information at hand, this method of estimating is satisfactory for a quick preliminary estimate and for a check on a more detailed estimate.

There are many variables to be considered when estimating wiring costs. Among these variables are the kind of wiring (conduit or cable), code used, local ordinances, kind and quality of materials specified, material costs, labor costs, labor regulations, kinds and quality of labor, and difficulties pertaining to the particular job.

The cost per outlet varies considerably: $5 and up for open work and $7 and up for conduit work (Table 14-1). At the present time, the average cost per outlet for conduit work in Middle Western medium-sized cities is $7 to $15. When the number of outlets per room or per building is high, the cost per outlet is usually less than when the number of outlets is low.

To determine the number of outlets in a room or building, it is necessary to count all the floor and wall plugs, all the switches, and all the wall and ceiling light fixtures required. The total cost is equal to the estimated cost per outlet multiplied by the number of outlets. The costs of leads from transformer or from the electrical company's lines to the building, main switches, main fuse boxes, etc., must be included in the estimate as an extra item.

A more accurate approximate method is to list all the different kinds of outlets separately and to use different prices for each kind of outlet and class of work. The outlets listed should include the meter outlet; main switches; fuse boxes; wall- and ceiling-fixture outlets; wall two-way, three-way, and four-way switches; and floor and base plugs. Each of these is called an outlet and is listed and priced accordingly. In addition, such items as doorbells, push buttons, and bell transformers are listed and priced. Approximate costs are given in Table 14-1. Costs of light fixtures are not included.

Wiring or Rough Electrical Work

Materials Takeoff Before beginning the takeoff, the estimator should make a careful study of the particular layout so that he can have a good general idea of the job. Sometimes the layout can be modified so as to save on the materials and labor or to improve the general wiring plan.

The materials takeoff for the rough work is a complete list of all the materials required. The items included may be conduit, fittings, wire, outlet boxes, cutout boxes, switches, plugs, fuse boxes, main switches, meter boards, knobs, circuit breakers, panel boards, and main switch and fuse boxes. Other items may be bells, buzzers, annunciators, bell transformers, dry batteries, push buttons, bell wire, etc. On some jobs there may be special items such as switchboards, large panel boards, dynamos, motors, and other electrical machinery. These special items should be included in the estimate for the rough work if installed at the time the rough wiring is done. If installed at the same time as the fixtures, these special items should be included in the finish estimate. Note if any of the special items require extra work such as foundations and insulation.

Table 14-1 *Approximate* 1967 *Prices of Electrical Outlets*

Outlet	Romex cable: No ground	Romex cable: With ground	BX cable	Flexible conduit	Thin-wall light conduit	Rigid heavy conduit
Electric wiring 110 or 220 volt						
Main switch	$4–$10	$6–$12	$ 7–$13	$ 9–$15	$ 8–$14	$10–$17
Main switch and circuit breaker	$8–$13	$9–$16	$11–$18	$12–$22	$10–$18	$14–$25
Small circuit breaker and box	$7–$10	$8–$12	$ 9–$13	$10–$16	$ 9–$14	$11–$18
Cutout box	$4–$ 9	$5–$11	$ 6–$12	$ 7–$14	$ 6–$12	$ 8–$16
Meter (less meter)	$4–$ 8	$5–$10	$ 6–$11	$ 7–$13	$ 5–$12	$ 7–$15
Switch, two-way	$4–$ 7	$4–$ 8	$ 5–$ 9	$ 5–$10	$ 5–$ 9	$ 6–$11
Switch, three-way	$4–$ 8	$5–$ 9	$ 5–$10	$ 6–$11	$ 6–$10	$ 7–$12
Switch, four-way	$5–$ 9	$5–$10	$ 6–$11	$ 7–$12	$ 7–$12	$ 8–$14
Fixture outlet only	$4–$ 6	$4–$ 7	$ 4–$ 8	$ 5–$11	$ 5–$10	$ 6–$12
Plug, base or floor	$4–$ 6	$4–$ 7	$ 4–$ 8	$ 5–$11	$ 5–$10	$ 6–$12
Average electric outlet	$5–$ 7	$5–$ 8	$ 6–$ 9	$ 8–$14	$ 7–$12	$ 9–$16
Electric bell or buzzer	......	......	......	$ 4–$ 7	$ 4–$ 7	$ 5–$ 8
Bell, push button or switch	......	......	......	$ 5–$ 9	$ 5–$ 9	$ 6–$10
Bell transformer	......	......	......	$ 5–$ 9	$ 5–$ 9	$ 6–$10
Telephone outlet	......	......	......	$ 4–$ 8	$ 4–$ 7	$ 5–$ 9
TV antenna outlet	......	......	......	$ 6–$10	$ 5–$ 9	$ 7–$12

A satisfactory method is to begin at the entrance of the service wires or leads and list the service wires, switch, cabinet, panel board, meter board, cutout box, fuse box, etc. Then each floor of the building can be taken in order, and all items on each floor listed. The estimator must include all vertical conduit and wiring between floors as well as the horizontal wiring. If desired, each room of each floor can be taken separately. Sometimes an estimator starts with a circuit as it leaves the main switch box, itemizes all materials connected with that circuit, and then takes all the remaining circuits in order. When going over the plans in this manner, it is necessary to measure or estimate the lengths of the circuits.

The use of specially devised blanks or forms is an aid in accurate estimating, as they tend to help the estimator to keep his details in a logical and orderly manner and they also can be arranged to serve as a reminder list if all the various items are shown on them. Some estimators make a practice of using a good reminder list to help them in avoiding errors and omissions.

After the materials takeoff is completed and checked, all the items can be priced, their costs estimated, and the totals found for the materials cost. Careful attention should be given to the buying so that the materials costs will be a minimum. It is often advisable to obtain prices from more than one firm. When the cost of any material is computed, the cost of freight, insurance, drayage, breakage, waste, and handling should be included so as to obtain the materials cost delivered at the job.

Materials The wire used for electrical work is insulated copper wire of various diameters. The wire may be solid or stranded. Rubber-insulated wire is used indoors and weatherproof outdoors. There are many types of insulation suitable for different purposes. Two or more insulated wires may be woven or twisted together. A cable usually consists of two or more insulated wires in parallel and all covered with extra insulation.

The units for measuring wire are the foot for length, the mil or gauge number for diameter, and the circular mil for area. A mil is 1/1,000 in. The circular mil is the area of a circle 1 mil in diameter. Table 14-2 gives information concerning insulated wires. Wires smaller than No. 12 or 14 should not be used for electric lighting and power rough work. Smaller wires are often satisfactory for bell and buzzer wiring. Wires may be

Table 14-2 *Insulated Solid Wires. Diameters, Areas, and Carrying Capacity*

B & S gauge number	Diameter solid wire, mils (1 mil = 0.001 in.)	Area, cir mils	Carrying capacity		Approximate price 1967 per 1,000 ft	
			Rubber insulation, amp	Other insulation, amp	Rubber-covered indoor	Weather-proof outdoor
18	40.3	1,624	3	5	$ 12	$ 17
16	50.8	2,583	6	10	13	19
14	64.1	4,107	15	20	15	22
12	80.8	6,530	20	25	20	28
10	101.9	10,380	25	30	30	40
8	128.5	16,510	35	50	47	60
6	162.0	26,250	50	70	75	92
5	181.9	33,100	55	80		
4	204.3	41,740	70	90	125	155
3	229.4	52,630	80	100		
2	257.6	66,370	90	125	175	185
1	289.3	83,690	100	150		
0	325.0	105,500	125	200		
00	364.8	133,100	150	225		
000	409.6	167,800	175	275		
	447.2	200,000	200	300		
0000	460.0	211,600	225	325		

either solid or stranded. There are various types and kinds of electric cable on the market, each cable usually containing two wires (sometimes one only and sometimes three or more). These cables are comparatively expensive, the price varying with the number and size of the wires and the insulation and armor.

Wire is usually estimated in lineal feet with details such as size (gauge), insulation, and solid or stranded noted. The allowance for waste, splicing, soldering, etc., may vary from about 5 to 10 percent or so. Some estimators allow an extra foot for each outlet, 2 ft for each main switch, and 3 to 5 ft for each main circuit.

The conduit used may be rigid (heavy), thin-wall (light), or flexible conduit. Rigid conduit is made of light galvanized-steel pipe and may be heavy (outdoor and indoor) or light (indoor only). The heavy conduit cannot be bent and requires threading and many fittings. The light, rigid conduit can be bent and does not require threading. Flexible conduit is made of two spirally wound interlocking metal strips. Electric cable is like flexible conduit except that the conduit is made in place over the conductors (usually two or three) so that both conduit and conductors can be installed together.

Conduit of various kinds is usually estimated in lineal feet, with diameter, kind, and other details being noted. The allowance for waste should be small, say 2 to 5 percent.

Tables 14-3 to 14-7 contain information in regard to armored cables, conduits, and fittings. Prices of various electric-wiring materials are given in Table 14-8. Prices in Table 14-6 are for quantity lots.

Table 14-3 *Approximate* 1967 *Prices of Electric Cable per* 1,000 *Ft*

Wire, gauge	14	14	12	12	10	10
Number of wires	2	3	2	3	2	3
BX cable	$90	$120	$120	$160	$210	$280
Romex cable with no ground wire	40	80	50	100		
Romex cable with ground wire	55	100	65	125		

Table 14-4 *Approximate* 1967 *Prices for Fittings for Electric Cables*

Steel connectors, small sizes, per 100	$ 8–$16
Steel connectors, large sizes, per 100	$20–$30
Steel straps, per 100	$ 2–$ 6
Plastic straps, per 100	$ 7–$10
Fiber bushings, per 100	$ 3–$ 5

Table 14-5 *Approximate* 1967 *Prices for Flexible Conduit and Fittings*

Size, in.	½	¾	1	1¼	1½	2
Price per 100 ft of conduit	$15	$20	$40	$50	$65	$85
Price per 100 fittings	9	13	18	24	35	45

Table 14-6 *Approximate* 1967 *Price per* 100 *Ft of Conduit or per* 100 *Fittings*

Item	Rigid lightweight (thin-wall) conduit and fittings, size, in.					
	½	¾	1	1¼	1½	2
Conduit, 10-ft lengths	$ 9	$13	$18	$ 26	$ 33	$ 45
Couplings	17	25	38	60	85	115
Connectors	15	22	32	58	82	115
Adapters	8	13	17	30	35	65
Elbows	30	40	50	65	80	135
Straps	3	4	5	6	8	9
Entrance heads and L's	50	60	90	120	210	400

Item	Rigid heavyweight conduit and fittings, size, in.							
	½	¾	1	1¼	1½	2	3	4
Conduit, 10-ft lengths	$21	$28	$ 40	$ 52	$ 65	$ 85	$165	$ 250
Couplings	10	13	18	23	30	45	150	250
Elbows	35	45	70	95	125	180	580	1,200
Locknuts	2	3	4	5	6	10	30	75
Bushings	3	5	8	11	14	25	55	120
Straps	2	2	3	4	5	6	10	15
Entrance caps and L's	60	75	100	150	200			

Attention is called to the fact that the prices given are approximate only and for the year 1967. Prices will vary considerably from time to time. Prices will also vary with the make, grade, and quality of any article. Consequently, the estimator should secure the actual local prices in force at the time he is preparing the estimate.

Rigid heavy-weight conduit is available in larger sizes up to 6 in. in diameter. Rigid heavy conduit up to about 1¼ in. can be bent to a fairly large radius (1 ft or so) with suitable tools.

For special items for electrical rough work, the estimator should consult the local dealers and secure their prices.

For large pieces of apparatus, such as engines, generators, motors, and motor-generator sets, it is advisable to secure bids from the factory for the

Table 14-7 *Number of Wires and Minimum Size of Conduit*

Size of wire B & S gauge	Number of wires in one circuit: 1	2	3	4	5	6	7	8	9
	Minimum size of conduit, in.								
14	½	½	½	¾	¾	1	1	1	1
12	½	½	¾	¾	¾	1	1	1	1¼
10	½	¾	¾	1	1	1	1¼	1¼	1¼
8	½	¾	1	1	1	1¼	1¼	1¼	1½
6	½	1	1¼	1¼	1½	1½	2	2	2
5	¾	1¼	1¼	1¼	1½	2	2	2	2
4	¾	1¼	1¼	1½	2	2	2	2	2½
3	¾	1¼	1¼	1½	2	2	2	2½	2½
2	¾	1¼	1½	1½	2	2	2½	2½	2½
1	¾	1½	1½	2	2	2½	2½	3	3
0	1	2	2	2	2½	2½	3	3	3
00	1	2	2	2½	2½	3	3	3	3½
000	1	2	2	2½	3	3	3	3½	3½
0000	1¼	2	2½	2½	3	3	3½	3½	4

complete apparatus, delivered and erected, together with foundations, connected ready for service. If the prices of the apparatus are f.o.b. factory or nearest railway station, the estimator can estimate the cost of the apparatus installed on the job according to the method given in Chap. 16 on Heavy Machinery. In addition to the cost f.o.b. railway station, the estimator will have to estimate the cost of unloading and transporting the apparatus to the site, erecting the apparatus on its foundations, lining up, and making all necessary wiring connections. In some cases, the cost of building the foundations will need to be included. Often the factory may provide the electrical machine, and the contractor who installs it must provide the wiring materials, switches, fuses, switchboard panels, etc., as well as furnishing all labor required for the complete installation with machine ready to run.

Labor The labor of electrical wiring or rough work is performed by electricians with or without the help of apprentices or helpers. A team of two electricians or of one electrician and one helper is a good working combination, especially on small jobs. Medium-sized and large jobs may require several electricians and helpers working under the direction of a foreman.

The labor required per outlet or for a certain item of work will vary

Table 14-8 *Various Electric-wiring Materials and Approximate* 1967 *Prices*

Item	*Price*
Main switches, up to 60 amp, each	\$ 5–\$ 20
Fuse cabinets, 60 amp or less, each	\$ 5–\$ 20
Combination main switch and fuse cabinets, 60 amp or less, each	\$10–\$ 40
Fuseless cutout boxes, each	\$ 6–\$ 20
Outlet, switch, receptacle, boxes, galvanized, per 100	\$25–\$ 40
Outlet box covers, galvanized, per 100	\$10–\$ 20
Bakelite outlet boxes, per 100	\$25–\$ 40
Bakelite outlet box covers, per 100	\$12–\$ 20
Switches, toggle, two-way, per 100	\$20–\$ 40
Switches, toggle, three- and four-way, per 100	\$30–\$ 45
Switches, mercury, two-way, per 100	\$60–\$ 90
Switches, mercury, three- and four-way, per 100	\$90–\$125
Convenience outlets, single and duplex, per 100	\$25–\$ 45
Bar hangers, per 100	\$12–\$ 30
Fuses, round glass, 10 to 30 amp, per 100	\$ 4–\$ 8
Fuses, cartridge, 15 to 30 amp, per 100	\$ 6–\$ 9
Fuses, cartridge, 35 to 60 amp, per 100	\$12–\$ 18
Wall plates, per 100	\$10–\$ 25
Receptacles, per 100	\$15–\$ 30
Bracket insulators, two or three wires, each	\$ 2–\$ 4
Service insulators, one wire per 100	\$25–\$ 50
Friction and rubber tape, 100 oz	\$ 4–\$ 6
Bell, buzzer, annunciator wire, per M ft	\$ 7–\$ 10
Bell transformers	\$ 2–\$ 6
Bells and buzzers, per 10	\$ 4–\$ 12
Push button, per 10	\$ 3–\$ 5

with the type of work, the working conditions, and the skill and inclination of the workers. The amount of time required for the rough work per outlet may vary from less than 1 hr to 3 or 4 hr. The labor per outlet will vary with the kind of wiring, being least with electric cable and greatest with heavy rigid conduit work. The labor will also vary with the size of conduit and the size of the wires, the larger sizes requiring more labor per 100 ft.

The approximate labor-hours required per outlet are given in Table 14-9.

Note that any one of the larger sizes of rigid conduit will usually carry wires for more than one outlet. Consequently, the labor-hours per outlet will not vary appreciably with the size of rigid conduit.

The approximate labor-hours for installing various items of rough electrical work are given in Table 14-10. The labor values given for installing conduit includes all labor for fittings such as threading pipe, making joints, and installing outlet boxes for wall and ceiling fixtures, switches, and for floor and base plugs, but not for pulling wire. The labor

Table 14-9 *Approximate Labor-hours for Installing an Outlet in Rough Electrical Work*

Kind of work installed	*Labor-hours per outlet*
Electric cable and fittings, Nos. 14, 12, and 10 gauge wire	0.8–2.0
Flexible conduit, and wire, ½ and ¾ in	0.8–2.0
Lightweight conduit, fittings, and wire:	
½ and ¾ in	1.0–2.0
1 and 1¼ in	1.1–2.2
Heavy-weight rigid conduit, fittings, and wire:	
½ and ¾ in	1.2–2.3
1 and 1¼ in	1.2–2.5
1½ and 2 in	1.3–2.7
2½ and 3 in	1.3–3.0

Table 14-10 *Approximate Labor-hours for Installing Conduit, Cable, and Wire*

Conduit and wire	Labor-hours per 100 ft	Conduit and wire	Labor-hours per 100 ft
Installing heavy rigid conduit, and fittings, including outlet boxes:		Installing flexible conduit and fittings:	
		½, ¾ and 1 in	5–10
½ and ¾ in	8–14	1¼, 1½ and 2 in	7–14
1 and 1¼ in	10–17	Pulling and installing wire in conduit:	
1½ in	12–20		
2 in	15–25	18 and 16 gauge	0.8–1.5
2½ in	21–35	14 and 12 gauge	1.0–1.8
3 in	26–45	10 and 8 gauge	1.5–2.5
4 in	32–55	6 gauge	2.2–3.5
5 in	40–65	4 gauge	2.8–4.5
6 in	55–90	3 gauge	3.5–5.5
Installing lightweight conduit and fittings, including outlet boxes:		2 gauge	4.5–6.5
		1 gauge	5.5–7.5
		Installing electric cable and fittings:	
½, ¾ and 1 in	6–10		
1¼, 1½ and 2 in	8–14	No. 14, 12, and 10 gauge wire	4–12

values given for wire pulling include the labor required for pulling the wire through the conduit and for soldering and taping ends in outlet boxes.

The wages of electricians vary considerably at different times and in different parts of the United States. They may range from $4 to $7.50 per hour with an average (1967) of about $4 to $6.50 per hour. Wages of helpers will be about half those of electricians, and will range from about $3 to $5 per hour.

Diagrams 14-1, 14-2, and 14-3 (Appendix F) can be used for estimating labor costs of rough electrical work and wiring per outlet (Table 14-9), per item (Table 14-11), per 100 ft of conduit (Table 14-10), and per 100 ft of wire (Table 14-10) and for various other items when the hourly wage is known and the hourly labor output can be reasonably assumed.

Table 14-11 *Approximate Labor-hours for Installing Various Items in Rough Electrical Work*

Item installed	*Labor-hours each*
Meter	0.5–1.0
Main switch and circuit breaker or box (up to 200 amp)	0.8–2.0
Circuit breaker or fuse box (200 amp or less)	0.5–2.0
Combined switch and circuit breaker or fuse box (200 amp or less)	1.0–2.5
Cabinet (200 amp or less)	1.0–3.0
Making conductor connections, as in a cabinet	0.2–0.4
Outlet boxes, for ceiling and wall fixtures	0.2–0.4
for service switches, floor and base plugs*	0.15–0.3

* Small service switches, floor and base plugs, and their covers and plates are usually installed with the finish work.

Equipment The equipment required for rough electrical work will vary with the type of work. Heavy rigid conduit work will require pipe cutters, hack saws, threading tools, reamers, and other small tools needed for conduit work. Light rigid conduit work, which can be bent and which does not require threading, will not require threading tools. Other small tools needed will include hand and electric drills, saws, hammers, chisels, gasoline torches, soldering outfits, fishing wire, pliers, and hand lines. Many firms also include tape and solder with the equipment. Ladders, stepladders, sawhorses, and planks may be needed for scaffolding.

The cost of equipment for a team of two men will vary with the size of the job and the amount of work required. About $5 to $20 per team per job should be sufficient, including transportation.

Overhead and Profit The general overhead for rough electrical work may range from about 5 to 15 percent of the sum of materials, labor, and equipment costs. When based on labor costs alone, the percentage may range from about 10 to 25 percent. Workmen's compensation insurance, social-security tax, etc., costs are included in the labor costs.

The profit charged on rough electrical work may range from about 6 to 15 percent of the sum of all other costs. However, if the contractor has estimated the costs of the materials at retail prices including their overhead and profit, profit on these items should not be included again.

Summary The estimate for wiring or for rough electrical work can be summarized to show total costs of materials, labor, equipment, overhead, and profit and also show unit costs per outlet. Sometimes the estimator computes the cost per 100 ft of conduit or per 100 ft of circuit as units. The summary may be as follows:

Item	Cost per outlet	Total cost
Materials..........	$	$
Labor.............		
Equipment........		
Overhead..........		
Profit.............		
Total...........	$	$

Illustrative estimate. Prepare an approximate estimate for the rough wiring for a new residence, and compute the estimated cost per outlet. The local electric company supplies the leads from the transformer to the house and supplies and installs the meter. There are to be 26 ceiling outlets, 14 wall outlets, 24 service switches, and 30 base plugs, giving a total of 94 outlets.

The materials takeoff gave the following information:

1 main switch and cutout box	=	$ 13
1 small circuit-breaker box, 6 circuits	=	12
32 outlet boxes with covers, at $0.50	=	16
94 switch, base plug, and receptacle boxes at $0.35	=	33
60 ft, ¾-in. thin-wall conduit (light weight) at $16 per 100 ft	=	10
Fittings for ¾-in. lightweight conduit, estimated	=	6
720 ft, ½-in. lightweight conduit at $12 per 100 ft	=	86
Fittings for ½-in. lightweight conduit, estimated	=	35
60 ft, ½-in. flexible conduit at $20 per 100 ft	=	12
Fittings for ½-in. flexible conduit, estimated	=	4
65 ft, No. 8 rubber-covered wire at $4.70 per 100 ft	=	3
65 ft, No. 10 rubber-covered wire at $3.00 per 100 ft	=	2
1,700 ft, No. 12 rubber-covered wire at $2.00 per 100 ft	=	34
Miscellaneous materials, estimate for job	=	15
Total materials	=	$ 281

The labor-hours required will be about as follows:

Installing main switch and cutout box	=	2 hr
Installing circuit-breaker box	=	2 hr
Installing 60 ft of ¾-in. lightweight conduit and fittings	=	6 hr
Installing 720 ft of ½-in. lightweight conduit and fittings	=	65 hr
Installing 60 ft of ½-in. flexible conduit	=	3 hr
Pulling 65 ft of No. 8 wire	=	1 hr
Pulling 65 ft No. 10 wire	=	1 hr
Pulling 1,700 ft of No. 12 wire	=	22 hr
Total labor-hours	=	102 hr

Labor costs with electricians at $5.70 per hour including payroll taxes and insurance	=	$ 581
Equipment cost for job, estimated	=	20
Overhead cost at 12% of labor cost	=	70
Profit at 10% of all other costs	=	95
Total estimated cost of rough electrical work	=	$1,047

This estimate can be summarized and unit costs computed on a basis of 94 outlets as follows:

Item	Cost per outlet	Total cost
Materials..........	$ 3.00	$ 281
Labor.............	6.18	581
Equipment........	0.21	20
Overhead..........	0.74	70
Profit.............	1.01	95
Total...........	$11.14	$1,047

Finish Electrical Work

Materials and Materials Takeoff The materials usually included in finish electrical work are ceiling fixtures, wall brackets and fixtures, service switches and plates, floor- and base-plug receptacles and plates. Ceiling and wall fixtures and brackets are usually selected by the owner. The price of a fixture may vary from about $2 to several hundred dollars. The cost of wall brackets may vary from $2 up to $20 or $25.

Service switches, service receptacles, and plates are comparatively inexpensive. Service receptacles and switches (two-way) without cover plates may cost 20 to 50 cents each for ordinary types. Three-way and four-way service switches without cover plates may cost 30 cents to $1 each. Convenience outlets, single and duplex, may cost about 20 to 75 cents each without cover plates. Cover plates may cost about 10 cents and up depending on material, style, finish, etc.

There are a great variety of rosettes, receptacles, sockets, covers, etc., on the market. Prices range from about 30 cents and up for each.

Lamp sockets of various types will cost about 30 up to about 80 cents for ordinary types.

Drop cord, lamp cord, fixture cord, etc., come in various styles, coverings, and colors. Prices range from about $3 to $6 per 100 ft. Most cord contains two stranded copper wires of No. 18 gauge, though cord containing larger wire (No. 16 or 14 gauge) is available at slightly higher prices.

Fixture wire, insulated, No. 18 gauge, costs about $1.50 per 100 ft. The No. 16 gauge costs a little more.

Please note that prices given for materials are approximate prices for 1967. Prices used in actual estimates should be obtained from dealers' catalogues.

The materials takeoff for finish electrical work should list each item of ceiling fixtures, wall fixtures and brackets, service switches with plates, plugs and other receptacles with plates, rosettes, covers, drop cord, sockets, etc. Finish, style, catalogue numbers, and other details should be noted that will aid in pricing the various items.

Labor The labor required for installing fixtures will vary somewhat depending on the size of the fixture, working conditions, and skill of the electrician. Most fixtures come ready wired so that the installing consists of installing the fixture in place and connecting the fixture wires to the circuit wires. Sometimes the fixture must be wired and labor and materials furnished for this work.

Table 14-12 gives approximate labor-hours required for installing various types of fixtures, switches, plugs, and service receptacles ready for operation.

Table 14-12 *Approximate Labor-hours Required for Installing Electric Fixtures*

Item	Labor-hours each	Item	Labor-hours each
Wall light or bracket (one or two lights)	0.20–0.40	Rosette with drop cord and socket	0.15–0.40
Ceiling fixture (one or two lights)	0.25–0.50	Two-way service switch	0.10–0.25
Ceiling fixture (three-four, or five lights)	0.30–0.70	Three-way service switch	0.20–0.50
Wiring fixtures in addition (per light or per socket)	0.10–0.25	Four-way service switch	0.25–0.75
		Base- or floor-plug outlet (cover plates included)	0.10–0.25

Labor wages for electricians vary from about $4 to $7.50 per hour (1967). Helpers' hourly wages (1967) may vary from about $3 to $5 per hour.

Diagram 14-4 (Appendix F) can be used for estimating the labor costs of finish electrical work when the labor wage per hour is known and the labor output can be reasonably assumed.

Equipment The equipment required for installing fixtures will include the electrician's ordinary hand tools with tape and small soldering outfit. Stepladders may be needed for installing ceiling and high wall fixtures. The cost of equipment for an ordinary installation job may range from $3 or $4 up to about $15 or so for each pair of electricians, depending upon the work and magnitude of the job.

Overhead and Profit Overhead costs for finish electrical work are usually based on labor costs. The percentage may range from 10 to 20 for ordinary overhead.

Prices of materials for finish electrical work are usually retail prices and include all shop overhead and profit. If the prices do not include shop overhead and profit, these cost items must not be omitted. Shop overhead on fixtures, etc., may vary from 10 to 20 percent of the cost.

The profit on finish electrical work may range from about 8 to 20 percent of the costs of labor, equipment, and labor overhead. Shop profit on fixtures, etc., may range from 10 to 50 percent. This profit is usually included in the retail price.

Summary The estimates for finish electrical work can be tabulated to show total costs and costs per outlet. The material costs per outlet may not have much significance because of the variation in the prices of the fixtures. The prices of the service switches and service outlets do not vary so much. Labor, equipment, and overhead costs per outlet are worth computing.

The estimate can be tabulated as follows. If desired, the materials costs can be divided by separating the fixture costs from the other material costs.

Item	Cost per outlet	Total cost
Materials............	$	$
Labor...............		
Equipment..........		
Overhead...........		
Profit...............		
Total.............	$	$

Illustrative estimate. Prepare an approximate estimate for furnishing and installing all fixtures, service switches, service outlets, etc., for the new residence for which the wiring or rough electrical work was estimated in the illustrative estimate in the section on Wiring or Rough Electrical Work in this chapter. It is assumed that the owner has selected the fixtures listed from the dealer's shop and that the price of the fixture includes all shop overhead and profit.

Materials takeoff and cost:

Basement:			
3 switches with plates	at $0.60	=	$ 2
5 drop lights with rosette, cord, and sockets	at $1.60	=	8
First floor:			
3 five-light ceiling fixtures	at $32	=	96
1 three-light ceiling fixture	at $20	=	20
4 one-light ceiling fixtures	at $ 7	=	28
5 small one-light ceiling fixtures	at $ 3	=	25
14 wall brackets	at $ 3	=	42
11 switches with plates	at $0.60	=	7
1 three-way switch with plate	at $0.80	=	1
20 base plugs with plates	at $0.40	=	8

Second floor:		
7 two-light ceiling fixtures	at $12 =	84
5 switches with plates	at $0.60 =	3
1 three-way switch with plate	at $0.80 =	1
10 base plugs with plates	at $0.40 =	4
Attic:		
1 switch	at $0.60 =	1
1 rosette with socket	at $1.60 =	2
Total cost of all fixtures		= $332

Labor cost of installations. For purposes of installation the various fixtures may be grouped as follows:

4 five- and three-light ceiling fixtures at 0.5 hr	=	2.0 hr
16 one- and two-light ceiling fixtures at 0.35 hr	=	5.6 hr
6 rosette, drop cord, and socket ceiling fixtures at 0.25 hr	=	1.5 hr
14 wall brackets at 0.25 hr	=	3.5 hr
2 three-way switches at 0.35 hr	=	0.7 hr
22 switches at 0.15 hr	=	3.3 hr
30 base plugs at 0.15 hr	=	4.5 hr
94 total, for a check	Total =	21.1 hr

Labor costs, say 21 hr at $5.70 per hour including payroll taxes and insurance	= $120
Equipment, say $5	= $ 5
Overhead, at 11% of labor costs	= $ 13
Profit, at 10% of all other costs except materials (Shop overhead and profit is included in prices of fixtures)	= $ 14
Total estimated cost	= $484

This estimate can be summarized and unit costs computed on a basis of 94 outlets as follows:

Item	Cost per outlet	Total cost
Materials	$3.53	$332
Labor	1.28	120
Equipment	0.05	5
Overhead	0.14	13
Profit	0.15	14
Total	$5.15	$484

Total or Combined Electrical Estimates

Combined Electrical Estimates On some jobs, the electrical contractor estimates both the rough and finish work and submits a bid for the entire job. When preparing an estimate for the complete job, the estimator usually estimates the rough and finish work separately and then adds the two estimates to find the costs of the complete or combined work.

Illustrative estimate. For an example of an estimate for a combined or complete electrical job, the following tabulation has been prepared, using the estimated costs found in the preceding estimates for rough and finish work. Note that the shop overhead and profits on fixtures are included in the material costs.

Item	Wiring or rough work	Fixtures or finish work	Total
Materials	$ 281	$332	$ 613
Labor	581	120	701
Equipment	20	5	25
Overhead	70	13	83
Profit	95	14	109
Total	$1,047	$484	$1,531
Cost per outlet (94 outlets)	$11.14	$5.15	$16.29

CHAPTER FIFTEEN

Painting, Papering, and Glazing

Painting

Painting The cost of painting will depend in general upon the kind of surface (wood, steel, plaster) to be painted, the form of surface (flat, grooved, paneled, etc.), the kind of paint used, the number of coats, the particular coat (priming, second, finish, etc.), and the skill and inclination of the workmen.

Painting is usually estimated by units of area of 1 sq ft, 1 sq yd, or 1 sq (100 sq ft). The unit of 1 sq seems to be the most practical and popular.

Surfaces that may have varying unit costs should be taken off separately. When doors and windows are listed, no deductions should be made for the glass, as the extra labor cost of painting the narrow strips is often more than the saving in paint. It is customary to consider each linear foot of wood trim as 1 sq ft, because there is but little difference in the time required to paint trim 1 ft wide or less.

The kind of surface (wood, metal, plaster, etc.) should always be noted. The nature of the surface is important because the spreading capacity of any paint, varnish, or enamel varies and the amount of labor required also varies for different kinds of surfaces.

The amount of paint used on any given surface will also depend on how thoroughly the paint is brushed in. Consequently, there will be a variation in the amount of paint that will be used by different painters to cover a given surface.

The following procedure for the takeoff is recommended. The number of coats and the kinds of coats (priming, second, finish, size, filler, etc.) should be noted. Many estimators prefer to divide the painting into exterior and interior work.

Exterior work:

- Walls and siding, area in square feet
- Trim, length in lineal feet, with width noted
- Doors by total areas in square feet
- Windows by number and kind and total areas in square feet
- Blinds by number and areas in square feet
- Cornice, length in lineal feet, with girth noted
- Balusters and posts, by number of pieces, or length in lineal feet
- Shingles by square feet of shingled area when paint is applied by a brush, or by number of shingles when dipped
- Steel or iron roofing, by area in square feet

Interior work:

- Wall and ceiling surfaces in square feet, noting kind of surface and treatment required
- Floor surface in square feet with treatment noted
- Interior trim by length in linear feet with width noted
- Stairs rails (posts, rails, balusters) by number of pieces or length in linear feet
- Doors by number, kind, and total areas in square feet
- Windows by number, kind, and total areas in square feet
- Piping by length in lineal feet
- Radiators by square feet of radiation

Structural-steel work:

- By tonnage or area, noting kinds and approximate sizes of members, such as columns, beams, girders, and purlins and of small, medium, and large sizes

When estimating the number of square feet of surface area per ton of structural steel, the following approximations can be used:

	Sq ft per ton
Small members, beams, column purlins, etc.	350–450
Medium members	250–350
Large members	175–250
Small trusses and built-up frames	350–450

Materials The paint specified may be a brand of ready-mixed paint, or the contractor may be required to mix the paint at his shop, using different materials in certain proportions. In many instances, the ready-mixed paints need to be thinned on the job. Hence, the materials required may include ready-mixed paint, linseed oil, turpentine, varnish, shellac, putty, oil, and drier.

Table 15-1 gives approximate quantities for 100 gal of paint, using white lead as a base or pigment. Proportions will vary somewhat, according to the ideas of the painters and the use of the paint.

Table 15-1 *Approximate Quantities of Ingredients for* 10 *Gal of Paint*

Paint	White lead, lb	Linseed oil, gal	Turpentine, gal	Drier, pt
Exterior:				
Prime coat	100–120	4–5	1.0–2.0	1–1.5
Second coat	120–160	3–5	1.5–2.5	0.5–1.5
Finish coat	120–160	4–7	0.2–0.3	0.5–1.5
Interior:				
Prime coat	100–120	3–5	2–3	1–2
Second coat	130–160	2–4	1.5–2.5	1–2
Finish, flat	160–180	0	2–6	0–0.5
Finish, gloss	160–170	3–4	0–0.5	0.5–1.0

To produce shades or colors, 3 to 15 lb of coloring matter will be required for each 10 gal of paint. About ½ pt of oil or turpentine may be needed with each pound of coloring matter.

Mixed paint will weigh about 15 or 16 lb per gallon, and white lead about 36 lb per gallon. Weights of other materials per gallon are varnish, 8 to 9 lb; linseed oil, 7.5 to 8 lb; turpentine, about 7 lb.

The proportions used in glue sizing are 1 lb of glue to about 2 gal of water.

For a preparatory coat on new concrete or stucco surfaces, a solution of 2 or 3 lb of zinc sulfate to 1 gal of water can be used.

The amount of surface that can be covered by a gallon of paint depends on the brushing, the porosity of the surface, the thinness of the paint, and the temperature. Except for metal surfaces, the first coat applied will be absorbed much more rapidly than the succeeding coats. Consequently, a larger amount of paint is needed for the first coat. In general, zinc paint will cover more surface than an equal amount of lead paint, but the lead paint covering is more effective and durable. Up to 10 percent of drier can be used to hasten the drying time of the first coat, but driers should be used sparingly in other coats, since the driers tend to reduce the wearing qualities of the paint. Glue size costs about 5 percent as much as good paint, but the size will not adhere nearly so well as a first coat of thin paint.

About 1 lb of putty will be needed for each 1,000 sq ft of surface.

Tables 15-2 and 15-4 give approximate covering capacities of painting materials on some surfaces.

Diagram 15-1 (Appendix F) can be used for finding the cost of painting materials for one coat per 100 sq ft (or 100 lin ft) when the costs of the materials are known and the coverage can be reasonably assumed.

Table 15-2 *Approximate Covering Capacity of Painting Materials (Square Feet per Gallon per Coat)*

Painting material	Metal, any coat	Wood		Plaster, brick cement, and stucco	
		First coat	Other coat	First coat	Other coat
White lead and oil	400–500	300–400	400–500	200–300	300–500
Zinc white and oil	500–600	400–500	500–600	250–350	400–600
Good ready mixed paint (exterior)	500–600	400–500	500–600	300–400	400–600
Enamel	500–600	400–500	500–600	350–450	400–600
Floor enamel	....	400–600	500–700	400–550	500–700
Interior flat paint	....	400–500	500–600	300–450	400–550
Interior gloss paint	....	250–350	300–450	200–300	300–400
Varnish	500–600	400–500	500–600	350–450	400–600
Shellac	....	600–700	....	400–600	
Stain	....	300–500			
Wall primer	....	....	....	300–500	
Wood paste filler	....	250–350			
Glue size	....	300–500			
Varnish size	....	....	....	300–500	
Zinc sulfate	....	....	....	200–400	
Calcimine (Kalsomine)	....	....	....	200–400	
Latex-base paint	....	200–350	....	200–300	

Prices (1967) of painting materials are approximately as given in Table 15-3.

For calcimine, 5 lb of material and 2 qt of water will make 1 gal. One pound of calcimine size will make ½ to ¾ gal. Paste wood filler if bought in pounds requires about 5 lb per 250 or 300 sq ft. Three pounds of zinc sulfate are mixed with 1 gal of water. Five pounds of cold-water paint are used per gallon of water.

Table 15-3 *Prices of Painting Materials*

Exterior paint	\$4.00–\$ 7.00 per gallon
Boiled linseed oil	3.00– 4.00 per gallon
Turpentine	1.50– 3.00 per gallon
Latex-base paint	3.50– 6.00 per gallon
Putty	0.15– 0.30 per pound
Enamel	4.00– 8.00 per gallon
Varnish	5.00– 8.00 per gallon
Shellac	3.00– 5.00 per gallon
Gloss and flat paint, interior	4.00– 7.00 per gallon
Wood paste filler	0.25– 0.45 per pound
Wall primer	3.50– 4.50 per gallon
Brick and stucco vinyl	4.00– 5.50 per gallon
Zolatone	4.00– 6.00 per gallon
Calcimine, regular	0.15– 0.25 per pound
Patch plaster	0.10– 0.20 per pound
Calcimine (Kalsomine) size	0.30– 0.55 per pound
Dye stain	3.50– 6.00 per gallon
Oil stain	3.50– 6.00 per gallon
Zinc sulfate	0.40– 0.70 per pound
Cold-water paint	0.15– 0.35 per pound
Aluminum paint	4.00– 6.00 per gallon
Graphite paint	3.00– 5.00 per gallon
Ferric oxide paint	3.00– 5.00 per gallon
Red lead, dry	0.15– 0.35 per pound
White lead, in oil	0.30– 0.45 per pound
Drop cloth	5.00– 11.00 each

Labor The labor-hours required to apply a coat of paint to 100 sq ft of surface will depend upon the surface, the paint, the weather, the location of the surface (floor wall, high ceiling, etc.), and the skill and inclination of the workman (Table 15-5).

Exterior painting should not be done in damp or wet weather. Interior painting can be done in damp or wet weather when required, provided that the surfaces can be kept clean and dry. Painting should not be done in very cold and freezing weather. However, interior painting can be done in cold weather if artificial heat is provided. The surface painted per hour per man may vary from 20 to 200 sq ft with brushes, from 60 to 300 sq ft with a roller, and from 100 to 500 sq ft with a sprayer. Conse-

Table 15-4 *Approximate Covering Capacity of Painting Materials (Gallons per 100 Sq Ft for One Coat)*

Painting material	Metal, any coat	Wood: First coat	Wood: Other coat	Plaster, brick, cement, and stucco: First coat	Plaster, brick, cement, and stucco: Other coat
White lead and oil	0.20–0.25	0.25–0.35	0.20–0.25	0.35–0.50	0.20–0.35
Zinc white and oil	0.17–0.20	0.20–0.25	0.17–0.20	0.30–0.40	0.17–0.25
Good ready-mixed paint (exterior)	0.17–0.20	0.20–0.25	0.17–0.20	0.25–0.35	0.17–0.25
Enamel	0.17–0.20	0.20–0.25	0.17–0.20	0.22–0.30	0.17–0.25
Floor enamel	....	0.17–0.25	0.14–0.20	0.18–0.25	0.14–0.20
Interior flat paint	....	0.20–0.25	0.17–0.20	0.22–0.35	0.18–0.25
Interior gloss paint	....	0.30–0.40	0.22–0.35	0.35–0.50	0.25–0.35
Varnish	0.17–0.20	0.20–0.25	0.17–0.20	0.22–0.30	0.17–0.25
Shellac	....	0.14–0.17	....	0.17–0.25	
Stain	....	0.20–0.35			
Wood paste filler	....	0.30–0.40			
Wall primer	....	....	....	0.20–0.35	
Glue size	....	0.20–0.35	....	0.20–0.35	
Varnish size	....	0.20–0.35	....	0.20–0.35	
Zinc sulfate	....	....	....	0.25–0.50	
Calcimine (Kalsomine)	....	....	....	0.25–0.50	
Latex-base paint	....	0.30–0.50	....	0.35–0.50	

quently, in order to arrive at a fair labor estimate, the estimator must know his workmen and must analyze carefully all details of the job. Labor wages for painters may vary from $3.50 to $7 per hour, depending upon time, locality and men available. Payroll taxes and insurance may vary from 8 to 15 percent of the labor wage for most work and from 10 to 20 percent of the labor wage for painting structural steel after erection.

Table 15-5 gives the approximate labor required for painting various surfaces.

The labor required for painting structural steel, one coat, is approximately as follows:

	Hours per ton
Heavy-weight members	0.5–0.9
Medium-weight members	0.7–1.4
Lightweight members	1.0–2.0
Small trusses and built-up frames	1.0–2.0

Diagram 15-2 (Appendix F) may be used for finding labor costs per 100 sq ft of surface (or 100 lin ft) for painting with one coat when the labor wage is known and the labor output can be reasonably assumed.

Table 15-5 *Approximate Labor Required for Painting Different Surfaces Using Brushes*

Kind of work	Sq ft per hr for one coat	Hr per 100 sq ft for one coat
Exterior work:		
Walls, siding and shingles	150–200	0.50–0.70
Doors, windows, blinds	120–180	0.55–0.85
Steel or iron roofing	160–220	0.45–0.65
Trim, cornice, water table, balusters, posts, rails, etc. (in lineal feet)	100–180	0.55–1.00
Stucco	120–150	0.65–0.85
Interior work:		
Trim, filling and puttying (all trim in lineal feet)	60–90	1.10–1.65
Staining	140–200	0.50–0.70
Shellacking	120–180	0.55–0.85
Sand papering	60–90	1.10–1.65
Rubbing	60–90	1.10–1.65
Varnishing	120–180	0.55–0.85
Flat painting	120–180	0.55–0.85
Enameling	100–150	0.65–1.00
Floors, filling and puttying	120–180	0.55–0.85
Staining	150–200	0.50–0.65
Shellacking	150–200	0.50–0.65
Varnishing	140–200	0.50–0.70
Waxing	150–200	0.50–0.65
Polishing	100–150	0.65–1.00
Sanding (hand)	40–80	1.25–2.50
Sanding (machine)	100–200	0.50–1.00
Plaster walls and ceilings, sizing	200–250	0.40–0.50
Calcimining	200–250	0.40–0.50
Painting	120–160	0.60–0.85
Stenciling (in lineal feet)	15–30	3.35–6.65
Brick and concrete work, oiling, sizing, sulfating, etc	120–180	0.55–0.85
Painting	120–150	0.65–0.85
Structural-steel work	150–200	0.50–0.65

Diagram 15-3 (Appendix F) can be used for finding labor costs of painting structural steel.

Equipment The equipment used on a painting job will include brushes and hand tools, ladders, planks, and light, movable scaffolding. When a builtup scaffold is required, this should be estimated separately in regard to materials and carpenters' time. The cost of sanding, rubbing, and cleaning materials is usually included with the equipment costs. If an electric sander or polisher is used, the cost of these machines should be included. A roller set costs about $5 and may last for a few jobs. A good small paint sprayer may cost from $70 to $200 and may last about two years.

The cost of ladders, light scaffolding, brushes, and small tools (including transportation) may vary from about $10 to $30 for a comparatively small job or per gang of three to five painters. When special scaffolding is required, this cost will be increased considerably. In some cases, the cost of a built up scaffold may be equal to the cost of applying one coat of paint.

Overhead and Profit Overhead charges for painting are usually based on labor costs but can be based on the sum of labor and material costs. When based on labor costs alone, the percentage for overhead may vary from about 10 to 20 percent for ordinary interior and exterior work on buildings. The percentage may vary from 8 to 15 percent when based on the sum of material and labor costs. For painting structural steel after erection, the overhead may be more.

Profit may range from about 8 to 15 percent of all other costs for ordinary painting jobs and from 10 percent and up for structural-steel jobs.

Painting Estimates The estimate for a complete painting job should include the following:

Materials. Paint, putty, oil, turpentine, size, varnish, shellac, etc.
Labor. All painters. Also foremen, helpers, apprentices, if any
Equipment. All equipment costs
Overhead. Based on labor costs or on sum of material and labor costs
Profit.

The summary or tabulation should show unit costs per 100 sq ft (or per 100 lin ft for trim) as well as total costs.

Illustrative estimate. Prepare a complete estimate for the interior paint job for a new six-room house. A study of the plans and specifications shows the following:

First Floor	*Second Floor*
2 outside doors	8 inside doors
4 inside doors	8 single windows
4 double windows	600 ft trim
4 single windows (2 in kitchen)	700 sq ft flooring
440-ft trim (42 ft kitchen)	1 stairs
700 sq ft flooring	1 medicine cabinet
1 stairs	1 linen closet
1 mantel	1 towel closet
1 broom closet	
2 kitchen cupboards	

All floors to be sanded, filled, and given 3 coats varnish
First-floor doors and windows and trim, stain, sandpapered, and 2 coats varnish, except kitchen which is to have prime coat, 2 flat coats, and 2 enamel

First-floor stairs same as floor and trim
First-floor walls and ceiling, 1 coat sizing and 2 coats flat oil paint
Kitchen equipment and mantel are finished at mill
Second-floor doors and windows and trim, 1 coat primer, 2 coats flat and 2 coats enamel
Second-floor stairs, stained and varnished as floors
Second-floor walls and ceilings left as they are
Medicine cabinet, towel closet, and linen closet finished at mill
Prices of materials are as follows:

Floor paste	$0.40 per pound
Floor varnish	6.50 per gallon
Stain, oil	5.00 per gallon
Trim varnish	6.50 per gallon
Wall sizing (prime)	4.00 per gallon
Wall paint	5.50 per gallon
Trim, flat paint	5.50 per gallon
Trim, enamel	6.50 per gallon
Materials for sanding	7.00 for job

Floor sander, rental charge, $3 per day plus $4 transportation
Wages of painters, $4.50 per hour plus 10 percent for payroll taxes and insurance
Allowance for other equipment for job, $40
Overhead, 11% of labor costs
Profit, 10% of all other costs

Solution: Materials. Perhaps the simplest way is to consider the different items.

Floors, 1,400 sq ft
  Paste, 1 coat, at 0.35 gal per 100 sq ft = 4.90 gal
  Varnish, 3 coats at 0.20 gal per 100 sq ft = 3 × 14 × 0.20 = 8.40 gal
Stairs (72 + 60 = 132 sq ft)
  Paste, 1 coat at 0.35 gal = 0.47 gal
  Varnish, 3 coats at 0.20 gal = 0.80 gal
Stair rails and balusters (50 lin ft)
  Stain 1 coat at 0.25 gal per 100 lin ft = 0.13 gal
  Varnish 2 coats at 0.20 gal per 100 lin ft = 0.20 gal
First-floor trim (less kitchen) (398 lin ft)
  Stain, 1 coat at 0.25 gal = 1.00 gal
  Varnish, 2 coats at 0.20 gal = 1.60 gal
First-floor doors, finish, 2 on one side, 3 on two sides (say 250 sq ft)
  Stain, 1 coat at 0.25 gal = 0.63 gal
  Varnish, 2 coats at 0.20 gal = 1.00 gal
First-floor windows, 4 double on one side, 2 single on one side (230 sq ft)
  Stain, 1 coat at 0.25 gal = 0.58 gal
  Varnish, 2 coats at 0.20 gal = 0.92 gal
First-floor kitchen doors, 2 on one side (65 sq ft)
  Flat paint, 3 coats at 0.20 gal = 0.40 gal
  Enamel paint, 2 coats at 0.25 gal = 0.35 gal

First-floor kitchen windows, 2 single on one side (50 sq ft)
  Flat paint, 3 coats at 0.20 gal = 0.30 gal
  Enamel paint, 2 coats at 0.25 gal = 0.25 gal
First-floor kitchen trim, 42 lin ft
  Flat paint, 3 coats at 0.20 gal = 0.25 gal
  Enamel paint, 2 coats at 0.25 gal = 0.20 gal
First-floor walls and ceilings (1,680 + 700 = 2,380 sq ft)
  Sizing (wall prime), 1 coat, at 0.25 gal = 6.00 gal
  Flat paint, 2 coats, at 0.30 gal = 14.30 gal
Second-floor trim, 600 lin ft
  Flat paint, 3 coats at 0.20 gal = 3.60 gal
  Enamel paint, 2 coats at 0.25 gal = 3.00 gal
Second-floor doors, 8 on two sides (250 sq ft)
  Flat paint, 3 coats at 0.20 gal = 1.50 gal
  Enamel paint, 2 coats at 0.25 gal = 1.25 gal
Second-floor windows, 8 single on one side (200 sq ft)
  Flat paint, 3 coats at 0.20 gal = 1.20 gal
  Enamel paint, 2 coats at 0.25 gal = 1.00 gal

Materials summary:

Floor paste, 5.37 gal, say 5.5 gal at 5 lb per gallon at $0.40	= $ 11		
Floor varnish, 9.20 gal, say 9.25 gal at $6.50	= 60		
Trim, door, and window stain, 2.34 gal, say 2.50 gal at $5.00	= 13		
Trim, door, and window varnish, 3.72 gal, say 3.75 gal at $6.50	= 24		
Trim, door, and window flat paint, 7.25 gal at $5.50	= 40		
Trim, door and window enamel, 6.05 gal, at $6.50	= 39		
Wall size or prime, 6.00 gal at $4.00	= 24		
Wall paint, 14.30 gal, say 14.5 gal at $5.50	= 80		
Turpentine 2 gal, and oil 1 gal, for thinning, say	= 9		
Materials for sanding	= 7		
Putty, at 1 lb per 1,000 sq ft, say 6 lb	= 2		
Total		= $ 309	

Labor, painters at $4.50 × 1.10 = $4.95 per hour

Floors, 1,400 sq ft, sand, fill, and varnish 3 coats = 10.5 + 10.5 + 3 × 8.4	= 46.2 hr	= $229
Stairs, 132 sq ft, sand, fill, and varnish 3 coats = 2.0 + 1.0 + 3.0	= 6.0 hr	= 30
Trim (including stair rail) 450 lin ft, sand twice, putty, stain, varnish 2 coats = 11.70 + 6.0 + 6.30	= 24.0 hr	= 119
Doors and windows, 480 sq ft, sand twice, putty, stain, varnish 2 coats = 12.5 + 6.0 + 7.2	= 25.7 hr	= 127
Trim, 642 ft, sand twice, putty, primer, flat paint 2 coats, enamel 2 coats = 17.0 + 7.0 + 9.0 + 10.5	= 43.5 hr	= 215

Doors and windows, 565 sq ft, sand twice, putty, primer, flat paint 2 coats, enamel 2 coats = 14.5 + 7.0 + 8.0 + 9.0	= 38.5 hr	= 190	
Walls, 2,380 sq ft, size, paint 2 coats = 11.5 + 33.5	= 45.0 hr	= 223	
Total labor	= 228.9 hr		= $1,133
Equipment: General, $40 plus sander $10			= 50
Overhead: 11% of $1,133			= 125
Profit: At 10% of $1,617			= 162
Total estimate			= $1,779

If desired, this estimate can be checked by using Diagram 15-1 for materials and Diagram 15-2 for labor.

This painting estimate has been worked out in more detail, perhaps, than most estimators would consider necessary. Prices will vary in different localities at different times. Coverage by paint and rates of work will vary considerably also.

Papering

Papering The cost of papering will depend upon the cost of the wallpaper and paste, the cost of preparing the walls and ceilings, the area to be covered, the work required for papering including the work around openings and matching, and the wages, skill, and inclination of the workmen.

On new work, the plastered walls must be given a coat of sizing before the wallpaper is applied. The cost of sizing can be found as directed in the articles on Painting. The unit of measurement for sizing is usually the square of 100 sq ft.

The materials and labor for papering are usually estimated per roll of paper instead of per square yard or per square of 100 sq ft. The width of a roll of paper is usually 18 in., though some styles may have larger widths. The length of a single roll is 24 ft, or 8 yd, and of a double roll 48 ft, or 16 yd. Borders are of varying widths and come in rolls. Borders can be estimated per linear yard or per roll.

Materials

Sizing. If the plaster walls and ceilings must be given a coat of sizing, 1 gal of glue sizing suitable for this purpose will cover 500 to 1,000 sq ft of surface and cost about 50 cents to $1 per gallon.

Wallpaper. In the cheaper work where the strips can be pieced, the number of single rolls can be found by first obtaining the gross area of the walls and ceiling, deducting the area of the openings, and dividing the net area by 36 (square feet in a roll) to obtain the number of single rolls. If the paper used on the ceiling is not the same as that on the walls, then walls and ceiling must be figured separately.

With the more expensive work where strips cannot be lapped, the number of rolls required for the walls can be found by first finding the perimeter of the room in feet and dividing this by 1.5 to obtain the number of strips. The number of strips that can be cut from one roll is found by dividing the length of the roll (single or double) by the length of a strip (height of room). The total number of strips divided by the number of strips that can be cut from one roll will give the number of rolls required for the walls. The rolls required for the ceiling can be found in the same manner. Only half the area of openings of 20 sq ft or larger should be deducted to allow for cutting and fitting around openings.

For waste caused by matching, allow

5 percent for paper with no figured pattern
10 percent for paper with a small figured pattern
15 percent for paper with a large figured pattern

Double rolls should be used, as the waste caused by cutting and matching is less than with single rolls.

The length of border is found by measuring the perimeter of the ceiling in yards. The border is usually estimated in yards but can be estimated in rolls.

Prices of wallpaper per double roll may vary from about $1.50 to $5 or more. Average costs are usually less than $3 a double roll. Border costs range from about 10 cents per yard and up.

Paste. The amount of paste required will vary with the area and weight of the rolls. Paste may cost about 30 cents to $1.00 per gallon. One gallon of paste will be sufficient for about three double rolls of heavy-weight paper, four double rolls of medium-weight paper, and five or six double rolls of lightweight paper.

Labor If plastered walls and ceilings must be sized, the labor required for this purpose will vary from about 0.25 to 0.50 hr per 100 sq ft.

The time required to apply a roll of paper will depend upon the weight of the paper, length of roll, the pattern (matching required), the number of openings, and the skill and ability of the paperhanger.

Table 15-6 gives the approximate amount of labor required for double

Table 15-6 *Approximate Labor Required for Hanging Paper*

Weight of paper	Double rolls per hr		Hr per double roll	
	Walls	Ceiling	Walls	Ceiling
Light	1.5–4.0	1.2–3.5	0.25–0.70	0.30–0.85
Medium	1.0–2.5	0.8–2.0	0.40–1.00	0.50–1.25
Heavy	0.7–2.0	0.5–1.5	0.50–1.40	0.65–2.00

rolls (48 ft or 16 yd long). The labor required for applying border rolls will be the same as for other rolls of equal length. Labor wages may range from about \$3.50 to \$7 per hour. Payroll taxes and insurance may vary from 8 to 15 percent of the labor wage.

Diagram 15-4 (Appendix F) can be used for finding the labor cost per roll for paper hanging when the hourly wage is known and the rate of work can be reasonably assumed.

Equipment The equipment required for papering will include stepladders and planks for scaffolding, cutting planks, paste buckets, brushes, shears, and other hand tools. The equipment costs for a small job may range from \$5 to \$25, including transportation. On large jobs, the equipment costs may be about \$10 to \$30 for each gang of about three to six paperhangers.

Overhead and Profit Overhead costs, based on labor, may range from 10 to 20 percent or from 8 to 15 percent when based on the sum of material and labor costs. Profit may range from 8 to 15 percent of the sum of all other costs.

Papering Estimates Papering estimates should include the costs of all materials, labor, equipment, overhead, and profit charged to the particular job. Estimates may show the cost per roll as well as total costs.

Illustrative estimate. Prepare an estimate for papering the living room of a residence.

Size of room is 12 × 20 by 8.5 ft high.
Openings are

2 triple windows 5 by 8 ft each
2 single windows 3 by 5 ft each
1 double French door, 5 by 7 ft

The room has an 8-in. oak baseboard at the floor and a combined oak cove and picture molding 4.5 in. wide around the ceiling. The walls have not been painted or calcimined, so that sizing will be needed. Wallpaper selected has medium-sized pattern and costs \$4 per double roll. Ceiling paper costs \$3.50 per double roll, and border costs \$3 per roll of 16 yd. Paperhanger's wage is \$4 per hour plus 40 cents for payroll taxes and insurance.

Solution: Materials.

Walls:

Perimeter = 64 ft. Strips = $64 \div 1.5 = 43$
Deduct for openings, $\frac{1}{2}(2 \times 5 + 2 \times 3 + 5) = 10.5$, or about 7 strips
Net 36 strips. Length of strip = 8.5 ft less 1 ft (trim) or about 7.5 ft

If pattern matches easily, 6 strips can be cut from 1 double roll 48 ft long, otherwise only 5 strips. 38 strips $\div$ 6 = $6\frac{1}{3}$ rolls, or may need 7.

Border = 64 ft. or 1⅓ double rolls. Will need 2

Ceiling. Strips crosswise. 20 ft ÷ 1.5 = 14 strips needed

Length of strip = 12 ft less about 8 in., or 11.33 ft

1 double roll would give 4 strips if pattern matches readily, otherwise only 3 strips

Rolls needed = 14 ÷ 4 = 3.5. Will need 4

Materials cost:

Walls 7 × $4	=	$28
Border 2 × $3	=	6
Ceiling 4 × $3.50	=	14
Paste, say 2 gal	=	2
Sizing, say 2 gal	=	2
Total		$52

Labor:

For sizing, about 64 × 8.5 + 12 × 20 − [2 × 5 × 8 + 2 × 3 × 5 + 5 × 7] or 544 + 240 − 145 = 639 sq ft

Assume 0.40 hr per 100 sq ft, time required will be 0.40 × 639 ÷ 100 = 2.55 hr

Labor for papering walls at 0.70 hr per roll gives 0.70 × 7 = 4.9 hr

Ceiling at 0.80 per roll gives 0.80 × 4 = 3.2 hr

Border at 0.70 per roll gives 0.70 × 2 = 1.4 hr

Total labor = 12.05 hr, say 12 hr

Labor cost = $4.40 × 12 = $53

Equipment costs will be low, say about $3 for transportation and depreciation

Overhead, say 12% of labor costs, or $6

Profit, say 10% of all other costs, or about $11

Summary:

Materials	$ 52
Labor	53
Equipment	3
Overhead	6
Profit	11
Total	$125

Or a cost of $125 ÷ 13, or $9.62 per roll

Glazing

Glazing The cost of glazing will depend upon the total area to be glazed, size of panes, quality of glass, price, difficulties of installing the glass in the frames, and the wages, skill, and inclination of the workmen.

For approximate estimates, the total area to be glazed with each quality or kind of glass is taken off and figures assumed for the costs of materials and labor per square of 100 sq ft.

For more detailed estimates, the takeoff shows the number and size of panes of each quality of glass, together with notes relating to the work on the particular job. Then the materials cost is computed. Labor output is assumed, and labor costs computed.

Most sash used for windows and doors in ordinary building construction comes ready glazed. However, when steel sash is used in industrial buildings, the glass is usually installed after the sash is in place and there is little danger of breakage. If large steel sash were glazed before erection, they would be very difficult to handle and the breakage would probably be large. Most steel sash is made so that the glass can be installed from the inside of the building. When glass must be installed from the outside, some form of swing scaffold can be used.

Materials For accurate estimates, the estimator should go over the plans and specifications and count the number of panes of each size and kind. Then the required glass should be listed as to size (dimensions) and kind. The kind may be single strength, double strength, plate, wire, ribbed, etc. The size is usually given in inches as 8 by 10, 10 by 16, 12 by 18, 14 by 20, etc. From 3 to 6 percent should be allowed for breakage.

Window glass is usually packed in boxes containing approximately 50 sq ft. Hence, two boxes are required for each square of 100 sq ft. There are a great many sizes. Widths vary from 6 to 12 in. by 1-in. intervals and from 12 up to about 30 in. by 2-in. intervals. Lengths vary from 8 to about 40 in. by 2-in. intervals and above 40 in. by larger intervals. Manufacturers and dealers rarely stock all the various sizes. Hence, their catalogues and stock lists should be consulted for information regarding sizes and grades, and number of panes per box. Table 15-7 gives information about some of the various sizes, areas, and number of panes or pieces per box. To get the number of panes in a box, divide 50 by the area of one pane in square feet.

Single-strength glass can be used in the smaller sizes. Double strength should be used for panes larger than about 3 sq ft in area.

Prices of glass vary with the thickness (single strength, double strength), clearness or grade, kind (ordinary, plate, wire, corrugated, etc.), dimensions (length and breadth), and areas of panes. Approximate prices (1967) per box of 50 sq ft vary from about $9 and up for single-strength B quality glass, and from about $11 per box and up for double-strength B quality glass. Glass of A quality costs about 20 percent more than glass of B quality. Plate glass may cost from about $2 per square foot and up, depending on the size, thickness, and quality.

About 1 lb of putty is required for every 5 to 7 lin ft of stopping or puttying. Some estimators allow 1 lb of putty for every 2 sq ft of glass, or about 25 lb of putty per box of glass. For glazing steel sash, "steel-sash" putty should be used. This putty is provided at reasonable prices

Table 15-7 *Glass Sizes, Areas, and Number of Panes per Box of* 50 *Sq Ft*

Size, in.	Net area of 100 panes, sq ft	Panes per box	Size, in.	Net area of 100 panes, sq ft	Panes per box
6 by 8	33.3	150	20 by 24	333	15
7 by 9	43.7	115	20 by 28	389	13
8 by 10	55.5	90	20 by 34	472	11
8 by 12	66.7	75	20 by 40	555	9
			20 by 46	638	8
10 by 12	83.3	60			
10 by 14	97.2	52	22 by 28	358	12
10 by 16	111.1	45	22 by 34	519	10
			22 by 40	611	8
12 by 16	133	38	22 by 46	703	7
12 by 20	167	30			
12 by 24	200	25	24 by 28	467	11
			24 by 34	567	9
14 by 16	155	32	24 by 40	667	8 (7)
14 by 20	194	26	24 by 46	767	7 (6)
14 by 24	233	22	24 by 52	867	6
14 by 28	272	19 (18)			
			26 by 34	613	8
16 by 20	222	23	26 by 40	722	7
16 by 22	244	21	26 by 46	831	6
16 by 24	267	19	26 by 52	939	5
16 by 28	311	16			
16 by 34	378	13			
18 by 20	250	20			
18 by 22	275	18			
18 by 24	300	17			
18 by 28	350	14			
18 by 34	425	12			

by the manufacturers of the sash. The price of putty may vary from about 15 to 30 cents per pound. Table 15-8 gives the approximate amount of putty required for glazing.

Diagram 15-5 (Appendix F) can be used for estimating costs of glass when the price per box is known. Curves are shown for glass with breakage allowances of 5 and 10 percent and for glass plus breakage allowances plus an allowance of $5 per box for putty.

Labor Labor costs vary greatly because of the difficulties found on the particular job, the weather, and the skill and inclination of the workers. A good average rate of work is about 100 panes, 14 by 20 in. in size, per 8-hr working day. Some jobs may run as low as 50 a day, and other jobs may run as high as 200 per day. More time is usually required

Table 15-8 *Pounds of Putty Required per* 100 *Panes and per Box of Glass*

Glass size, in.	Pounds putty		Glass size, in.	Pounds putty	
	Per 100 panes	Per box		Per 100 panes	Per box
8 by 12	56	42	16 by 34	155	20
10 by 16	75	34	20 by 28	150	19.5
12 by 20	93	28	20 by 40	185	16.5
14 by 20	100	26	24 by 38	160	17.5
14 by 28	125	24	24 by 46	215	15
16 by 22	115	24			

(about 10 percent) to glaze monitors and roofs than to glaze side walls. Compared with summer work, winter work may require up to 20 percent more time, depending upon the severity of the weather. The larger the size of the light, the more the time required per light for glazing.

Table 15-9 gives the approximate number of hours required by skilled workers for glazing 100 lights or panes. Summer weather and side walls are assumed.

Table 15-9 *Labor Required for Glazing Steel Sash*

Glass size, in.	Panes, per hour	Labor-hours per 100 panes	Glass size, in.	Panes per hour	Labor-hours per 100 panes
10 by 16	12–17	6–8	16 by 22	9–12	8–11
12 by 18	11–17	6–9	24 by 34	4–6	18–27
14 by 20	10–14	7–10	34 by 46	3–5	22–30

Glazing can also be estimated per square of 100 sq ft. Average rates of work are about as follows:

Table 15-10 *Labor-hours per Square of* 100 *Sq Ft of Glazed Surface*

Size of panes	Square feet per labor-hour	Labor-hours per square
Small, 1.5 sq ft or less	14–25	4.0–7.0
Medium, 1.5 to 4.0 sq ft	18–33	3.0–5.5
Large, 4.0 sq ft or more	22–40	2.5–4.5

Glazing is often done by painters, and prevailing hourly wages apply. In some places, glazing is done by a special crew of glaziers who receive comparatively high pay and who have a high hourly output. Sometimes the glazing is paid for on a piecework basis with varying rates for panes of different sizes. Wages of skilled workmen may vary from about $3.50 to $7 per hour.

Diagrams 15-6 and 15-7 (Appendix F) can be used for estimating the labor cost of glazing.

Steel sash are usually erected by structural-steel or iron workers and not by the glazier. The painting or calking of the steel sash may be done by steelworkers, masons, or handy men and rarely by painters or glaziers.

Equipment The cost of equipment for a glazing job is usually small. The equipment needed may consist of a swinging scaffold, ladders, stepladders, putty knives, and glass cutters. Transportation will be needed for scaffolds and ladders. The glaziers can move the swinging scaffold and ladders as needed. An allowance of about $10 to $20 is usually sufficient for a small gang of about three to five men.

If built-up scaffolds are required, these should be estimated separately. The estimate should include all materials and all labor needed for erection, moving, and dismantling.

Overhead and Profit Overhead is usually based on labor costs and may range from 10 to 25 percent. When based on the sum of materials and labor, the percentage may range from 7 to 15 percent.

The percentage allowed for profit may vary from 8 to 20 percent of the sum of all other costs.

Glazing Estimates Glazing estimates should include the costs of all materials delivered at the job (including freight, insurance, truckage, handling, storage, and breakage), and the costs of labor, equipment, overhead, and profit.

If desired, estimates may be subdivided to show costs per sash, costs per light or pane (of each size), or costs per 100 sq ft of glass.

Illustrative estimate. Estimate the costs of materials, labor, equipment, overhead, and profit for the following job of glazing steel sash. All lights are double-strength (DS) A grade clear glass.

Materials takeoff

Size of lights, in.	No. of lights	Sq ft
10 by 16	250	278
12 by 18	300	450
14 by 20	672	1307
20 by 22	154	471

Cost of lights = \$17 per box of 50 sq ft delivered at job
Cost of sash putty = \$0.16 per pound
Labor wage = \$5 per hour plus \$0.50 per hour for payroll taxes and insurance
Equipment costs assumed as \$30 for the job
Overhead assumed as 13% of labor cost
Profit assumed as 12% of all other costs

Materials (allow 3 to 5% for breakage):

250 lights, 10 by 16 in., 45 per box, say 6 boxes			
300 lights, 12 by 18 in., 34 per box, say 9 boxes			
672 lights, 14 by 20 in., 26 per box, say 27 boxes			
154 lights, 20 by 22 in., 16 per box, say 10 boxes			
Total lights	= 52 boxes at \$17	= \$ 884	
Putty: 75 × 2.50 =	190 lb		
90 × 3.00 =	270 lb		
100 × 6.72 =	672 lb		
125 × 1.54 =	193 lb		
Total =	1,325 lb at \$0.16	= \$ 216	
Materials total		= \$1,100	
Materials cost for glass can be checked by the area method.			
Total area = 2,506 sq ft plus 4% for breakage gives 2,600 sq ft, or 52 boxes at \$17			= \$ 884
Putty at 26 lb per box = 1,350 lb at \$0.16			= 216
Total materials			= \$1,100

Diagram 15-5, with 4 per cent breakage and an allowance of \$4 per box for putty, gives a cost of about \$44 per 100 sq ft.
Total materials cost = \$44 × 25.06 sq = \$1,100

Labor at \$5.50 per hour:

10 by 16 glass at 7 hr per 100, 250 panes	= 17.5 hr	
12 by 18 glass at 7.5 hr per 100, 300 panes	= 22.5 hr	
14 by 20 glass at 8.5 hr per 100, 672 panes	= 57 hr	
20 by 22 glass at 10 hr per 100, 154 panes	= 15.5 hr	
Total labor	= 112.5 hr	= \$ 619

Or, estimating about 4.5 hr per 100 sq ft,
Labor-hours = 4.50 × 25.06 sq = 113 hr, or \$622

By using Diagram 15-6, the cost per square, for a wage of \$5.50 per hour and 4.50 hr per square, is \$24.75. \$24.75 × 25.06 sq gives \$620.

Equipment, assumed	= \$ 30
Overhead, 13% of \$619 (labor)	= 80
Profit, 12% of \$1,829	= 219
Total estimated cost	= \$2,048

Or a cost of about \$82 per 100 sq ft of glazed surface.

CHAPTER SIXTEEN

Heavy Machinery

Estimating Machinery The estimation of the costs of heavy machinery is a special problem in the handling, transporting, and installing of comparatively large and heavy objects. Nearly every problem of this kind includes such items as the following:

Machinery cost, usually f.o.b. factory. Sometimes the manufacturer allows or prepays freight to destination.

Freight to most convenient railway siding, if shipped by railroad, including demurrage and other extra charges.

Freight to site if shipped by truck.

Transporting, erecting, operating, and taking down of special equipment required for handling heavy machinery.

Unloading from railway car and transporting from railway car to site or unloading from truck at site.

Installing the machinery in place, which may include erection, setting, lining up, and grouting.

Miscellaneous work, which may include the construction of foundations and the removal and replacing of building walls and roofs.

On some jobs, the owner or general contractor assumes the cost of the machinery f.o.b. destination and a subcontractor cares for the handling, transportation from railway siding to site, and installation at the site. When estimates are prepared for the subcontractor, the materials cost (except that for miscellaneous materials) is omitted. Insurance is cared for in the overhead costs.

When preparing an estimate for heavy machinery, the estimator must have information concerning the kinds, sizes, and weights of the machinery; the method of transportation from factory to destination; the kinds of equipment the contractor may have or may be able to secure by rental or purchase; the ability and skill of the foreman and crew; the methods that can be satisfactorily and economically used for handling, transporting, and installing this particular kind of machinery; state and local regulations regarding this kind of work, especially such items as permits, allowable loads on streets, and available routes; the amounts and kinds of the miscellaneous work that must also be done.

In almost every city of medium or large size, there are available trucks, trailers, rollers, special four-wheel trucks or dollies, jacks, winches, cribbing, gin poles, derricks, cranes, and other tools. If the contractor's equipment is not suitable for the work, the equipment needed can be rented. Sometimes the work of unloading the machinery and transporting it to the job site can be economically sublet to a local company that specializes in the moving of heavy and bulky objects.

Whenever practical, the estimator should consult with men who have had experience in loading and unloading, moving, and installing heavy machinery. From these men, the estimator can obtain considerable valuable information in regard to the equipment and gangs which can be used, the methods which can be economically followed, and the approximate time required.

When preparing a cost estimate for heavy machinery, the estimator should be able to visualize in detail all the methods and operations in the order that they may be used.

The estimator should be sure to allow for time lost by delays caused by bad weather, accidents, and breakdowns.

The cost of unloading at railway siding, transporting and erecting heavy machinery may range from $5 to $20 per ton and up, depending upon the size, weight, and the ease with which the machinery can be placed or erected and the work required. Large heavy pieces will usually cost less per ton. Large and light pieces will cost more per ton. Field riveting and bolting as of stacks, and lining up, as of engines, dynamos, and shafting, will increase the costs.

Materials For estimating purposes, the estimator should know the kinds, sizes, and weights of the machinery; the cost at the factory; and the transportation costs up to the time the shipment is turned over to the contractor. In addition to the first cost and freight, there may be demurrage costs, insurance on materials in transit, inspection at factory, expense of tracing, and interest on investment. Most of these latter items, except demurrage, are often included in overhead expenses.

On many jobs, miscellaneous materials may be required, such as bolts and washers, materials for making grout, or masonry materials as the case may be. The costs of these miscellaneous materials delivered at the job should be estimated and preferably included in the materials cost.

Labor The labor required may include the following:

Handling. Unloading from railway car to ground or to truck, trailer, or dollies, and unloading at the site. The machinery may be unloaded from the railway by means of a crane, derrick, or gin pole or by means of jacking, bulling, or skidding on small rollers as the case may be.

Transportation. Extra labor is often required to help in the transportation of the machinery to the site. The machinery may be moved on skids and small rollers, on heavy four-wheel trucks or dollies, on large-capacity autotrailers, or on large-capacity trucks. The extra labor required will vary with the particular job.

Installation. The labor required will depend on the kind, size, and weight of the machines, the equipment provided, and the methods used. This labor may include the labor of erection as for a steel smokestack, the labor of setting or placing a machine (such as an engine or dynamo) in position, lining up and plumbing the machine after setting, fastening the machine in position by bolting, riveting, or grouting.

Equipment. Labor will always be required for moving, erecting, operating, taking down, loading, and unloading equipment of various kinds.

Miscellaneous. On many jobs, labor may have to be provided for constructing foundations, setting anchors and anchor bolts, removing and replacing walls and roofs, etc.

The size of the gang will vary with the machine to be installed, the equipment used, and the kind of work. Usually, a gang will contain about 4 to 10 men under a capable foreman. The gang may be composed of foreman, skilled labor (truck driver, crane operators, steelworkers, masons, etc.), and unskilled labor, depending on the kind of work. When conditions permit it, as many unskilled men should be used as practical. If the men in the gang are used to working together under the particular foreman, there may be a considerable saving in labor time on work of this kind when compared with a green or inexperienced gang. There is a skill or knack of handling heavy machinery that can be acquired only by experience.

The wages of the men may vary about as follows:

Foreman	$4.00 to $9.00 per hour
Skilled labor	$3.00 to $6.00 per hour
Unskilled labor	$2.00 to $5.00 per hour

Payroll taxes and insurance may vary from about 12 to 25 percent of the labor wage.

The labor of handling, transporting, and installing heavy machinery can be most conveniently estimated in labor-hours per ton, with details noted in regard to machinery, equipment, and working conditions. Sometimes a heavy machine can be handled at less cost per ton than a light machine.

The labor required for the equipment can be estimated in labor-hours required for each kind of equipment.

The labor required for miscellaneous work can be estimated in labor-hours for the particular job according to the kind and quantity of work to be done.

Tables 16-1 to 16-4 give the labor-hours per ton required for doing various kinds of work. The values given are necessarily very approximate and must be so considered. It is practically impossible to give accurate information because of the differences in working conditions, variations in the skill and ability of the foreman and men, variation in equipment, and differences in kinds of machinery.

Table 16-1 *Approximate Labor-hours for Heavy Machinery*

Item	*Labor-hours*
Bulling and moving or turning up to 8 or 10 ft	1–2 per ton
Jacking up or down, and placing or removing cribbing	0.5–1.0 per ton per foot of height
Jacking up and installing skids	1–2 per ton
Moving on skids and small rollers	0.5–1.5 per ton per 100 ft
Unloading, from car to ground, from car to truck or trailer, from truck or trailer to ground, with use of derrick or gin pole	1–6 per ton
Erecting with use of derrick or gin pole (steel stack for example)	2–12 per ton
Setting, plumbing, etc. (steam boiler, for example)	1–3 per ton
Lining up, roughly	0.5–3 per ton
Lining up, accurately	3–15 per ton
Grouting	0.10–0.25 per ton
Installing guy wires on stack	1–3 per guy wire
Riveting, hand-driven field rivets	12–25 per 100 rivets
Jacking up or down, and placing or removing large dolly trucks	1–5 per ton

Table 16-2 *Approximate Labor-hours Required for Unloading, Erecting, and Lining Up Certain Kinds of Machinery*

Kind of machinery	Work	Labor-hours per ton
A. Large and bulky machinery such as smokestacks, shells, and tanks weighing about 1 to 5 tons per piece. (Smokestack and like types requiring riveting and guying.)	Unload car to truck or trailer or to ground	2–8
	Unload truck or trailer to ground	1.5–6
	Erect only	3–10
	Erect and rivet sections	6–20
	Erect and line up	5–15
	Erect on ground, hand rivet, guy	8–25
	Erect on roof, hand rivet, guy	10–35
B. Heavy but not bulky machinery such as boilers, compressors, engines, and motors weighing 2 to 10 tons per piece.	Unload car to truck or trailer or to ground	2–7
	Unload truck or trailer to ground	1–5
	Set and roughly line up	2–6
	Accurately line up when required	3–18
C. Heavy but not bulky machinery such as engines, boilers, and motors, weighing 10 to 50 tons per piece.	Unload car to trailer or to ground	2–6
	Unload trailer to ground	1–5
	Set and roughly line up	2–5
	Accurately line up when required	3–15

The values given do not include any labor for equipment, such as labor of unloading, erecting, rigging, taking down, and loading derricks or gin poles, but do include labor of arranging ordinary cribbing and blocking.

Labor Diagrams. Diagram 16-1 (Appendix F) can be used for estimating the labor cost of handling and installing heavy machinery, and Diagram 18-4 (Appendix F) for estimating the labor cost for equipment.

Equipment The equipment used for handling, transporting, and installing heavy machinery will depend upon the needs of the particular job, the methods that can be used, the equipment available (either owned or rentable), and what the contractor thinks should be used. There is, consequently, a great variety of combinations of kinds and sizes or capacities of equipment that may be considered. In this text, some of the equipment that can be used will be listed. The reader is referred to texts on construction methods for the proper selection of equipment and its use.

To find the equipment costs for a given job, all equipment used must be listed with the costs and the total found. The cost per ton equals the total equipment cost divided by the tonnage of the machinery.

Table 16-3 *Approximate Labor-hours Required for Equipment*

Item	*Labor-hours*
Handling cribbing and timbers, per 100 pieces	1–5
Gin pole or guy derrick, large, 15- to 30-ton capacity:	
Unload and erect	40–100
Take down and load	15–50
Gin pole or guy derrick, small, 5- to 15-ton capacity:	
Unload and erect	20–50
Take down and load	8–20
Stiffleg derrick, large, 25- to 60-ton capacity:	
Unload and erect	150–300
Take down and load	80–150
Stiffleg derrick, medium or small, 10- to 30-ton capacity:	
Unload and erect	60–150
Take down and load	30–80
Hand derrick, small, 1- to 5-ton capacity:	
Unload and erect	15–40
Take down and load	8–20

Table 16-4 *Approximate Extra Labor Required with Equipment*

Item	*Extra labor*
With each autotruck	1 driver, 1 or 2 helpers
With each autotruck trailer	1 to 3 helpers
With each set (3 or 4) of heavy dolly trucks	2 to 4 helpers
With each hand windlass	2 to 5 helpers
With each crane or derrick (power)	1 foreman, 1 operator, and 4 to 10 helpers
For handling cribbing or timbers	2 or more
For jacking, bulling, cribbing, etc. (About 2 plus 1 to 3 for each 10 tons of weight)	1 foreman, and 2 to 12 helpers

Various types of equipment that may be used for handling, transporting, and installing heavy machinery are:

Handling, loading, and unloading:
- Cribbing and timbers
- Skids and small rollers
- Jacks
- Pinch, claw, and crow bars
- Dollies, small
- Gin poles
- Shear legs and A frames
- Stationary cranes and derricks
- Crawler and auto cranes
- Hand- and power-operated cranes and derricks
- Winches
- Block and tackle
- Wire and manila rope
- Small tools

Transporting:
- Autotrucks
- Autotrucks and trailers
- Heavy four-wheel dollies
- Skids and rollers
- Winches
- Block and tackle
- Wire and manila rope

- Various small and hand tools
- Lighters
- Tugs
- Floating cranes and derricks

Erection or installation:

- Cribbing and timbers
- Jacks
- Gin poles
- Shear legs and A frames
- Cranes and derricks, both hand and power
- Crawler and auto cranes
- Block and tackle
- Wire and manila rope
- Various small and hand tools
- Ladders
- Scaffolds
- Slings
- Welding equipment
- Riveting equipment
- Air compressors
- Steelworkers' tools
- Grouting equipment
- Mason's tools
- Roofers' equipment
- Tractors

Rental costs may be per hour, per day, per week, or per month with certain minimum charges. The daily rate may be based on a day of 8 hr or less. The hourly rate may be one-eighth of the daily rate. The weekly rate may be three to four times the daily rate. The monthly rate may be about three times the weekly rate. Rental rates start at the time the equipment is turned over to the renter, either at the job site or f.o.b. the nearest railway siding. Rental rates may include machine and fuel and major but not minor repairs and may or may not include an operator. They usually do not include transportation and freight.

The best method of estimating equipment costs is first to estimate the number of working hours that each piece of equipment will be used. Then the total cost of one piece of equipment for the job would include hourly ownership or rental cost times number of hours used on the job, plus transportation costs to and from the job, plus maintenance, and plus labor costs of operation, erection, takedown, etc. (if these labor costs are not included in the labor estimate).

Information concerning ownership and rental costs of various pieces of equipment is given in Appendixes B and C.

The equipment costs per ton of heavy machinery is a suitable unit to use for approximate estimates when, and only when, the estimator has adequate information. Naturally the costs per ton of machinery will vary considerably with the different kinds of equipment used, the sizes and weights of the machinery, the skill of the foreman and crew and the transportation costs of the equipment to and from the job. The labor costs of erecting, operating, moving, taking down, loading, and unloading equipment are usually included in the labor estimate.

Table 16-5 gives approximate equipment costs per ton of heavy machinery for different kinds of equipment. The values given do not include

Table 16-5 *Approximate Equipment Costs per Ton of Heavy Machinery (Transportation and Labor Costs Are Not Included)*

Item	*Cost per ton*
Handling, loading, and unloading:	
Cribbing and timbers	$0.10–$1.50
Skids and small rollers or dollies	$0.20–$2.00
Jacks	$0.05–$0.25
Hand tools, hammers, bars, etc	$0.10–$1.00
Small gin poles, shear legs, A frames, etc	$0.10–$3.00
Hand derricks and cranes	$0.10–$3.00
Medium and large stationary cranes and derricks	$0.20–$4.00
Block and tackle, and wire and manila rope	$0.10–$1.00
Crawler cranes	$0.20–$4.00
Transportation:	
Autotrucks ($2 for load and unload + $0.20 per ton mile)	$2.20–$6.00
Autotrucks and trailers (ton costs about same as for autotrucks)	$2.35–$7.00
Heavy four-wheel dollies (set of 3 or 4), per mile	$0.20–$1.00
Skids and rollers or small dollies	$0.25–$1.00
Winches	$0.20–$1.50
Block and tackle, and wire and manila rope	$0.10–$1.00
Erection:	
See items under Handling, loading, and unloading	
Ladders, slings	$0.10–$0.50
Scaffolds	$0.10–$2.50

transportation costs of equipment to and from the job and labor costs of loading and unloading, erecting, moving, operating, and taking down the equipment unless otherwise specified.

The range in costs as given in Table 16-5 is quite large for several reasons. For example, one piece of equipment with a minimum rental price of $5 per day may be required for handling a machine of 2 tons weight, giving a cost of $2.50 per ton. On another job, a similar piece of equipment may be used to handle several machines in a day with a total tonnage of, say, 50 tons. This would give a cost of but 10 cents per ton. An example would be a crane or derrick unloading machinery at a railway siding.

The cost of transportating heavy machinery with autotrucks or with autotrucks and trailers may be determined approximately by the formula

$$\begin{pmatrix}\text{Cost per}\\ \text{ton of}\\ \text{machinery}\end{pmatrix} = \begin{pmatrix}\text{cost of loading}\\ \text{and unloading}\\ \text{per ton}\end{pmatrix} + \begin{pmatrix}\text{cost per}\\ \text{ton mile}\end{pmatrix} \times \begin{pmatrix}\text{miles hauled}\\ \text{loaded}\end{pmatrix}$$

Loading and unloading costs may range from 75 cents to $3 per ton, and the cost per ton-mile may range from about 7 to 30 cents.

Diagrams. Diagrams 16-2, 18-4, and 18-5 can be used for estimating equipment costs, and Diagram 16-3 (Appendix F) for estimating transportation costs.

Overhead and Profit Overhead costs may be comparatively large because of the job hazard. The overhead costs may be based on the sum of material, labor, and equipment costs; based on labor costs alone; or based on material, labor, and equipment costs using different rates or percentages for each.

When based on labor costs alone, overhead costs may range from about 10 to 40 percent of the labor costs.

When different rates are used for materials, labor, and equipment, approximate variations are

From 2 to 10 percent of the materials cost
From 10 to 30 percent of the labor cost
From 3 to 10 percent of the equipment cost

The profit charged on a heavy machinery job is usually fairly high. Average rates may vary from about 8 to 25 percent of the sum of all other costs.

Heavy-machinery Estimates The estimate for heavy machinery should include:

Materials. Cost of machinery, freight, and miscellaneous materials
Labor. All labor costs of handling, transporting, and installing the machinery
Equipment. All equipment costs
Overhead. Usually based on labor costs
Profit. Based on sum of all other costs

The estimate may be in more detail if desired, and unit costs per ton of machinery may be computed.

Some estimators arrange their estimate differently by separating the handling and transportation and installation costs about as follows:

Materials. Cost of machinery, freight, and miscellaneous materials
Handling and transportation. From railway car to job site
  Labor. Cost of all labor used for this purpose
  Equipment. Cost of all handling and transporting equipment
Installation, erecting, setting, lining up, grouting, etc.
  Labor. Cost of all labor of installation
  Equipment. Cost of all equipment used for installation
Overhead. As a separate item or subdivided under materials, handling and transportation, and installation
Profit. On sum of all other costs

Illustrative estimates. *Steam boiler.* Prepare an estimate of the cost of unloading from a railway car a return tubular steam boiler weighing 12 tons, transporting the boiler 3 miles to the job site, and erecting the boiler on its foundations. As the walls and roof of the boiler house have not been built, it will not be necessary to make an opening in the walls. Boiler is mounted on skids and was shipped on a flatcar.

Boiler is to be unloaded from flatcar to trailer, transported 3 miles to job site, unloaded, and placed in position. The following data are assumed:

Owner's cost of truck and trailer = \$3.15 per hour
Miscellaneous cribbing, timbers, small rollers, tools, etc., including transportation = \$12 per job
Workmen = \$3 per hour
Foreman = \$5 per hour
Payroll taxes and insurance = 18% of labor wage
Overhead = 14% of labor cost
Profit = 12% of all other costs
First day
Unloading, car to trailer
Place rollers under skids, place timber and cribbing, load on trailer, place cribbing, etc., on truck
Use a gang of 1 foreman and 6 laborers, one as truck driver.
Hourly wage of gang, including payroll taxes and insurance, = (\$5 + \$3 × 6) 1.18 = \$27.14
Estimate about 2.5 hr per ton, or 2.5 × 12 = 30 labor-hours, or about 4.3 gang-hours for a gang of 7
Trucking 3 miles to job site may require about 1 hr
Unloading, trailer to timbers on ground, about 1.5 hr per ton, or 1.5 × 12 = 18 labor-hours, or about 2.6 gang-hours
This will take a day of 8 hr
Second day
Skid and roll boiler to foundation, lower and set on foundation, hang doors, clean up and remove all equipment. This is estimated to take about 2.25 hr per ton, or 2.25 × 12 = 27 labor-hours. Or about 4 gang-hours
Total gang-hours for both days = 12
Summary (no materials cost):

Item	Cost per ton	Total cost
Labor, \$27.14 × 12	\$27	\$326
Equipment, \$3.15 × 12 + \$12	4	50
Overhead, 14% of \$326	4	46
Profit, 12% of \$422	4	51
Total	\$40	\$473

NOTE: In 1967 two similar jobs in two states cost $450 and $515.

Smokestack. Prepare an estimate of the cost of unloading from a railway car to trailer a steel smokestack 3 ft in diameter and 50 ft high, transporting the smokestack 3 miles to the job site, riveting the sections together (the stack comes in three sections), and erecting the stack on the roof of the boiler house. Weight of stack is about 3 tons. The stack is to be erected by using a suitable gin pole. The contractor has the requisite equipment.

The following data are assumed:

Gin pole and rigging, owner's cost per hour = $0.35
Truck, 5 ton, with power winch mounted back of seat, hourly cost = $2.50
Two-wheel trailer, 5 ton, suitable for moving stack sections, gin pole, other equipment, hourly cost = $0.25
Equipment for cold riveting and rivets, job cost, estimated = $17
Miscellaneous tools and equipment, job cost estimated = $10
Contractor acts as foreman, hourly wage = $5
Five workmen, one acting as truck driver, hourly wage = $3
Payroll taxes and insurance = 18% of labor wage
Overhead = 14% of labor cost
Profit = 12% of all other costs

Stack sections can be unloaded from flatcar to trailer and from trailer to ground by hand labor without use of gin pole. Gin pole is used for erection only.

	Labor-hours
Stack labor:	
Unloading stack from car to trailer, say 4 labor-hours per ton or section, giving	12
Unloading at job site, a little less, say	10
Cold-riveting sections together, about 90 rivets at 20 hr per 100 rivets, say	18
Erecting stack, including raising to a vertical position, raising to roof of boiler house, and setting with aid of gin pole and winch on truck, say about 9 labor-hours per ton or	27
Guying 4 guys, say	3
Trucking, assume 1 hr with gang	6
Total for stack	76
Equipment labor:	
Loading gin pole and other equipment at contractor's yard	4
Unloading at job	3
Erecting and rigging gin pole	16
Takedown and load, gin pole	8
Unload again at contractor's yard	3
Trucking 2 hr with gang	10
Total for equipment	44
Total for job	120
Gang-hours for a gang of 6 men	20

Assume 20 working hours.

Hourly wage of gang, including payroll taxes and insurance,
= ($5 + $3 × 5)1.18 = $23.60

Total labor cost	= $23.60 × 20	= $472
Equipment cost:		
Truck	= $2.50 × 20 = $50	
Trailer	= $0.25 × 20 = 5	
Gin pole	= $0.35 × 20 = 7	
Riveting and rivets	= 17	
Miscellaneous	= 10	
Total equipment cost		= 89
Overhead at 14% of $472		= 66
Profit at 12% of $627		= 75
Total		= $702

Summarizing (no materials cost) for 3-ton stack:

Item	Cost per ton	Total cost
Labor	$157	$472
Equipment	30	89
Overhead	22	66
Profit	25	75
Total	$234	$702

CHAPTER SEVENTEEN

Asphalt Walks, Driveways, and Pavements

Asphalt work for walks, driveways and pavements is usually sublet to a contractor specializing in this type of work. This contractor or his estimator is capable of preparing detailed and accurate estimates. Consequently, the general contractor or his estimator rarely needs to prepare a detailed estimate for asphalt work, but he should have a good general knowledge of estimating so that he can check and determine the reasonableness of the detailed bids submitted by the asphalt contractors.

Classification The asphalt work covered in this chapter can be classified according to traffic (light, medium, or heavy), use (walks, playgrounds, driveways, parking lots, streets, roadways, etc.), or construction (surface treatment, cold mix, or hot mix). The classification according to traffic appears to be the best. Most of the new construction is by the hot-mix method. The cold mix is used for patching and temporary repairs, and the surface treatment for country roads where the traffic is light, for resurfacing, and sometimes for stabilizing the base.

The Asphalt Institute suggests that soils on which asphalt concrete is to be placed may be classified as follows.

Good to Excellent. Good subgrade soils will retain a substantial amount of their load supporting capacity when wet, such as clean sands, sandy gravels and soils free of detrimental amounts of plastic materials.

Excellent subgrade soils are unaffected by moisture or frost, such as well-graded sands and gravels.

Medium. Medium subgrade soils which retain a moderate degree of firmness under adverse moisture conditions, such as loams, silty sands and sand-gravels containing only moderate amounts of clay and fine silt.

Poor. Poor subgrade soils which become soft and plastic when moist, such as soils containing appreciable amounts of clay and fine silt, and some of the coarse silts and sandy loams.

The weight of 1 sq ft of a compacted asphalt concrete course should be 12 lb or a little more per inch of thickness, corresponding to a density of 88 percent. This density may be considered as being achieved when the weight of a fully loaded truck of 8 tons of payload capacity does not make indentations of more than 1 in. in depth. City, county, and state highway departments usually require a density of 92 percent or more for streets and highways.

The weight of 1 sq ft of a well-compacted subbase material should be 9.5 lb, or a little more, per inch of thickness.

With some soils, subdrainage must be provided by constructing crushed-rock or tile drains. With other soils, an insulation course, consisting of a compacted layer of stone screenings, granulated sand, or other good sand, may be required.

Surface drainage can be obtained by giving the finished wearing surface a slope of 1/8 to 1/4 in. per foot.

The soil should be sterilized in all areas to be paved so as to prevent the growth of vegetation.

Forms are usually needed for asphalt walks and drives, and they are usually left in place. The forms may be of wood, concrete, or steel. Sometimes an earth or asphalt-concrete shoulder is used instead of forms.

Construction The construction of an asphalt concrete paving job may include the following items:

The subgrade must be properly prepared and compacted by rolling with a roller of not less than 5 tons. Required subdrains should be installed. Sometimes the subgrade is prepared by a different contractor.

The soil should be sterilized.

Then the insulation course (if required) should be placed and compacted by rolling.

The subbase course (if required) is placed and compacted.

The base courses (if required) are placed and compacted. These courses should consist of crushed rock, gravel, slag, decomposed granite, or other suitable material uniformly spread and compacted by a power roller weighing 5 tons or more. The compacted thickness of any one layer should not exceed 4 in.

Prime, or tack, coats should be applied if required.

The asphalt concrete is placed on the base course. No compacted layer should be more than 3 or less than 1 in. in thickness. A standard paving machine should be used if warranted by the amount of work. When hand methods are used, the mix should be shoveled in place from a truck or dumped on a dump board and then shoveled in place. Excessive raking should be avoided. When the asphalt concrete is placed in two layers, the upper layer is the wearing course or layer.

In the hot-mix method the aggregate can be heated to a temperature of 300 to 375°F, then screened and combined to give the desired grading. The asphalt cement is heated to about 275 to 375°F, and then the asphalt and aggregate are thoroughly mixed in a mixer. The hot asphalt concrete is discharged from the mixer into trucks and hauled to the job, where the mix is spread in uniformly thick layers by machine or hand methods. Each layer is compacted to the desired density by rolling with steel rollers weighing 5, 10, or more tons as specified.

Items Included in an Estimate The following are some of the items that may need to be included (or considered) in the estimate:

Excavation and fill. Usually done by another contractor
Soil sterilization
Drainage
Leveling of subgrade
Insulation course
Base and subbase courses
Primer coat
Forms
Asphalt-concrete course or courses
Cleaning up

Some other items which may affect the estimate for a large job are climate and weather conditions, maintaining of traffic around the job, rent of land for the mixer plant, cost of transporting asphalt and aggregates to the mixer plant, cost of transporting hot mixes to the job, and special specification and contractural requirements.

Units of Measurement The units of measurement are usually the square foot or square yard of surface area, with the compacted thickness of each of the courses and other details given. However, many estimators use the weight in tons of the materials (base and subbase courses, asphalt concrete, etc.) placed and compacted. Unit costs may be computed per square foot or per square yard of surface area and also per ton of asphalt concrete and base course materials. The materials used should be measured by weight in pounds or tons, and not by volume.

Asphalts The asphalts used can be classified as road oils, slow-curing (SC) asphalts; cutback asphalts, medium-curing (MC) or rapid-curing (RC); emulsified asphalts; asphalt cements; natural asphalts; asphalt primers; etc.

Present prices (November, 1967) are approximately as follows for materials purchased in tank-car lots f.o.b. refinery. Transportation costs must be added.

Paving asphalt, less than 80 penetration, per ton	\$20–\$33
Cutback asphalt (MC), per ton	\$19–\$30
Road oil, per ton	\$20–\$33
Road oil, per gallon	\$0.08–\$0.14
Asphalt emulsion (RS-1), per gallon	\$0.10–\$0.18
Asphalt binders and fluxes:	
Per ton	\$20–\$33
Per gallon	\$0.10–\$0.22

Materials in drums may cost from 50 to 100 percent more.

The amount of asphalt required for a hot mix may vary from 4 to 7 percent of the weight of the aggregate.

For prime coats, the rate of application is about 0.15 to 0.25 gal per square yard.

For various surface treatments, the rate of application may vary from about 0.20 to 0.55 gal per square yard of surface per application, depending upon the kind of asphalt used.

Diagram 17-1 can be used for finding the cost of asphalt per ton of asphalt concrete.

Aggregates The aggregates used in hot mixes include sand, gravel, and crushed rock of good quality. They must be proportioned in size to give the desired density. This requires screening (and sometimes washing) and remixing of the various sizes. The maximum size of the aggregate should not be more than about two-thirds of the compacted thickness of any layer or course of asphalt concrete.

Typical specifications for aggregate gradings are given in Table 17-1. The gradings are based on the use of U.S. Standard square-opening sieves.

If a subbase course is required, this course should be constructed of a pit run or an unwashed stream-bar gravel, a crusher run material, or a blend of commercial rock products.

The aggregate should conform to the following gradings:

Passing a 2-in. sieve	100%
Passing a 1-in. sieve	70–100%
Passing a No. 4 sieve	35–70%
Passing a No. 200 sieve	3–11%

The cost of aggregates will vary considerably in different localities. The cost of base course materials, such as decomposed granite, bank run gravel, slag, etc., may range from about 50 cents to $2 per ton at the job. The cost of aggregates suitable for asphalt-concrete mixes may range from about $1.25 to $4 per ton at the mixing plant.

Asphalt-concrete Mixes Different asphalt-concrete mixes and different sizes and gradings of aggregates should be used for different kinds of asphalt-concrete work

The proportions of asphalt to aggregate, and the sizes and gradings of the aggregates for asphalt-concrete mixes are given in Table 17-1. The data in this table agree with the present-day recommendations of the Asphalt Institute.

Table 17-1 *Asphalt Content and Size and Grading of Aggregates for Asphalt-concrete Mixes*

Sieve size	Percentages passing sieves			
	Mix A	Mix B	Mix C	Mix D
1 in.	100			
¾ in.	75–100	100		
½ in.	......	80–100	100	100
⅜ in.	45–70	70–90	80–100	80–100
No. 4	30–50	50–70	55–75	65–80
No. 8	20–35	35–50	35–50	50–65
No. 30	5–20	18–29	18–29	25–40
No. 50	3–12	13–23	13–23	
No. 100	2–8	8–16	8–16	10–20
No. 200	0–4	4–10	4–10	3–10
Percent by weight of aggregate	Asphalt content of mix			
	4.0–5.5	4.5–6.0	5.0–6.5	5.5–7.0

Mix A should be used as a binder and surface course only where heavy truckloads are expected and in regions of low rainfall.

Mix B is a general all-purpose mix for heavy-duty construction of driveways and parking areas.

Mix C should be used where a smoother-textured surface is desired, such as in residential driveways and service stations.

Mix D is intended for use on playgrounds and walks.

If an asphalt-concrete base is desired instead of an untreated rock base, mix A is recommended for that purpose.

The weight of the aggregate required for an asphalt-concrete mix may be taken equal to the weight of the mix (145 lb per cubic foot) to allow for losses in screening, proportioning and other waste.

Labor The labor required for any given job will vary with the kind of work (walks, pavements, etc.), the size of the job, the kind and amount of equipment used, and the requirements of the particular job.

For a small contractor, who constructs walks, driveways, and small parking areas and who buys his asphalt concrete from a mixing plant, the average crew may consist of

One foreman
One roller operator
One or more truck drivers, as required
Three to six laborers

For a medium contractor, who specializes in medium-sized jobs, such as medium- or large-sized parking areas, asphalt work in subdivisions, street pavements a few blocks or so in length, but not in large street and highway jobs, and who buys his asphalt concrete from a mixing plant, the crew may consist of

One foreman
One or two roller operators
One motor grader or bulldozer operator
One or more truck drivers as required
Six to twelve laborers, organized in one or more gangs

If he uses a regular paving machine, he will also need a distributor operator and a spreader operator, but he will not need so many laborers.

For a large contractor, who specializes in highway pavements and extensive street-paving projects and who operates his own mixing plant, his mixer plant and paving crews may include

Mixer plant:

One plant operator
One crane (clamshell) operator
One assistant plant operator
Two to six laborers

Paving crew:

One foreman
One or two machine operators
Two roller operators
One motor grader or bulldozer operator
Four or more truck drivers as required
Three to six laborers

These are typical crews, and they may need to be increased or decreased according to the requirements of the particular job.

The output of any crew will vary considerably, depending upon the skill and inclination of the foreman and workers and the job requirements. The following are approximate values and do not include transportation labor (hours of truck drivers).

On *small* jobs, such as walks, driveways, and parking areas, and using rollers:

Spreading and compacting insulation, base and subbase courses, 50 to 100 labor-hours per 100 tons of materials

Spreading and compacting asphalt-concrete courses, 70 to 150 labor-hours per 100 tons of materials

On *large* jobs, such as streets and highways, and using rollers, graders, and paving machines:

Spreading and compacting insulation, base and subbase courses, 5 to 9 labor-hours per 100 tons of materials

Spreading, compacting and finishing asphalt-concrete courses, 7 to 12 labor-hours per 100 tons of materials

Mixing plants, 1 to 3 tons per batch or 100 to 200 tph, 5 to 8 labor-hours per 100 tons of asphalt concrete

Wages of workmen vary from time to time and in different parts of the country. Approximate hourly wages (1967) are foremen \$5 to \$8, machine operators \$4 to \$6, truck drivers \$3.50 to \$5, semiskilled laborers \$3 to \$4.50, and ordinary laborers \$2.50 to \$4.50.

Payroll taxes and insurance may vary from 8 to 15 percent of the labor wage.

Diagram 17-3 can be used for finding the labor costs of asphalt-concrete walks and driveways.

Equipment The equipment for a small contractor may include a small (2- or 3-ton) tandem steel-wheel roller, a 5-ton roller, one or two trailers for hauling rollers, one or more dump trucks, a soil stabilizer, two or three hot irons, a dump board, shovels, rakes, brooms, tampers and other small tools.

The equipment needed by a medium contractor will include the items needed by a small contractor, plus a 10-ton roller, more dump trucks, more small equipment (as his crew is larger), and perhaps a paving machine of a capacity up to 100 tph or more if the amount of work justifies the use of such a machine.

A large contractor will require the same equipment that a medium contractor does but more of it. He may need one or more 10-, 12-, or 15-ton rollers; a water truck; a general-purpose truck; a larger paving machine; a mixing plant complete; and other miscellaneous equipment.

The equipment required for an asphalt-concrete mixing plant may include cold-aggregate storage bins and feeder bins, asphalt storage tanks, fuel-oil storage tank, cold and hot elevators, driers, screens, dust collector, hot-aggregate storage bins, batcher, mixer (1 to 3 tons per batch), asphalt pump and engine, oil heaters, car heaters, truck heaters, crawler crane and clamshell bucket (1 to 2 cu yd), and other miscellaneous equipment and a small office building.

The capacity of batch-type mixing plants may range from 500 to 8,000 lb per batch and from 30 to 60 batches per hour. The capacity of continuous-type mixing plants may range from 15 to 200 tph.

The total cost of an asphalt concrete mixing plant may vary from about $100,000 to $200,000 or more. The cost of moving and setting up such a plant may vary from about $1,000 to $6,000 or more.

The equipment ownership and operating costs can be computed according to the method explained in Chap. 18, Construction Plant and Equipment, and in Appendix B, Equipment Costs. The contractor or estimator should use his own figures when such are available.

Diagram 17-4 can be used for finding the equipment costs for asphalt-concrete walks and driveways.

Overhead and Profit Overhead costs may range from about 5 to 15 percent of the cost of materials, labor, and equipment or from 10 to 30 percent of the labor cost alone.

Profit may vary from about 8 to 15 percent of the sum of all other costs.

Summary The cost items for an asphalt-concrete job can be tabulated and summarized in different ways. Some large contractors like to prepare separate estimates for the labor and equipment costs for the mixing plant, transportation (trucking), and paving. Other contractors do not care to separate these items in the summary.

For a large contractor, the tabulation may be as follows;

Item	Unit cost per sq yd	Unit cost per ton	Total cost
Materials..................	$	$	$
Labor.....................	$	$	$
Mixer plant $			
Trucking $			
Paving $			
Equipment................	$	$	$
Mixer plant $			
Trucking $			
Paving $			
Overhead..................	$	$	$
Profit.....................	$	$	$
Total..................	$	$	$

For a "small" contractor, the tabulation may be simpler:

Item	Unit cost per sq yd. or per sq ft	Unit cost per ton	Total cost
Materials	$	$	$
Labor	$	$	$
Equipment	$	$	$
Overhead	$	$	$
Profit	$	$	$
Total	$	$	$

Estimating Asphalt Walks, Driveways, and Parking Areas In addition to the information given in the preceding articles of this chapter, the information given in this article and in Table 17-2 may be of help to the estimator when preparing estimates for asphalt walks, driveways, and parking areas.

The price of asphalt concrete delivered at the job varies in different localities, for different lengths of haul, and for different mixes. Prices of asphalt concrete (1967) delivered at the job may range from about $5 to $8 or more per ton.

Table 17-2 *Suggested Minimum Thicknesses for Asphalt Driveways and Parking Lots**

Subgrade soil (see article on Classification for a description of soils)	Thicknesses of asphalt-concrete courses, in.		Thicknesses of combined asphalt-concrete and granular-base courses, in.		
	Surface course	Base course	Surface course	Base course	Granular base
For passenger cars					
Good to excellent	1	2½			
Medium	1	3½	1	2	3
Poor	1	5	1	2	6
For light and medium trucks and passenger cars					
Good to excellent	1½	3½			
Medium	1½	5½	1½	3	5
Poor	1½	7½	1½	3	9

* The data in Table 17-2 are the recommendations of the Asphalt Institute. The same asphalt-concrete mix that is used for the surface course may be used for the base course. A 2-in. course of coarse sand or crushed stone should be placed between the base course and a poor soil. Asphalt concrete weighs about 145 to 150 lb per cubic foot.

Prices of asphalt-concrete walks (1967) may range from about $0.20 to $0.50 per square foot depending on the size of the job, thickness of asphalt concrete, base course, and other conditions.

The thickness of an asphalt-concrete walk may vary from about 2 to 4 in. Many authorities specify a minimum thickness of 3 in. A 2-in. layer of sand may be placed on the soil before the asphalt concrete is placed.

Table 17-3 gives the quantities of materials required for 100 sq ft of surface area for asphalt walks, driveways, and parking areas.

Table 17-3 *Tons of Materials Required per 100 Sq Ft of Surface Area of Asphalt Walks, Driveways, and Parking Areas*

	Asphalt concrete	Granular material	Subgrade material
	Weight of compacted material, lb per cu ft		
	145	115	95 to 110
Thickness, in.	Tons per 100 sq ft of surface area		
1	0.63	0.50	0.42–0.48
1½	0.94	0.75	0.63–0.72
2	1.25	1.00	0.83–0.96
2½	1.57	1.25	1.04–1.20
3	1.88	1.50	1.25–1.45
3½	2.20	1.75	1.45–1.70
4	2.50	2.00	1.67–1.92
5	3.13	2.50	2.10–2.40
6	3.75	3.00	2.50–2.90
7	4.40	3.50	2.90–3.35
8	5.00	4.00	3.35–3.85
9	5.65	4.50	3.75–4.35
10	6.25	5.00	4.20–4.80

An estimate for an asphalt concrete walk, driveway, or parking area may be summarized as described in the Summary of this chapter.

Illustrative estimate for asphalt driveways and parking area. Prepare a cost estimate for the driveways and parking area for an apartment building. Total area is 6,600 sq ft. Soil is medium. Contractor is to provide a 3-in. base course of decomposed granite and a 3-in. wearing course of asphalt concrete. Forms will be left in place, and 180 lin ft are required. Decomposed granite costs 90 cents per ton and asphalt concrete $5.75 per ton delivered at the job and roughly spread by dump trucks.

Work Plan. This is a two-day job. On the first day the subgrade will be checked, soil sterilized, forms erected along driveway, and base course spread, rolled, and compacted. On the second day, the asphalt concrete will be spread, rolled, and compacted.

Materials:

Forms. 180 lin ft of 2 by 6's plus stakes and nails	= $ 25	
Lumber cost at $120 per 1,000 fbm		
Soil-sterilization chemicals	= 5	
Base course, Table 17-3, 0.90 tons per square		
66 squares × 0.90 tons × $0.90	= 54	
Asphalt concrete, Table 17-3, 2.50 tons per square		
66 squares × 2.50 tons × $5.75	= 949	
Total materials		= $1,033

Labor:

1 foreman, 16 hr at $4 per hour	= $ 64	
1 roller operator and truck driver, 16 hr at $3.50 per hour	= 56	
4 laborers, 16 hr each = 64 hr at $3 per hour	= 192	
Total wages	= $312	
Payroll taxes and insurance, 10% of wages	= 31	
Total labor		= $ 343

Equipment: Costs are computed by method given in Appendix B. Job will need a truck, a trailer for transporting the roller, a soil sterilizer, a 5-ton tandem steel-wheel roller, and hot irons, shovels, rakes, and other small tools.

Truck, 4 hr at $3.00 per hour	= $12	
Trailer, 2 hr at $0.50 per hour	= 1	
Sterilizer, 1 hr at $1 per hour	= 1	
Roller, 14 hr at $1.50 per hour	= 21	
Small tools for the job	= 8	
Total equipment		= $ 43
Overhead: Allow 20% of labor cost		= 69
Profit at 10% of all other costs		= 149
Total estimated cost		= $1,637

If desired, this estimate can be summarized as follows:

Item	Unit cost per sq ft	Total cost
Materials	$0.16	$1,033
Labor	0.05	343
Equipment	0.01	43
Overhead	0.01	69
Profit	0.02	149
Total	$0.25	$1,637

Diagrams 17-2, 17-3, and 17-4 can be used for finding the costs per 100 sq ft of surface area for asphalt concrete, labor, and equipment.

Estimating Asphalt-concrete Pavements The construction of an asphalt-concrete pavement or highway is usually a large job and the work is done by a large contractor, as mentioned in the articles on Labor and Equipment in this chapter. He will probably have his own asphalt concrete mixing plant and all other equipment needed for a large job.

The pavement design and specifications will have been prepared by a city, county, or state engineering department.

In addition to the information given in the preceding articles of this chapter, the following items may be helpful to the estimator.

The unit of measurement for a paving job is usually the square yard of surface area, with the thickness, mix, and other items given.

The materials are usually estimated and priced by the ton, though the cubic yard or another unit of measurement may be used.

Table 17-4 *Tons of Materials for 100 Sq Yd of Surface Area for Asphalt-concrete Pavements*

	Asphalt concrete	Granular material	Subgrade material
	Weight of compacted material, lb per cu ft		
	145	115	95–110
Thickness, in.	Tons per 100 sq yd of surface area		
1	5.63	4.50	3.75–4.30
1½	8.45	6.75	5.65–6.45
2	11.25	9.00	7.50–8.65
2½	14.1	11.25	9.40–10.80
3	16.9	13.50	11.25–12.95
3½	19.7	15.75	13.15–15.10
4	22.5	18.0	15.0–17.3
5	28.1	22.5	18.8–21.6
6	33.8	27.0	22.5–25.9
7	39.4	31.5	26.3–30.2
8	45.0	36.0	30.0–34.5
9	50.7	40.5	33.8–38.8
10	56.3	45.0	37.5–43.1
11	61.9	49.5	41.3–47.4
12	67.5	54.0	45.0–51.8
14	78.8	63.0	52.5–60.4
16	90.0	72.0	60.0–69.0
18	101.3	81.0	67.5–77.6

In estimating the weights of asphalt and aggregate required for an asphalt-concrete mix, the weight of asphalt required is usually given as a percentage by weight of the aggregate, as in Table 17-1. The weight of the aggregate required may be taken equal to the weight of the asphalt concrete mix (145 lb per cubic foot) to allow for losses in screening, proportioning, and other waste.

Table 17-4 gives the quantities of materials required per 100 sq yd of surface area of asphalt-concrete pavements. These quantities should allow of a small amount of waste.

The cost of asphalt pavements will vary greatly, depending on the thickness of the asphalt concrete, the base and subbase courses needed, specification requirements, weather, job conditions, and the costs of materials, labor, equipment, and overhead.

Present prices (1967) may vary from about \$2.50 to \$5 per square yard. The cost of asphalt concrete in place may vary from about \$7 to \$13 per ton.

The estimate for an asphalt-concrete paving job may be prepared and summarized as described in the Summary article of this chapter.

Illustrative estimate for an asphalt-concrete pavement. Prepare an estimate of the total cost, cost per ton, and cost per square yard of surface area for mixing and placing an asphalt-concrete highway pavement 26 miles long, 24 ft wide, and 9 in. thick (7-in. lower course and 2-in. wearing course) on a specially prepared base. The asphalt cement shall equal 5.20 percent of the weight of the combined aggregate. The aggregate shall be screened and graded to meet specification requirements. The asphalt concrete is to be rolled and compacted with tandem steel-wheeled rollers to a density of not less than 92 percent as determined by the Proctor method.

Work Plan: A job of this size justifies the use of two pavers and one large or two medium-sized mixing plants. Contractor has two pavers with a rated capacity of 100 tph each, two mixing plants with a capacity of 100 to 120 tph each, two 12-ton and two 10-ton steel-wheel rollers, a fleet of 10-ton heavy-duty diesel trucks, a 3-ton stake service truck, and all needed small tools and miscellaneous equipment. The mixing plants can be set up alongside a railway siding about 1 mile from the center of the job.

Materials: Suitable aggregates, bank run sand and gravel, can be purchased f.o.b. railway cars on the siding for \$2.50 per ton. Asphalt costs \$24 per ton f.o.b. cars on the siding.

$$\text{Surface area of paving} = \frac{26 \times 5{,}280 \times 24}{9} = 366{,}100 \text{ sq yd}$$

$$\text{Aggregate, Table 17-4} = \frac{50.7 \times 366{,}100}{100} = 185{,}600 \text{ tons}$$

This amount will allow about 5 to 6 percent for waste in screening, etc.

Aggregate cost = \$2.50 × 185,600 = \$464,000

Asphalt = 185,600 × 95% × 5.20% = 9,170 tons
(5 percent of the aggregate was deducted for waste)
Asphalt cost = $24 × 9,170 = $220,080
Total materials cost = $684,080

Labor: The number of working hours on this job will depend upon the capacity of the two pavers, which is 200 tph.

$$\text{Working hours} = \frac{185{,}600 + 9{,}170}{200} = 974$$

Two paving crews, 2 foremen at $6 per hour	=	$12.00
2 machine operators at $5 per hour	=	10.00
2 helpers at $4 per hour	=	8.00
4 roller operators at $4.50 per hour	=	18.00
6 laborers at $3 per hour	=	18.00
Hourly wage of 2 paving crews	=	$66.00
Payroll taxes and insurance, 10%	=	6.60
Hourly cost of 2 paving crews	=	$72.60

Cost of 2 paving crews for the job = $72.60 × 974 = $70,712

Two mixing plant crews, 2 plant operators at $5 per hour	=	$10.00
2 assistants at $4 per hour	=	8.00
2 crane operators at $4.50 per hour	=	9.00
6 laborers at $3 per hour	=	18.00
Hourly wage of 2 mixing plant crews	=	$45.00
Payroll taxes and insurance, 10%	=	4.50
Hourly cost of 2 mixing crews	=	$49.50

Cost of 2 mixing crews for the job = $49.50 × 974 = $48,213

Trucking crew: If one paver starts at one end of the job and works toward the center and the other paver starts at the center of the job and works toward the other end, the number of trucks required will be constant throughout the job. Round-trip time to the end of the job will be about 68 min (load, travel loaded, unload, return, and delays). Round-trip time to the center of the job will be about 14 min. Average time equals 41 min.

Number of trucks required = $\frac{200}{10} \times \frac{41}{60}$ = 13.67, or 14 trucks

Hourly wage of trucking crew = $4 × 14	=	$56.00
Payroll taxes and insurance, 10%	=	5.60
Hourly cost of trucking crew	=	$61.60

Cost of trucking crew for the job, $61.60 × 974 = $59,998

Superintendent's crew, 1 superintendent at $8.00 per hour	=	$ 8.00
1 office man at $4.00 per hour	=	4.00
1 service truck driver at $4.00 per hour	=	4.00
Hourly wage of superintendent's crew	=	$ 16.00
Payroll taxes and insurance, 10%	=	1.60
Hourly cost of superintendent's crew	=	$ 17.60
Cost of superintendent's crew for the job = $17.60 × 974	=	$17,143

Total labor cost = $196,066

Equipment: Equipment costs will be estimated by the method described in Appendix B, Equipment Costs, but using the contractor's own figures.

Mixing plants:

One plant, capital investment = $118,750

Contractor's hourly ownership cost = 0.036%

Hourly ownership cost = $118,750 × 0.036% = $42.75

Hourly operating cost (fuel, oil, minor repairs) = 46.50

Hourly cost of ownership and operation, one plant = $89.25

Charge cost of transportation one way and erection and dismantling to this job, one plant = $2,285

Cost of two mixing plants for this job = $89.25 × 974 × 2 + $2,285 × 2 = $178,429

Pavers: Contractor's books show ownership and operating costs of $5.25 per hour for each. Transportation, erection and dismantling for the job cost $247 for each

Cost of two pavers for this job = $5.25 × 974 × 2 + $247 × 2 = $ 10,721

Rollers: Two 12-ton and two 10-ton rollers. Contractor's books show ownership and operating costs of $2.15 per hour for each. Transportation cost for each is $85

Cost of 4 rollers for this job = $2.15 × 974 × 4 + $85 × 4 = $ 8,716

Dump trucks: Fourteen 10-ton heavy-duty diesel trucks. Ownership and operating cost per truck is $4.65 per hour. Transportation cost for each truck is $35

Cost of 14 dump trucks for this job = $4.65 × 974 × 14 + $35 × 14 = $63,897

Service truck: One 3-ton stake truck costs $2.55 per hour. Transportation cost for this truck is $25

Cost of the service truck for this job = $2.55 × 974 + $25 = $ 2,509

Cost of superintendent's car = $1.75 per hour

Transportation cost = $20

Cost for the job = $1.75 × 974 + $20 = $ 1,725

NOTE: Mixing plants are shipped by railway freight. Pavers, rollers, tools, and miscellaneous equipment are hauled by the contractor's trucks.

Office building. Estimate for the job = $ 500

Washroom and toilets. Estimate for the job = $ 400

Miscellaneous tools and equipment. Estimated at = $ 1,600

Total equipment cost = $ 268,487

Overhead: Assume 8% of material, labor, and equipment costs = \$1,148,643 × 8% = \$ 91,891

Profit: Assume 10% of all other costs = \$1,240,534 × 10% = \$ 124,053

Total estimated cost of the job = \$1,364,587

This estimate can be summarized as follows, with unit costs based on 366,100 sq yd of pavement surface and 185,600 tons of asphalt concrete in place.

Item	Unit cost per sq yd	Unit cost per ton	Total cost
Materials	\$1.87	\$3.69	\$ 684,080
Labor	0.54	1.06	196,066
Equipment	0.73	1.45	268,497
Overhead	0.25	0.50	91,891
Profit	0.34	0.67	124,053
Total	\$3.73	\$7.37	\$1,364,587

CHAPTER EIGHTEEN

Construction Plant and Equipment

Definitions The term *construction plant* includes all the temporary buildings and other structures, machinery, and tools assembled and used for any construction job.

The term *construction equipment* is considered by some authorities as synonymous with construction plant. Other authorities consider equipment as referring to an individual machine or set of tools.

In this text, the terms construction plant and construction equipment have been considered as synonymous. The complete construction plant or complete construction equipment is considered as including all the plant or equipment used. An individual machine is referred to as one piece or part of the plant or equipment.

Plant Economics The most economical construction plant for any job is the one that will do the work in the required time and for the least total cost. In general, the use of larger and more complete equipment will result in a saving of labor costs, and vice versa. However, the total combined cost (plant, labor, overhead, etc.) should govern the choice of plant. In regard to economy, the fundamental requirement is to select, arrange, and use the men, equipment, and materials so that the work will be done in the most efficient manner, especially in regard to costs. Hence, the construction equipment or plant must be selected, arranged, and used so that it will function most efficiently in regard to total costs. The equipment must be carefully selected for the requirements of the particular job. It must be arranged so that it can be worked efficiently, and then it must be operated in an efficient manner. All this requires that the contractor, superintendent, or estimator have a good knowledge of construction methods in general and also of the detailed methods applicable to the job under consideration.

Some of the general principles that should be considered are as follows:

Selection and arrangement of equipment suitable for the job, consideration being given to kind of work, amount of work, characteristics of site, equipment available, etc.

No piece of equipment should be used on any job unless this use will be economical, especially in regard to costs.

Whenever the topography of the site permits, gravity should be utilized for handling materials.

Capacities of machines should be sufficient for the work, so that neither overloading nor idling is required. That is, for efficient operation, machines should be worked as near their rated capacities as possible.

Machines should be located so that they can be used efficiently and also so that the number of moves will be a reasonable minimum.

The tendency toward accidents and delays should be reduced whenever possible. A stock of spare parts kept on the job will tend to reduce the time of delays caused by breakdowns in machinery.

Remember that the sum of all items of unit cost must be a minimum.

Construction Plant Required for a Job The methods followed in various construction jobs are partly standardized according to the kind of work, amount of work, and location. Construction equipment is also standardized to some extent, different machines being developed for different kinds of work. Consequently, the contractor and his estimator should have a good knowledge of suitable methods and equipment for any particular job.

The following will give a general idea in regard to the different equipment required for various jobs:

Earth excavation may need power shovels, cranes, derricks, grab buckets, plows, picks, shovels, scrapers, etc.

Rock excavation may need explosives, drills, and other tools.

Materials transportation may require motor trucks, trailers, tractors, wheelbarrows, wagons, belt conveyors, and unloading and loading equipment.

Piling may require pile drivers, hammers, templates, caps, points, etc.

A concrete building may require woodworking machinery for forms, bar benders, and cutters, and bins for storing aggregates, cement sheds, proportioning equipment, mixers, towers, and hoists, barrows, chutes, belts, pumps, etc.

A structural-steel office building may require guy or stiffleg derricks, auto or crawler cranes, power hoists, riveting equipment, etc.

An industrial-type steel building may need a gin pole or tractor crane, riveting equipment, etc.

A concrete highway may need excavating machinery, road-working machinery, autotrucks, tractors, concrete pavers, etc.

The equipment required for various kinds of work has been discussed in most of the preceding chapters. For further information, the reader is referred to books on construction methods and plant.

Plant Design Construction-plant design includes the selection and arrangement of the construction plant or equipment needed for any particular construction job.

Before the construction plant is selected and the arrangement or layout planned, the site should be visited and conditions studied. Notes should be taken in regard to the location of the work; kind of soil; availability of power, light, water, and sewer; local ordinances and regulations; highways and other transportation facilities; railway yards; material dealers; banking facilities; local labor supply and conditions; boarding and rooming places for workmen; etc. Sketches are often helpful and should be made. In some instances, topographic maps are needed.

Having the various details concerning the site in mind and knowing the kind and amount of construction work to be done, the contractor or his estimator must be able to visualize the work to be done, starting with the preliminary operations and following through to the final operations. Then the various pieces of equipment can be selected and the arrangement or layout prepared.

The equipment selected for any particular job will depend on what is thought to be desirable, what the contractor has available, and what he may reasonably expect to acquire by purchase or rental. Each piece of equipment must be suitable for the work to be done. The type and capacity of each machine must be such that it can be operated economically in connection with the other machines and the workmen on the job. That is, all pieces of equipment should be of such size and capacity that the plant is "balanced." By a balanced plant is meant one in which each machine will be operating efficiently and at practically full capacity without any idleness or delays.

Sometimes the contractor has all the desired equipment, and at other times he must purchase or rent additional equipment. Whether the contractor should purchase new or secondhand equipment or rent it may depend upon his finances and the particular job. If the job is a comparatively small one, or if a certain kind of equipment is needed that may be rarely used afterward, the contractor may find it advisable to rent the equipment. However, if the needed equipment is such that the contractor may find use for it on other jobs, he may find it advisable to purchase the equipment either new or secondhand. Sometimes good standard secondhand equipment can be purchased at reasonable prices.

After the equipment has been selected, a layout should be prepared showing the arrangement of the equipment. The layout may be a map or a sketch showing the location of each major piece of equipment and all the probable moves. Notes can be made explaining the reasons for the locations of the machines on the map and their probable moves. If possible, all the construction plant should be located at the site, as it is then easier to superintend the work of this equipment. However, on some jobs, some equipment may be needed at the railroad siding, other equipment at the contractor's shop or yard, and the remaining equipment at the job site. Conditions vary on different jobs, and the equipment must be arranged to the best advantage for the job in question. The equipment should be arranged so that the work will progress from one machine to another without unnecessary travel or delays.

When selecting the equipment and planning the layout, consideration should be given to the following:

1. Plant should be balanced. Capacities of machines should be such that neither overloading nor idling of machines is necessary.

2. Power supply should be sufficient at all times. A small excess of power is advisable.

3. Power supply should be arranged so that a breakdown of one machine will not cause others to be shut down.

4. A reasonable stock of spare parts should be kept on hand so as to reduce the time lost by minor breakdowns.

5. Supplies of fuel, water, lubricants, and accessories should be sufficient at all times.

6. Stock piles should be placed as close to place of final use as practicable. The material having the greatest bulk should be placed the closest.

7. If power equipment is used for handling materials, handling by manpower should be kept at a minimum.

8. Each tower, derrick, hoist, etc., should be located so that it will serve as large an area as practicable and also so that the moves will be a minimum.

9. Mixers, saws, batchers, etc., should be placed so that their output can be delivered as closely to the work as is possible.

10. Gravity should be utilized for handling materials and parts of the work whenever practicable.

11. Sufficient small tools of various kinds should be provided.

12. The plant layout should be such that the total amount of work, as measured in foot-pounds of work or energy, should be a minimum.

Plant Operation When selecting equipment and planning the plant layout, some thought must be given to the operation of the plant. The man who designs the plant must have a knowledge of the operation of the plant so that he can select the proper equipment and plan a correct layout.

The economical operation of the plant requires careful planning of the work as well as the careful selection and arrangement of the plant machinery. On large jobs, it is advisable to have a planning department on the job under the supervision of the superintendent of construction. The planning department on a reinforced concrete job, for example, takes care of the supplies of materials and the order of work for forming, steel bending and placing, concreting, form removal, and surface finishing. The function of a planning department in construction is like that of the planning department in a factory.

A brief discussion of job management is given in Chap. 22 on Construction Management. For more complete discussions, the reader is referred to texts on construction-job management, job superintendence, engineering management, and engineering economics.

Equipment Costs The cost of any piece of equipment on a construction job will usually include three or four of the following:

1. Ownership costs
2. Rental costs
3. Transportation costs to and from the job
4. Erection and dismantling costs
5. Operating costs

No one piece of equipment on a construction job will be subject to all five of these cost items. If the contractor owns the equipment, there will be no rental costs, and if he rents, there will be no ownership costs. Most equipment will have some transportation costs. Some equipment, like autotrucks, will have no erection and dismantling costs. Some equipment, like scaffolds, will have no operating costs. Other equipment, like a power hoist with tower, ropes, and pulleys, will have operating costs and erection and dismantling costs.

The total equipment costs on a job will be the sum of all of the equipment-cost items that apply.

When equipment costs are estimated for any job, a simple tabulation can be prepared. The different pieces of equipment can be listed in the first column, and the various cost items (and totals) listed in other columns, as desired.

Equipment-ownership Costs The ownership costs of a piece of equipment will include the following three items, which are usually expressed as a percentage of the capital investment or first cost of the equipment.

1. Depreciation
2. Major repairs and overhaul
3. Interest, insurance, taxes, and storage

The capital investment, or first cost, of a piece of equipment will include such items as cost f.o.b. factory, transportation, unloading, assembling, testing, and preparing the equipment for use.

Depreciation depends upon the useful life of a piece of equipment, usually expressed in years, or on the number of hours it can be worked before it is worn out. Idle equipment should not depreciate, though it may become obsolete. The total depreciation is 100 percent of the capital investment less salvage or scrap value. The salvage or scrap value of a piece of worn-out contractor's equipment is usually very small (often less than 2 percent) and rarely needs consideration in estimating. The total number of working hours will vary considerably, depending on the kind of equipment, ruggedness, job conditions, and care used in operation.

Depreciation can be computed in many ways, such as straight-line, sum-of-digits, declining-balance, etc., but the total depreciation is 100 percent of the capital investment less salvage or scrap value.

Major repair and overhaul costs are usually assumed to vary with the kind of equipment and the number of working hours, though accidents, job conditions, and care in operation will have their effect. Total major repair and overhaul costs can be expressed as a percentage of the capital investment in the equipment.

Interest, taxes, and insurance are usually expressed as a percentage of the current value (capital investment less depreciation) of a piece of equipment. Storage costs are usually based on the capital investment and do not vary with the age of the equipment. Interest, insurance, taxes, and storage costs continue as long as the contractor has the equipment, no matter whether it is working or not.

In Appendix B, Equipment Costs, the approximate costs per working hour are listed for several typical pieces of equipment. These costs are expressed as a percentage of the capital investment and are based on an average of the total number of working hours for the particular piece of equipment. When the percentages for interest, insurance, and taxes were computed, allowance was made for the depreciated value computed by the straight-line method. It is thought that the method of computing ownership costs per working hour used in Appendix B is simple and easy for the estimator to understand and use.

Diagrams 18-1, 18-2, and 18-3 can be used for computing and checking equipment-ownership costs.

As equipment-ownership costs will vary in different localities and for different contractors, each contractor should keep his own records and use his own data for estimating.

Some estimators base ownership costs on the time that the equipment is assigned to the job (calendar months, weeks, or days). Other estimators use actual working days (of 8 hr) or working hours. The working hour is preferable, especially as it is used in the estimate of labor costs and equipment operating costs.

Illustrative estimate on ownership costs. Estimate the ownership cost per working hour for a 10-ton diesel dump truck with a capital investment of \$16,000. Assume the following:

Useful life = 5 years or 10,000 working hours
Salvage or scrap value = 5%
Total depreciation = 100 − 5 = 95%
Total major repairs and overhaul, estimated = 65%
Total interest, insurance, taxes, and storage = 45%

Interest, insurance, and taxes are estimated at 10 percent per year of the current value using straight-line depreciation for a 5-year life, or $(1.00 + 0.80 + 0.60 + 0.40 + 0.20) \times 10\% = 30\%$ total. The storage cost is estimated at 3 percent per year of the capital investment or a total of 15 percent for 5 years.

Total interest, insurance, taxes, and storage costs = 45%

$$\text{Cost per working hour} = \frac{\$16{,}000}{10{,}000} \times \frac{(95 + 65 + 45)}{100} = \$3.28$$

Diagram 18-1, for a total percentage of 205 and 10,000 working hours, gives a value per working hour of 0.0205 percent of the capital investment.

Diagram 18-3, for 0.0205 percent per working hour and a capital investment of \$16,000, gives an ownership cost of about \$3.30 per working hour.

Rental Costs The total cost of a rented piece of equipment will include such items as rental rates, time rented, transportation, erection and dismantling, moving, operation, and minor or field repairs. Some idea of 1967 rental rates and what they include can be obtained from Appendix C. In some special cases the rental rates may include such items as operation costs, operator's wages, transportation, and minor (as well as major) repairs.

Rental rates may be quoted by the month, week, or day. The rental cost is found by multiplying the rental rate by the total time that the equipment is held available for the job. This is the total elapsed time measured in calendar months, weeks, or days with no deductions for Sundays, holidays, bad weather, or other idle time.

Rentals are usually based on one shift of 8 hr or less per day. Rental prices are usually increased when the equipment is operated more than 8 hr a day, the rental charge being increased about 50 percent for each additional 8-hr shift. When the additional shift is less than 8 hr, a proportional charge is made.

Sometimes the rental rates are based on ownership costs. In such cases, an allowance must be added for general overhead, profit, and any other expense items, in addition to those stated, which must be paid for by the owner.

Rental costs per working hour can be found by dividing the total estimated rental costs by the estimated number of working hours that the equipment will be used on the job.

Equipment-transportation Costs All equipment used on a job must be transported to and from that job. On a small job, the equipment transportation costs may be only a few dollars, while on a large job (such as a paving job), the equipment-transportation costs may amount to $1,000 or more. Equipment-transportation costs will vary with the kind and amount of equipment, distance, method of transportation used, and other variables. When equipment is rented, the transportation costs for this equipment must be added. Transportation costs of equipment are important, are usually appreciable, and should always be allowed for by the estimator.

Equipment-erection and -dismantling Costs The equipment costs of erection and dismantling will vary with the kind, size, and number of pieces used on the job. Some equipment, such as trucks, hand tools, etc., do not require erection or dismantling. For other equipment, such as portable buildings, scaffolding, towers and hoists, portable bins, etc., the costs of erection and dismantling may be considerable. On some large jobs some of the equipment must be moved, and this moving may require some erection and dismantling expense.

Diagrams 18-4 and 18-5 can be used in the estimate of the labor costs of erecting and dismantling equipment, as well as such other labor costs as loading, unloading, moving, etc., of equipment.

Equipment-operation Costs The costs of operating construction equipment may be based on the working week, day, or hour. Many estimators prefer the working hour. The operating costs include such items as fuel, oil, grease, electric power, supplies, small parts, and minor or field repairs. Operation costs usually do not include wages of operator, transportation costs, erection and dismantling costs, or any of the items included in ownership costs.

Equipment-operating costs will vary with the kind, type, and size of the equipment; job conditions; and care and skill of the operator. For power equipment, the operating cost may be based on the rated brake horsepower of the machine.

A gasoline engine may require from 0.06 to 0.075 gal of gasoline per brake horsepower-hour when operating at full rated capacity. However, the engine will usually operate at 50 to 85 percent of full rated capacity during an average working hour. Gasoline may cost from 25 to 45 cents per gallon (including tax) at the job. The costs of oil, grease, supplies, small parts, and minor repairs for gasoline-engine-operated equipment may vary from about 0.3 to 0.7 cent per brake horsepower-hour. Diagram 18-6 can be used in the estimate of operating costs per working hour for gasoline-engine-powered equipment.

A diesel engine may require from 0.04 to 0.06 gal of diesel fuel oil per brake horsepower-hour when operating at full rated capacity. However, the engine will usually operate at 50 to 85 percent of full rated capacity during an average working hour. Diesel fuel oil may cost from 15 to 30 cents per gallon (including tax) at the job. The cost of oil, grease, supplies, small parts, and minor repairs for diesel-engine-operated equipment may vary from about 0.3 to 0.6 cent per brake horsepower-hour. Diagram 18-7 can be used in the estimate of operating costs per working hour for diesel-engine-powered equipment.

An electric motor, assuming it to operate at an efficiency of about 75 percent, will require about 1 kw of electricity to develop 1 bhp at the shaft. The cost of electric power at the job may vary from about 1.5 to 4 cents per kilowatthour. Additional costs for electric-motor-driven equipment may vary from about 0.2 to 0.4 cent per brake horsepower-hour. Diagram 18-8 can be used in the estimate of operating costs per working hour for electric-motor-powered equipment.

Table 18-1 is a summary of the range of the estimated costs per brake

Table 18-1

Item	Gasoline	Diesel	Electric
Fuel or power costs	\$0.015–\$0.033	\$0.006–\$0.018	\$0.015–\$0.040
Other costs	\$0.003–\$0.007	\$0.003–\$0.006	\$0.002–\$0.004
Total costs	\$0.018–\$0.040	\$0.009–\$0.024	\$0.017–\$0.044

horsepower-hour for gasoline-, diesel-, and electric-power-operated equipment.

Some kinds of equipment which have no power unit, such as tractor-drawn scrapers and graders, truck-drawn trailers, etc., do have operating costs for oil, grease, supplies, small parts, and minor repairs. Estimated operating costs per working hour for some kinds and types of this equipment are given in Appendix B.

The labor costs of equipment handling can be found by Diagrams 18-4 and 18-5 when the hourly crew cost is known and the hourly labor output can be assumed.

Illustrative estimates on operating costs. The following three estimates are included to show how operating costs for construction equipment can be estimated.

Diesel-powered Shovel. Estimate the operating costs per working hour for a 1 cu-yd diesel-powered shovel with an 85-hp engine. Diesel fuel oil costs 16 cents per gallon at the job, including tax. The shovel is estimated to operate at 70 percent of full rated capacity during an average working hour. The engine consumes 0.045 gal of fuel oil per brake horsepower-hour. Other costs are estimated at 0.4 cent per brake horsepower-hour.

Fuel cost per working hour	= 0.045 × 85 × 70% × \$0.16	= \$0.429
Other costs per working hour	= 85 × 70% × \$0.004	= 0.238
Total operating cost per working hour		= \$0.667

Diagram 18-7 gives about 67 cents per working hour.

Gasoline-powered Shovel. Estimate the operating costs per working hour for a 1-cu-yd gasoline-powered shovel with a 90-hp engine. Gasoline costs 26 cents per gallon (including tax) at the job. The shovel is estimated to operate at 70 percent of full rated capacity during an average working hour. The engine consumes 0.06 gal of gasoline per brake horsepower-hour. Other costs are estimated at 0.5 cent per brake horsepower-hour.

Fuel cost per working hour	= 0.06 × 90 × 70% × \$0.26	= \$0.983
Other costs per working hour	= 90 × 70% × \$0.005	= 0.315
Total operating cost per working hour		= \$1.298

Diagram 18-6 gives about \$1.30 per working hour.

Electric-powered Hoist. Estimate the operating costs per working hour for an electric-motor-driven double-drum hoist. Motor is 40 hp. Hoist is estimated to be operating at full capacity about 80 percent of the time. Electricity costs 2½ cents per kilowatthour at the job. Other costs are estimated at 0.2 cent per brake horsepower-hour. Assume that 1 kw of electricity is needed to give 1 bhp at the shaft.

Power cost per working hour	= 40 × 80% × \$0.025	= \$0.800
Other costs per working hour	= 40 × 80% × \$0.002	= 0.064
Total operating cost per working hour		= \$0.864

Diagram 18-8 cannot be used to check this estimate because the motor is assumed to be operating at 80 percent, instead of 70 percent, of full rated capacity.

Small-tools Costs On almost every construction job, the estimator must estimate the costs of the small tools used. Such small tools are wheelbarrows, jacks, dollies, hand trucks, wrenches, crowbars and pinch bars, hammers, sledges, picks, shovels, hose, lanterns, ropes, saws, etc.

On most jobs, the mechanics have all their small tools. On such jobs, the contractor may need a chest of miscellaneous tools. When the contractor has to supply all the small tools, the kind and number of each should be listed and their cost estimated. The common practice is for the estimator to allow a lump sum for the cost of the small tools for the job.

The life of small tools is usually taken as one year. Some small tools wear out in a few months, and others may last for years. The losses due to breakage and disappearances are comparatively large.

Apportionment of Equipment Costs There are several ways of apportioning plant and equipment costs. For estimating purposes, the cost computations are usually based on time.

When any piece of equipment is used on two or more kinds of work, the cost for each kind of work can be estimated. An example of the use of a piece of equipment for more than one kind of work would be the use of a tower and hoist in building construction for hoisting the lumber, concrete, brick, and various other materials.

Unit costs, based on any desired unit, can be found by dividing the total cost by the number of units. There are many units used by estimators. Some examples are cost per square foot of area; cost per cubic foot; cost per cubic yard; cost per square (100 sq ft); cost per lineal foot; cost per 1,000 brick, block, or tile; cost per 1,000 fbm of lumber; cost per pound or per ton for steel; cost per fixture for plumbing; cost per outlet in electrical work; etc.

Construction-plant Procedure The procedure that may be followed in selecting, arranging, and planning the use of construction plant and equipment and in estimating plant and equipment costs can be summarized as follows:

1. Visit the site and visualize the work.
2. Select the equipment and plan the arrangement or layout.
3. Plan the work and the operation of the equipment.
4. Estimate the time required, both total time and working time.
5. Estimate equipment costs.
6. Check all computations.
7. Repeat procedure for each of other practical plant designs.
8. Compare costs, and select the most practical design.

CHAPTER NINETEEN

Overhead and Profit

Overhead Overhead costs usually include all the general costs of operating and maintaining a contracting business plus such job costs that are not included in materials, labor, and equipment costs. Overhead costs can be divided into general overhead and job overhead costs.

General overhead costs include all overhead costs that cannot be charged to any particular job. Such costs are continuous, though they may vary somewhat with the amount of business done or with the number and size of the jobs.

Job overhead costs include all overhead costs that can be readily charged to the job but which cannot be charged to labor, materials, or equipment.

Total overhead costs may vary from 5 to 20 percent of the sum of the materials, labor, and equipment costs or from 10 to 40 percent of the labor cost alone, depending upon the locality, kind of job, and items included.

General Overhead The general overhead costs may vary from almost nothing, as for a small contractor who maintains no office, up to

8 to 15 percent of the total business done. Average percentages may vary from about 2 to 8. The following list will give an idea of the various items that can be included in general overhead costs. Almost any job will include several of these items, though no one job may include all of them.

General Office Costs

Rent	Interest
Heat	Taxes
Light	Salaries of general officers
Telephone	Wages of estimators
Telegraph	Wages of draftsmen
Stationery and supplies	Wages of clerks
Furniture and fixtures	Wages of stenographers
Advertising	Wages of secretaries
Entertainment	Wages of bookkeeper
Traveling	Legal expenses (general)
Automobile	Consulting fees (general)
Insurance	

Shop and yard costs and plant and equipment costs are usually not included in overhead costs but are charged directly to plant and equipment. Plant and equipment costs are discussed in the preceding chapter and in Appendix B.

Job Overhead Job-overhead costs may vary considerably, depending upon the particular job, the kind of work, the preliminary expenses involved, the difficulties encountered when doing the work, and the costs of settling up after the work is completed. The job-overhead costs may vary from 4 or 5 up to 10 percent or more of the sum of material, labor, and equipment costs or from about 10 to 20 percent of the labor costs alone. The following list will give an idea of the various items that can be included in job-overhead costs. The job-overhead costs for any one job will usually include several of the items. The costs of bonds and insurance are discussed in more detail in following articles. Payroll taxes and insurance costs, which are based on the job payroll, are usually added to the labor costs and are not included in the job-overhead costs. Items marked with an asterisk (*) are often included with the equipment costs.

Job-overhead Costs

*Job-office building	Light
*Job-office furniture	Power
Job-office stationery	Water
Job-office supplies	*Construction buildings
Telephone	*Equipment for buildings
Telegraph	*Tool and materials sheds
Heat	*Water closets

*Temporary enclosures
Contingencies
Engineering
Police protection
Fire protection
Property protection
Surveys
Transportation
  Railway
  Trucking
*Temporary walks
*Ladders
*Temporary stairs
Approval of plans
Permits (see list)
Bonds (see list)
Insurance (see list)
Interest
Taxes
Removing and replacing
  Pavements
  Sewers and drains
  Water and gas pipes
  Poles and wires
  Conduit
Removing rubbish
Cleaning up

Approval of Plans. Before the contractor can start work, the plans and specifications must first be approved by some such organization as the building department, fire department, industrial or labor commission, board of health, department of public safety, department of education, various boards and commissions, planning board, or zoning commission as the case may be. In many instances, fees are charged by the various boards concerned, and these fees are usually paid by the contractor.

Permits. On most jobs, the contractor is required to secure and pay for one or more permits such as:

To build
To cross sidewalk
To use part of street and walk
To erect shelters over sidewalk
To store gasoline and other materials
To construct temporary construction buildings
To maintain sanitary facilities for workmen
To obtain gas, water, electricity, etc.
To open street for gas, water, sewer pipes, etc.
To repair pavement or walks
To use city dumps
To transport heavy loads over city streets

Bonds. Some of the bonds that a contractor may have to provide include surety or guarantee bonds, proposal or bid bond, contract bond, maintenance bond, and fidelity bonds. These bonds are described briefly in the article on Bonds.

Insurance. The various kinds of insurance in which a contractor is interested are many. Some of the kinds of insurance are required by law, other kinds are required by the owner, and some other kinds are desirable for the protection of the contractor. Some of the more general kinds are described briefly in the article on Insurance. Some base rates for Workmen's Compensation Insurance in force in 1967 are given in Appendix E.

Some of the kinds of insurance are as follows:

General	*Miscellaneous*
Workmen's liability	Automobile
Public liability	Vehicle
Property damage	Boiler
Fire	Wind
Windstorm	Rain
Owner's contingent	Flood
Contractor's contingent	Earthquake
Workmen's compensation	Plate glass
Old age	Theft
Unemployment	Payroll
Social security and Medicare	Forgery
Structure	Elevator
Various contingencies	Use
	Occupancy

Interest. On most construction jobs, there may be interest payments, which payments should be included in the overhead.

Taxes. If the contractor, instead of the owner, is required to pay taxes on a structure (say an office building, for example) during the construction period, such taxes should be included as an overhead cost.

Contingencies. On practically every construction job, there will be certain cost items that were not foreseen (and in some cases could not be reasonably predicted) when the estimates were prepared. Most contractors include an allowance for contingencies in their estimate of overhead costs. On ordinary jobs, the percentage allowed for contingencies may be comparatively small.

Apportionment of Overhead There are various ways of apportioning the general overhead costs to the various jobs. No doubt the best way would be to charge the actual costs to each job, provided that such costs could be accurately estimated in each case.

The more common methods are to divide general overhead costs among the various jobs according to the estimated costs (materials, labor, and equipment) of the jobs or according to the estimated labor costs. In these methods, a certain percentage is added to the sum of the costs of materials, labor, and equipment or to the labor cost to care for the overhead costs. The percentage allowed is based on past costs and experiences and also to some extent on the probable business for the current year.

In some instances, estimators have divided the general overhead costs according to the number of jobs. In other instances, estimators have estimated the overhead costs according to the length of time required to complete a particular job.

The job-overhead costs can be estimated in detail and included with the estimates of materials, labor, and equipment when the total estimated cost is prepared. Though this method is, perhaps, the most accurate method, the two common methods are to use selected percentages and to base the job-overhead costs either on the sum of materials, labor, and equipment costs or else on labor costs alone. When the percentages are selected, consideration is given to such things as the kind of job, magnitude of work, probable difficulties to be encountered, and costs on previous jobs.

Many contractors do not divide their overhead costs into general- and job-overhead costs but consider all overhead costs together. Then, for any particular job, they allow either a selected percentage of the sum of materials, labor, and equipment costs or a percentage of labor costs alone. The percentage may vary from about 5 to 25 of the sum of materials, labor, and equipment costs or from about 10 to 50 of labor costs alone.

Bonds There are many types and kinds of bonds. Only a few of the more common ones will be briefly described in this article. Unless otherwise specified, the contractor will be expected to pay the cost of the bonds. Such costs should be included in his overhead costs.

Bid Bond. A bid or proposal bond is a bond submitted by the contractor with his bid as evidence that he will enter into the contract if the contract be tendered to him. Such a bond may take the place of a certified check. When the contract is awarded to a particular contractor, his bid bond or certified check is held until he signs the contract and submits his contract or performance bond. Then the bid bond or certified check is returned to him. The cost of a bid bond is nominal, usually being about $5 to $25.

Contract Bond. A contract or performance bond is a bond submitted by the contractor after he has signed the contract. The purpose of such a bond is to guarantee the faithful fulfillment of the contract by the contractor. Contract bonds are of two kinds, construction and material supply. These bonds are sometimes called surety or guarantee bonds.

A construction bond guarantees the faithful performance of the construction work by the contractor. The amount of the bond may vary from about 20 to 100 percent of the contract price. The cost or premium of such a bond varies from time to time and also with the kind of work. The premium is usually based on the contract price rather than upon the amount of the bond. For ordinary construction work, which may be completed in about 2 years or less, the premium is about 1.5 percent of the contract price. For longer times, an additional premium of about 0.75 percent per year is charged. The premium for street and highway work varies from about 0.75 to 1.5 percent per year. Probably a premium of about 1 to 1.5 percent per year would be the average for most work.

A material supply bond is given by a material dealer to ensure the proper deliveries of materials in regard to time, quantity, and quality. The premiums on these bonds may vary from about 0.25 percent of the contract price for materials delivered from stock to 0.75 percent for materials specially designed or ordered.

Maintenance Bond. Sometimes the contractor is required to maintain the building, pavement, or other structure for a period of time after its completion. The maintenance bond is submitted by the contractor as a guarantee that he will care for the maintenance work needed. The premium on a maintenance bond can be computed on the contract price or upon the value of the bond. When based on contract prices, the premium rates may vary from about 0.15 to 0.75 percent.

Fidelity Bond. A fidelity bond is one given by an employee to his employer to protect the employer from loss of money through any negligent action of the employee. Each employee who may be required to handle any considerable sum or sums of money should be placed under bond. The premium for a fidelity bond is nominal. This premium is usually paid by the employer and included in his overhead costs.

Insurance Insurance can be secured to cover practically every operation, condition, or risk in almost every kind of construction work. The different kinds of insurance and the amount of each kind will depend upon the conditions of the particular job, the law, the contract requirements, and the desires and experiences of the contractor. The estimator should determine the kinds, amounts, and other details of the kinds of insurance required by law in the state or locality in which the particular construction work is to be done.

A number of different kinds of insurance are listed in the article of this chapter on Job Overhead. The estimator should consult a reliable insurance agency for information concerning correct insurance rates for the locality in which any particular job is located.

No attempt will be made in this article to describe all the kinds of insurance that can be secured. Only a few of the more common kinds of insurance will be briefly described.

Workmen's Compensation Insurance. The rates charged for this insurance are based upon the labor payroll. The rates vary in different states, for different kinds of work, at different times, and for different contractors according to their experience (accident) ratings. Information on these rates is given in Appendix E. Note that they vary from about 2 percent or less up to as much as 30 percent. A contractor should be careful to classify his workmen correctly and to keep his accident rate low.

Contractor's Public Liability Insurance. This insurance is taken by the contractor as a protection from losses caused by accidents to property and to any person or persons not in his employment. The rates are usually based upon the labor payroll and may vary with the payroll, kind of structure, policy limits, and the amount of the policy. Minimum limits are usually $5,000 per person and $10,000 per accident for bodily injury and $5,000 per accident with $25,000 aggregate for property damage. Higher limits are desirable. Rates may range from about 0.25 to 0.70 percent of the total payroll for bodily injuries and from about 0.15 to 0.30 percent of the total payroll for property damage. The total cost usually ranges from about 40 to 95 cents per $100 of payroll.

Worker's Health and Welfare Assessments. In many of the larger cities and in other places where the labor unions are strong, the labor contracts often include a special assessment, or benefit, to care for the worker's health and welfare. These assessments are paid by the contractor and are added by him to his labor costs. These assessments may vary from about 5 to 70 cents per labor-hour. The contractor may pay these assessments to a holding company (such as a bank or other agency), to an insurance company, or to a union welfare fund.

Automobile and Vehicle Insurance. If a contractor owns and operates automobiles, trucks, and other vehicles, he should take out insurance covering these vehicles and their operation. Rates may vary from about $55 and up per vehicle, depending on the conditions in the particular policy selected.

Property-damage Insurance. The contractor on some jobs should take out insurance against claims arising from possible damage to other property. Such insurance is especially desirable when the proposed work may cause damage to an adjoining structure, such as damage caused by the settlement of the foundations of a structure on an adjoining lot when the excavation is made for a new building. The rates for property-damage insurance will vary with the risks involved.

Contingent Insurance. Contingent insurance can be taken out by the contractor to protect the owner during the construction period. Such insurance is called owner's contingent insurance. Rates are usually based upon the kind and cost of the job.

Another kind of contingent insurance is that taken out by the contractor to protect him from claims resulting from any acts of his subcontractors. This insurance is called contractor's contingent insurance. Rates are usually based upon the kind and cost of the job.

Fire Insurance. The fire insurance should cover the structure itself (if inflammable), scaffolding, sheds, temporary structures, and all materials, equipment, and tools on the property either in the structure, in temporary structures and sheds, or upon the lot. Fire insurance may cost 30 to 90 cents per year per $100 of valuation.

Wind and Rain Insurance. Such insurance, often called tornado or windstorm insurance, should be carried if there is any possibility of the structure being damaged by wind and rain. Rates for this kind of insurance are usually less than fire insurance rates.

Structure or Blanket Insurance. Some insurance companies issue what is known as structure or blanket insurance in which one policy covers all the protection required and which formerly required several policies of various kinds. When taking out structure or blanket insurance, the contractor must inform himself of all the different hazards covered and the amounts. Rates for blanket insurance are usually appreciably less than the sum of the rates for the different kinds of insurance considered separately. This insurance is often called *packaged* insurance.

Payroll Taxes The Federal government and several states levy payroll taxes for various purposes. The two payroll taxes levied by the Federal government and two payroll taxes levied by the state of California are described in this article. The contractor's costs for these taxes are usually included in his labor costs.

Social Security Tax (FICA) and Medicare. Under this act assessments are made on the wages paid the workmen. The employer and the workmen each pay part of the tax. The assessments against the employer in 1967 are 4.4 percent of all wages paid. The assessments against the employee in 1967 are 4.4 percent of all wages received, the assessments being deducted from the employee's wages. These percentages apply to all salaries and wages paid employees up to a total of $6,600 per year per person.

Federal Unemployment Tax (FUTA). Under this act assessments equal to 0.30 percent are made on all salaries and wages paid by the employer. These assessments are paid by the employer; the worker pays nothing.

California State Unemployment Insurance Tax. In California the employer is assessed 3.50 percent (1966) on all salaries and wages paid employees. The employees are not assessed.

California State Unemployment Compensation and Disability Tax. This tax (UCD) is 1 percent (1966) of all salaries and wages paid the employees and is assessed against each employee. The employer deducts this tax from the employee's wage.

Profit The amount of profit is usually expressed as a percentage of the sum of the estimated costs of materials, labor, equipment, and overhead. The percentage allowed for profit will vary according to the kind of the job, the amount involved, and the time required for completion.

Usually a larger percentage is allowed for smaller and more hazardous jobs. The percentage allowed for profit will also depend upon the contractor's desire for the particular work, what he thinks is fair, and what he thinks he can get.

The percentage allowed for profit may vary from about 5 to 25 percent. For small-sized and hazardous jobs, 20 to 25 percent may be used, 15 to 20 percent for jobs of medium size and which are somewhat less hazardous, 10 to 15 percent for large jobs when the hazards are not great, from 5 to 10 percent for very large jobs when the hazards are small.

CHAPTER TWENTY

Complete Estimates

Complete Estimates A complete cost estimate, especially from an owner's viewpoint, contains many items other than those included in the main and subcontracts. Many people think that the total cost of a structure, say an office building for example, includes only the costs of the land and of the main and subcontracts. There are many other cost items to be included.

A discussion of the costs of several of these items is given in the following articles of this chapter. An attempt has been made to include all the main items in the discussion. However, there may be jobs on which other items should be included. Likewise, on many jobs all items mentioned may not be included.

Land There are several items of cost other than the first cost or price of the land which an owner may be required to pay. The cost of the land will include one or more of the following items:

First Cost. Price paid to former owner.

Real Estate Commissions. These are usually, but not always, paid by the seller.

Survey. A survey of the land by a responsible land surveyor is often advisable to determine if the actual boundaries agree with those given in the deed.

Deed. The deed should be examined by a competent lawyer to see if it has been correctly drawn.

Escrow Company. If available, a good escrow company should be used. This company will see that all requirements of the transaction are met, that the deed and other papers are properly prepared, that the deed is recorded, will make all payments, and will secure a title insurance policy or an abstract of title, etc. The escrow fees are usually paid by the seller.

Title Insurance. If possible, the buyer should secure a title insurance policy guaranteeing him a clear title to the property and protecting him against loss up to a certain amount (usually the purchase price). The premium for this policy is usually paid by the seller.

Abstract of Title. When a title insurance policy is not available, the seller should provide the buyer with an abstract of title or other proof of ownership of the land. This abstract should be examined by a competent lawyer to see if it is correct and that the seller has a clear title to the land, and can transfer a clear title to the buyer.

Zoning. Check on zoning restrictions, if any.

Back Taxes. Determine if all regular taxes have been paid.

Special Taxes. Such as sewer, water, and street-improvement taxes. Determine number and amount of installments to be paid.

Legal Expenses On practically every construction job, except possibly the smallest and simplest, the owner should have a competent lawyer examine and pass on all contracts and agreements before the owner signs them. Legal advice is important on all items that involve possible litigation. The owner often retains the same lawyer or firm of lawyers to look after all the legal affairs connected with any particular construction job.

Architectural and Engineering Fees These fees are often divided into two parts, one for the preparation of designs, plans, and specifications and the other for supervision during construction. The fees charged for designs, plans, and specifications may range from about 2 to 12 percent of the estimated cost of the structure, depending upon the work involved. The fees charged for supervision of the work during construction may vary from 2 to 15 percent, depending on the particular job and the amount of supervision (general or detailed) desired.

Main Contract Sometimes the main contract includes all the work of construction. On other jobs, only the general or main parts of the job are included and subcontracts are let for the specialized work. For example, on a building job, the main contract may include all the work to be done by the general and subcontractors or it may include only the work to be done by the general contractor. The owner should know what work is to be done by the main or general contractor and what work by each of the subcontractors. He must also know whether or not the general contract includes work by the subcontractors.

Subcontracts On many construction jobs, the specialized work is let to qualified subcontractors. These subcontractors may work under the general contractor and be responsible directly to him, or they may not be under the main contractor but may be responsible directly to the architect, engineer, owner, or owner's agent, as the case may be.

For example, in the construction of a building, subcontracts may be let for the plumbing, heating and air conditioning, electric wiring, and painting and decorating.

For another example, in the construction of a bridge a subcontractor may build the foundations and piers, another may build the approaches, and the main contractor constructs the bridge on the prepared foundations.

Extras On practically every job, there will be some extras and contingencies that are not covered in the main and subcontracts. These contracts usually include clauses in regard to extras and their cost. On many construction jobs, the extras may cost 2 to 10 percent or more of the sum of the main and subcontracts. Consequently, the owner should consult with his architect and engineer and make a reasonable allowance for extras for the particular job.

Permits and Inspection Fees On many jobs, permits must be obtained for doing certain kinds of work and for making connections such as for water and electricity. Many kinds of construction work require inspections by various government, state, and local officials and inspectors.

The cost of all permits and inspections will be usually charged against the owner, and he will be required to pay them unless provision for their payment by the contractors concerned is provided for in the main and subcontracts.

Surety Bonds On most large jobs, each contractor is required to give a surety bond for the successful completion of his work according to the contract and specifications. The individual contractors are usually required to pay for these bonds. However, if the contract in question does not state that the contractor is to pay the premium, the contractor can pay it and add the cost as an extra.

Financing When the owner does not have sufficient funds available to pay for the entire costs of construction, he must make arrangements to obtain the needed funds from some other source. The bank, trust company, or finance company involved usually makes a charge for the service of securing the necessary funds. This charge usually ranges from about 2 to 5 percent of the funds provided, though the charge may be more in times when money is difficult to obtain, when the amount required is more than about half to two-thirds the value of the job, when the owner's credit is bad, or when the apparent risk is greater.

Sometimes the general contractor arranges for the financing as well as for the construction of the job. When this is the case, the general contractor includes a financing charge with his bid.

Insurance Though the contractors who do the work are usually required to provide all insurance that may be deemed necessary, the owner is also held responsible in many cases. Consequently, it may be advisable for the owner to take out insurance, such as public liability, personal injury, fire, tornado, or property damage, for his own protection. On a large building construction job, the cost of such insurance may be an appreciable item.

Interest Interest charges during construction should always be allowed. When the time required is a year or more and the amounts paid the contractors before the final completion of the job are large, the construction interest charges may be considerable. For example, if the time required to construct a $1,000,000 building was 2 years and the contractors received partial payments up to 90 percent of the cost, the interest charges during the construction period would be approximately $43,000, by assuming interest rates at 5 percent per annum.

Taxes On practically every private construction job that extends over one tax-assessing and -paying time, some taxes will have to be paid. The amount of taxes, on a new building for example, will depend on the tax rate, the assessed valuation of the land, and the assessed valuation of the portion of the structure erected at the time the assessment was made. This particular valuation of the structure will include all materials actually installed in the structure and may or may not include part of or all the materials delivered at the site but not yet incorporated in the structure.

Miscellaneous On most large jobs, there will be a number of miscellaneous expenses that an owner may have to pay. Some such items are stenographic services, transportation to and from the job for owner's inspections, and entertainment (if the owner deems it necessary to entertain the architect, engineer, contractor, or others). Though not large, the sum total of such items should be allowed for.

Summary The items that may be included in a complete estimate can be summarized somewhat as follows:

Land
Legal expenses
Architectural and engineering fees and commissions
Main contract

Subcontracts
Extras
Permits and inspection fees
Surety bonds
Financing costs
Insurance, owner's
Interest, during construction
Taxes, during construction
Miscellaneous items
Total

CHAPTER TWENTY-ONE

Approximate Estimates

Approximate Estimates The only safe and sure method of estimating costs is to take off the actual quantities of materials, judge the labor-hours required, and use the prevailing prices for labor and materials. However, many experienced extimators, when estimating the costs of certain types of structures, make use of shortcuts or approximations (1) to obtain the approximate cost in a short time, (2) to check roughly the costs found by more accurate methods, and (3) to compare the costs of like structures erected at other times and places.

Costs estimated by a shortcut or approximate method are approximate only. Their value depends upon the judgment, skill, and experience of the estimator; upon the care with which the estimates are prepared; and upon the correctness of the prices used.

Estimating Building Costs by the Square-foot-of-floor-area Method The method of estimating by the square foot of floor area is applicable to such structures as office buildings, schools, mills, warehouses, factories, hospitals, churches, stores, residences, and garages. This method is useful in comparing the costs of different buildings of the same type or kind where the floor area is important, as in office buildings and factories. This method is also useful in determining the costs per person, as in schools, hospitals, and churches.

There are several variations of the method for estimating building costs by the square feet of floor area. One variation is to allow a certain price per square foot for each floor, including the basement, attic, and roof. Another variation is to use different unit cost prices for the basement floor, other floors, and roof. A third variation is to use different unit cost prices for the different floors and have the price for the lower floor include the cost of the basement and foundations. The price for the upper floor should include the cost of the roof. A fourth method is to use the same unit cost price for all usable floors, omitting the roof area or both the roof and basement areas. This fourth method is preferred by many estimators. Consequently, it is important to state just what is included in the unit cost price used in the estimate.

The following illustrative example shows the effect on the computed cost per square foot by using the different methods.

Illustrative estimate. A certain building is 40 by 60 ft in size and consists of a basement; first, second, and third floors; and attic. The cost of the building was $115,100.

Compute the cost per square foot of the building:

1. Based on total area of three floors, basement, attic, and roof.
2. Based on total area of three floors, basement, and roof, assuming basement to cost 60 percent of the cost of the first floor, attic floor to cost 40 percent of the cost of the first floor, roof to cost 50 percent of the cost of the first floor, and first, second, and third floors to have equal costs.
3. Based on area of the three floors, assuming the cost of the first floor to include cost of basement and to be 1.60 times the cost of the second floor, and assuming the cost of the third floor to include cost of attic floor and roof and to be 1.90 times the cost of the second floor.
4. Based on total area of the three floors only.

Solution 1:

	Sq ft
Basement area	2,400
First-floor area	2,400
Second-floor area	2,400
Third-floor area	2,400
Attic-floor area	2,400
Roof (horizontal) area	2,400
Total	14,400

$$\text{Cost per square foot} = \frac{\$115{,}100}{14{,}400} = \$8.00$$

Solution 2:

	Sq ft
Basement area, equivalent = 2,400 × 0.60	1,440
First floor	2,400
Second floor	2,400
Third floor	2,400
Attic floor, equivalent = 2,400 × 0.40	960
Roof, equivalent = 2,400 × 0.50	1,200
Total equivalent area	10,800

Cost per square foot, first, second, and third floors = \$115,100/10,800 = \$10.66
Cost per square foot, basement = \$10.66 × 0.60 = \$6.40
Cost per square foot, attic = \$10.66 × 0.40 = \$4.26
Cost per square foot, roof = \$10.66 × 0.50 = \$5.33

Solution 3:

	Sq ft
First floor, equivalent = 2,400 × 1.60	3,840
Second floor	2,400
Third floor, equivalent = 2,400 × 1.90	4,560
Total equivalent	10,800

Cost per square foot, second floor = \$115,100/10,800 = \$10.66
Cost per square foot, first floor = \$10.66 × 1.60 = \$17.06
Cost per square foot, third floor = \$10.66 × 1.90 = \$20.25

Solution 4:

	Sq ft
First-floor area	2,400
Second-floor area	2,400
Third-floor area	2,400
Total area, three floors	7,200

Cost per square foot = \$115,100/7,200 = \$16.00

Estimating Building Costs by the Cubic-foot-of-volume Method The method of estimating building costs by the cubic foot of volume is more accurate, in general, than the method of estimating costs by the square foot of floor area. Buildings of different types and different kinds of construction require different cost prices per cubic foot.

The best way of estimating costs by the cubic-foot method is to find the total volume of the building in cubic feet and multiply this volume by a selected price per cubic foot for this particular class and type of building.

Another way is to use a certain unit cost per cubic foot for the parts of the building that are more expensively finished and to use lower prices for the parts that are more cheaply or less completely finished. For example, in a residence one unit price would be used for the basement, attic, and attached garage and another unit price for the living room, dining room, kitchen, and bedrooms.

A third way is to consider only half or two-thirds of the volume of the basement, attic, and attached garage, when computing the volume of a residence, and then to use the same cost price per cubic foot for all parts of the building.

The American Institute of Architects, Washington, D.C., in the *AIA Document* 239, recommends the following:

The cubic content (cube or cubage) of a building is the actual cubic space enclosed within the outer surfaces of the outside or enclosing walls and contained between the outer surfaces of the roof and 6 in. below the finished surfaces of the lowest floors.

The preceding definition requires the cube of dormers, penthouses, vaults, pits, enclosed porches, and other enclosed appendages to be included as a part of the cube of the building. It does not include the cube of courts or light shafts open at the top, or the cube of outside steps, cornices, parapets, or open porches or loggias.

The following items shall be listed separately:

1. Cube of enclosed courts or light shafts open at the top, measured from outside face of enclosing walls and from 6 in. below the finished floor or paving to top of enclosing walls
2. Cube of open porches measured from outside face of wall, outside face of columns, finished floor, and finished roof

It is recommended that the following items also be listed separately:

1. Square feet of area of all stoops, balconies, and terraces
2. Memoranda or brief description of caissons, piling, special foundations, or features, if any

The following illustrative example shows how the costs per cubic foot can be found by two methods. The first method is recommended.

Illustrative estimate. A certain building is 40 by 60 ft in size and consists of basement; first, second, and third floors; and attic. The heights, including floor thicknesses, are 8.5, 10, 9, and 9 ft for the basement, and first, second, and third floors, respectively. The average height of the attic is 7.5 ft. The cost of the building was $115,100.

1. Compute the cost per cubic foot based on the total volume of the building in cubic feet.

2. Compute the cost per cubic foot, allowing 60 percent for the basement, full value for the three stories, and 50 percent for the attic.

Solution 1: Cubical contents:

	Cubic feet
Basement = 40 × 60 × 8.5	20,400
First floor = 40 × 60 × 10	24,000
Second floor = 40 × 60 × 9	21,600
Third floor = 40 × 60 × 9	21,600
Attic = 40 × 60 × 7.5	18,000
Total	105,600

Cost per cubic foot = \$115,100/105,600 = \$1.10

Solution 2: Equivalent cubical contents:

	Cubic feet
Basement = 40 × 60 × 8.5 × 0.60	12,240
First floor = 40 × 60 × 10	24,000
Second floor = 40 × 60 × 9	21,600
Third floor = 40 × 60 × 9	21,600
Attic = 40 × 60 × 7.5 × 0.50	9,000
Total equivalent	88,440

$$\text{Cost per cubic foot} = \frac{\$115{,}100}{88{,}440} = \$1.30$$

Panel Method for Buildings In the panel method of estimating building costs, the costs of each kind of the "panels" (wall and floor) are computed separately and then these costs are multiplied by the respective number of panels of each kind. There are about as many variations of the panel method as there are different kinds of buildings.

The panel costs usually include the items of materials, labor, and equipment. Heating and air conditioning, plumbing, wiring, painting, and decorating are added later as separate items, though painting and decorating can be included in the panel costs. Overhead and profit for the whole structure are usually added later as separate items.

Some estimators think the panel method is better than the square-foot-of-floor-area method or the cubic-foot-of-volume method because the panel method takes into consideration the variations in the ratios of floor and roof areas to wall areas that practically always occur with changes in the shapes and sizes of buildings.

Panel Method for Residences and Small Buildings of Frame Construction The panel method for estimating building costs can be applied to residences and other small buildings of frame construction by obtaining the cost per square foot or per square of 100 sq ft for the floors (basement, first, second, attic, etc.) and for the roof, allowing for such items as the overhang, eaves, and valleys. For the exterior walls and partitions, the costs can be computed per square foot or per square of 100 sq ft of wall area. However, for wooden-frame construction most estimators prefer to consider a wall panel as the area whose width is equal to the distance center to center of studs (usually 16 in.) and whose height is equal to that of one story, often from 8 to 10 ft. There are several ways of allowing for the windows and doors in the wall panels. The simplest way is to compute the cost of a plain wall panel and add certain amounts for each window and door. Another way is to compute the cost of a wall panel containing a window or a door, as the case may be.

Wall panels containing doors and windows usually cost more than plain wall panels.

Items such as stairs, steps, and porches must be added as separate items.

When the cost for each kind of panel is known, it is only necessary to count the number of panels of each kind and compute the total costs. For estimating residences, it is usually advisable to include overhead and profit in the panel costs. Painting and decorating, heating, plumbing, wiring, and electric fixtures are usually added as separate items.

The contractor can prepare cost estimates for typical wall, floor, and roof panels of different kinds of construction, using different kinds and qualities of materials, and then can use these panel costs when preparing preliminary estimates for customers. Changes in the prices of materials and in wages must be allowed for as they occur.

Panel Method for Industrial and Commercial Buildings The panel method can be used for large buildings such as warehouses, factories, and office buildings of different kinds of construction such as wooden-mill construction, steel-mill construction, steel-frame buildings, and reinforced concrete construction.

The panel estimate can be divided into the following items:

1. Columns including foundations and excavation
   a. Exterior
   b. Interior
2. Floors, including basement floor and excavation
3. Roof
4. Trusses, such as roof trusses
5. Exterior walls, including doors and windows
   a. Basement walls, including footings and excavation
   b. Wall panels for each story above basement
   c. Parapet walls, as around roof
6. Interior or partition walls, including doors
7. Stairs and stair walls
8. Elevators
9. Heating and air conditioning
10. Plumbing

11. Electrical work
12. Sprinkler system (if any)
13. Painting and decorating
14. Contingencies
15. Overhead
16. Profit
17. Summary or total estimate

When the estimate is prepared, the items listed need not necessarily be taken in the order given. Some items can be omitted or combined with others, and other items added as desired.

For example, in the estimate of the cost of a two-story and basement warehouse, 60 by 100 ft in area with columns spaced 20 ft on centers, there may be

16 exterior columns
8 interior columns
15 basement-floor panels
15 first-floor panels
15 second-floor panels
15 roof-floor panels
16 exterior-basement-wall panels
32 exterior-wall panels
16 parapet-wall panels
18 partition-wall panels
7 sets of stairs
1 elevator
1 sprinkling system for three floors

Some of the floor panels may be modified because of stairs and elevator. Allowances must be made for heating, ventilation, plumbing, electrical work, painting, and other items and contingencies as required by the plans and specifications. Overhead and profit must be included.

Bay Method for Buildings When a building consists of several equal bays, the cost of the building can be estimated by estimating the cost of one bay and then multiplying this cost by the number of bays.

Allowances must be made for end walls, entrances, and other items that are not included in a typical bay. For example, an exterior corner bay will have two exterior walls, an exterior side bay will have one exterior wall, all bays may have one or more interior walls or partitions, some bays may have stairs and stair walls, and some may contain elevator shafts. Special items not included in a typical bay may be roof trusses, other trusses or girders, elevators, extra columns and beams, and also any omissions of columns, beams, and slabs.

The bay method of estimating buildings is usually not so accurate as the panel method but may be more accurate than the square-foot-of-floor-area or cubic-foot-of-volume method.

Comparative Costs When unit costs of structures are compared, consideration must be given to the use of the structure (bridge, residence, store, warehouse, etc.), kind of construction (truss bridge, wooden-frame residence, steel-frame building, reinforced concrete building, etc.), quality of construction (cheap, medium, good), locality (city, village, country, eastern, southern, etc.), and year and time of year. If the classification idea is carried further and buildings only are considered, the buildings can be classified as follows:

Use of Occupancy. Residence, apartment, store, warehouse, office, school, hospital, theater, garage, etc.

Kind of Construction. Wooden frame, siding, brick or stone veneer, stucco, steel frame, reinforced concrete, steel mill, wooden mill, brick, stone, etc. Insulated, noninsulated. Fireproof, fire-resisting, semi-fire-resisting, non-fire-resisting

Quality of Construction. Cheap, medium, good, very good

Foundations. Kind of soil, soft, medium, hard, loam, sand, gravel, rock, under water. Simple or combined footings, mat or raft foundations, bearing or sheet piling, piers, open well or caisson

Location. City, village, country, eastern, western, southern, northern. Wet, medium, or dry climate

Time. Year. Summer or winter

Other kinds of structures can be classified in somewhat similar ways.

The unit on which the comparative costs are based must be suitable for the kind of structure considered and the use to which the structure is to be put. Some examples follow:

Bridges, span in feet

Pavements, area in square yards

Excavation, cubic yards

Concrete, cubic feet, or cubic yards

Buildings, square feet of floor area or cubic feet of volume

Schools, per pupil, per square foot of floor area

Office buildings, per office, per room, per square foot of rentable floor space

Apartment buildings, per apartment, per room, per square foot of floor area, per cubic foot of volume

Theaters, per seat

Stadium, per seat

Grain elevators, per bushel of capacity

Storage warehouse, per cubic foot of storage space

Manufacturing buildings, per unit of daily output, per square foot of available floor space

Automobile factory, per car

Power plant, capacity in horsepower or kilowatts

Hospitals, per bed

Hotels, per room

Store, per square foot of rentable floor area

Range of Unit Costs Unit costs will vary considerably with buildings of different occupancies, kinds of construction, sizes and shapes, quality of construction, locations, and times.

Table 21-1 gives approximate ranges of costs per square foot and per cubic foot for several kinds of buildings.

Table 21-1 *Approximate* 1967 *Costs of Buildings per Square Foot and per Cubic Foot*

Kind of building	Cost per sq ft	Cost per cu ft
Frame residences:		
Cheap construction	\$ 9–\$13	\$0.55–\$1.05
Medium construction	\$12–\$20	\$0.90–\$1.30
Good construction	\$17–\$27	\$1.10–\$1.70
Very good construction	\$24–\$45	\$1.40–\$2.65
Apartment buildings, wood frame:		
Medium construction	\$ 8–\$15	\$0.70–\$1.30
Good construction	\$14–\$25	\$1.10–\$2.00
Office buildings	\$17–\$55	\$1.00–\$3.50
Stores	\$12–\$42	\$0.65–\$2.40
Banks and trust companies	\$30–\$75	\$1.80–\$4.60
Schools	\$15–\$45	\$0.80–\$3.50
Hospitals	\$22–\$50	\$1.45–\$3.35
Hotels	\$18–\$55	\$1.30–\$3.20
Industrial buildings:		
Steel mill, one or one and one half stories	\$10–\$17	\$0.25–\$0.60
Wood mill, two to six stories	\$ 9–\$27	\$0.60–\$1.70
Steel frame, multistoried	\$18–\$33	\$1.15–\$2.20
Reinforced concrete, multistoried	\$18–\$35	\$1.15–\$2.30

The cost per square foot is usually found by dividing the total cost by the total floor area. When the total floor area is computed, the outside dimensions of the walls are used and no deductions are made for openings.

The cost per cubic foot is usually found by dividing the total cost by the volume of the building in cubic feet. The number of cubic feet are usually computed by the method recommended by the American Institute of Architects and described previously in this chapter.

Distribution of Costs There are many ways of distributing or breaking down the total cost of a structure. The cost may be divided into materials, labor, equipment, overhead, and profit for the entire structure, or the costs may be distributed according to the different kinds of work, such as foundations, framing, exterior and interior finish, mechanical trades, and painting and decorating for a building.

Approximate costs, expressed as percentages of the total, for materials, labor, overhead and equipment, and profit for several different kinds of construction work are given in Table 21-2 (United States Construction Census Reports).

Table 21-2 *Approximate Costs in Percentage of Total Costs*

Type of construction	Percentage of total costs			
	Materials	Labor	Overhead and equipment	Profit
All classes	43.8	30.7	18.0	7.5
All building	43.7	31.5	15.3	9.5
Building:				
Commercial	41.2	32.0	16.7	9.9
Manufacturing	51.9	22.8	14.4	10.9
Residential	45.2	29.0	14.7	11.1
Highway	41.8	25.7	23.3	9.2
Street paving	42.3	24.7	23.0	10.0
Grading	8.8	39.3	40.3	11.6
Bridge and culvert	43.0	29.8	19.7	7.5
Sewer, gas, water, conduit	37.2	33.5	20.6	8.7
Dam and reservoir	32.2	32.0	22.8	13.0
Water works	43.5	27.7	21.3	7.5
Dredging, river and harbor	24.9	29.8	37.5	7.8
Levee	16.4	32.5	41.7	9.4
Railroad	28.3	37.2	24.1	10.4
Foundation	35.3	35.8	22.5	6.4
Subways	32.5	34.7	18.9	13.9

The costs, and percentages of total costs for various items, will vary considerably with the kind and quality of construction, kind of building, and specification requirements. The cost distributions given here are included to give an approximate idea of the various percentages.

Table 21-3 gives the cost distributions, in percentages of the total costs, for two frame residences: one a full two-story frame residence with wood siding and a basement under the entire building and erected in a cold climate and the other a one-story stucco frame residence with no basement and erected in a warm climate.

Table 21-4 gives the cost distributions for some large buildings with steel and reinforced-concrete frames.

The cost-distribution percentages vary considerably for different kinds of buildings, erected in different locations, and for different occupancies. There appears to be no general rule that the estimator can follow when estimating cost-distribution percentages. For more complete and detailed information in regard to building costs and other construction costs, the reader is referred to such publications as the *Engineering News-Record.*

Table 21-3 *Cost Distributions in Percentage of Total Cost for Two Frame Residences*

Item	Two-story siding with basement, cold climate	One-story stucco, no basement, warm climate
Excavation	1.5	None
Concrete work	10.0	5.0
Carpentry and millwork	42.0	45.0
Lathing and plastering	10.0	16.0
Sheet metal	1.5	1.5
Tile work	3.0	3.0
Plumbing	9.0	10.0
Heating and air conditioning	11.0	7.0
Electrical	4.0	5.0
Painting	7.0	6.0
Finish hardware	1.0	1.5
Total	100.0	100.0

Cost Indexes At the present time, there are many cost indexes available which give information in tabular or graphical form or both relating to various construction costs. These costs are usually given in percentage form and are based on the costs of some certain year which are considered as 100 percent. By the study and use of the cost indexes and by consideration of construction trends and price trends, an experienced estimator is enabled to bring the cost data for any year up to the present time. Then this cost data can be used in preparing estimates.

Perhaps the best known of the cost indexes are those published monthly in the *Engineering News-Record.* Many of the cost indexes are based on the year 1913, considering that year as 100 percent. If desired, any other year may be used as a base. No attempt will be made in this article to list or discuss all the cost indexes available.

Table 21-4 *Cost Distribution, in Percentage of Total Cost, for Some Typical Buildings*

Building	Industrial	Steel mill	Office	Office	Apartment	Apartment
Frame	Steel	Steel	Steel	Concrete	Steel	Concrete
Outside walls	Block	Metal	Brick	Brick	Brick	Stucco
Stories	One	One	5–10	5–10	4–8	3–8
Item	Percentage of total cost					
Foundations	12–18	15–23	4–9	5–9	6–9	4–7
Structural frame	13–19	17–22	15–21	12–18	14–20	11–18
Masonry	13–17	5–8	3–5	4–8	3–7	5–9
Roofing*	9–13	10–22	1–3	1–3	1–4	1–4
Partitions	1–3	0–2	7–11	7–11	7–11	8–14
Insulation	.....	.....	1–4	1–4	1–3	1–3
Lath and plaster	.....	.....	5–8	5–8	4–7	4–8
Sash, doors, and glazing	3–7	9–16	9–14	9–13	1–4	2–4
Floor covering	.....	.....	2–5	2–5	5–7	5–7
Painting	2–4	1–3	1–2	1–2	1–3	1–3
Plumbing	2–4	2–4	4–6	4–6	7–12	8–12
Heating and air conditioning	5–9	5–9	10–14	10–13	14–20	13–18
Electrical	6–11	3–7	8–12	8–12	7–12	7–11
Elevators	.....	.....	6–10	6–10	2–6	2–7
Sprinklers	7–12					
Architect's fee	3–5	3–5	4–6	4–6	4–7	4–7

* Roofing item for the steel-mill buildings includes metal wall covering.

Table 21-5 gives some cost indexes for the years 1962 to 1966, inclusive, with year 1962 = 100. These cost indexes represent average values for some twenty larger cities in the United States and they do not apply to any particular city or locality. The purpose of this table is to show the general trends in these cost indexes for the years 1962 to 1966, inclusive.

In addition to the various cost indexes that are published by different authorities, the estimator should study trends in the prices of construction and building materials and machinery and also the trends in construction and building wages. Much information is given in the *Engineering News-Record* and other publications in regard to price and wage trends. This information is valuable, comprehensive, and instructive.

Table 21-6 gives information for the years 1962 to 1966, inclusive, for construction skilled and common labor wages and the costs of the three basic construction materials: steel, cement, and lumber. The wages given are average hourly wages in some 20 cities. Materials costs are structural steel, base price per hundredweight (average of three mills);

Table 21-5 *Average Cost Indexes for General Construction, Building Construction, Construction Materials, Construction Machinery, and Skilled and Common Construction Labor Wages*

Year	Construction cost	Building cost	Construction materials cost	Construction machinery cost	Construction labor wages	
					Skilled	Common
1962	100.0	100.0	100.0	100.0	100.0	100.0
1963	103.2	102.3	100.2	102.5	103.8	104.0
1964	107.5	105.0	101.4	104.5	108.7	109.2
1965	111.4	108.0	102.5	108.0	113.7	115.5
1966	119.0	113.3	105.4	112.0	122.5	125.5

Table 21-6 *Prices of Structural Steel, Cement, Lumber, and Wages of Construction, Skilled and Common Labor**

Year	Structural steel mill base per cwt	Cement CL lots f.o.b. city per bbl	Fir 2 by 4, S4S CL f.o.b. city per M fbm	Pine 2 by 4, S4S CL f.o.b. city per M fbm	Construction labor wage per hr	
					Skilled	Common
1962	$5.50	$4.06	$116	$108	$4.35	$2.95
1963	5.70	3.98	117	109	4.55	3.10
1964	5.70	3.98	117	108	4.75	3.25
1965	5.70	3.87	120	109	4.95	3.45
1966	5.85	3.89	120	112	5.35	3.75

* The data in this table have been collected from different sources.

cement, bulk, per barrel, carload lots f.o.b. city; and lumber, fir and pine, 2 by 4, S4S, per 1,000 fbm, f.o.b. city.

The estimator (if he desires) may prepare his own tables and diagrams showing recent changes and trends (say for the last five or ten years) in cost indexes or in the total costs, materials prices, equipment costs, and wages of skilled and common labor. Such tables and diagrams may be prepared for one or more kinds of construction (buildings of different types, highways, sewers, etc.) and for any large city or major locality.

Other information which may be useful is that relating to building permits issued in cities; construction expenditures for different kinds of construction, both public and private, in different sections of the country; and domestic money rates.

The estimator should study unit bid prices for different kinds of construction in various parts of the country so that he can obtain information concerning prices bid by various contractors.

CHAPTER TWENTY-TWO

Construction Management

An experienced contractor once said that there were two important steps in the successful performance of a construction job.

1. The work must be carefully planned and scheduled.
2. The work must be kept up to schedule by means of thorough supervision aided by adequate records of the progress.

Planning the Work

Tentative Plan As soon as the contract is signed, the contractor must plan on starting the work on the date set and on finishing the work on or before the date set for completion. An experienced contractor does not have to do much planning for a small job because he already knows what to do and when to do it. For a large job, the planning must be more systematically and carefully done. Records must be kept in detail, as the contractor should not rely on his memory for all the various details.

The first step in planning a large job is to prepare a rough or tentative plan. In this plan, the contractor should list the different parts or kinds of work that are to be done, the approximate time needed for each part, and the approximate dates on which each kind of work should be started and finished.

After these approximate dates have been listed, the contractor must determine in a general way the various kinds and sizes of construction equipment required. He will note what equipment he has available and what equipment he may need to purchase or rent.

Then the classes of labor (various skilled and unskilled) needed can be listed, the approximate date or times that they will be required noted, and the number of each class estimated. The contractor will plan in a general way on how many men of his permanent organization he will use on the job and on how many new men he will need to hire. He will also note which men he will use for superintendents and general foremen.

The contractor will note approximately when and where the orders for materials should be placed and about when the deliveries should be made and in what quantities.

Consideration should be given to the subcontractors and their work. Approximate dates of starting and completion of work should be noted in each case.

Some thought at this time should be given to the overhead for the job and to how this is to be cared for.

After the contractor has prepared his tentative or rough plan, he is ready to proceed with the preparation of the detailed plans.

Visiting the Site Before starting the detailed plans for the work, the contractor and his superintendent should visit the site again. They should preferably be accompanied by the owner or the owner's representative, the architect, or the engineer as the case may be.

The job should be discussed in a general way. Notes should be taken concerning the opinions and suggestions of the owner, architect, engineer, or their representatives. Such notes will be of value if differences of opinion and arguments should develop later.

After going over the site with the owner, architect, or engineer, the contractor and his superintendent or foreman should go over the site again and discuss the construction work in detail. Such things as the location of the office shanty, sheds, tower, and major pieces of equipment can be tentatively decided upon and noted. Sketches should be made when desirable.

Before leaving the locality, the contractor may take additional notes to supplement the information already obtained in previous visits. Additional information may be needed in regard to such items as local ordinances and regulations, permits, banking facilities, light, power, water, sanitation, roads and streets, railway siding, trucking, local material dealers, local labor supply, and boarding and rooming facilities.

Planning Plant and Equipment After the site has been visited and studied, the plant can be planned. First, the work must be visualized in detail, keeping in mind the conditions at the site, the kinds and amounts of work to be done, the equipment that may be used, and that available.

Then suitable equipment can be selected. The machines should be such that they will work well together and that the plant will be balanced. The selection of equipment will depend on what the contractor has available and on what he can readily procure for the job.

After the equipment has been selected, the layout must be carefully planned so that the various pieces will work most efficiently. Sketches and notes should be made illustrating the layout.

Often several arrangements or layouts should be investigated so that the most suitable one can be found. Sometimes machines of different kinds and sizes (capacities) should be considered when planning different layouts.

On some jobs, equipment time and work schedules are advisable. Such a schedule should show the various pieces of major equipment, the kinds of work they are to be used upon, and the dates or times each piece of equipment is to be available for use on each kind of work. The schedule should be set up in simple tabular or graphical form. Further information on time and work schedules is given in the article on Time and Work Schedules in this chapter.

Further information about construction plant and equipment is given in Chap. 18 on Construction Plant and Equipment and in the various chapters on estimating. A study of the publications of the Associated General Contractors and other organizations is well worth while.

Sequence of Trades Before the labor for a job can be planned in detail, consideration must be given to the order of sequence for trades and gangs on the job. The order of sequence will depend upon the particular job. It is usually not necessary for any one operation to be wholly completed before another operation is started. However, the trades and gangs should be scheduled so that they will have the least interference with one another and so that they will cause the minimum of delays. On almost any job, there will be some interference among gangs and some delays, but proper planning will reduce this interference and these delays to a minimum.

How closely one gang can follow another will depend on the particular job. For example, on a concrete paving job it may work best to keep the grading crew at least two days ahead of the concrete crew and to keep the roller one day ahead. On another job, the grading crew may be but one day ahead of the concrete gang. When erecting a reinforced concrete building, the form crew should be two days ahead and the steel crew one day ahead of the concrete gang. If the form crew is kept but one day ahead of the concrete gang, the steel crew will find it difficult to place the steel after the form crew and keep ahead of the concrete gang at the same time.

In building work, the rough plumbing, steam fitting, wiring, and ventilating should be installed as soon as the rough framework of the building is available to work upon. The finished plumbing, heating, and ventilating work should usually be installed after the completion of the finished plaster, floor, and marble work. Interior finish, such as wood trim, flooring, and glazing, should not be installed until after the plaster is hard and dry. Painting and interior decorating usually come last. After the plasterwork is finished, the building should be cleaned up and kept clean thereafter, as this is the best method of protecting finished surfaces.

Planning Labor The kinds and amounts of labor will vary with different jobs. For any particular job, the different kinds and amounts of work can be noted and the classes of skilled labor required for each kind of work noted. Then consideration being given the time estimated for doing each kind of work and the amount of each kind of work, the number of laborers of each class of skilled labor and of unskilled labor can be estimated. If the number of laborers required for any one kind of work is four or more, they can be organized as a gang and a foreman provided. When the times for doing different kinds of work are planned, consideration must be given to the proper sequence of work and trades.

After the different classes of laborers and the number of each class required have been decided upon, a labor time and work schedule can be prepared. Such a schedule may be in a simple tabular or graphical form as desired. The schedule should show the following:

1. The different kinds of work
2. The dates and times that this work is to be performed
3. The number of skilled and unskilled laborers required for each kind of work
4. A summary showing for each day the total number of laborers on the job as well as the number of each class

When labor is planned, consideration should be given to the men in the contractor's permanent organization, old employees who may be available, labor that may be available locally, and labor that may need to be imported from another locality. Boarding and rooming places will be needed for imported labor. If local facilities are not sufficient, the contractor may need to provide board and lodging.

Consideration should also be given to construction trends, labor trends, and wages. The amount of construction varies from time to time and in different localities. Likewise, the labor supply and labor wages will vary from time to time and in different localities. In general, when there is a large amount of construction work in any locality, the labor supply will be low, the wages will be high, and the labor efficiency will be low.

Labor wages depend not only upon the amount of work and the labor supply available but also upon several other items such as union rules and regulations, government regulations, general prosperity of the country or locality, work available in other fields (offices, factories, industries, stores, etc.), and currency inflation. In general, currency inflation will cause an increase in wages, but wages will rarely keep pace with the inflation and will lag behind. The opposite is true when there is a currency deflation. Information concerning labor wages and trends can be found in such publications as the *Engineering News-Record.*

Planning Materials When materials are planned, consideration should be given to such things as prices, price trends, time to buy, amounts to buy, dealers or manufacturers, transportation, deliveries, amounts to be kept on hand, checking, testing, insurance, and follow up.

The correct procedure in buying construction materials is to purchase these materials so that they will arrive at the job in proper order, in the correct amounts, and at the right times. The materials must also conform to all specification requirements.

An expert buyer must be familiar with the markets, prices, factories (producers), shipping rates, time required for transportation of materials from producer to job, tracing of shipments, and the reliability of different producers.

Some producers are vary careful to see that their products are of the proper quality and are shipped at the desired times and in the correct amounts. Other producers are not so careful in regard to the uniformity of quality of their products, and they also may be careless in their shipments, shipping either too soon or too late and not always in the correct amounts.

When the market prices are remaining stationary, only the necessary quantities of the materials can be purchased from time to time to keep the work progressing in the most efficient manner. Contracts for the materials can be made a reasonable time in advance. Discounts for prompt payments (like 2 percent off in 10 days) should always be taken advantage of when the contractor's finances permit.

When the market prices are falling, materials should be purchased only as needed and in as small quantities as possible. Sometimes some risk may be run of delaying the work a short time in order to secure better prices on the materials.

When the market prices are rising, the materials for a job should be purchased as soon as possible, due consideration being given to the contractor's finances. If not purchased outright, the materials should be contracted for in advance so that the contractor will not have to pay higher prices later.

Some consideration must be given to the way in which prices are quoted. Some examples are f.o.b. manufacturing plant, f.o.b. nearest railway siding, delivered at job, carload lots or less than carload lots, crated or uncrated, minimum quantities, extra prices for small quantities, odd sizes, special deliveries, discounts from list prices, discounts for prompt payments, etc.

Much information concerning materials prices and price trends can be obtained from such publications as the *Engineering News-Record.*

Some consideration should be given to the method of transporting the materials from the manufacturer or producer to the job. The materials may be transported by railway, boat, or truck. The expense of transportation, probable time required including delays, and the reliability of the transportation company should be considered.

When planning on deliveries, some thought must be given to the total amount of each material, the amounts of each material to be kept on the job (such as a few day's supply, a week's supply, or the entire supply), the time or date that the material will be needed, and the rate at which each material will be used. Materials should not arrive at the job too soon or in too large quantities so that other work will be interfered with or that the materials will need much rehandling.

Some thought must also be given in regard to the overhead work required in checking purchases, testing materials, insurance, checking shipments, tracing of shipments during transportation, unloading, storage, watching, etc.

On most jobs, it is advisable to contract for all materials as to price, quality, quantity, delivery, etc., before the work starts on the job. When practical, some allowance should be made for variations in quality, quantity, and delivery because the job conditions and needs may vary after work starts. Sometimes clauses are included regarding unavoidable delays. Other clauses may be included regarding changes in prices.

Materials time schedules should be prepared for nearly all medium- and large-sized jobs. Such a schedule can be prepared in a simple tabular or graphical form. The schedule should show such information as kind of materials, total amounts of each kind, delivery dates for each kind, and amounts of each kind to be delivered on each date. Schedules may also show dates of first delivery and minimum amounts of each kind of material to be kept on hand at the job after the first delivery date and until all the required material is used up. Some schedules may also show prices delivered at the job.

Planning Overhead Every construction job will require more or less overhead. Before work is started on any particular job, some plan should be provided for handling the necessary overhead work for that job.

In the main office, the overhead work for the particular job may be assigned to certain clerks, draftsman, and stenographers under the general supervision of one of the executives of the company. They will prepare drawings; keep records; care for correspondence; check job reports; look after materials orders and transportation; trace shipments; check on payrolls, insurance, and social security; care for permits; etc. A reminder or tickler file is often useful. Plans must be made for inspecting and keeping records of the materials to be furnished and the work to be done by the subcontractors.

On the job, advance plans should be made in regard to the superintendent, general foreman, other general employees, records, reports, etc., as may be required for all necessary overhead work at the job. Provision must be made for keeping track of the work of the subcontractors on the job.

Planning Subcontracts On most jobs, all subcontracts should be let before work starts. The subcontractors will have full information given in their contracts as to the materials they are to furnish, the work they are to do, and other contractural relations they are to fulfill. The subcontractors should also know the approximate dates they are to start and finish their work, and some information should be given in regard to the rate of progress.

A time and work schedule should be prepared for all work to be done by subcontractors. This may be a separate schedule, or it may be included with the general job schedule. The time and work schedule for the subcontractors should show the kinds of work to be done, the names of the subcontractors, the dates for starting and finishing the work, and notes regarding the rate of progress and other details.

Planning the Management of the Job The general planning of the management of a job includes the bringing together and coordinating of all the various plans mentioned in the preceding articles of this chapter. The items considered are tentative or rough plan, visit to site, plant plan, labor plan, materials plan, overhead plan, and subcontract plan.

Many large contractors have planning departments whose duty is to plan the work. This department is usually kept separate from the executive or construction departments because they have found that better results are secured when the planning is separated from the doing.

The work of a planning department can be briefly outlined as follows:

1. Preliminary work plans (before actual work starts on the job)
   a. Work in main office
   b. Work in job office

2. Construction work plans (after actual work starts on the job)
   a. Work in main office
   b. Work in job office

It is the work of the planning department to prepare all time and work plans for the job in question. The plans are prepared after consultation with company executives and superintendents. The plans must be flexible so that they can be changed from time to time if necessary.

In addition to preparing the plans, the planning department keeps records as to the progress and costs of the work, suggests changes in plans when advisable, reports all discrepancies between plans and actual work to superiors, may make investigations as to causes of discrepancies between plans and actual work, and may make suggestions relating to improvement of the work, especially in regard to progress and costs.

When properly and carefully done, the work of the planning department is of great service to the contractor in giving him information in regard to the rate of progress of the different kinds of work on the job, and the costs of this work and in giving him an opportunity to correct errors in planning and execution and to regulate the progress of the work.

Time and Work Schedules On medium-sized and large jobs, it is advisable to provide time and work schedules for the convenience of the main office, the superintendent, and the foremen. Such a schedule lists the different construction operations, the estimated dates that each operation should start and finish, and the actual dates. Such a schedule, together with progress reports and charts (if the job is a large one), enables the contractor or engineer to note whether the work is progressing as planned and to observe which items are ahead or behind the schedule.

Certain construction operations and trades should follow each other in regular order and without interference. Confusion, with a resulting loss of efficient work, often occurs when two comparatively large gangs are scheduled to work on the same part of a job at the same time. Two small gangs may frequently work on the job at the same time (such as plumbers and electricians doing rough plumbing and wiring in a residence) without interference. In general, the paint gang should come last (except for priming-coat work) on any section of the job.

It is not necessary for any one operation to be wholly completed before another operation is started, but the work should be so planned that the different operations will not interfere with each other. For instance, on concrete-paving work, the excavation gang should be about a half a day or a day in advance of the roller and the concreting gang should follow about a day behind the roller. This allows the work to go forward efficiently, prevents interference between gangs, and permits the general superintendent or contractor to speed up any gang that is lagging.

On large jobs, delivery schedules are often provided for all the materials used, so that there will be no delays due to lack of materials and neither will there be a surplus of materials to clutter up the work. Labor schedules are sometimes made showing the numbers of each of the different classes of laborers required on the job for each day.

Time and work schedules usually show a list of the various work items, the estimated dates for starting and finishing, and the actual dates of starting and finishing. However, on some large jobs, like the erection of a large multistoried office building, for example, where there are many subcontracts and where plans are made separately for the work in the main office and the work on the job, the time and work schedules are more elaborate and detailed. The main-office schedule would show such dates as when architect's drawings are to be ready, subcontracts must be let, and all details must be ready for subcontractors. Some companies provide separate time and work schedules for the main office work, their work on the job, the work of subcontractors on the job, their materials, and their labor. The time and work schedules are usually prepared in tabular form, though a graphical form is sometimes used, especially when the time and work schedule is combined with a progress report.

Illustrative time and work schedules. A few illustrations of simple time and work schedules will be given in this article. The reader is referred for further information to texts specializing in construction methods and management.

One-course Paving Job:

TIME AND WORK SCHEDULE
TYPE OF WORK: One-course concrete pavement
LOCATION: Highway 147. SUPT. Conway

No.	Item	Estimated dates		Actual dates	
		Start	Finish	Start	Finish
1	Minor excavation				
2	Rolling subgrade				
3	Forming				
4	Concreting and finishing				
5	Curing				
6	Removing forms				
7	Cleaning up				
8	Finishing shoulders				
9	Completion, schedule time				
10	Completion, contract time				

Garage Building:

TIME AND WORK SCHEDULE

TYPE OF BUILDING: Garage

LOCATION: 2463 First St. SUPT. Hanson

No.	Item	Estimated dates		Actual dates	
		Start	Finish	Start	Finish
1	Excavation				
2	Make and erect forms				
3	Bend and place steel				
4	Mix and place concrete				
5	Remove forms				
6	Finish concrete surface				
7	Brick work				
8	Sash, frames, and trim				
9	Glass and glazing				
10	Roofing and flashing				
11	Plumbing				
12	Steam fitting (heating)				
13	Wiring				
14	Cleaning up				
15	Painting				
16	Schedule time of completion				
17	Contract time of completion				

CONTRACTOR'S JOB SCHEDULE

TYPE OF BLDG.: Steel-frame 7-story office

LOCATION: 211 Grand Ave. SUPT. Fowler

No.	Item	Estimated dates		Actual dates	
		Start	Finish	Start	Finish
1	Wrecking				
2	Excavation				
3	Drains and water				
4	Foundations				
5	Basement walls				
6	Steel erection				
7	Floors				
8	Stairs				
9	Roofing				
10	Walls, masonry				
11	Exterior doors and windows				
12	Partitions				
13	Lath and plaster				
14	Interior doors				
15	Finish flooring				
16	Other interior finish				
17	Cleaning up				
18	Etc.				
19	Schedule time for building				
20	Contract time for building				

SUBCONTRACTOR'S JOB SCHEDULE
TYPE OF BLDG.: Steel-frame 7-story office
LOCATION: 211 Grand Ave. SUPT. Fowler

No.	Item	Estimated dates		Actual dates	
		Start	Finish	Start	Finish
101	Elevators				
102	Plumbing, rough				
103	Plumbing, finish				
104	Heating and air conditioning, rough				
105	Heating and air conditioning, finish				
106	Electrical, rough				
107	Electrical, finish				
108	Sheet-metal work				
109	Painting and decorating				
110	Etc.				
111	Schedule time for building				
112	Contract time for building				

Multistoried Steel-frame Office Building. The time and work schedules for such a job might include contractors' office schedule, contractor's job schedule, subcontractor's job schedule, materials schedule, plant schedule, and a labor schedule. Only the contractors and subcontractor's job schedules are given in this article.

Managing the Job

Starting the Work All necessary preliminary work, such as discussed in the first part of this chapter, should preferably be completed before the actual work on the job is started. That is, the work will have been well planned in advance. The plant will have been designed and secured, labor supply will be arranged for, materials' orders will have been given and deliveries decided upon, all necessary permits will have been obtained, subcontracts will have been let (or be ready to let) and subcontractors informed as to their work, and all other preliminary work cared for.

A day or so before the work on the job is to start, the boss (contractor or superintendent) should visit the site again with a foreman and spend as much time as necessary checking over plans previously made and planning the actual work.

On the day the work is to start, the superintendent should be present with a foreman and a gang of men to assist in placing the construction equipment as it arrives; erect job office, tool shed, and other temporary buildings; see that telephone, light, and water are provided for; and care for supplies of materials as they arrive and need to be stored or piled. One or more copies of the plans and specifications for the job and the various time and work schedules should be kept on the job at all times.

Work in the Main Office The superintendent or foreman on the job cannot accomplish much unless the main office does its share of the work. At the main office, some one man should be responsible for each particular job. This man should preferably be an experienced officer of the company. He should keep track of all matters pertaining to this particular job and should know whether these matters relate to the superintendent, foremen, laborers, subcontractors, material men, owners, or architects. It is expected that he will consult other officials of the firm whenever advisable.

If the job is a large one, a superintendent should be placed in charge of the job and all correspondence relating to the job should pass through his hands. He should give all orders to foremen and material men and should be kept informed of what is going on even though he does not transact all the business relating to the job. The superintendent may divide his time between the main office and the job.

A full set of plans and specifications and copies of all time and work schedules should be kept in the main office at all times. If the job is a large one, an extra set or two of plans and specifications should be provided so that a set can be loaned to material men or subcontractors from time to time. It is usually advisable to keep the plans and specifications flat in a large shallow drawer. One or more drawers, properly labeled, should be provided for each job. Copies of all letters and memorandums relating to the job should be kept on file.

Separate contracts should have been drawn up with each material dealer and subcontractor, binding them to provide the materials and do the work as desired. Care should be exercised in setting the dates for providing materials and work so that there will be no delays and the work will be finished on or a little ahead of the date of completion. Schedules for material men and subcontractors should be kept on file, and these men should be checked frequently to see if they are living up to their contracts.

If special detailed drawings are to be provided from time to time by the architect, they should be provided when needed. It is often advisable to write the architect, calling attention to the drawings or data needed at specified dates and insisting that he provide them on time.

Work on the Job The office man in charge of the particular job should visit the job occasionally so that he can be informed of what is going on. The boss or superintendent (if he does not spend full time at the job) should visit the job preferably once or twice daily. He should spend enough time on the job with the foremen to see that the work is progressing favorably and that delays are avoided.

When the boss visits the job, it is a good plan for him to go over all parts of the job and observe everything that is being done by the company workmen and by the subcontractors.

It is a good plan for the superintendent to go over the work in detail each day with the foremen and discuss matters with them. The superintendent should give instruction to the foremen regarding the work, settle details, make suggestions regarding the work, and note things to be taken up with the owner or architect. The superintendent should be careful to avoid giving direct orders to the workmen, as this duty should be left to the foremen.

The superintendent should be on the lookout for improper and inefficient work at all times. He should be careful to observe if the materials on hand are sufficient and if material deliveries are satisfactory. It is advisable for the superintendent (and the foremen as well) to keep a daily notebook record, or diary, concerning the work.

The superintendent should keep the work progressing according to the planned time and work schedule as nearly as possible. He should consult with the main office regarding all changes in the time and work schedule.

The main office will undoubtedly require several daily and weekly reports in regard to the work. The superintendent should see that all required reports are prepared by himself, foreman, timekeepers, and clerks and sent to the main office.

When work is being done by subcontractors, the superintendent should see that they do their work in accordance with the plans and specifications and also in accordance with the time and work schedules. Discrepancies in materials and workmanship should be reported to the main office. If necessary, work should be stopped. The superintendent should make arrangements so that the subcontractors can proceed with their work in an efficient manner and without unnecessary interference from (or with) other workmen.

The superintendent should consult the architect or engineer frequently and secure needed instructions in advance concerning all doubtful details of the work. This procedure will avoid tearing down and doing over certain parts of the work at the contractor's expense.

The superintendent should be a good finisher as well as a good starter. He must be careful not to let his energy, persistence, and enthusiasm diminish as the work progresses. He must endeavor at all times to keep the work up to schedule in regard to time, quality, quantity, and costs. Most men can start a piece of work more easily than they can finish it.

Reports and Records Most large construction companies require that daily and weekly reports be sent from the job to the main office, describing in detail the work that has been done. The daily reports should give information about materials received, used, and on hand;

numbers and kinds of laborers and their work; amounts of different kinds of work done; work done by subcontractors; and any other information deemed of value. These reports may be prepared by the superintendent, foremen, timekeepers, or clerks as required and preferably signed by the superintendent.

In addition to the daily and weekly reports and records, a daily diary should be kept by the proper official in which all essentials relating to the particular job are noted. Photographs of the work may be taken from time to time and copies included in the diary.

The main office, by means of the daily reports and records, is enabled to keep in close touch with the job and to take any needed steps in regard to materials and labor. The daily reports should be checked by a competent official to provide against possible dishonesty and errors.

Weekly and monthly summary reports may be prepared as desired by using the information given in the daily reports and records.

Progress Reports and Charts So that the owner, contractor, architect, or engineer can clearly and easily visualize the amount of work done, the rate of progress of the job, and the cost of the work, progress charts are prepared and kept up to date. A progress chart is a graphical representation of the progress of the work on the job in question. These charts are prepared from the data given in the progress reports and must be kept up to date. A progress chart may be simple or complicated and may show a few or many details, as the case may be. The charts should not be too complicated or contain too much detail if they are to be read and understood by the average person.

Some companies prepare their progress charts in blueprint form with the estimated quantities of work and time shown as on the original tracing and the actual progress filled in each week by a colored crayon or full heavy line. When issued weekly, those charts are of value in indicating the progress of the work.

There are a great many kinds of progress charts. Almost every large construction company has its own type of charts. Most charts show progress of work in terms of time rates of different kinds of work. Some charts show comparisons between total estimated and actual costs, and other charts show unit costs. Only a few types of charts will be discussed in this chapter.

On certain jobs requiring large amounts of work of certain kinds and extending over comparatively long times, such as large excavation jobs or large concrete jobs, it is often advisable to prepare a unit-cost chart showing how the actual costs compare with the estimated costs. Figure 22-1 shows such a chart prepared for an excavating job. The horizontal dashed line shows the estimated cost, and the zigzag full line shows the actual costs which were computed at the end of each week and for that particular week. When the full line is *below* the dashed line, the actual costs are *less* than the estimated costs and the work is being done at a profit greater than that estimated. When the full line is *above* the dashed line, the actual costs are *more* than the estimated costs and the work is being done at a profit less than that estimated and perhaps at a loss.

An easily understandable form of progress chart is an outline sketch of the structure with the progress made each month (or week) shown by different colors. An example of this form of progress chart would be an elevation of a building or bridge with the progress made each month indicated by different colors.

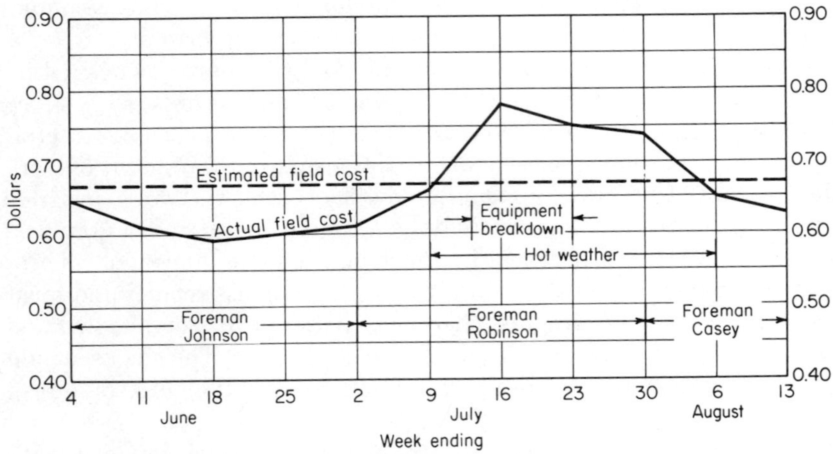

Figure 22-1 Estimated and actual unit costs.

A popular form of chart for showing the actual weekly progress of the work is one in which the various items of work are listed in the vertical left-hand column, with the allotted times extending horizontally to the right across the remainder of the sheet. Such a chart is easy to read and understand.

An example of this type of chart is shown in Fig. 22-2. This particular chart shows a time-work schedule for a concrete construction job involving excavation, assembling and erecting forms, bending and placing steel reinforcement, mixing and placing concrete, removing forms, and surface finishing. The cross-hatched lines show the estimated time, and the full heavy lines show the actual time required. Each of the horizontal work lines are divided into five parts as indicated by the short vertical lines. Each vertical line indicates 20 percent. Note that this chart is for the completed job. Such a chart can be readily lengthened and elaborated on to include as many operations as desired.

The work completed on this particular job up to June 12 is shown in Fig. 22-3. Note that in this progress chart, the form work, steelwork, and concrete mixing and placing are lagging a little behind schedule.

If desired, progress charts can be prepared showing costs instead of rates of doing work.

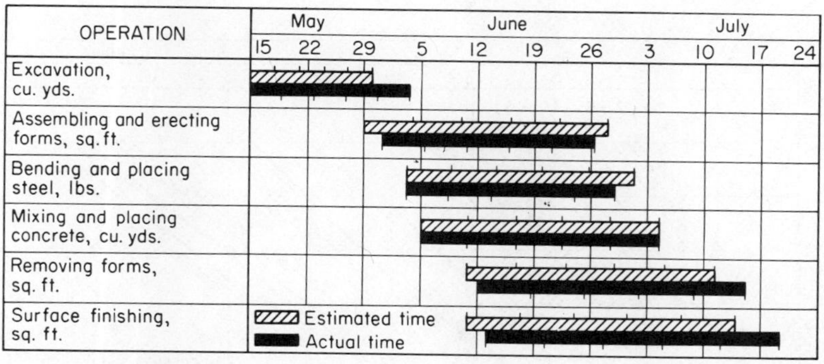

Figure 22-2 Time and work schedule and progress chart for a completed job.

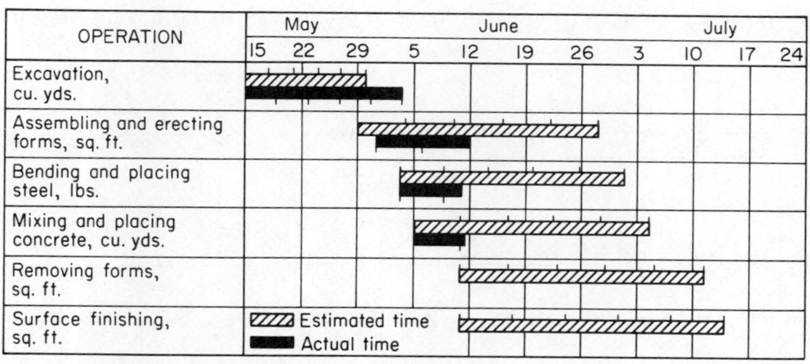

Figure 22-3 Time and work schedule and progress chart for a partially completed job.

The same data shown in Fig. 22-2 are shown in another manner in Fig. 22-4. In this chart, the light lines indicate the estimated times of doing the work and the actual times are shown by the heavy lines. Note that the actual excavation exceeded the estimated quantity (because of a bank cave-in) and the other actual quantities agreed with the estimated quantities. In regard to the actual time required, the excavation, form removal, and surface finishing each required a longer time than was estimated and the assembling and erection of forms and steel placing required less time. A chart of this type can be lengthened to include as many operations as desired.

Figure 22-5 shows the work completed on this particular job up to June 12. This chart shows the same data as is given in Fig. 22-3, but plotted in the same manner as in Fig. 22-4.

Another variation of the chart shown in Fig. 22-4 is to use a light black line for showing the relation between estimated time and quantity of work

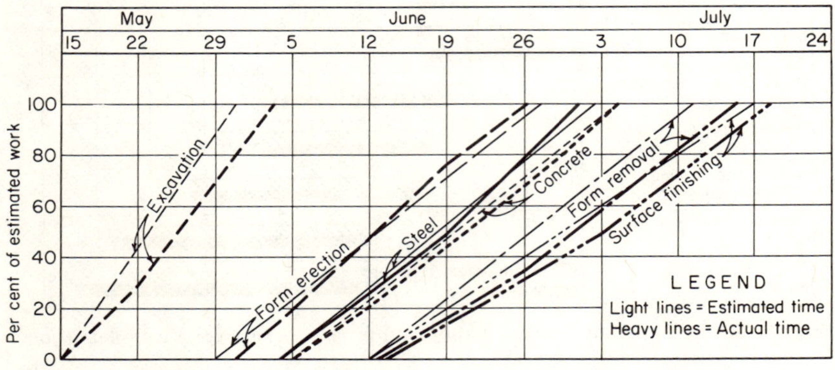

Figure 22-4 Time and work schedule and progress chart for a completed job.

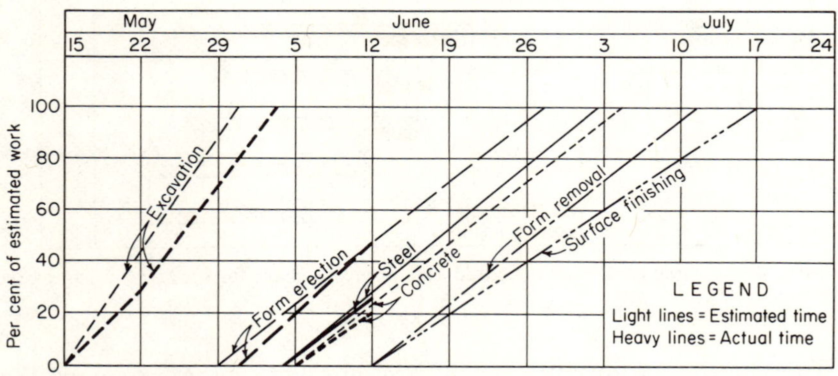

Figure 22-5 Time and work schedule and progress chart for a partially completed job.

in percentages, a heavy black line for showing actual time required, and a red line for costs. If, for any operation, the red (cost) line agrees with the heavy black line, the estimated and actual costs agree; if the red line is below the heavy black line, the costs are less than were estimated; and if the red line goes above the heavy black line, the costs are more than were estimated for the corresponding quantity of work. Note that the quantity of work and costs will be expressed in the form of a percentage of the estimated total of each. Actual quantities and costs can be written on the chart adjacent to the corresponding points on the lines if desired.

Managing a Small Construction Job The work of managing a small construction job, like a small store, a one-family residence, or a garage, need not be planned in so much detail as a large construction job. However, some general method of procedure is desirable.

The preliminary work after the contract is signed would include the signing of the subcontracts, the placing of material orders, the securing of all needed permits, and a general plan of construction.

For an example, the construction of a one-family wood-frame residence will be considered.

When work is started, the first step would be the staking out of the building according to the plans and specifications. Then the tool shed could be erected and the equipment for excavating could be brought to the job.

While the excavating was being done, the equipment for concrete work could be brought to the job and lumber for form work and concrete materials should be delivered.

After the basement walls have been formed and the concrete poured, lumber for framing should be delivered and the carpenter crew organized. The framing should start as soon as the concrete walls are strong enough to carry the loads and permit the removal of forms.

While the rough carpenter work is being completed, the rough work for heating, plumbing, and wiring can be started and completed when the carpenter crew is putting on the roof.

Then the lathing and plastering crews can do their work. At this time, the carpenter crew can do the exterior finish work. The concrete floor in the basement can be poured and finished. Sheet-metal work can be installed as convenient.

After the plastering work is completed, the building should be cleaned. Finished carpenter work can be started as soon as the plaster is dry enough. Finish plumbing and heating work can now be installed. Electrical finish work can be installed except light fixtures and outlet plates, which should wait until after the painting is completed.

At about this time, the yard can be graded and the required concrete walks and drives built.

When the floors are sanded, the painters can start work and work as rapidly as conditions permit. Light fixtures, outlet plates, and other miscellaneous finish work can now be done, and the building prepared for delivery to the owner. Needless to state, the premises should be cleaned up before final delivery to the owner.

The contractor on a small job will need to keep complete records in regard to construction equipment, materials, labor, and overhead costs and also some records in regard to the work done by subcontractors. Although the records should be complete, they need not be kept in so elaborate and detailed form as for a large construction job.

APPENDIX A

Reminder List

Many estimators use a reminder list so that they will not omit any important items when they prepare construction cost estimates. A separate list should be made for each kind of construction work such as a wooden-frame apartment building, a reinforced concrete office building, a structural-steel building, a steel highway bridge, an asphalt pavement, etc.

It is practically impossible to prepare a reminder list that is absolutely complete. Consequently, the estimator should not depend too much on any reminder list and should not neglect to study the plans and specifications carefully for small details, supplementary notes, and inconspicuous clauses.

The cost of a building, for example, is the sum of the costs of hundreds of items, both large and small. Some of these items may be overlooked by the estimator. A good reminder list will help him to avoid overlooking any important items.

The following reminder list is offered as an aid to the building estimator or contractor. This list has been compiled from lists used by the author and several building contractors and estimators.

Reminder List for Building Cost Estimates

1. ***General.*** See Chap. 19, Overhead and Profit, for more details.
   Main office expense: portion chargeable to the job.
   Job expenses, superintendent, timekeepers, clerks, etc.
   Temporary office building, tool and storage sheds, toilets.
   Telephone, light, heat, water, power.
   Office supplies, extra copies of plans and specifications.
   Permits: building, sidewalk, street, water, sewer, wrecking, other.
   Bonds: Check with list in Chap. 19.
   Insurance: Check with list in Chap. 19.
   Taxes: payroll, property, sales, excise, real estate, etc.
   Temporary enclosures, sidewalks, fences, ramps.
   Property protection: fire, watchmen, police, adjacent property.
   Cleaning rubbish, yard, floors, windows.
   Repair damage to adjacent property, streets, walks.
   Subcontractors, excavation, concrete, plumbing, heating and air conditioning, electrical, roofing, painting, glazing, sheet metal, etc.
   Contingencies: various.
   Profit: Be sure to include this item.
2. ***Equipment or Plant.*** See various chapters for equipment needed for different jobs.
   Costs of ownership, operation, transportation to and from the job, erection, moves, dismantling, rentals.
   Fuel, power, oil, grease, parts, minor repairs, supplies.
3. ***Excavation.*** See Chap. 3, Excavation.
   Wrecking old buildings, trees, shrubs: grubbing.
   General, special, backfill and tamping, boulders, rock, blasting, shoring, sheet piling, pumping, rough and fine grading.
4. ***Foundations.*** See Chap. 4, Piling and Bracing.
   Bearing piles, sheet piling, caissons, tests, borings, shoring, bracing, pumping, piers, wall footings, column footings, spread and raft footings, backfill, tamping.
   Pipes, water, gas, sewer, drains.
5. ***Concrete Forms.*** See Chap. 5, Concrete, the section on Forms.
   Wood, metal, special, shores, braces, ties, clamps.
   Footings, piers, wall, columns, beams, girders, floors, slabs, drop panels, spandrels, arches, domes, steps, platforms, special.
   Cleaning, oiling, wetting.
6. ***Concrete.*** See Chap. 5, Concrete, the sections on Concrete and Finishing.

Materials, cement, fine and coarse aggregates, water, admixtures, other.

Mixes: general and special, mortars, ready-mix concrete, pump-crete, cement gun.

Foundations, footings, piers, columns, beams, girders, slabs, floors, drop panels, walls, steps, cornice, lintels, water tables, copings, sills, walks, driveways, curbs, gutters, fills, other.

Special finishes: terrazzo, granolithic; grouting, bonding.

Curing, water, straw, canvas, cleaning, removing fins.

Cold-weather concreting, heating water and aggregates, curing, straw, canvas, salamanders, steam boiler and piping.

7. ***Reinforcing Steel.*** See Chap. 5, Concrete, the section on Reinforcement.

Sizes (bar numbers), specials, kind (grade), extras (size, quantity, milling, freight, trucking, testing, etc.), bends, hooks, spacers, chairs, ties, sleeves, wire mesh.

8. ***Masonry.*** See Chap. 6, Masonry.

Kind: brick, stone, block, tile, terra cotta, clay; cement, concrete.

Brick: common, face, fire, special, clay; cement, sand-lime.

Stone: rubble, random, range, coursed, ashlar, special cut stone, marble, limestone, sandstone, etc.

Block and tile: kind, sizes, special facing, special molding.

Terra cotta: kind, sizes, specials.

Mortars and mortar materials: admixtures, coloring.

Special work: lintels, water tables, coping, cornice, fireplaces, arches, pilasters, fireproofing, chimney flues, column and beam covering, platforms, steps, walks.

Anchors, dowels, bolts, ties.

9. ***Dampproofing and Waterproofing.*** See Chap. 7, Dampproofing and Waterproofing.

Foundations, floors, walls, external, interior.

Admixtures, paints, plaster, membranes, plastics, asphalt, tar, etc.

10. ***Wood Construction, Heavy.*** See Chap. 8, Wood Construction, the section on Heavy.

Lumber: kinds, grades, sizes, lengths, timbers, plank.

Columns, beams, girders, floors, roofs, cross bracing.

Hardware, column caps and bases, beam hangers and boxes, fish plates, tie rods, bolts, spikes.

Cleaning building and grounds.

11. ***Wood Construction, Light, Rough.*** See Chap. 8, Wood Construction, the section on Rough Work.

Framing, sills, plates, caps, studs, posts, beams, girders, joists, rafters, other.

Boarding, sheathing, flooring, roofing, insulation.

Walls, floors, roofs, partitions, dormers, skylights, gables.

Furring, grounds, bridging, caps, porches, steps, stoops.

Door and window frames, other.

Hardware, nails, bolts, hinges, plates, special, other.

Cleaning and removing rubbish.

12. ***Wood Construction, Light, Finish.*** See Chap. 8, Wood Construction, the section on Finish Work.

Exterior Finish: Siding, wall shingles, building paper, trim, molding, cornices, corner boards, water tables, porches, steps, stoops, skylights, dormer trim, exterior doors and windows, screens, storm doors and windows, blinds, shutters, exterior hardware.

Interior Finish: Flooring, paper, deadening felt, sanding, trim, base, chair and plate rails, picture molding, beam casing, wainscot, panels, quarter rounds, wood ceilings.

Stairs, stringers, treads, risers, quarter rounds, newels, rails, balusters, coves, etc.

Door and window frames, interior doors, porches, weather strip.

Cabinet work and millwork: cupboards, ironing boards, cabinets, linen closets, other closets, shelving, medicine cabinets, book cases, columns, mantels, breakfast sets, bars and stools, wardrobes, chests, drawers, blinds (venetian, etc.).

Interior finish hardware (for doors, windows, cabinets, etc.) such as hinges, bolts, lock sets, butts, bolts, stops.

Cleaning floors, removing rubbish.

13. ***Structural Steel.*** See Chap. 9, Structural Steel.

Sections, sizes, lengths, weights, quantities, specials.

Mill costs, size and quantity extras, other extras, shop drawings, fabrication, engineering, inspection, testing, shop painting, handling, freight, trucking, tracing, insurance.

Unloading and loading, erection, plumbing, riveting, bolting, welding, removing rust, painting.

Rivets, bolts, welding material, cross bracing, wind bracing, skylights, roofing, purlins, paint, clips, ties, plates, bars, rods.

Corrugated siding and roofing, steel sash and doors, glazing, hardware, miscellaneous items.

14. ***Ornamental and Miscellaneous Iron, Steel, and Bronze.***

Stairs, rails, treads, risers, column guards, wheel guards, fire escapes, platforms, ladders, lintels, doors and frames, chutes (mail, linen, garbage, etc.), bronze details, hardware.

15. ***Roofing and Flashing.*** See Chap. 10, Roofing and Flashing.

Paper, felt, composition roll, composition built-up, tile (clay, metal, etc.), canvas, shingles (wood, asphalt, tin, asbestos, slate, etc.), copper, aluminum, corrugated steel, flat or seamed steel, insulation, base, solder, rosin, charcoal.

Flashing, ridges, valleys, finials, nails, clips, miscellaneous.

16. Sheet Metal.

Flashing, cornice, valleys, finials, louvres, gutters, down spouts, ridges, ventilators skylights, ducts, chutes (linen, garbage, etc.), roofs, heating ducts (warm and cold air).

17. Lathing, Plastering, and Drywall Construction. See Chap. 11, Lathing, Plastering, and Drywall Construction.

Lathing: Wood, metal, plasterboard, etc.

Furring, grounds, corner bead, interior, exterior (stucco). Walls, ceilings, closets, partitions, special, etc.

Plastering: Exterior, interior, lime plaster, cement plaster, patent plaster, cement stucco, magnesite, special, molding, trim, etc.

Drywall: Gypsum wallboard, wood panels, tile, etc.

Furring, grounds, nailing strips, tape, paste, adhesive, nails, staples, clips.

18. Heating, Cooling, and Air Conditioning. See Chap. 12, Heating, Cooling, and Air Conditioning.

Steam and Hot Water: Pressure or vacuum, boiler, stoker, burner, oil tanks, smokestack, dampers, gauges, valves, pumps, accessories.

Piping: horizontal and vertical, insulation, air and other valves, joints, T's, elbows, hangers, unions, and other fittings.

Radiators, valves, air valves, floor plates, hangers, covers, etc.

Heat controls: gauges, pumps, thermostats, wiring, clocks.

Warm Air: Central or unit systems, "packaged deals," kind of fuel.

Central: furnace, stack, dampers, stoker, burner, tanks, fans, filters, humidifiers and dehumidifiers, warm- and cold-air ducts and risers, air washers, duct insulation, cold- and warm-air boxes and registers, accessories.

Thermostats, wiring, clocks, etc.

Unit: wall, floor, or ceiling, gas or electricity, piping, wiring, ducts, vents, thermostats, "packaged deals."

Air Cooling and Conditioning: Central or unit system. Separate or combined system. Several parts or "package deal." Water or refrigeration cooled. Boxes, fans, water pipes, electrical circuits, air filters and washers, humidifiers and dehumidifiers, ducts and pipes for conveying air, air outlets, inlets and returns, registers, grills, accessories, etc.

Thermostats, clocks wiring, and other controls.

Unit: room coolers, usually a "packaged deal," water coolers, refrigeration coolers, piping, wiring, thermostats (built in or separate).

19. **Plumbing.** See Chap. 13, Plumbing.

Rough Plumbing: Water, gas, and sewer lines; permits, trenches, meters, water pipes and fittings, gas pipes and fittings, sewer and soil pipes, vents, and fittings, caulking and caulking materials.

Fittings include joints, T's, elbows, hangers, traps, cleanouts, valves, plugs, tanks, etc.

Finish Plumbing: Laundry trays, water heaters, water softeners, gas stoves, laundry stoves, toilets, urinals, showers, bath tubs, foot baths, lavatories, kitchen sinks, shower stalls, garbage disposals, clothes washers and dryers, dishwashers, gas ranges, tanks, slop sinks, faucets, valves, traps, legs, and other accessories, medicine cabinets, shower doors or curtains.

20. **Electrical Work.** See Chap. 14, Electrical Work.

Rough: Wires (sizes, lengths, indoor, outdoor), main leads, circuit breakers, light and heavy conduit, armored and flexible cable, meters, small circuit breakers, fuse boxes, cutout boxes, transformers, switches (2-, 3-, 4-way), fixture outlets (floor, wall, and ceiling), outdoor outlets, plugs (floor, wall, ceiling), receptacles.

Bell wiring, transformers, outlets, TV antennas, wiring, outlets.

Special wiring for stoves, air conditioners, motors, etc.

Finish: Electric fixtures, all kinds; floor, ceiling, and wall plates; sockets, rosettes, drop cords, wall brackets, outdoor fixtures, time clocks.

Bell push buttons, buzzers, bells, chimes, etc., TV outlets.

Electric stoves, air conditioners (coolers), heaters, motors, fans, and other apparatus.

21. **Painting.** See Chap. 15, Painting, Papering, and Glazing.

Exterior: Priming coat, number of other coats, kind of paint, walls, siding, doors, windows, cornice, trim, dormers, gables, coves, water tables, shingles, ironwork, special, others.

Interior: Woodwork priming and other coats of filler, shellac, varnish, flat and gloss paint, enamel, rubbing, polishing.

Door, windows, mantels, cupboards, stairs, bookcases, other.

Walls, ceilings, sizing, number of coats, flat or gloss paint.

Floors, sanding, filler, shellac, varnish, number of coats, special floor finishes.

Radiators, radiator covers, registers, etc.

Special decorations and murals, ironwork.

22. **Papering.** See Chap. 15, Painting, Papering, and Glazing.

Sizing, paper, smoothing and cleaning walls and ceiling, removing old paper.

Special materials and work.

23. Glass and Glazing. See Chap. 15, Painting, Papering, and Glazing.

Glass sizes, kind (single or double strength, plate, wire, corrugated, prismatic, mirrors, special), putty.

Glazing windows, doors, skylights, wood or steel sash, caulking.

24. Fireproofing.

Columns, beams, girders, walls, floors, drop panels, doors, window shutters, other.

25. Sprinklers.

Pumps, tanks, pipes and connections, sprinkler heads, valves and other fittings, standpipes, fire-hose connections, etc.

26. Elevators.

Passenger, freight, sidewalk, etc.; number, kind, capacity.

Special requirements in regard to controls, doors, shafts, etc.

27. Special Finishes.

Marble, special tile, terrazzo, granolithic, magnesite, cork, rubber, inlay, plastic, ceramic, carving, etc.

28. Miscellaneous.

Flagpoles, signs, pneumatic tubes, mail chutes, vacuum cleaning systems, vaults, vault doors, time clocks, venetian blinds, window shades, engines, motors, generators, large blowers, awnings, etc.

APPENDIX B

Equipment Ownership and Operating Costs

The cost of owning and operating construction equipment is often a large and important factor in the cost of construction. As the use of mechanized and labor-saving equipment increases, equipment costs will become an increasingly larger factor in the cost of construction.

No attempt is made in this text to cover all kinds, types, and sizes of construction equipment. For more complete information, the reader is referred to publications specializing in the selection, uses, and costs of construction equipment.

Please note that the costs given in this appendix are *approximate* values which may or may not apply to any one contractor and his equipment. The contractor should use his own data when available.

Ownership Costs In this appendix, approximate ownership costs of equipment per working hour have been computed by a simple method that is sufficiently accurate for estimating purposes. The approximate ownership costs of equipment per working hour are expressed as a percentage of the capital investment, or first cost. The following assumptions were made:

The equipment has no salvage or scrap value.

Total depreciation equals 100 percent of the capital investment.

Total costs of the major repairs and overhaul are expressed as a percentage of the capital investment. These percentages are approximate and were obtained from various sources.

Total interest, insurance, taxes, and storage costs are expressed as a percentage of the capital investment. The storage cost is assumed at 3 percent of the capital investment per year. Interest, insurance, and taxes are assumed at 10 percent of the current value (capital investment less depreciation computed by the straight-line method) per year.

The total percentages for interest, insurance, taxes, and storage are computed by the formula

$$\text{Total percentage} = \tfrac{1}{2}(n + 1)10 + 3n$$

where n is the average life of equipment in years.

For example, for a 5-year life, the total percentage is

$$\tfrac{1}{2}(5 + 1)10 + 3 \times 5 = 30 + 15 = 45\%$$

Or, by using arithmetic, this total percentage is

$$(1.00 + 0.80 + 0.60 + 0.40 + 0.20)10 + 3 \times 5 = 30 + 15 = 45\%$$

Illustrative estimate on ownership costs. Estimate the ownership costs per working hour for a 1-cu-yd diesel crawler-type shovel, assuming the following:

Capital investment = \$26,300
Total working hours = 9,000 (in 5 years)
Total depreciation = 100% of capital investment
Total major repairs and overhaul = 70% of capital investment
Total interest, insurance, taxes, and storage for 5 years = 45% of capital investment

Total ownership cost per working hour

$$= \frac{\$26{,}300 \times (100 + 70 + 45)}{9{,}000 \times 100} = \$6.30$$

This cost can be checked by Diagrams 18-1 and 18-3.

Operating Costs The operating costs per working hour include only the costs of fuel, power, oil, grease, supplies, small parts, and minor (or field) repairs. The operating costs do not include any ownership costs; transportation, erection, and dismantling costs; wages of operators; general overhead; or profit. The operating costs given in this appendix are approximate. The contractor or estimator should use his own cost data when available.

The operating costs per working hour for power-operated equipment are computed as explained in Chap. 18, Construction Plant and Equipment, assuming the following:

Equipment will operate at an average of 70 percent of rated capacity. A gasoline engine will use 0.065 gal of gasoline per brake horsepower-hour. Gasoline costs 27 cents per gallon (including tax) at the job. Other items will cost 0.5 cent per brake horsepower-hour. The total operating cost per brake horsepower-hour will be

$$(\$0.27 \times 0.065 + \$0.005) \times 70\% = \$0.015785 \text{ (say } \$0.016)$$

A diesel engine will use 0.045 gal of diesel fuel oil per brake horsepower-hour. Diesel fuel oil costs 21 cents per gallon (including tax) at the job. Other items will cost 0.4 cent per brake horsepower-hour. The total operating cost per brake horsepower-hour will be

$$(\$0.21 \times 0.045 + \$0.004) \times 70\% = \$0.009415 \text{ (say } \$0.0095)$$

An electric motor, operating at about 75 percent efficiency, will require 1 kwhr of electricity for 1 bhp-hr. Electric power will cost 2.25 cents per kilowatthour at the job. Other items will cost 0.3 cent per brake horsepower-hour. The total cost per brake horsepower-hour will be

$$(\$0.0225 + \$0.003) \times 70\% = \$0.01785 \text{ (say } \$0.018)$$

Diagrams 18-6, 18-7, and 18-8 can be used for estimating or checking operating costs of power-driven equipment.

In some instances, operating costs per working hour are given for some kinds of equipment which have no direct power units. These costs include oil, grease, supplies, small parts, and minor repairs.

Illustrative estimate on operating costs. Estimate the operating cost per working hour for a 1-cu-yd diesel crawler shovel with an engine of 100-hp rated capacity.

$$\text{Cost per working hour} = \$0.0095 \times 100 = \$0.95$$

Diagram 18-7 gives a cost per working hour of about 95 cents.

Abbreviations. The following abbreviations are used in the tables of this appendix.

ac	alternating current	in.	inch
app	approximate	ins	insurance
ave	average	int	interest
bbl	barrel	kw	kilowatt
bhp	brake horsepower	kwhr	kilowatthour
cfm	cubic feet per minute	lb	pounds
comp'r	compressor	maj	major
cu ft	cubic feet	mat'l	materials
cu yd	cubic yards	min	minute
dc	direct current	ovhl	overhaul
dep	depreciation	own	ownership
dsl	diesel	psi	pounds per square inch
elec	electric	pwr	power
eng.	engine	rpr	repairs
ft	foot or feet	sq ft	square foot
ft-lb	foot-pound	sto	storage
gal	gallon	tax	taxes
gas	gasoline	tot	total
hp	horsepower	wkhr	working hour
hp-hr	horsepower-hour	yr	years
hr	hour		

Approximate Equipment-ownership and -operating Costs

Equipment	App life		App tot cost, percent of capital investment					Operating cost per wkhr
	Yr	Wkhr	Tot dep	Tot maj rpr ovhl	Tot int ins tax sto	Tot own cost	Own cost per wkhr	
Air compressor, 100 psi:								
Gas, portable:								
60–185 cfm	4	5,000	100	60	37	197	0.0394	$ 0.30–$ 0.90
200–350 cfm	5	6,000	100	70	45	215	0.0358	$ 1.00–$ 1.90
Dsl, portable:								
105–185 cfm	4	5,000	100	60	37	197	0.0394	$ 0.25–$ 0.50
200–650 cfm	5	6,000	100	70	45	215	0.0358	$ 0.55–$ 1.40
Air tools:								
Drills:								
Drifters	4	5,000	100	40	37	177	0.0354	
Wagon	7	10,000	100	70	61	231	0.0231	
Grinders	3	3,500	100	30	29	159	0.0455	
Hammers, all kinds	3	3,600	100	30	29	159	0.0442	
Hose and couplings	2	3,000	100	20	21	141	0.0470	
Spades, tampers, vibrators	3	4,000	100	30	29	159	0.0398	
Automobiles:								
Light	2	4,500	100	35	21	156	0.0347	$ 0.65–$ 0.85
Medium	3	6,000	100	45	29	174	0.0290	$ 0.90–$ 1.10
Heavy	4	7,500	100	55	37	192	0.0256	$ 1.10–$ 1.30
Bituminous equipment:								
Brooms	5	7,000	100	85	45	230	0.0329	$ 0.30–$ 0.45
Distributors, no truck, 600- to 2,000-gal tank	5	7,000	100	85	45	230	0.0329	$ 1.30–$ 1.50
Distributors with truck, 600- to 1,500-gal tank	5	7,000	100	85	45	230	0.0329	$ 2.60–$ 3.25
Mixing plant, gas, dsl, or elec:								
Batcher, 15–150 ton per hr, 500- to 5,000-lb pugmill	6	6,500	100	85	53	238	0.0366	$ 7.00–$50.00
Continuous, 15–200 ton per hr	6	6,500	100	85	53	238	0.0366	$ 7.00–$65.00
Traveling crawler, gas or dsl, 1–2 cu yd per min, 100–200 tons per hr	5	6,000	100	85	45	230	0.0383	$40.00–$70.00
Heater:								
Tank car or truck	5	6,500	100	75	45	220	0.0399	$ 0.20–$ 0.55

Approximate Equipment-ownership and -operating Costs (CONTINUED)

Equipment	App life		App tot cost, percent of capital investment					Operating cost per wkhr
	Yr	Wkhr	Tot dep	Tot maj rpr ovhl	Tot int ins tax sto	Tot own cost	Own cost per wkhr	
Bituminous equipment: (*Contd.*)								
Surface, 8–24 sq ft	5	6,500	100	75	45	220	0.0339	\$ 0.15–\$ 0.35
Kettles, portable, oil burner, 50–600 gal	6	7,000	100	65	53	218	0.0311	\$ 0.20–\$ 0.55
Pavers, spreaders, finishers	4	6,500	100	60	37	197	0.0303	\$ 0.45–\$ 0.75
Boilers, steam, 150 psi:								
Locomotive, 25–200 hp	10	12,000	100	100	85	285	0.0238	\$ 0.25–\$ 2.00
Vertical, 20–100 hp	7	9,000	100	80	61	241	0.0268	\$ 0.20–\$ 1.00
Buckets:								
Clamshell, ¼–4 cu yd	5	6,000	100	80	45	225	0.0375	
Concrete bottom dump ½–4 cu yd	5	6,000	100	75	45	220	0.0367	
Dragline:								
⅜–1½ cu yd	4	5,500	100	60	37	197	0.0358	
2–15 cu yd	5	6,000	100	70	45	215	0.0358	
Turnover, ball and pivot, ⅜–2 cu yd	5	6,000	100	90	45	235	0.0392	
Buildings, portable	5	8,000	100	75	45	220	0.0275	
Communicating systems, radio	4	5,500	100	45	37	182	0.0331	
Concrete equipment:								
Bar benders and cutters	5	8,000	100	70	45	215	0.0268	
Batcher, portable	5	6,000	100	65	37	202	0.0342	
Batcher and weigher, 1–4 mat'l, to 1- to 4-cu-yd batch	4	5,000	100	65	37	202	0.0404	
Bins: steel, aggregate:								
Portable, 8–70 cu yd	5	6,000	100	75	45	220	0.0367	
Knockdown, 50–200 cu yd	6	7,000	100	85	53	238	0.0340	
Bins, steel, cement:								
Portable, 100–300 bbl	5	6,000	100	75	45	220	0.0367	
Knockdown, 250–1,000 bbl	6	7,000	100	85	53	238	0.0340	
Carts:								
Steel wheels, 6–11 cu ft	3	4,500	100	60	29	189	0.0420	
Rubber tires, 6–12 cu ft	3	4,500	100	75	29	204	0.0454	

Approximate Equipment-ownership and -operating Costs (CONTINUED)

Equipment	App life		App tot cost, percent of capital investment					Operating cost per wkhr
	Yr	Wkhr	Tot dep	Tot maj rpr ovhl	Tot int ins tax sto	Tot own cost	Own cost per wkhr	
Concrete equipment: (*Contd.*)								
Power driven, 6–15 cu ft	3	4,500	100	75	29	204	0.0454	\$ 0.10–\$ 0.15
Cement gun machines:								
With nozzle and hose only	4	6,500	100	60	37	197	0.0303	
With gas engine and comp'r	4	6,500	100	60	37	197	0.0303	\$ 0.75–\$ 1.00
Column clamps	5	10,000	100	50	45	195	0.0195	
Concrete buckets:								
Bottom dump, ½–4 cu yd	5	6,000	100	75	45	220	0.0367	
Tower hoist, ⅓–4 cu yd	5	6,000	100	75	45	220	0.0367	
Concrete hoppers, floor:								
Single gate, ½–3 cu yd	5	7,000	100	100	45	245	0.0350	
Double gate, 2–5 cu yd	5	7,000	100	100	45	245	0.0350	
Floor shares, adjustable	4	6,000	100	65	37	202	0.0337	
Forms, steel, all kinds	4	7,000	100	80	37	217	0.0310	
Mixers, construction:								
Gas engine:								
3½S–6S	2	4,000	100	24	21	145	0.0363	\$ 0.07–\$ 0.14
11S–16S	3	4,500	100	36	29	165	0.0367	\$ 0.25–\$ 0.30
28S, 56S, 84S	4	5,000	100	48	37	185	0.0370	\$ 0.50–\$ 1.40
Elec motor:								
3½S–6S	4	6,000	100	48	37	185	0.0308	\$ 0.08–\$ 0.16
11S–84S	5	6,000	100	60	45	205	0.0342	\$ 0.55–\$ 1.60
Mixers, paving:								
Gas, wheel or crawler, 10E–40E	4	5,500	100	60	37	197	0.0358	\$ 1.05–\$ 1.75
Dsl, crawler, single or double, 27E–34E	5	6,500	100	65	45	210	0.0323	\$ 0.55–\$ 0.90
Mixers, truck type, gas eng.:								
1–2 cu yd	3	6,500	100	50	29	179	0.0276	\$ 1.30–\$ 2.00
3–5½ cu yd	4	7,500	100	60	37	197	0.0263	\$ 2.25–\$ 3.00
Mixer agitators, truck type, gas eng.:								
1½–2 cu yd	3	6,500	100	50	29	179	0.0276	\$ 1.30–\$ 2.00
3–6½ cu yd	4	7,500	100	60	37	197	0.0263	\$ 2.25–\$ 3.00

Approximate Equipment-ownership and -operating Costs (CONTINUED)

Equipment	App life		App tot cost, percent of capital investment					
	Yr	Wkhr	Tot dep	Tot maj rpr ovhl	Tot int ins tax sto	Tot own cost	Own cost per wkhr	Operating cost per wkhr
Concrete equipment: (*Contd.*)								
Mixers:								
Mortar and plaster, gas eng., 4–13 cu ft.	3	5,000	100	36	29	165	0.0330	$ 0.03–$ 0.10
Pugmill:								
Gas, 1,000- to 4,000-lb batch.........	4	5,000	100	60	37	197	0.0394	$ 0.13–$ 0.55
Dsl, 1,000- to 5,000-lb batch.........	4	5,000	100	60	37	197	0.0394	$ 0.10–$ 0.50
Pumps, concrete on skids with remixer and 1,000-ft pipe:								
Gas, single, 20–35 cu yd per hr..........	4	5,500	100	60	37	197	0.0358	$ 0.50–$ 1.00
Gas, double, 40–65 cu yd per hr..........	5	6,500	100	70	45	215	0.0331	$ 1.00–$ 1.35
Screeds, vibrating, gas pwr.................	4	5,500	100	60	37	197	0.0358	$ 0.04–$ 0.06
Spouting plant complete, without pump.......	4	6,000	100	50	37	187	0.0311	
Vibrators:								
Gas, 2–6 hp..........	3	4,500	100	30	29	159	0.0354	$ 0.04–$ 0.12
Electric, 2–6 hp......	3	4,500	100	30	29	159	0.0354	$ 0.05–$ 0.15
Washer, gravel, with screen, no power.....	5	6,000	100	70	45	215	0.0358	
Conveyors and loaders:								
Belt, 18–30 in.:								
Portable, no power...	4	5,000	100	65	37	202	0.0404	
Portable, with power.	4	5,000	100	75	37	212	0.0424	$ 0.05–$ 0.20
Stationary, no power.	6	7,000	100	85	53	238	0.0340	
Bucket loader, gas, 60–300 cu yd per hr.....	5	7,000	100	70	45	215	0.0307	$ 0.45–$ 2.20
Cranes:								
Crawler, gas power:								
4–12 tons, 10–12 ft radius.............	5	8,000	100	50	45	195	0.0244	$ 0.70–$ 1.35
12–40 tons, 12–20 ft radius.............	6	8,500	100	60	53	213	0.0251	$ 1.35–$ 2.10

Approximate Equipment-ownership and -operating Costs (CONTINUED)

Equipment	App life		App tot cost, percent of capital investment					Operating cost per wkhr
	Yr	Wkhr	Tot dep	Tot maj rpr ovhl	Tot int ins tax sto	Tot own cost	Own cost per wkhr	
Cranes: (*Contd.*)								
Crawler, dsl power:								
4–12 tons, 10–12 ft radius	5	8,000	100	50	45	195	0.0244	$ 0.35–$ 0.80
12–40 tons, 12–20 ft radius	6	8,500	100	60	53	213	0.0251	$ 0.80–$ 1.25
Truck, gas power, 4–25 tons, 10 ft radius	5	8,000	100	75	45	220	0.0275	$ 1.00–$ 2.25
Truck, dsl power, 4–30 tons, 10 ft radius	5	8,000	100	65	45	210	0.0263	$ 0.75–$ 1.65
Crushers, no power, all sizes:								
Cone or gyratory	10	12,000	100	100	85	285	0.0238	
Jaw or roller	6	8,000	100	70	53	223	0.0279	
Derricks, with winch, ropes, and load block, no power:								
Circle, swing, hand, ½–1½ tons	8	10,000	100	120	69	289	0.0289	
Guy, steel, 3–50 tons, 50- to 120-ft boom	12	16,000	100	140	101	341	0.0213	
Stiff-leg steel with or without sills, 5–50 tons, 50- to 100-ft boom	12	16,000	100	175	101	376	0.0235	
Stiff-leg wood, 1–25 tons	8	10,000	100	150	69	319	0.0319	
Draglines and excavators:								
Crawler with rope and bucket:								
Gas:								
½–¾ cu yd, 35-ft boom	4	8,000	100	65	37	202	0.0253	$ 1.20–$ 1.45
1–1½ cu yd, 40- to 45-ft boom	5	9,000	100	75	45	220	0.0244	$ 1.55–$ 2.00
1¾–2½ cu yd, 45- to 60-ft boom	6	10,000	100	85	53	238	0.0238	$ 2.40–$ 3.00
Dsl:								
½–¾ cu yd, 35-ft boom	4	8,000	100	65	37	202	0.0253	$ 0.65–$ 0.85

Approximate Equipment-ownership and -operating Costs (CONTINUED)

Equipment	App life: Yr	App life: Wkhr	App tot cost, percent of capital investment: Tot dep	Tot maj rpr ovhl	Tot int ins tax sto	Tot own cost	Own cost per wkhr	Operating cost per wkhr
Draglines and excavators: (*Contd.*)								
1–1½ cu yd, 40- to 45-ft boom.......	5	9,000	100	75	45	220	0.0244	$ 0.90–$ 1.25
1¾–4 cu yd, 50- to 75-ft boom......	6	10,000	100	85	53	238	0.0238	$ 1.50–$ 2.80
Electric:								
½–¾ cu yd, 35-ft boom...........	6	10,000	100	70	53	223	0.0223	$ 1.30–$ 1.60
1–1¾ cu yd, 40- to 50-ft boom......	8	12,000	100	90	69	259	0.0216	$ 1.70–$ 2.40
2–7 cu yd, 50- to 90-ft boom......	10	14,000	100	110	85	295	0.0211	$ 2.50–$ 7.00
Cable type with power, slack line, up to 500-ft span ½–4 cu yd......	5	7,000	100	75	45	220	0.0314	$ 0.75–$ 3.30
Drag scrapers with power, 300- to 400-ft span, ½–4 cu yd.....	5	7,000	100	75	45	220	0.0314	$ 0.75–$ 3.30
Drills:								
Augur, with power.....	3	4,500	100	30	29	159	0.0353	$ 0.06–$ 0.20
Core with power, 1–6½ in..................	3	4,500	100	30	29	159	0.0353	$ 0.06–$ 0.25
Elec generators and motors:								
Generators with power, 1–100 kw............	10	12,000	100	110	85	295	0.0246	$ 0.05–$ 2.50
Motors, 5–250 hp.......	10	12,000	100	90	85	275	0.0229	$ 0.13–$ 6.50
Elec-light plants, ac or dc:								
Gas power, 1–100 kw...	7	8,500	100	85	61	246	0.0290	$ 0.05–$ 2.50
Dsl power, 1–100 kw...	7	8,500	100	85	61	246	0.0290	$ 0.04–$ 1.70
Engines with clutch:								
Gas:								
1–25 hp............	6	8,000	100	80	53	233	0.0291	$0.023 per bhp-hr (1–25 hp and 25–300 hp)
25–300 hp...........	8	10,000	100	90	69	259	0.0259	
Dsl:								
10–25 hp............	6	8,000	100	75	53	228	0.0285	$0.0135 per bhp-hr (10–25, 30–150 and 175–500 hp)
30–150 hp...........	8	10,000	100	90	69	259	0.0259	
175–500 hp..........	10	12,000	100	100	85	285	0.0238	

Approximate Equipment-ownership and -operating Costs (CONTINUED)

Equipment	App life		App tot cost, percent of capital investment					
	Yr	Wkhr	Tot dep	Tot maj rpr ovhl	Tot int ins tax sto	Tot own cost	Own cost per wkhr	Operating cost per wkhr
Graders:								
Elevating, 36- to 48-in. belt	5	7,000	100	90	45	235	0.0336	$ 0.20–$ 0.50
Gas, 3–12 tons, 8- to 10-ft blade	5	8,000	100	75	45	220	0.0275	$ 0.50–$ 1.70
Dsl, 4–15 tons, 9- to 12-ft blade	5	8,000	100	75	45	220	0.0275	$ 0.40–$ 1.10
Towed, 6- to 12-ft blade	5	8,000	100	90	45	235	0.0294	
Hammer, drop and pile:								
Drop hammers, all sizes	6	8,000	100	75	53	228	0.0285	
Leads, all sizes	6	8,000	100	75	53	228	0.0285	
Pile hammers, air or steam:								
Single acting:								
3,000–8,000 ft-lb	5	7,000	100	60	45	205	0.0293	
10,000 ft-lb and over	6	7,500	100	70	53	223	0.0297	
Double acting:								
100–2,000 ft-lb	4	6,000	100	60	37	197	0.0328	
2,000–8,000 ft-lb	5	7,000	100	65	45	210	0.0300	
10,000 ft-lb and over	6	7,500	100	70	53	223	0.0297	
Differential acting:								
3,500–8,000 ft-lb	5	7,000	100	65	45	210	0.0300	
10,000 ft-lb and over	6	7,500	100	70	53	223	0.0297	
Dsl, all sizes	5	7,000	100	80	45	225	0.0321	$ 0.23–$ 0.85
Hoists:								
Hand hoist, 1–12 tons	8	12,000	100	75	69	244	0.0203	
Gas:								
One drum, 4–100 hp	5	8,000	100	70	45	215	0.0269	$ 0.07–$ 1.60
Two drums, 25–100 hp	6	9,000	100	60	53	213	0.0237	$ 0.40–$ 1.65
Three drums, 40–100 hp	6	9,000	100	60	53	213	0.0237	$ 0.64–$ 1.65
Elec:								
One-drum, 4–100 hp	6	9,000	100	65	53	218	0.0242	$ 0.08–$ 1.80
Two-drum, 20–100 hp	7	10,000	100	70	61	231	0.0231	$ 0.36–$ 1.85
Three-drum 40–100 hp	8	11,000	100	75	69	244	0.0222	$ 0.72–$ 1.85
Hoppers:								
Floor, 1 or 2 gates, with or without legs, ½–2½ cu yd	5	8,000	100	90	45	235	0.0294	

Approximate Equipment-ownership and -operating Costs (CONTINUED)

Equipment	App life		App tot cost, percent of capital investment					
	Yr	Wkhr	Tot dep	Tot maj rpr ovhl	Tot int ins tax sto	Tot own cost	Own cost per wkhr	Operating cost per wkhr
Hoppers: (*Contd.*)								
Tower, ½–3 cu yd......	5	8,000	100	90	45	235	0.0294	
Truck mixer, with legs 3–4 cu yd............	4	6,500	100	50	37	187	0.0288	
Hose, all kinds...........	2	3,000	100	20	21	141	0.0470	
Jacks:								
Hydraulic, 5–100 tons..	8	10,000	100	120	69	289	0.0289	
Screw and ratchet, 10–25 tons.................	5	8,000	100	25	45	170	0.0213	
Trench, 12–24 in.......	5	8,000	100	60	45	205	0.0256	
Loaders. See conveyors and tractor attachments								
Pipe and connections, black and galvanized, 1–4 in.................	4	5,000	100	20	37	157	0.0314	
Pit and quarry equipment, portable, with power, bins, conveyors, crushers, screens, etc.								
Crushing, screening and loading:								
One small crusher......	4	5,000	100	60	37	197	0.0394	No data
One or two large crushers..................	6	7,000	100	70	53	223	0.0319	No data
Plows and scarifiers:								
Disk, furrow, and rooter plows..............	4	6,000	100	80	37	217	0.0362	
Scarifier attachment....	4	6,000	100	80	37	217	0.0362	
Post drivers, hole diggers, and pullers...........	3	4,000	100	40	29	169	0.0423	
Power tools:								
Elec small tools........	3	5,000	100	30	29	159	0.0318	$ 0.02–$ 0.06
Elec table saws.........	4	6,000	100	40	37	177	0.0295	$ 0.03–$ 0.08
Gas table saws, 10- to 20-in. blade.........	4	6,000	100	60	37	197	0.0328	$ 0.03–$ 0.07
Gas chain saws, 18- to 84-in. cut............	3	4,500	100	45	29	174	0.0387	$ 0.06–$ 0.15
Pumps, centrifugal, portable:								

Approximate Equipment-ownership and -operating Costs (CONTINUED)

Equipment	App life		App tot cost, percent of capital investment					Operating cost per wkhr
	Yr	Wkhr	Tot dep	Tot maj rpr ovhl	Tot int ins tax sto	Tot own cost	Own cost per wkhr	
Pumps: (*Contd.*)								
Gas, 1½–10 in., 4,000–240,000 gph	5	6,000	100	90	45	235	0.0392	$ 0.04–$ 1.85
Elec, 1–8 in. 2,500–150,000 gph	6	7,000	100	90	53	243	0.0347	$ 0.06–$ 2.50
Riveting equipment:								
Compressors. See air compressors								
Hammers. See air tools								
Forges	3	5,000	100	45	29	174	0.0348	$ 0.25–$ 0.75
Rollers:								
Hand, steel, 250–1,000 lb	5	8,000	100	60	45	205	0.0256	
Sheep's foot, 4–8 ft wide	4	7,000	100	60	37	197	0.0281	
Tamping, rubber tires, 9–13 wheels	3	5,000	100	90	29	219	0.0438	
Tandem:								
Gas, 2-21 tons	7	12,000	100	85	61	246	0.0205	$ 0.30–$ 1.60
Dsl, 2–21 tons	7	12,000	100	85	61	246	0.0205	$ 0.20–$ 0.75
Three-wheel:								
Gas, 5–13 tons	7	12,000	100	85	61	246	0.0205	$ 0.70–$ 1.40
Dsl, 5–13 tons	7	12,000	100	85	61	246	0.0205	$ 0.35–$ 0.75
Rope:								
Manila, ½–1½ in	2	3,000	100	40	21	161	0.0537	
Wire:								
½–1 in	2	3,500	100	40	21	161	0.0460	
1¼–2¼ in	3	4,000	100	50	29	179	0.0448	
Salamanders, coal or oil	5	6,000	100	70	45	215	0.0358	
Scales:								
Truck, 10–50 tons	8	11,000	100	75	69	244	0.0222	
Wheelbarrow and cart	8	11,000	100	75	69	244	0.0222	
Scrapers:								
Slip, 3–5 cu ft	2	3,000	100	30	21	151	0.0503	
Wheel, 6–40 cu ft	4	5,000	100	60	37	197	0.0394	
Drag, 20–70 cu ft	4	5,000	100	50	37	187	0.0374	
Self-load, no power, rubber tires, 2 or 4 wheels, 2–36 cu yd struck, 3–44 cu yd heaped	5	9,000	100	70	45	215	0.0239	$ 0.20–$ 1.00

Approximate Equipment-ownership and -operating Costs (CONTINUED)

Equipment	App life		App tot cost, percent of capital investment					Operating cost per wkhr
	Yr	Wkhr	Tot dep	Tot maj rpr ovhl	Tot int ins tax sto	Tot own cost	Own cost per wkhr	
Shovels (and backhoes):								
Crawler:								
Gas:								
⅜, ½, ¾ cu yd....	4	8,000	100	60	37	197	0.0246	$ 1.15–$ 1.45
1, 1¼ cu yd.......	5	9,000	100	75	45	220	0.0244	$ 1.55–$ 1.80
1½, 2, 2½ cu yd...	6	10,000	100	85	53	238	0.0238	$ 2.00–$ 2.65
Dsl:								
⅜, ½, ¾ cu yd....	4	8,000	100	60	37	197	0.0246	$ 0.55–$ 0.85
1, 1¼ cu yd.......	5	9,000	100	75	45	220	0.0244	$ 0.85–$ 1.05
1½, 2, 2½ cu yd...	6	10,000	100	85	53	238	0.0238	$ 1.10– $1.95
Elec:								
¾, ½, ¾ cu yd....	5	9,000	100	60	45	205	0.0228	$ 1.20–$ 1.65
1–5 cu yd..........	8	12,000	100	90	69	259	0.0216	$ 1.70–$ 4.00
Truck mounted:								
Gas, ⅜–1 cu yd........	4	7,500	100	60	37	197	0.0263	$ 1.30–$ 1.90
Dsl, ⅜–1 cu yd........	4	7,500	100	60	37	197	0.0263	$ 0.70–$ 1.20
Surveying instruments....	10	15,000	100	50	85	235	0.0157	
Towers, steel, single or double:								
Angle steel, 50–200 ft...	5	8,000	100	40	45	185	0.0231	
Tube steel, light, 25–200 ft 3,000-lb load.......	5	8,000	100	50	45	195	0.0244	
Tube steel, heavy, 40–300 ft, 5,000-lb load..	5	8,000	100	40	45	185	0.0231	
Portable, 10-ft sections 20–50 ft high.........	5	8,000	100	50	45	195	0.0244	
Sheaves and swivels, 12–18 in................	2	3,500	100	30	21	151	0.0431	
Tractors:								
Crawler:								
Gas:								
20–50 hp..........	3	6,000	100	50	29	179	0.0298	$ 0.32–$ 0.80
50–110 hp.........	4	8,000	100	65	37	202	0.0253	$ 0.80–$ 1.70
Dsl:								
20–50 hp..........	4	8,000	100	60	37	197	0.0246	$ 0.20–$ 0.50
50–200 hp.........	5	9,000	100	70	45	215	0.0239	$ 0.50–$ 1.85
Four-wheel, rubber tires:								
Gas:								
10–35 hp..........	3	6,000	100	50	29	179	0.0298	$ 0.16–$ 0.60

Approximate Equipment-ownership and -operating Costs (CONTINUED)

Equipment	App life		App tot cost, percent of capital investment					Operating cost per wkhr
	Yr	Wkhr	Tot dep	Tot maj rpr ovhl	Tot int ins tax sto	Tot own cost	Own cost per wkhr	
Tractors: (*Contd.*)								
35–75 hp	4	8,000	100	65	37	202	0.0253	$ 0.60–$ 1.25
Dsl:								
35–75 hp	4	8,000	100	60	37	197	0.0246	$ 0.35–$ 0.70
75–300 hp	5	9,000	100	70	45	215	0.0239	$ 0.70–$ 2.85
Two-wheel, rubber tires, dsl, 100–240 hp	5	9,000	100	70	45	215	0.0239	$ 0.95–$ 2.25
Tractor attachments:								
Angle dozers, all sizes	4	7,000	100	70	37	207	0.0296	
Bulldozers, all sizes	4	8,000	100	70	37	207	0.0259	
Front-end loaders, all sizes	4	8,000	100	80	37	217	0.0271	
Pipe layers, all sizes	4	8,000	100	70	37	207	0.0259	
Winches, towing, all sizes	4	7,000	100	70	37	207	0.0296	
Tractor and trailer units:								
Dsl power, all sizes, two-wheel with cable scraper or dump trailer, four-wheel with cable scraper or dump wagons	5	9,000	100	75	45	220	0.0244	$0.0125 per tractor hp
Gas power, all sizes, four-wheel with cable scraper or dump wagons	5	9,000	100	75	45	220	0.0244	$ 0.020 per tractor hp
Trailers, platform type:								
Drop frame, 10–30 tons	5	6,000	100	50	45	195	0.0325	
Reversible, 5–15 tons	4	5,000	100	40	37	177	0.0354	
Trench braces 24–70 in	5	9,000	100	60	45	205	0.0228	
Trenching machines:								
Ladder type:								
Gas, 10–26 ft	5	8,000	100	75	45	220	0.0275	$ 0.80–$ 1.80
Dsl, 10–26 ft	5	8,000	100	75	45	220	0.0275	$ 0.45–$ 0.95
Wheel type:								
Gas, 8.5 ft	4	7,500	100	70	37	207	0.0276	$ 1.05–$ 1.55
Dsl, 8.5 ft	4	7,500	100	70	37	207	0.0276	$ 0.55–$ 0.80

Approximate Equipment-ownership and -operating Costs (CONTINUED)

Equipment	App life		App tot cost, percent of capital investment					Operating cost per wkhr
	Yr	Wkhr	Tot dep	Tot maj rpr ovhl	Tot int ins tax sto	Tot own cost	Own cost per wkhr	
Trucks:								
Gas pickup, ½–1 ton...	3	6,000	100	50	29	179	0.0298	$ 0.75–$ 1.00
Gas stake:								
½–1 ton.............	3	6,000	100	50	29	179	0.0298	$ 0.75–$ 1.00
2–5 tons.............	4	7,000	100	60	37	197	0.0281	$ 1.10–$ 1.40
Gas dump, 2–5 tons....	4	7,000	100	60	37	197	0.0281	$ 1.10–$ 1.40
Gas heavy-duty dump, 7–18 tons............	5	9,000	100	60	45	205	0.0228	$ 1.50–$ 3.00
Dsl stake, 2–10 tons....	5	8,000	100	60	45	205	0.0256	$ 0.35–$ 1.40
Dsl dump, 2–5 tons.....	5	8,000	100	60	45	205	0.0256	$ 0.35–$ 0.80
Dsl heavy-duty dump, 7–30 tons............	6	10,000	100	60	53	213	0.0213	$ 0.80–$ 3.20
Wagons, dump:								
Crawler, side dump or bottom dump........	5	10,000	100	75	45	220	0.0220	
Four-wheel, rubber tires.	5	10,000	100	85	45	230	0.0230	
Welding equipment, portable:								
Acetylene torch, all sizes................	4	7,500	100	50	37	187	0.0250	$ 0.30–$ 1.25
Elec welding machines:								
Ac arc, 100–1,000 amp..............	5	9,000	100	60	45	205	0.0228	$ 0.15–$ 1.25
Dc arc:								
Gas, 200–600 amp..	5	9,000	100	70	45	215	0.0239	$ 0.50–$ 1.50
Dsl, 200–600 amp..	5	9,000	100	50	45	195	0.0217	$ 0.30–$ 0.90
Elec motor, 200–600 amp............	5	9,000	100	60	45	205	0.0228	$ 0.30–$ 1.20
Transformer, rectifiers, 200–600 amp	6	10,000	100	60	53	213	0.0213	$ 0.35–$ 0.85
Transformers, 200–600 amp................	5	10,000	100	50	45	195	0.0195	$ 0.25–$ 0.75
Wheelbarrows, steel wheel or rubber tire, 3- to 5-cu-ft tray...........	2	4,000	100	30	21	151	0.0378	

APPENDIX C

Construction-equipment Rental Rates

The rental rates given in this appendix are *approximate* only. The rental rates will vary with the kind, type, size, and capacity of the machines and with the time, locality, and particular rental agency. Also the rental rate for a newer and more efficient machine may be more.

The rental rates given in this appendix were obtained from various sources and represent *approximate* rates (1967) for some typical pieces of construction equipment, without including all types and sizes.

Rental rates are usually based on one 8-hr shift per day, 40 hr per week, or 175 hr per month. When a machine is operated more than 8 hr in one day, the rental rate may be increased about 50 percent for each additional 8-hr shift. If the additional shift is less than 8 hr, a proportional charge may be made. The rental rates are usually for the machine "bare," without fuel, lubricants, or operators, or for repairs and maintenance charges.

The rental time usually starts when the equipment leaves the lessor's (owner's) warehouse or yard and continues until the equipment is returned to the lessor's warehouse or yard.

The lessor (owner) is responsible only for the repairs caused by normal wear and tear. The lessee is responsible for all costs of transportation, operation, repairs (except normal wear and tear), maintenance, fuel, cleaning, insurance, sales or use taxes, fees, etc.

Approximate 1967 *Construction-equipment Rental Rates*

Equipment	Monthly	Weekly	Daily
Air compressors, bare, 100 psi:			
Gas, portable, 60–315 cfm	\$ 180–\$ 450	\$ 60–\$ 150	\$ 15–\$ 45
Dsl, portable, 200–600 cfm	\$ 400–\$ 900	\$ 140–\$ 300	\$ 45–\$100
Elec, portable, 60–465 cfm	\$ 140–\$ 420	\$ 45–\$ 140	\$ 12–\$ 35
Air tools:			
Drills:			
Drifters	\$ 36–\$ 90	\$ 12–\$ 30	\$ 4–\$ 10
Wagon	\$ 300–\$ 360	\$ 100–\$ 120	\$ 25–\$ 30
Grinders and sanders:			
Floor	\$ 33–\$ 54	\$ 12–\$ 20	\$ 3–\$ 6
Cement, terrazo	\$ 90–\$ 180	\$ 30–\$ 60	\$ 8–\$ 15
Hose and couplings, 50-ft length	\$ 6–\$ 15	\$ 3–\$ 8	\$ 2–\$ 3
Jackhammers	\$ 50–\$ 120	\$ 20–\$ 40	\$ 5–\$ 10
Spades and tampers	\$ 36–\$ 45	\$ 12–\$ 15	\$ 4–\$ 5
Asphalt and tar heating kettles	\$ 80–\$ 100	\$ 27–\$ 33	\$ 7–\$ 10
Blowers, fans, and ventilators	\$ 45–\$ 63	\$ 15–\$ 21	\$ 5–\$ 7
Boilers, steam, 5–100 hp	\$ 200–\$ 800	\$ 75–\$ 275	\$ 25–\$100
Brooms, road	\$ 125–\$ 210	\$ 45–\$ 75	\$ 15–\$ 25
Bulldozer attachments for crawler tractors:			
35–100 drawbar hp	\$ 165–\$ 270	\$ 55–\$ 90	\$ 18–\$ 30
101–200 drawbar hp	\$ 285–\$ 400	\$ 95–\$ 130	\$ 35–\$ 45
201–265 drawbar hp	\$ 400–\$ 500	\$ 135–\$ 165	\$ 45–\$ 55
Concrete accessories:			
Buckets, ½–4 cu yd	\$ 65–\$ 180	\$ 30–\$ 60	\$ 10–\$ 20
Carts:			
Rubber tired	\$18	\$9	\$3
With power	\$ 150–\$ 175	\$ 50–\$ 60	\$ 17–\$ 22
Column clamps, per 10	\$ 5–\$ 7		
Chutes	\$ 15–\$ 25	\$ 5–\$ 10	\$ 2–\$ 3
Finishers, 36 or 48 in	\$ 70–\$ 125	\$ 30–\$ 50	\$ 8–\$ 14
Hoppers, discharge	\$ 5–\$ 6	\$ 2–\$ 3	\$ 1–\$ 2
Mixers:			
⅓–½ sk	\$ 50	\$17	\$ 5
1 sk	\$135	\$45	\$15
2 sk	\$180	\$60	\$20
3 sk	\$225	\$75	\$25
Shores and jacks, per 10	\$ 4–\$ 6		
Vibrators:			
Air	\$72	\$24	\$ 8
Gas	\$90	\$30	\$10
Elec	\$80	\$27	\$ 9
Conveyors, portable, 20, 30, or 40 ft	\$ 240–\$ 375	\$ 80–\$ 125	\$ 20–\$ 35
Cranes, dsl crawler, no bucket:			
6–10 tons, 10–12 ft radius	\$ 900–\$1,250	\$ 300–\$ 400	\$100–\$150
14–30 tons, 12–20 ft radius	\$1,500–\$2,450	\$ 500–\$ 800	\$180–\$275
10–24 tons, to 45 ft radius	\$4,100–\$5,300	\$1,400–\$1,800	\$450–\$600
Yard, gas, 3–10 tons	\$ 450–\$ 650	\$ 140–\$ 225	\$ 35–\$ 50

Approximate 1967 *Construction-equipment Rental Rates* (CONTINUED)

Equipment	Monthly	Weekly	Daily
Floor polisher, elec	$ 30–$ 60	$ 10–$ 20	$ 3–$ 6
Floor sander, elec	$ 40–$ 60	$ 14–$ 20	$ 5–$ 7
Graders, rubber-tired, conventional transmission, Midwest:			
Dsl:			
To 20,000 lb	$ 650–$ 800	$ 215–$ 265	$ 70–$ 80
20,000–30,000 lb	$1,080–$1,300	$ 365–$ 425	$115–$140
Graders, rubber-tired, West Coast:			
Gas:			
Light	$380	$125	$ 40
Medium	$610	$205	$ 67
Heavy-duty	$775	$260	$ 85
Dsl:			
Light	$475	$160	$ 53
Medium	$645	$215	$ 70
Heavy-duty	$965	$325	$110
Hoists:			
Gas:			
Single drum, 10–110 hp	$ 75–$ 335	$ 25–$ 125	$ 10–$ 45
Double drum, 20–120 hp	$ 115–$ 500	$ 40–$ 175	$ 15–$ 65
Elec:			
Single drum, 10–220 hp	$ 70–$ 600	$ 25–$ 200	$ 10–$ 75
Double drum, 20–220 hp	$ 90–$ 750	$ 35–$ 270	$ 15–$100
Air, single drum, 1,200–3,500 lb	$ 90–$ 225	$ 30–$ 75	$ 10–$ 25
Hose, coupled:			
Suction, 2, 3, or 4 in., 20 ft	$ 12–$ 35	$ 4–$ 12	$ 2–$ 5
Discharge, 2, 3, or 4 in., 20 ft	$ 6–$ 20	$ 2–$ 7	$ 1–$ 3
Fire hose, 1½, 2 or 2½ in., 50 ft	$ 12–$ 27	$ 4–$ 10	$ 2–$ 3
Jacks:			
Hydraulic, 5–100 tons	$ 14–$ 120	$ 5–$ 40	$ 2–$ 15
Ratchet, 12–22 tons	$ 18–$ 32	$ 6–$ 11	$ 2–$ 4
Screw	$9	$3	$1
Trench	$3	$2	$1
Loaders, front end, wheel type, ⅓–2 cu yd	$ 325–$1,800	$ 110–$ 600	$ 30–$150
Mixers, portable:			
Cement, ⅓–½ sk, gas power	$ 45–$ 60	$ 15–$ 20	$ 5–$ 7
Concrete:			
1–3 sk, gas power	$ 135–$ 225	$ 45–$ 75	$ 15–$ 25
3–9 cu ft, mixer only	$ 70–$ 105	$ 25–$ 40	$ 8–$ 14
8–16 cu ft, power charge, mixer only	$ 130–$ 270	$ 45–$ 95	$ 15–$ 35
Mortar and plaster	$ 100–$ 130	$ 35–$ 45	$ 7–$ 12
Paint-spray equipment, 1 hp	$ 60–$ 90	$ 20–$ 30	$ 6–$ 10
Trailer unit	$ 180–$ 240	$ 60–$ 80	$ 15–$ 25
Power plants, elec:			
3–25 kw	$ 60–$ 240	$ 20–$ 80	$ 6–$ 25
25–75 kw	$ 275–$ 500	$ 90–$ 160	$ 25–$ 45

Approximate 1967 *Construction-equipment Rental Rates* (CONTINUED)

Equipment	Monthly	Weekly	Daily
Pumps without hose:			
Centrifugal, self-priming:			
Gas:			
2 in., 7–10 M gph	$ 60–$ 75	$ 20–$ 25	$ 7–$ 10
3 in., 15–20 M gph	$ 75–$ 90	$ 25–$ 30	$ 8–$ 10
4 in., 30–40 M gph	$ 120–$ 135	$ 40–$ 50	$ 13–$ 18
6 in., 90 M gph	$ 165–$ 200	$ 55–$ 70	$ 18–$ 24
8 in., 125 M gph	$ 240–$ 280	$ 80–$ 95	$ 27–$ 32
Dsl:			
6 in., 90 M gph	$ 330–$ 390	$ 110–$ 130	$ 35–$ 45
8 in., 125 M gph	$ 500–$ 575	$ 170–$ 195	$ 55–$ 65
Diaphragm, 5 hp:			
3 in., single action	$ 60–$ 90	$ 20–$ 30	$ 7–$ 10
4 in., single action	$ 75–$ 100	$ 25–$ 40	$ 9–$ 13
4 in., double action	$ 135–$ 150	$ 45–$ 60	$ 15–$ 20
Sludge	$ 75–$ 90	$ 25–$ 30	$ 8–$ 11
Rivet hammers	$45	$15	$5
Rivet sets	$60	$20	$7
Rollers, road:			
Gas:			
Two-wheel tandem, 2–5 tons	$ 240–$ 300	$ 80–$ 100	$ 18–$ 25
Three-wheel, 8–13 tons	$ 375–$ 550	$ 125–$ 185	$ 40–$ 60
Dsl, three-wheel, 9–13 tons	$ 570–$ 620	$ 195–$ 210	$ 65–$ 70
Rubber-tired towed, 9 or 13 wheels	$ 175–$ 200	$ 43–$ 65	$ 14–$ 16
Sheep's-foot	$ 125–$ 300	$ 45–$ 95	$ 15–$ 32
Saws:			
Chain:			
Gas	$ 135–$ 165	$ 65–$ 75	$ 17–$ 24
Elec	$ 75–$ 95	$ 45–$ 55	$ 12–$ 18
Hand portable:			
Air	$ 75–$ 90	$ 25–$ 30	$ 8–$ 12
Elec, 6- to 12-in. blade	$ 25–$ 55	$ 12–$ 20	$ 4–$ 7
Table, fixed or tilting:			
Gas	$ 70–$ 110	$ 27–$ 40	$ 9–$ 13
Elec	$ 70–$ 110	$ 27–$ 40	$ 9–$ 13
Scrapers and carryalls, no power, heaped:			
5–10 cu yd	$ 360–$ 510	$ 120–$ 170	$ 40–$ 55
10–17 cu yd	$ 610–$ 725	$ 200–$ 245	$ 65–$ 80
18–29 cu yd	$ 940–$1,260	$ 330–$ 420	$105–$140
30–37 cu yd	$1,500–$1,800	$ 500–$ 600	$165–$200
Shovels and back hoes:			
Gas:			
⅜ and ½ cu yd	$1,000–$1,250	$ 330–$ 415	$110–$135
¾ and 1 cu yd	$1,350–$1,800	$ 450–$ 575	$150–$190
Dsl:			
⅜ and ½ cu yd	$ 950–$1,250	$ 325–$ 425	$110–$145

Approximate 1967 *Construction-equipment Rental Rates* (CONTINUED)

Equipment	Monthly	Weekly	Daily
¾ and 1 cu yd	\$1,500–\$2,000	\$ 525–\$ 670	\$175–\$225
1¼ and 1½ cu yd	\$2,100–\$2,850	\$ 700–\$ 950	\$235–\$320
2 and 2½ cu yd	\$3,900–\$4,600	\$1,330–\$1,575	\$435–\$525
3 and 3½ cu yd	\$5,900–\$6,750	\$1,950–\$2,250	\$650–\$750
Surveying instruments:			
Levels	\$35	\$12	\$ 4
Transits	\$90	\$35	\$10
Tractors:			
Gas, 4 wheel	\$ 270–\$ 300	\$ 90–\$ 100	\$ 20–\$ 30
Dsl, crawler, gear drive:			
25–60 drawbar hp	\$ 420–\$ 720	\$ 140–\$ 240	\$ 45–\$ 80
60–115 drawbar hp	\$ 810–\$1,350	\$ 270–\$ 450	\$ 90–\$150
120–185 drawbar hp	\$1,400–\$2,050	\$ 465–\$ 675	\$155–\$225
200–270 drawbar hp	\$2,200–\$3,000	\$ 725–\$1,000	\$240–\$330
Dsl crawler with torque converter:			
120–185 net eng. hp	\$1,500–\$1,800	\$ 500–\$ 600	\$165–\$200
190–330 net eng. hp	\$2,250–\$3,600	\$ 750–\$1,200	\$250–\$400
Tractor and scraper units, 2-wheel tractor and 2-wheel scraper:			
7–18 cu yd, heaped	\$1,500–\$2,600	\$ 500–\$ 875	\$170–\$300
18–27 cu yd, heaped	\$2,800–\$3,500	\$ 950–\$1,250	\$330–\$400
Trenchers:			
Ladder:			
Gas, 18–36 in., 4- to 8-ft depth	\$ 700–\$3,425	\$ 230–\$1,160	\$ 75–\$400
Dsl, 18–36 in., 4- to 8-ft depth	\$1,050–\$3,700	\$ 350–\$1,240	\$115–\$400
Wheel:			
Gas, 24–36 in., 5- to 6-ft depth	\$ 600–\$1,900	\$ 200–\$ 625	\$ 65–\$215
Dsl, 24–36 in., 5- to 6-ft, depth	\$ 630–\$2,350	\$ 210–\$ 800	\$ 70–\$270
Trucks:			
Fork lift, construction-type, rubber-tired, 4,000–6,000 lb, 16- to 20-ft lift	\$ 560–\$ 600	\$ 140–\$ 180	\$ 30–\$ 50
Pickup and stake, West Coast	\$ 150–\$ 250	\$ 50–\$ 80	\$ 10–\$ 20
Rear dump, single axle:			
7–15 tons	\$1,800–\$2,250	\$ 600–\$ 750	\$200–\$250
16–25 tons	\$2,600–\$3,500	\$ 800–\$1,200	\$275–\$400
Welders:			
Acetylene set	\$45	\$15	\$5
Gas, 200–300 amp	\$ 90–\$ 120	\$ 30–\$ 40	\$ 8–\$ 12
Elec, 200–400 amp	\$ 50–\$ 90	\$ 20–\$ 35	\$ 7–\$ 12
Dsl, 75 kw	\$ 500–\$ 600	\$ 165–\$ 200	\$ 55–\$ 75
Wheelbarrows, rubber-tired	\$9	\$3	\$1

APPENDIX D

Building Cost Estimate Summary

The following building cost estimate summary form is similar to one used by a large construction company that specializes in the construction of buildings of various types (such as offices, apartments, stores, schools, banks, hotels, factories, etc.) ranging in value up to a million dollars or more.

This company prepares its own estimates for all parts of the job, even though several parts of the job are usually let to subcontractors.

For the parts of the job that are to be let to subcontractors, the company prepares an estimate of the cost of each part so that the company estimates may be compared with the bids of the subcontractors.

For the parts of the job that are to be done by the company, the company estimates the costs of materials, labor, and equipment for each part.

The overhead costs are estimated for the entire job. These costs are itemized. The items are divided into general and job overhead costs about as described in Chap. 19.

Profit is usually estimates as a percentage of the sum of all other costs for the entire job. This percentage may vary with the kind and size of the job, job hazards, time required for completion, how and when partial payments are made, and the desire of the company for the work.

Extras are listed as they occur and their estimated costs noted. Almost every job has one or more extras.

After the completion of the job, the actual costs are compared with the estimated costs.

Building Cost Estimate Summary

Owner	Job no.
Phone	Date
Address	Building type
Architect	Floor area
Phone	Volume
Address	Location

Item	Estimated Cost	Actual Cost	Remarks
1. Clearing site			
2. Excavation			
3. Fill			
4. Foundations			
5. Concrete work			
6. Masonry:			
a. Brick			
b. Stone			
c. Block			
d. Tile			
e. Other			
7. Dampproofing and waterproofing			
8. Structural steel:			
a. Framing			
b. Other			
9. Ornamental and misc. metal work			
10. Rough wood construction			
11. Finish wood construction			
12. Mill work			
13. Roofing and flashing			
14. Sheet metal work			
15. Insulation and sound proofing			
16. Lathing and plastering:			
a. Plasterboard			
b. Wire lath			
c. Plaster			
17. Stucco:			
a. Stucco mesh, etc.			
b. Stucco			
18. Drywall construction			
19. Rough plumbing			
20. Finish plumbing			
21. Steam heating			
22. Hot-water heating			
23. Warm-air heating			
24. Air cooling			
25. Combined air heating, cooling, and conditioning			
26. Rough electrical work			

Building Cost Estimate Summary (CONTINUED)

Item	Estimated Cost	Actual Cost	Remarks
27. Finish electrical work			
28. Glass and glazing			
29. Painting			
30. Papering			
31. Fireproofing			
32. Fire protection:			
a. Pipes and valves			
b. Sprinklers			
33. Elevators:			
a. Passenger			
b. Freight			
34. Walks and drives			
35. Curbs and gutters			
36. Pavements			
37. Landscaping			
38. Miscellaneous			
39. Cleaning up			
40. Overhead:			
a. General			
b. Job			
41. Sum of items 1 to 40 inc			
42. Profit			
43. Total cost of job			
44. Extras:			
a.			
b.			
c.			
d.			
e.			

APPENDIX E

Hourly Wage Rates and Compensation Insurance Rates For Construction

Hourly wage rates vary greatly from time to time, for different kinds of construction work, in different localities, and on different jobs. Labor wages are often less in small and medium-sized urban areas than in the large cities.

When preparing an estimate of labor costs, the estimator should always check the labor wage rates prevailing in the particular locality for the different kinds of construction work involved.

The cost of compensation insurance varies greatly among different classes of work and also among different states. The rates are based upon experience and data collected over a period of years and are changed from time to time. The base rates are subject to adjustment, either up or down, depending upon whether the individual contractor's experience rating is bad or good.

Careful payroll classification can afford an appreciable saving in insurance costs.

When preparing estimates, the estimator should always consult the insurance companies to ascertain the rates which are in effect at the time and in the locality in which the work is to be done and also which are in effect for the particular contractor and his experience rating.

The following table of base rates for important construction operations gives the rates in effect on July 1, 1967, and shows the variation in these base rates between the states and between different kinds of work. The data in this table were abstracted, with permission, from the *Engineering News-Record.*

The data given in this appendix are for illustrative purposes only and should not be used when preparing actual cost estimates.

The reader is referred to the *Engineering News-Record* for information on materials prices, labor wages cost trends, and actual costs of different kinds of construction work.

Hourly Wage Rates—Base rate + fringe benefits
(*Monthly Market Quotations by ENR Field Reporters February* 1, 1968)*

Classification	Atlanta $ + ¢	Baltimore $ + ¢	Birmingham $ + ¢	Boston $ + ¢
Common labor:				
Building	2.80 + 10	3.10 + 25	2.95	3.85 + 30
Heavy const.	2.80 + 10	2.90 + 15½	2.85 + 10	3.55 + 30
Skilled building trades:				
Bricklayers	4.90 + 58	5.15 + 25	5.00	5.60 + 50
Carpenters	4.55 + 10	4.54 + 14½	4.25 + 1½	5.15 + 35
Struct. iron workers	4.80 + 29	5.36 + 41	4.91	5.64 + 36
Plasterers	4.65	4.50 + 15	4.08	5.15 + 70
Electrical workers	5.00 + 40	5.15 + 15	4.65 + 30	5.90 + 60½
Steamfitters	5.15 + 29	5.03 + 32	4.95 + 46	5.65 + 34
Equipment operators:				
Hoisting engineers	4.30 + 30	5.06 + 41	4.20 + 10	5.48½ + 39
Tractor operators†				
34–49 hp	3.90 + 30	4.21 + 39	4.20 + 10	5.21½ + 39½
70–100 hp	4.55 + 30	4.66 + 40	4.20 + 10	5.21½ + 39½
Over 100 hp	3.90 + 30	4.66 + 40	4.20 + 10	5.21½ + 39½
Tractor scraper, 15–16 cy	3.90 + 30	4.66 + 40	4.20 + 10	5.21½ + 39½
Power cranes & shovels				
½ cy	4.45 + 30	5.06 + 41	4.20 + 10	5.58½ + 39½
¾ cy	4.55 + 30	5.06 + 41	4.20 + 10	5.58½ + 39½
Motor grader operator	3.90 + 30	4.06 + 39	4.20 + 10	5.46½ + 39½
Air comp. operator	2.82 + 30	4.26 + 40	4.20 + 10	4.68 + 39½
Air tool operator	3.02 + 10	3.00 + 15½	3.00 + 10	3.80 + 30
Truck drivers:				
1½ ton	3.05	3.00 + 15½	2.95	3.80 + 28
Under 4 cy	2.85	3.00 + 15½	3.15	3.80 + 28
Over 4 cy	3.10	3.00 + 15½	4.20 + 10	3.80 + 28

Classification	Chicago $ + ¢	Cincinnati $ + ¢	Cleveland $ + ¢	Dallas $ + ¢
Common labor:				
Building	4.12½ + 37½	4.10 + 15	4.77 + 40	2.47½ + 12½
Heavy const.	4.12½ + 37½	3.97½ + 15	4.73 + 25	2.47½ + 12½
Skilled building trades:				
Bricklayers	5.95 + 43	5.35 + 19	6.06 + 50	4.65 + 25
Carpenters	5.45 + 42	4.90 + 45	5.33 + 98	4.50 + 1
Struct. iron workers	5.85 + 59½	5.27½ + 65½	6.10 + 36	4.40 + 24
Plasterers	5.30 + 39½	4.95 + 21	6.06	4.26¼ + 57½
Electrical workers	5.40 + 75½	5.40 + 21.8	5.51 + 75	4.82½ + 32
Steamfitters	5.65 + 39	4.95 + 58½	5.51 + 75	4.80 + 30

Hourly Wage Rates—Base rate + fringe benefits (CONTINUED)

Classification	Chicago $ + ¢	Cincinnati $ + ¢	Cleveland $ + ¢	Dallas $ + ¢
Equipment operators:				
Hoisting engineers	5.20 + 50	4.93½/5.01	5.67 + 34	4.65
Tractor operators†				
34–49 hp	4.95 + 50	4.93½ + 34	5.67 + 34	1.80/2.62½
70–100 hp	4.95 + 50	4.93½ + 34	5.67 + 34	1.80/2.62½
Over 100 hp	5.30 + 50	4.93½ + 34	5.67 + 34	1.80/2.62½
Tractor scraper, 15–16 cy	4.95 + 50	4.93½ + 34	5.67 + 34	1.80/2.62½
Power cranes & shovels				
½ cy	5.75 + 50	4.93½ + 34	5.67 + 34	4.25
¾ cy	5.75 + 50	5.01 + 34	5.67 + 34	4.25
Motor grader operator	5.75 + 50	5.01 + 34	5.22 + 34	1.80/2.62½
Air comp. operator	4.80 + 50	5.01 + 34	5.29½ + 34	2.57½ + 12½
Air tool operator	4.35 + 37½	5.01 + 34	5.39½	2.57½ + 12½
Truck drivers:				
1½ ton	3.45 + 40	3.29 + 38.9	3.85 + 50	1.80/2.62½
Under 4 cy	3.90 + 25	3.34 + 38.9	3.85 + 50	1.80/2.62½
Over 4 cy	4.40 + 25	3.39 + 38.9	3.85 + 50	1.80/2.62½

Classification	Denver $ + ¢	Detroit $ + ¢	Kansas City $ + ¢	Los Angeles $ + ¢
Common labor				
Building	3.40	4.25 + 53	3.37 + 30	3.85 + 68½
Heavy const.	3.30	4.25 + 53	3.43 + 25	3.85 + 68½
Skilled building trades:				
Bricklayers	4.85 + 15	5.44 + 91	5.075 + 15	4.90 + 61
Carpenters	4.81½ + 20	5.06 + 91	4.55 + 17	5.09 + 71
Struct. iron workers	4.95 + 15	5.50 + 115	4.77½ + 15½	5.98 + 76
Plasterers	4.90 + 7½	5.32 + 68	4.75 + 17½	4.94 + 83
Electrical workers	5.07 + 17	5.40 + 102	5.40 + 37	5.70 + 60
Steamfitters	4.86 + 49	5.77½ + 88	5.20 + 45½	5.64 + 110
Equipment operators:				
Hoisting engineers	4.60 + 01	5.48 + 50	4.47 + 27½	5.41 + 105
Tractor operators†				
34–49 hp	3.60	5.48 + 50	3.67½ + 27½	5.41 + 105
70–100 hp	4.20	5.48 + 50	4.22½ + 27½	5.41 + 105
Over 100 hp	4.20	5.48 + 50	4.22½ + 27½	5.41 + 105
Tractor scraper, 15–16 cy	4.20	5.48 + 50	4.22½ + 27½	5.41 + 105
Power cranes & shovels				
½ cy	4.35	5.68 + 50	4.47 + 27½	5.41 + 105
¾ cy	4.35	5.68 + 50	4.47 + 27½	5.41 + 105

Hourly Wage Rates—Base rate + fringe benefits (CONTINUED)

Classification	Denver $ + ¢	Detroit $ + ¢	Kansas City $ + ¢	Los Angeles $ + ¢
Equipment operators (*Cont.*)				
Motor grader operator	4.20	5.48 + 50	4.22½ + 27½	5.41 + 105
Air comp. operator	3.95	4.72½ + 50	4.22½ + 27½	4.53 + 105
Air tool operator	3.68	4.43 + 53	3.46½ + 30	4.14 + 67½
Truck drivers:				
1½ ton	3.65	4.48 + 50	3.54½	4.50 + 80
Under 4 cy	3.80	4.13	3.54½	4.50 + 80
Over 4 cy	4.00	4.13	3.62	4.53 + 80

Classification	Minneapolis $ + ¢	New Orleans $ + ¢	New York $ + ¢	Philadelphia $ + ¢
Common labor:				
Building	4.00 + 15	3.03 + 10‡	4.95 + 85	3.60 + 25
Heavy const.	4.20	2.63 + 10‡	4.95 + 85	3.60 + 20
Skilled building trades:				
Bricklayers	4.85 + 62½	4.78 + 25	5.95 + 201	5.40 + 67
Carpenters	4.53 + 50	4.35 + 30	6.15 + 96	4.70 + 68
Struct. iron workers	5.10 + 15	4.85 + 20	6.55 + 128	5.55 + 70
Plasterers	4.55 + 65	4.20 + 20	5.75 + 130	5.23½ + 58
Electrical workers	4.95 + 74	4.95	5.76 + 152§	5.60 + 41
Steamfitters	5.09 + 47	4.93 + 40	5.54 + 167	5.64½ + 47
Equipment operators:				
Hoisting engineers	4.90 + 15	4.65 + 25	6.45 + 100	5.86/5.39 +40
Tractor operators†				
34–49 hp	3.78 + 15	4.65 + 25	5.72½ + 57.2	5.14 + 40
70–100 hp	4.68 + 30	4.65 + 25	5.82½ + 93	5.39 + 40
Over 100 hp	4.68 + 30	4.65 + 25	5.82½ + 93	5.39 + 40
Tractor scraper, 15–16 cy	4.78 + 30	4.65 + 25	5.80 + 58½	5.39 + 40
Power cranes & shovels				
½ cy	4.90 + 30	4.65 + 25	6.70 + 102	5.89 + 40
¾ cy	5.16 + 30	4.65 + 25	6.70 + 102	5.89 + 40
Motor grader operator	4.20 + 30	4.65 + 25	5.85 + 58½	5.99 + 40
Air comp. operator	4.00 + 30	3.82½ + 25	5.95 + 95	4.99 + 40
Air tool operator	4.15 + 15	3.11 + 10	5.50 + 75	3.60 + 20
Truck drivers:				
1½ ton	4.20	4.65 + 25	4.68½ + 29½	3.59 + 27½
Under 4 cy	4.30	4.65 + 25	4.68½ + 29½	3.59 + 27½
Over 4 cy	4.40	4.65 + 25	4.68½ + 29½	3.59 + 27½

Hourly Wage Rates—Base rate + fringe benefits (CONTINUED)

Classification	Pittsburgh $ + ¢	St. Louis $ + ¢	San Francisco $ + ¢
Common labor:			
Building	3.90 + 22½	4.37½ + 37	4.175 + 90
Heavy const.	3.69 + 32	4.37½ + 37	4.175 + 90
Skilled building trades:			
Bricklayers	6.07½ + 42½	5.50 + 103	6.00 + 143
Carpenters	5.65 + 678	5.57½ + 12	5.205 + 112
Struct. iron workers	5.45 + 39	5.37½ + 37	5.98 + 77½
Plasterers	5.43 + 35	5.45 + 17¾	5.00 + 114
Electrical workers	5.40 + 79	5.45 + 106½	6.91 + 41
Steamfitters	5.06 + 61	5.37½ + 163½	7.47½ + 147
Equipment operators:			
Hoisting engineers	4.79 + 22	5.32½ + 37	6.16 + 92
Tractor operators†			
34–39 hp	5.30 + 22	5.32½ + 37	5.76 + 92
70–100 hp	5.39 + 22	5.32½ + 37	5.76 + 92
Over 100 hp	5.39 + 22	5.32½ + 37	5.76 + 92
Tractor scraper, 15–16 cy	5.65 + 22	5.32 + 37	5.76 + 92
Power cranes & shovels			
½ cy	5.65 + 22	5.32½ + 37	6.00 + 92
¾ cy	5.65 + 22	5.32½ + 37	6.00 + 92
Motor grader operator	5.39 + 22	5.32½ + 37	5.53½ + 92
Air comp. operator	5.03 + 22	4.77½ + 37	4.81½ + 92
Air tool operator	3.74 + 32	4.37½ + 37	4.42½ + 90
Truck drivers:			
1½ ton	3.96 + 32	4.15 + 37	4.50½ + 98½
Under 4 cy	3.87 + 32	4.15 + 37	4.50½ + 98½
Over 4 cy	3.87 + 32	4.45 + 37	4.60 + 98½

Classification	Seattle $ + ¢	Montreal $ + ¢	Toronto $ + ¢
Common labor:			
Building	4.30 + 47	3.16 + 31	3.05 + 18.3
Heavy const.	4.30 + 47	3.16 + 31	3.30 + 19.8
Skilled building trades:			
Bricklayers	5.53 + 35	4.10 + 38	4.15 + 37
Carpenters	4.68 + 40½	3.89 + 37	4.00 + 34
Struct. iron workers	5.45 + 36	3.80 + 36	4.18 + 48
Plasterers	4.85 + 48	4.10 + 38	3.65 + 59
Electrical workers	5.47 + 41	4.15 + 39	4.40 + 40
Steamfitters	5.56 + 77	4.25 + 39	4.37 + 44

Hourly Wage Rates—Base rate + fringe benefits (CONTINUED)

Classification	Seattle $ + ¢	Montreal $ + ¢	Toronto $ + ¢
Equipment operators:			
Hoisting engineers	5.39 + 55	3.79 + 34	4.15 + 25
Tractor operators†			
34–49 hp	5.28 + 55	3.35 + 33	3.65 + 22
70–100 hp	5.65 + 55	3.69 + 35	3.65 + 22
Over 100 hp	5.65 + 55	3.69 + 35	3.65 + 22
Tractor scraper,			
15–16 cy	5.65 + 55	3.69 + 35	3.65 + 22
Power cranes & shovels			
½ cy	5.65 + 55	4.00 + 37	4.15 + 25
¾ cy	5.65 + 55	4.00 + 37	4.15 + 25
Motor grader operator	5.37 + 55	3.35 + 33	4.15 + 25
Air comp. operator	5.01 + 55	3.56 + 34	3.65 + 22
Air tool operator	4.30 + 47	3.56 + 34	3.65 + 22
Truck drivers:			
1½ ton	4.94 + 48	3.17 + 32	3.35 + 16¾
Under 4 cy	4.94 + 48	3.17 + 32	3.35 + 16¾
Over 4 cy	5.14 + 48	3.17 + 32	3.35 + 16¾

* ENR skilled labor average (bricklayers, carpenters, ironworkers): $5.660; ENR common labor average: $3.997. † Including bulldozer. ‡ Effective November, 1967. § Average for 6-hr day, including mandatory sixth hour at time and one-half.

Base Rates on Compensation Insurance for Construction—July 1, 1967*

Classification of work	Ala.	Alaska	Ariz.	Ark.	Cal.	Colo.	Conn.	Del.	D.C.	Fla.	Ga.	Hawaii	Ida.
Carpentry—1, 2 family residence	$1.95	$2.78	$6.44	$3.84	$5.69	$2.85	$3.38	$2.25	$2.93	$4.93	$2.74	$4.29	$2.87
Carpentry—3 stories or less	1.99	2.78	6.44	4.59	5.69	3.22	3.38	2.25	3.08	4.93	2.74	6.11	2.87
Carpentry—interior cab WK	.98	1.85	6.44	1.92	5.69	1.82	1.47	2.25	2.93	3.32	1.27	3.38	1.93
Carpentry—general	2.28	3.73	6.44	5.06	8.56	3.24	3.38	2.25	4.11	6.46	3.65	10.25	4.54
Chimney—construction brick con.	6.33	3.54	9.40	14.07	22.83	11.59	11.83	2.60	17.56	17.12	5.56	17.23	14.43
Concrete work—bridges, culverts‡	3.23	4.97	12.91	7.89	10.08	6.60	5.64	2.60	9.66	6.95	4.42	9.24	3.82
Concrete work dwellings—1–2 family	.95	4.99	8.87	1.99	2.65	2.99	2.03	2.60	4.13	5.25	1.02	3.69	2.30
Concrete work—N.O.C.	2.65	3.54	8.87	4.79	7.82	5.06	4.13	2.60	5.62	6.90	2.88	7.77	2.87
Concrete or cement work—floors, sidewalks	.95	3.24	8.87	2.07	2.36	1.74	2.53	1.15	2.83	3.84	1.59	2.73	2.35
Electrical wiring—inside	1.01	1.46	4.15	1.79	3.06	1.71	1.07	.75	2.23	2.59	1.36	3.12	2.12
Excavation work N.O.C	2.19	3.48	4.16	3.84	4.32	3.68	3.94	2.55	3.54	6.48	2.52	11.72	5.86
Excavation—rock	3.16	4.28	8.88	4.94	11.38	5.84	4.27	2.55	12.61	11.66	5.12	11.61	4.09
Glaziers	2.35	3.19	5.79	4.60	5.07	3.54	3.15	2.10	4.04	4.59	3.37	5.66	3.35
Insulation work	1.52	4.55	5.72	1.91	5.48	2.76	2.66	2.25	3.55	3.71	1.92	4.35	2.95
Lathing	1.17	2.91	3.39	2.57	4.95	2.05	1.55	1.20	2.42	2.35	1.59	3.79	2.27
Masonry	.95	4.37	6.49	1.99	5.16	2.99	2.30	2.15	4.44	3.59	1.37	3.69	2.30
Painting and decorating	1.78	3.74	5.81	2.70	5.46	2.42	2.37	2.30	3.17	3.45	1.88	3.49	2.50
Pile driving	6.88	8.44	19.20	10.62	15.13	11.28	2.91	7.50	15.52	12.84	5.64	12.50	13.18
Plastering	1.52	2.03	5.72	1.91	5.67	2.76	2.66	1.20	3.55	4.09	1.92	4.35	2.95
Plumbing	1.08	2.36	3.81	2.22	3.39	2.51	1.53	.84	2.08	3.17	1.83	2.88	1.42
Roofing	3.57	8.94	11.70	9.63	11.38	6.21	8.22	3.65	7.76	9.93	3.95	17.49	7.10
Sheet metal work—erection—installation and repair	1.72	2.01	5.05	3.12	3.56	2.74	2.18	3.65	2.98	3.87	2.30	4.76	2.23
Steel erection—doors and sash	1.63	5.95	2.74	3.18	5.19	3.45	3.23	3.35	3.06	4.96	1.78	3.77	3.20
Steel erection—interior ornament	1.63	5.95	2.74	3.18	5.19	3.45	3.23	3.35	3.06	4.96	1.78	3.77	3.20
Steel erection structure	3.62	14.05	36.06	8.50	18.40	12.42	10.19	3.35	24.26	14.62	8.39	31.47	7.24
Steel erection—dwelling 2 stories	3.05	8.23	13.38	5.26	14.33	7.62	5.96	3.35	8.76	9.82	5.21	11.58	6.16
Steel erection—N.O.C	3.62	14.05	36.06	5.26	15.23	7.77	10.19	3.35	21.64	12.26	7.21	20.93	7.24
Tile work—interior	1.03	.83	2.15	1.89	2.96	1.85	1.89	.71	4.26	4.23	1.09	2.65	1.42
Timekeepers and watchmen	1.73	4.88	2.52	3.08	3.79	2.91	3.96	‡	3.91	4.90	2.60	3.87	3.52
Waterproofing interior (brush)	1.78	3.74	5.81	2.70	5.46	2.42	2.37	2.30	3.17	3.45	1.88	3.49	2.50
Waterproofing—trowel interior	1.52	2.03	5.72	1.91	5.67	2.76	2.66	1.20	3.55	4.09	1.92	4.35	2 95
Waterproofing—trowel exterior	.95	4.37	6.49	1.99	5.16	2.99	2.30	2.15	4.44	3.59	1.37	3.69	2.30
Waterproofing pressure gum	2.65	3.54	8.87	4.79	7.82	5.06	4.13	2.60	5.62	6.90	2.88	7.77	2.87
Wrecking	7.98	12.71	27.15	26.77	21.27	24.12	12.04	6.35	18.17	27.39	9.38	32.61	21.90

* Rate/$100 payroll. Compiled by Herbert J. Jamison & Co., Insurance Advisers and Auditors, New York, N.Y. (Base rates in effect July 1, 1967—subject to increase or decease according to experience ratings).

† FA Specially rated—refer to company. ‡ Assign governing class. § Subscription to State Fund Mandatory; Industrial Commission, Carson City, Nev.; Workmen's Compensation Bureau, Bismarck, N.D.; The Industrial Commission of Ohio, Columbus, Ohio; Dept. of Labor and Industry, Olympia, Wash.; State Compensation Commission, Charleston, W.Va.; Workmen's Compensation, State Treasurer's Office, Cheyenne, Wyo.

Base Rates on Compensation Insurance for Construction—July 1, 1967 (CONTINUED)

Classification of work	Ill.	Ind.	Iowa	Kan.	Ky.	La.	Me.	Md.	Mass.	Mich.	Minn.	Miss.	Mo.
Carpentry—1, 2 family residence	$2.87	$2.38	$2.11	$2.90	$2.91	$5.07	$2.31	$3.26	$4.80	$4.23	$3.22	$4.36	$2.96
Carpentry—3 stories or less	2.87	2.47	2.54	2.90	3.19	6.60	2.31	3.26	5.83	4.23	3.22	4.36	2.96
Carpentry—interior cab WK	2.05	1.30	1.04	1.62	1.96	2.94	1.09	2.25	1.95	2.56	3.22	2.20	2.13
Carpentry—general	5.17	2.66	3.02	4.25	4.23	8.38	3.90	3.94	11.44	4.23	4.92	4.59	3.72
Chimney—construction brick con	14.75	8.76	7.57	13.04	12.12	24.20	7.67	13.67	12.30	19.46	15.87	15.23	14.39
Concrete work—bridges, culverts§	6.17	4.97	5.04	4.95	4.47	7.25	4.18	7.91	5.59	7.80	5.15	7.33	5.26
Concrete work dwellings—1–2 family	1.73	1.69	1.43	2.40	1.73	3.17	2.12	2.83	4.02	4.39	2.03	2.36	2.00
Concrete work—N.O.C	5.33	2.72	2.86	5.30	2.96	6.18	3.08	4.44	8.66	7.07	4.21	4.14	4.34
Concrete or cement work—floors, sidewalks	2.13	1.15	2.17	2.62	2.13	3.53	1.25	2.71	4.18	3.87	2.04	2.48	1.95
Electrical wiring—inside	1.85	1.15	1.01	2.18	1.40	2.88	.91	2.03	2.20	1.67	1.65	2.52	1.93
Excavation work N.O.C	3.19	2.33	2.42	3.51	3.28	6.38	2.13	2.80	3.63	5.11	3.19	5.01	4.62
Excavation—rock	13.88	5.29	4.34	5.15	3.49	17.49	4.24	6.61	4.58	9.31	3.86	10.78	6.20
Glaziers	3.62	2.58	2.70	3.11	3.01	6.44	3.05	4.98	5.31	4.07	4.63	5.49	3.59
Insulation work	3.31	1.57	1.40	2.52	2.00	5.74	2.05	2.16	3.74	2.73	3.53	2.61	2.37
Lathing	1.52	1.43	1.44	2.10	1.91	4.56	1.51	1.92	3.59	1.86	1.60	2.84	1.72
Masonry	2.70	1.59	1.43	2.40	1.73	3.17	2.12	2.91	5.08	3.21	2.13	2.36	2.23
Painting and decorating	3.42	1.87	1.90	2.81	2.34	4.25	2.25	3.16	3.11	4.90	2.14	3.10	2.55
Pile driving	9.09	6.11	5.33	9.70	9.12	18.43	3.82	7.13	10.65	12.42	11.77	12.97	7.06
Plastering	2.01	1.57	1.40	2.52	2.00	3.20	2.05	2.16	3.91	2.73	2.08	2.61	2.02
Plumbing	2.19	1.62	1.48	2.52	1.93	5.32	1.28	2.41	2.64	2.32	2.14	2.26	3.03
Roofing	7.65	5.55	5.37	9.25	6.64	7.77	5.64	6.98	23.21	11.00	8.00	6.61	6.57
Sheet metal work—erection—installation and repair	2.28	1.71	1.48	3.25	2.57	4.39	1.70	2.60	3.71	3.24	2.43	5.09	2.55
Steel erection—doors and sash	2.13	2.71	1.83	2.66	3.44	4.47	2.09	4.38	3.87	3.42	3.07	3.81	2.57
Steel erection—interior ornament	2.13	2.71	1.83	2.66	3.44	4.47	2.09	4.38	3.87	3.42	3.07	3.81	2.57
Steel erection structure	16.30	8.25	9.89	12.66	11.09	12.61	6.64	14.18	22.88	15.45	14.07	9.21	8.74
Steel erection—dwelling 2 stories	6.73	3.96	4.76	7.49	5.66	11.27	4.18	7.80	19.55	7.26	7.71	8.17	7.88
Steel erection—N.O.C	12.10	6.25	7.47	7.49	6.46	12.61	6.22	14.18	22.88	11.42	6.21	9.21	8.74
Tile work—interior	1.47	1.15	.86	1.80	1.13	2.28	1.02	1.68	3.43	1.78	1.78	2.29	1.26
Timekeepers and watchmen	3.88	2.54	2.27	3.47	2.85	6.38	2.23	3.65	7.34	4.47	3.07	3.49	4.28
Waterproofing interior (brush)	3.42	1.87	1.90	2.81	2.34	4.25	2.25	3.16	3.11	4.90	2.14	3.10	2.55
Waterproofing—trowel interior	2.01	1.57	1.40	2.52	2.00	3.20	2.05	2.16	3.91	2.73	2.08	2.61	2.02
Waterproofing—trowel exterior	2.70	1.59	1.43	2.40	1.73	3.17	2.12	2.91	5.08	3.21	2.13	2.36	2.23
Waterproofing pressure gum	5.33	2.72	2.86	5.30	2.96	6.18	3.08	4.44	8.66	7.07	4.21	4.14	4.34
Wrecking	14.41	10.12	14.35	12.56	17.65	25.56	18.55	26.39	19.99	32.13	14.65	13.83	19.22

† FA Specially rated—refer to company. ‡ Assign governing class. § Subscription to State Fund Mandatory: Industrial Commission, Carson City, Nev.; Workmen's Compensation Bureau, Bismarck, N.D.; The Industrial Commission of Ohio, Columbus, Ohio; Dept. of Labor and Industry, Olympia, Wash.; State Compensation Commission, Charleston, W. Va.; Workmen's Compensation, State Treasurer's Office, Cheyenne, Wyo.

Base Rates on Compensation Insurance for Construction—July 1, 1967 (CONTINUED)

Classification of work	Mont.	Neb.	Nev.§	N.H.	N.J.	N.M.	N.Y.	N.C.	N.D.§	Ohio§	Okla.	Ore.	Penn.
Carpentry—1, 2 family residence	$4.59	$2.08	$6.29	$1.81	$5.28	$4.01	$3.30	$2.54	$3.11	$2.51	$5.12	$4.24	$1.75
Carpentry—3 stories or less	4.59	2.57	6.29	1.81	3.85	4.01	3.20	2.54	3.11	2.60	6.41	4.00	1.75
Carpentry—interior cab WK	1.86	1.04	2.55	1.10	3.14	1.92	4.30	1.71	3.11	1.37	4.09	3.81	1.75
Carpentry—general	4.85	3.60	6.64	2.87	6.40	5.30	4.30	3.67	4.76	2.80	6.45	7.01	1.75
Chimney—construction brick con	17.42	7.85	23.87	8.29	15.39	12.64	22.20	10.27	15.35	9.23	21.72	15.22	1.75
Concrete work—bridges, culverts§	7.55	3.43	10.34	4.43	7.28	10.40	6.40	5.52	4.98	5.24	13.41	10.25	1.75
Concrete work dwellings—1-2 family	3.36	1.53	4.60	1.84	2.93	1.40	5.30	1.04	1.96	1.78	2.44	3.86	1.75
Concrete work—N.O.C	5.44	3.16	7.45	2.56	7.28	4.50	5.30	3.37	4.07	2.87	8.90	8.00	1.75
Concrete or cement work—floors, sidewalks	2.28	1.39	3.12	1.63	3.37	2.36	2.90	1.47	1.97	1.21	3.76	3.68	.85
Electrical wiring—inside	1.45	1.27	1.99	1.03	3.02	2.11	2.40	1.14	1.60	1.21	2.84	1.96	.67
Excavation work N.O.C	5.57	1.98	7.63	3.57	6.51	3.20	4.50	3.05	3.08	2.46	6.91	6.71	2.10
Excavation—rock	7.12	6.00	9.75	4.32	8.75	3.45	4.00	6.42	3.73	5.58	6.91	7.32	2.10
Glaziers	5.75	2.93	7.88	3.10	6.15	4.66	5.30	3.10	4.48	2.72	5.34	3.08	1.15
Insulation work	4.90	1.30	6.71	1.66	4.61	2.87	2.60	1.46	3.41	1.65	4.66	4.47	1.75
Lathing	2.51	1.39	3.44	1.45	3.57	1.68	1.75	1.73	1.55	1.51	2.84	4.81	.99
Masonry	3.36	1.53	4.60	1.84	5.02	2.87	3.60	1.04	2.06	1.68	3.87	6.48	1.50
Painting and decorating	5.31	1.81	7.27	2.55	7.57	2.92	4.00	2.08	2.07	1.97	3.79	5.88	2.15
Pile driving	15.10	6.57	20.69	4.10	9.14	12.91	7.00	4.86	11.38	6.44	15.68	15.44	2.60
Plastering	4.90	1.30	6.71	1.66	3.57	2.87	3.70	1.46	2.01	1.65	4.04	4.09	.99
Plumbing	2.99	1.05	4.10	1.26	3.39	2.34	2.50	1.46	2.07	1.71	3.60	3.05	1.15
Roofing	11.13	6.52	15.25	9.34	14.36	8.35	8.20	5.74	7.74	5.85	14.63	8.49	2.55
Sheet metal work—erection—installation and repair	3.66	2.51	5.01	1.30	3.48	3.11	3.60	1.90	2.35	1.80	4.36	2.88	2.55
Steel erection—doors and sash	3.56	1.99	4.88	2.03	6.45	2.89	3.80	1.95	2.97	2.86	4.52	5.84	5.00
Steel erection—interior ornament	3.56	1.99	4.88	2.03	†	2.89	3.80	1.95	2.97	2.86	4.52	5.84	5.00
Steel erection structure	19.33	4.69	26.48	9.23	15.99	16.07	11.20	10.56	13.61	8.70	25.20	10.78	5.00
Steel erection—dwelling 2 stories	9.83	4.69	13.47	4.69	8.08	8.75	4.30	5.22	7.46	4.17	14.33	8.37	5.00
Steel erection—N.O.C	18.60	4.69	25.48	8.23	11.02	16.07	10.80	8.06	6.01	6.59	18.74	10.93	5.00
Tile work—interior	1.55	.95	2.12	1.00	2.26	2.40	2.20	.93	1.72	1.21	2.50	2.45	1.10
Timekeepers and watchmen	3.97	2.17	5.44	2.17	3.96	3.69	2.90	2.58	2.97	2.68	4.51	8.37	‡
Waterproofing interior (brush)	5.31	1.81	7.27	2.55	7.57	2.92	4.00	2.08	2.07	1.97	3.79	5.88	2.15
Waterproofing—trowel interior	4.90	1.30	6.71	1.66	3.57	2.87	3.70	1.46	2.01	1.65	4.04	4.09	.99
Waterproofing—trowel exterior	3.36	1.53	4.60	1.84	5.02	2.87	3.60	1.04	2.06	1.68	3.87	6.48	1.50
Waterproofing pressure gum	5.44	3.16	7.45	2.56	7.28	4.50	5.30	3.37	4.07	2.87	8.90	8.00	1.75
Wrecking	24.12	13.38	33.04	11.87	30.37	16.95	†	8.25	14.17	10.67	27.36	35.99	13.45

† FA Specially rated—refer to company. ‡ Assign governing class. § Subscription to State Fund Mandatory; Industrial Commission, Carson City, Nev.; Workmen's Compensation Bureau, Bismarck, N.D.; The Industrial Commission of Ohio, Columbus, Ohio; Dept. of Labor and Industry, Olympia, Wash.; State Compensation Commission, Charleston, W.Va.; Workmen's Compensation, State Treasurer's Office, Cheyenne, Wyo.

Base Rates on Compensation Insurance for Construction—July 1, 1967 (CONTINUED)

Classification of work	R.I.	S.C.	S.D.	Tenn.	Tex.	Utah	Vt.	Va.	Wash §	W.Va. §	Wisc.	Wyo.
Carpentry—1, 2 family residence	$2.54	$4.01	$2.21	$3.64	$4.65	$2.69	$2.16	$1.64	$3.87	$2.91	$2.42	$2.37
Carpentry—3 stories or less	2.54	4.01	2.21	3.64	4.65	2.69	2.16	1.56	3.98	3.19	2.10	2.37
Carpentry—interior cab WK	1.72	1.94	1.04	1.80	2.34	2.69	1.00	1.05	1.27	1.96	2.42	2.37
Carpentry—general	3.27	6.81	2.44	4.17	6.51	2.69	2.46	2.02	5.11	4.23	4.18	2.37
Chimney—construction brick con	16.51	7.35	6.03	13.57	26.37	9.14	7.30	5.64	4.20	12.12	13.72	8.04
Concrete work—bridges, culverts §	3.30	8.95	4.36	6.01	6.49	4.09	4.56	3.08	5.78	4.47	4.44	3.60
Concrete work dwellings—1-2 family	2.71	1.94	1.47	2.22	4.14	1.99	1.60	.99	3.66	1.73	2.06	1.75
Concrete work N.O.C.	6.87	5.57	2.81	4.55	7.24	2.66	2.39	2.67	4.20	2.96	3.53	2.34
Concrete or cement work—floors, sidewalks	2.50	2.72	1.18	2.19	3.30	1.31	1.28	.99	1.94	2.13	2.24	1.15
Electrical wiring—inside	1.40	2.39	1.01	1.54	2.22	1.28	1.03	.77	1.95	1.40	1.80	1.13
Excavation work N.O.C	3.24	3.50	2.16	3.44	4.48	3.25	3.74	2.08	4.40	3.28	3.27	2.86
Excavation—rock	8.03	9.62	3.62	4.28	7.46	3.33	3.54	2.28	9.32	3.49	8.20	2.93
Glaziers	4.51	5.41	2.56	4.34	2.74	1.17	2.85	2.00	4.35	3.01	3.39	1.03
Insulation work	2.50	3.09	1.65	1.96	4.37	2.31	1.68	1.26	3.40	2.00	2.51	2.03
Lathing	2.59	2.92	1.25	2.55	1.69	1.66	1.32	1.08	3.30	1.91	1.74	1.40
Masonry	2.71	1.94	1.47	2.22	4.14	2.09	1.60	1.17	3.77	1.73	2.09	1.84
Painting and decorating	3.63	2.75	1.85	2.48	2.92	1.29	1.99	1.78	4.19	2.34	2.29	1.14
Pile driving	6.18	11.16	6.77	9.21	11.84	4.72	3.77	4.93	9.79	9.12	8.55	4.15
Plastering	2.50	3.09	1.64	1.96	3.90	2.31	1.68	1.26	3.40	2.00	1.63	2.03
Plumbing	1.71	2.29	1.04	2.09	3.38	1.55	1.26	1.15	1.72	1.93	1.95	1.36
Roofing	5.63	6.47	5.08	5.43	11.53	5.13	7.81	5.42	11.97	6.64	8.75	4.51
Sheet metal work—erection—installation and repair	2.09	3.50	1.87	2.45	4.25	1.61	1.59	1.56	3.75	2.57	2.29	1.42
Steel erection—doors and sash	3.22	2.85	1.79	2.86	3.55	1.55	1.89	1.56	10.56	3.44	3.09	1.36
Steel erection—interior ornament	3.22	2.85	1.79	2.86	3.55	1.55	1.89	1.56	10.56	3.44	3.09	1.36
Steel erection structure	12.75	12.83	9.01	8.03	14.47	6.55	10.59	5.98	15.86	11.09	9.64	5.76
Steel erection—dwelling 2 stories	8.50	8.73	3.53	5.69	14.47	5.15	4.94	3.23	7.51	5.66	5.41	4.53
Steel erection—N.O.C	12.42	8.77	3.53	8.03	6.41	6.55	6.43	4.84	10.56	6.46	12.39	5.76
Tile work—interior	1.66	1.75	.90	1.68	1.65	1.42	.97	.87	1.38	1.13	1.36	1.25
Timekeepers and watchmen	4.01	3.63	2.07	3.82	2.12	2.36	2.06	1.76	4.18	2.85	3.47	2.08
Waterproofing interior (brush)	3.63	2.75	1.85	2.48	2.92	1.29	1.99	1.78	4.19	2.34	2.29	1.14
Waterproofing—trowel interior	2.50	3.09	1.65	1.96	3.90	2.31	1.68	1.26	3.40	2.00	1.63	2.03
Waterproofing—trowel exterior	2.71	1.94	1.47	2.22	4.14	2.09	1.60	1.17	3.77	1.73	2.09	1.84
Waterproofing pressure gum	6.87	5.57	2.81	4.55	7.24	2.66	2.39	2.67	4.20	2.96	3.53	2.34
Wrecking	14.97	26.21	10.83	18.94	24.43	14.78	11.59	6.00	21.52	17.65	12.63	13.01

† FA Specially rated—refer to company. ‡ Assign governing class. § Subscription to State Fund Mandatory; Industrial Commission, Carson City, Nev.; Workmen's Compensation Bureau, Bismarck, N.D.; The Industrial Commission of Ohio, Columbus, Ohio; Dept. of Labor and Industry, Olympia, Wash.; State Compensation Commission, Charleston, W.Va.; Workmen's Compensation, State Treasurer's Office, Cheyenne, Wyo.

APPENDIX F

Diagrams

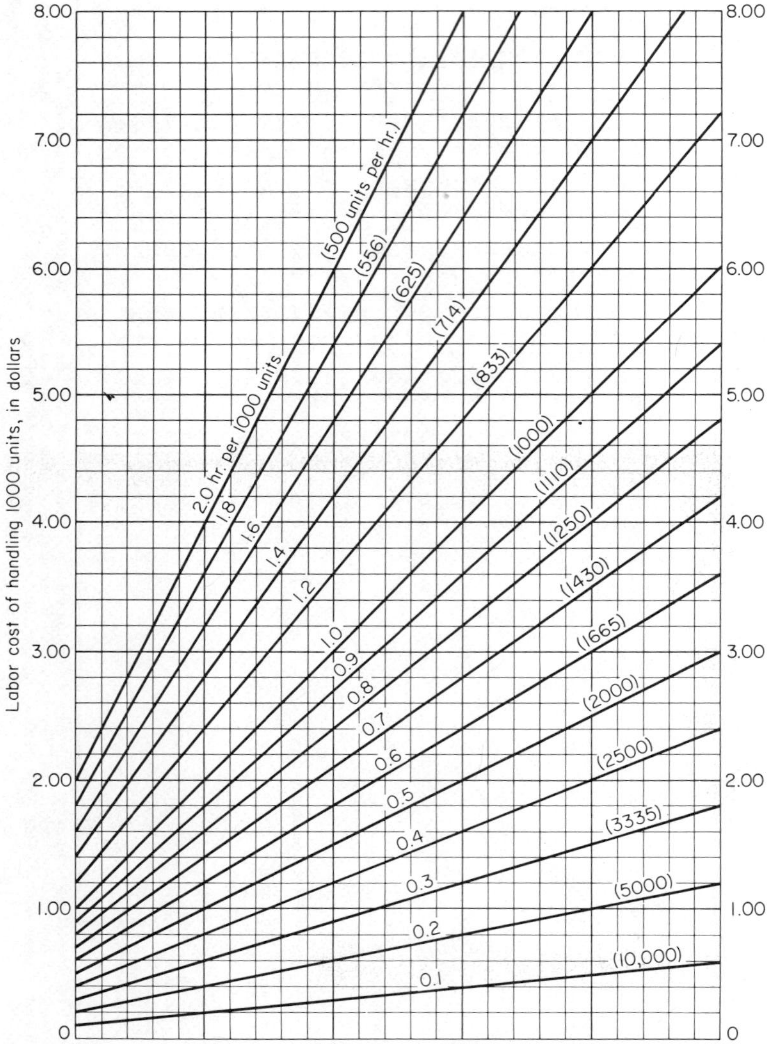

Diagram 2-1 Labor costs of handling 1,000 units of material, such as picking up materials, taking a few steps, and placing or piling.

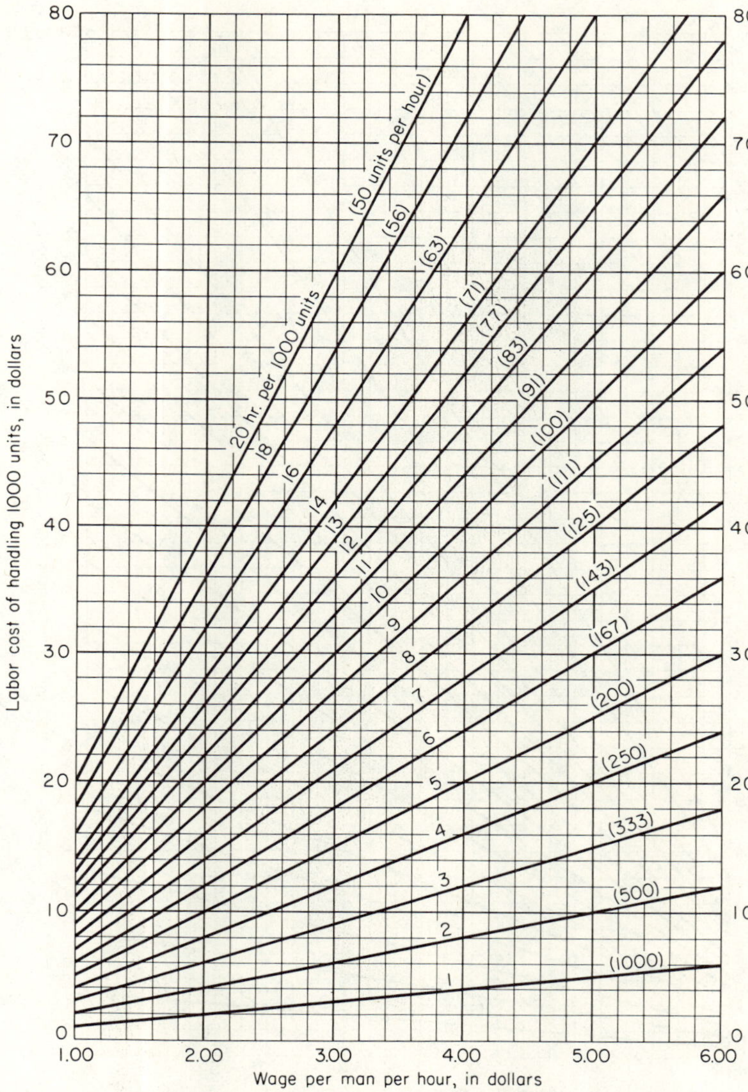

Diagram 2-2 Labor costs of handling 1,000 units of material, such as picking up materials, taking a few steps, and placing or piling.

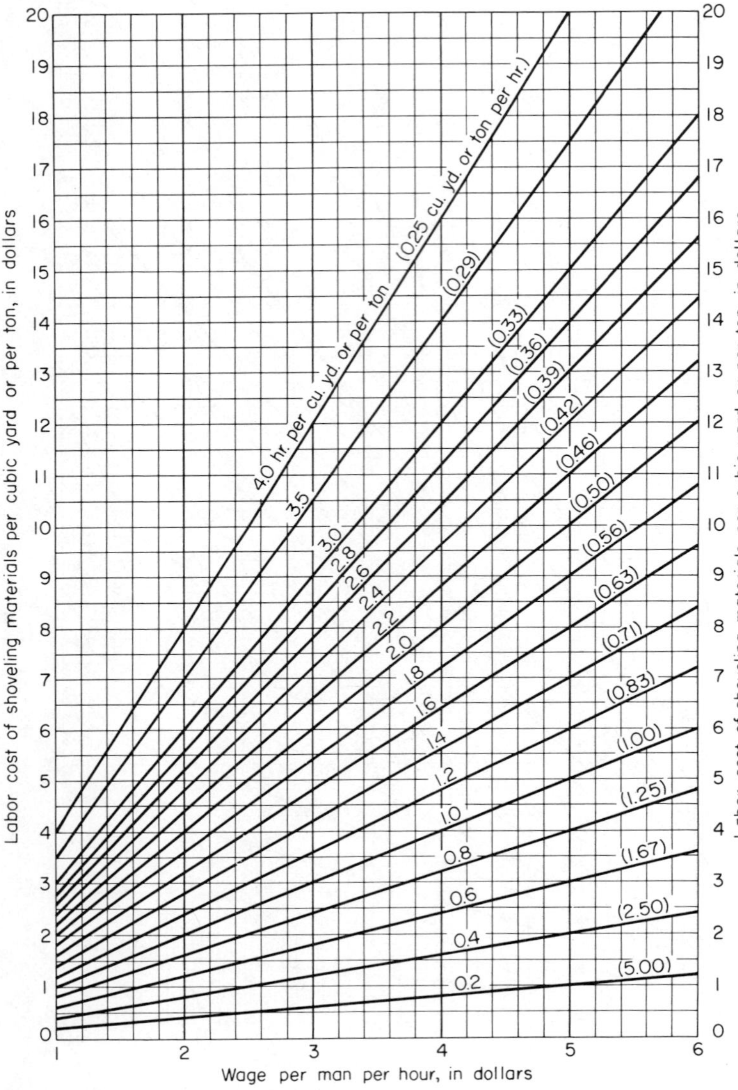

Diagram 2-3 Labor costs of shoveling materials per cubic yard or per ton.

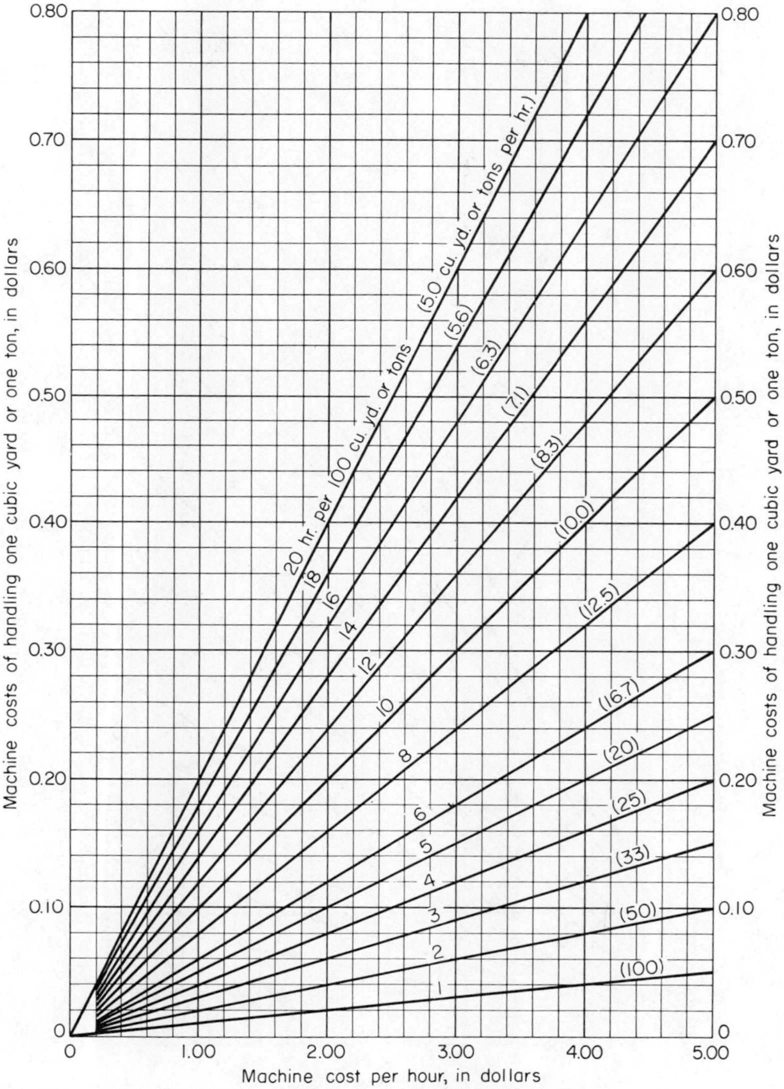

Diagram 2-4 Machine costs of handling materials per cubic yard or per ton. Machine costs per hour may or may not include cost of operator.

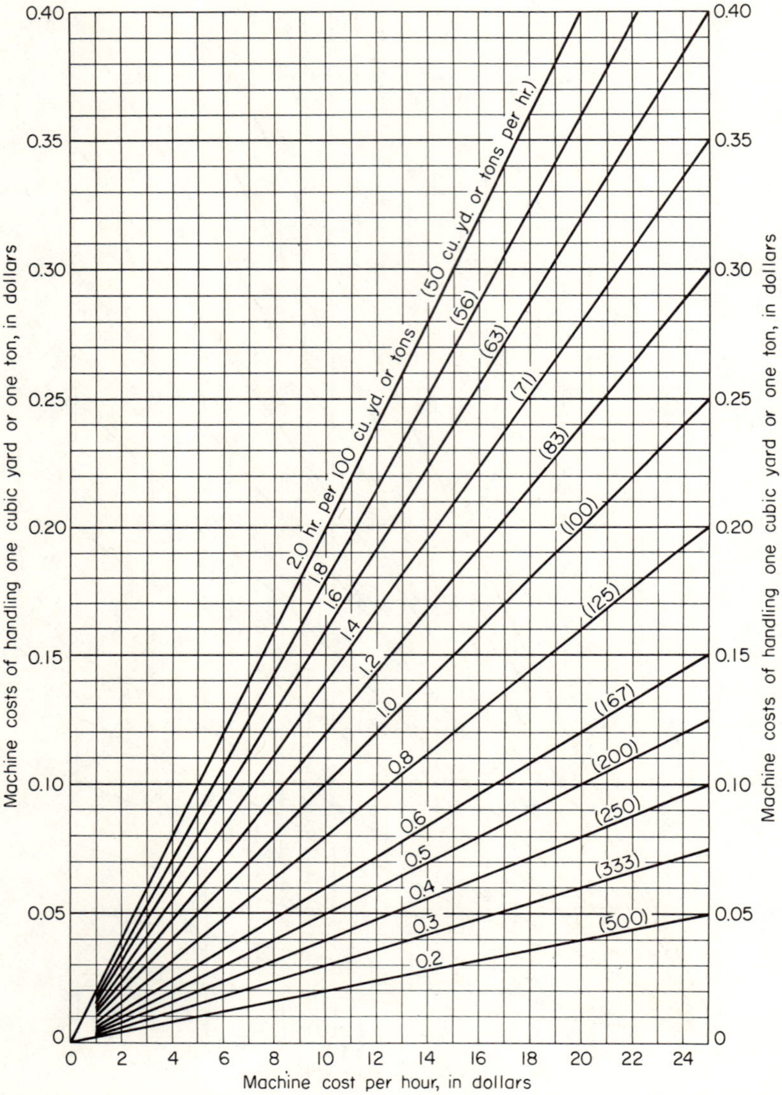

Diagram 2-5 Machine costs of handling materials per cubic yard or per ton. Machine costs per hour may or may not include cost of operator and crew.

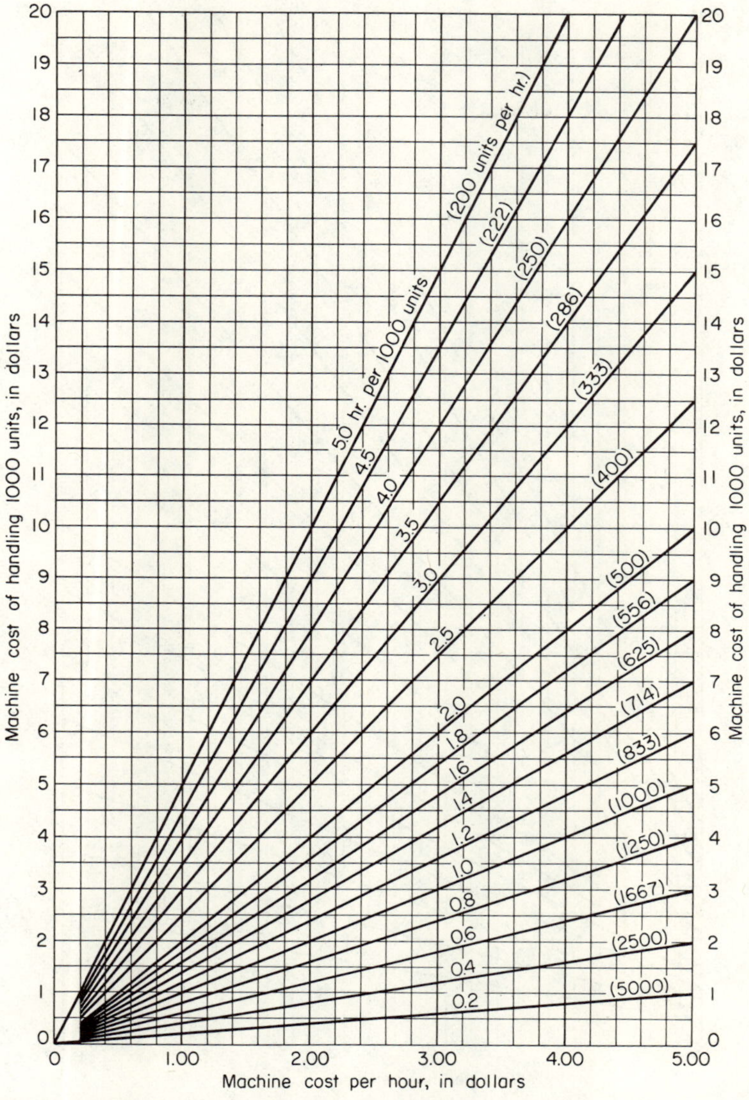

Diagram 2-6 Machine costs of handling 1,000 units of materials. Machine costs per hour may or may not include cost of operator.

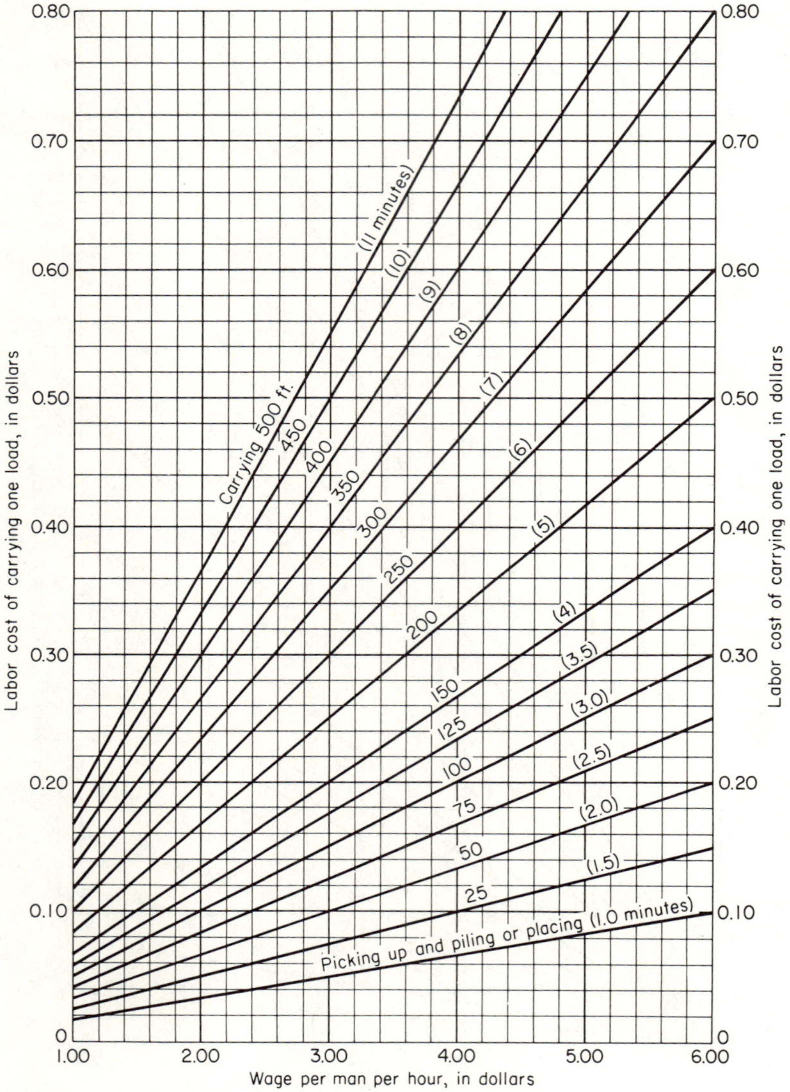

Diagram 2-7 Labor cost of carrying materials by hand.

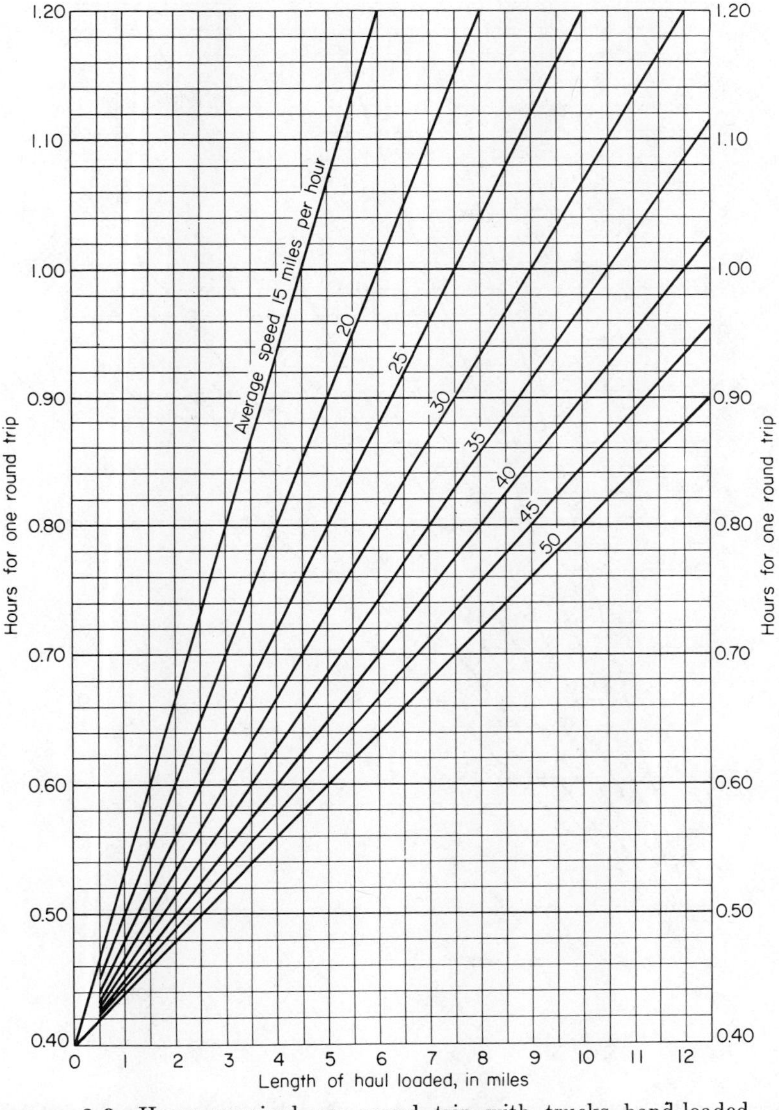

Diagram 2-8 Hours required per round trip with trucks hand-loaded. Time allowed for loading, unloading, and delays is 0.40 hr or 24 min.

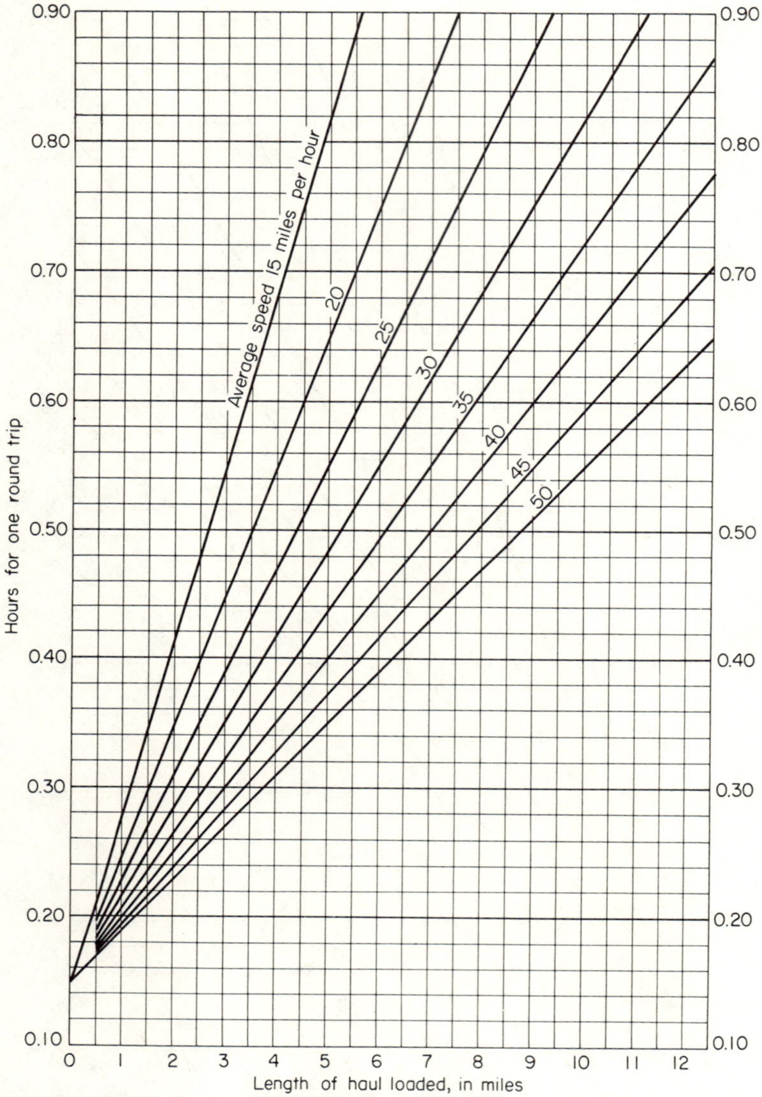

Diagram 2-9 Hours required per round trip with trucks power-loaded. Time allowed for loading, unloading, and delays is 0.15 hr or 9 min.

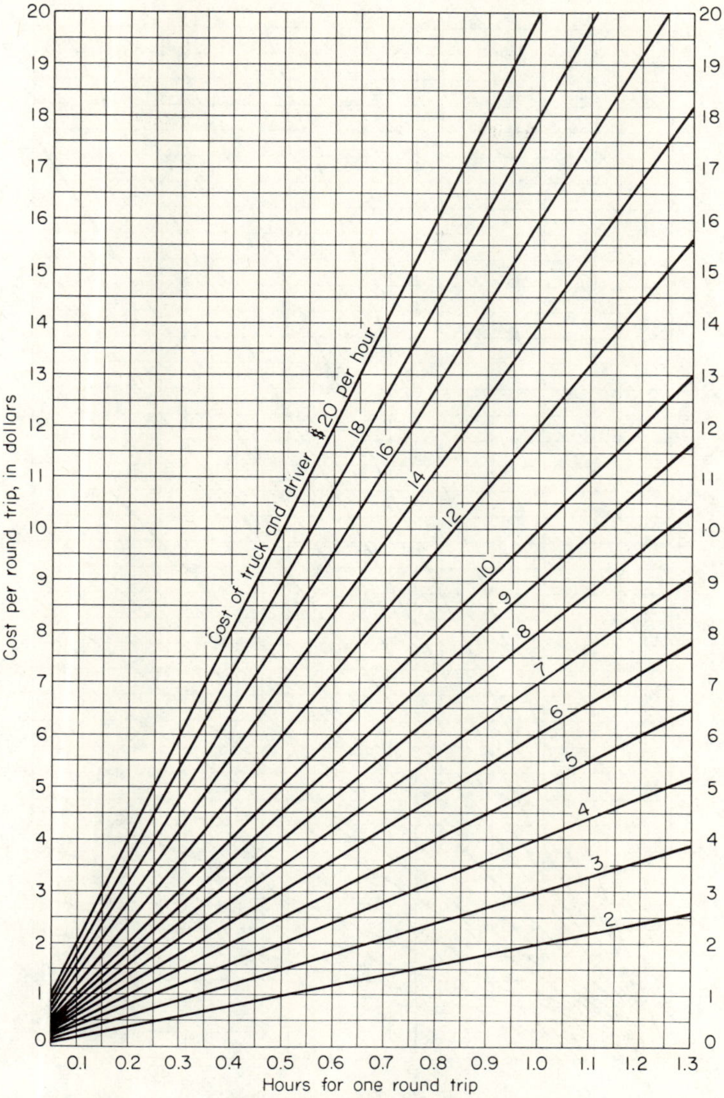

Diagram 2-10 Cost of one round trip in dollars of transporting materials by truck.

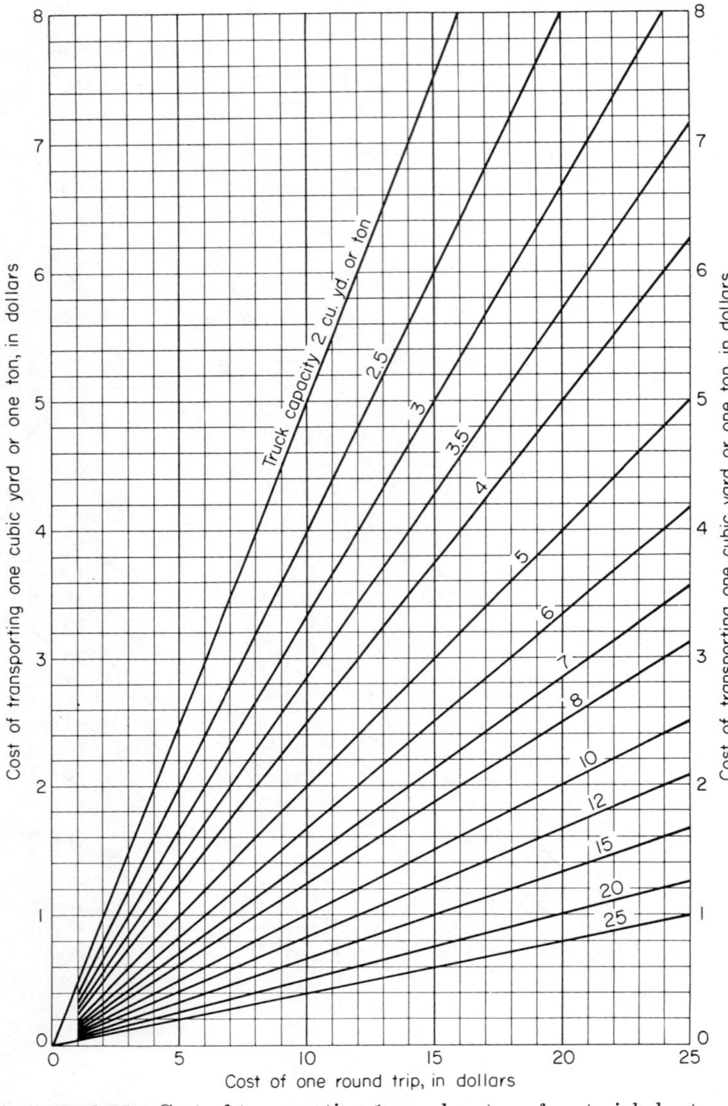

Diagram 2-11 Cost of transporting 1 cu yd or ton of materials by truck.

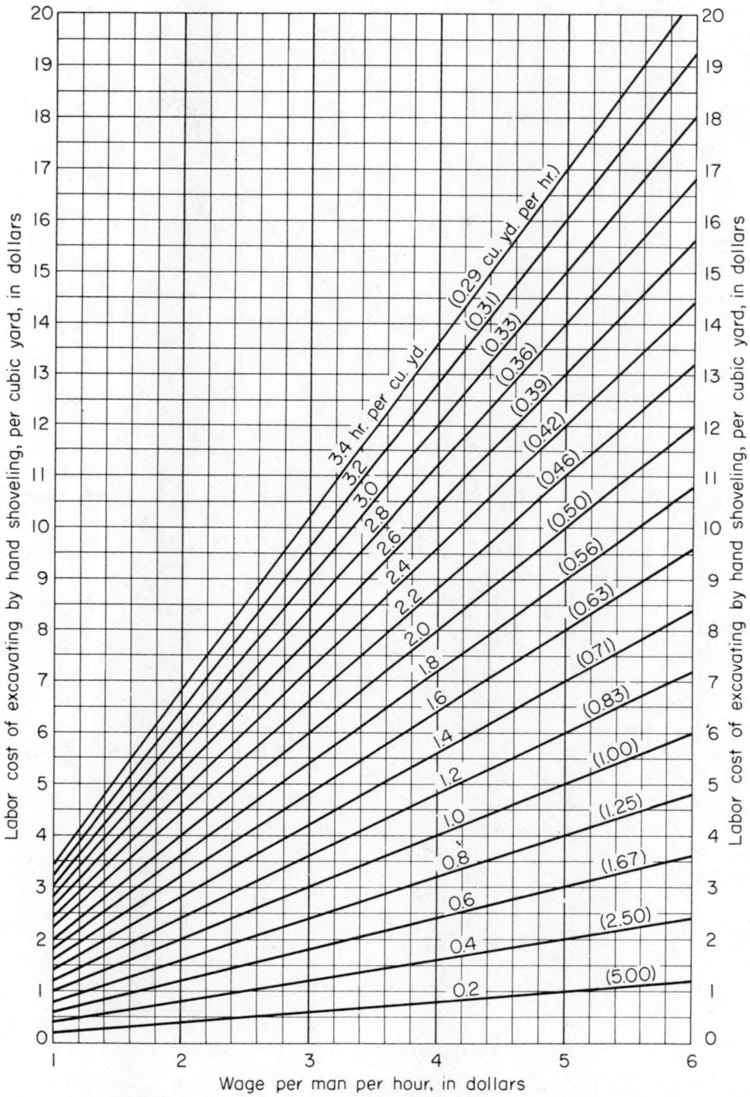

Diagram 3-1 Labor costs of excavating soil by hand shoveling. This diagram may be used for estimating costs of excavation by hand shoveling, trenching by hand labor, excavation by hand labor and wheelbarrows, backfilling by hand labor, and spreading soil by hand labor.

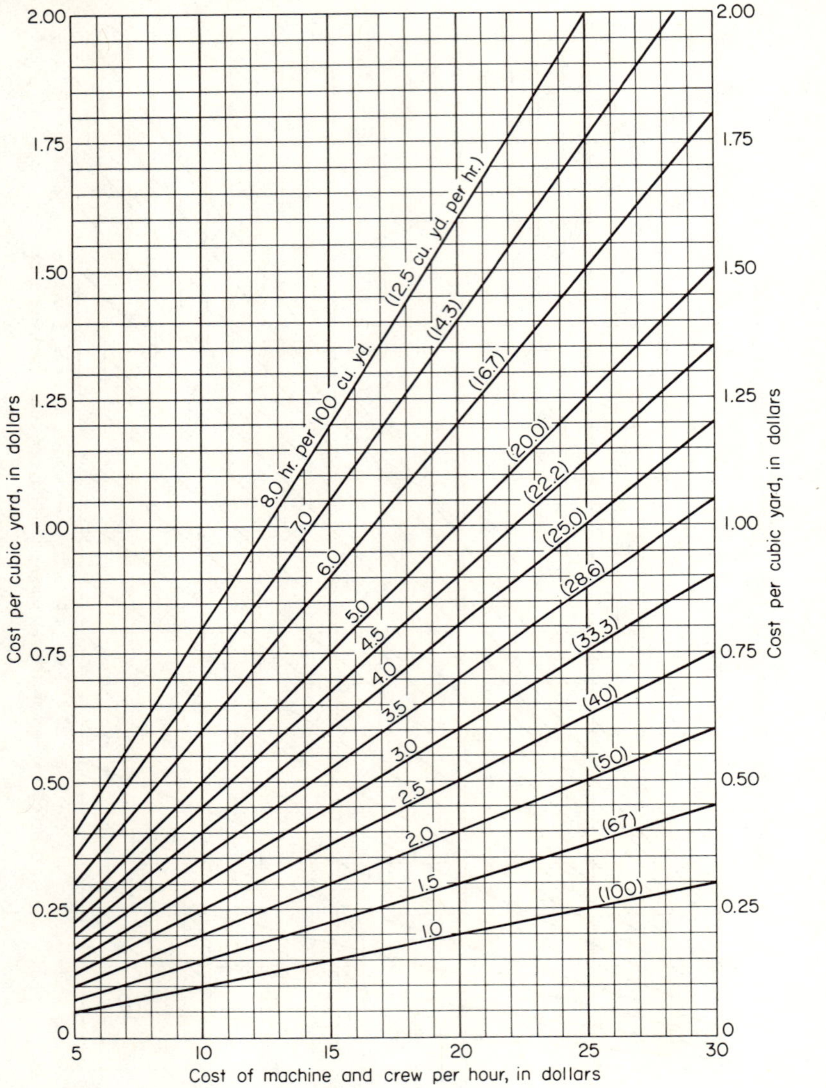

Diagram 3-2 Cost per cubic yard of trenching and backfilling by machine and crew. Cost of transporting the machine to and from the job is not included.

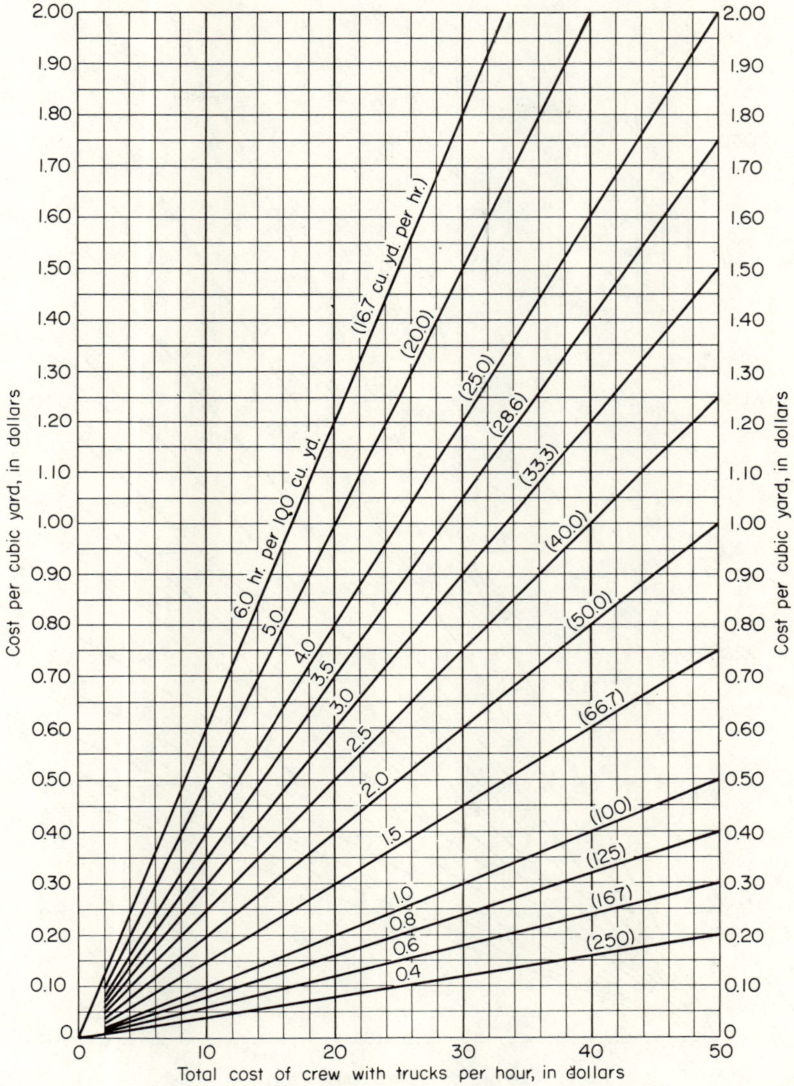

Diagram 3-3 Cost per cubic yard of excavating soil with hand labor and trucks.

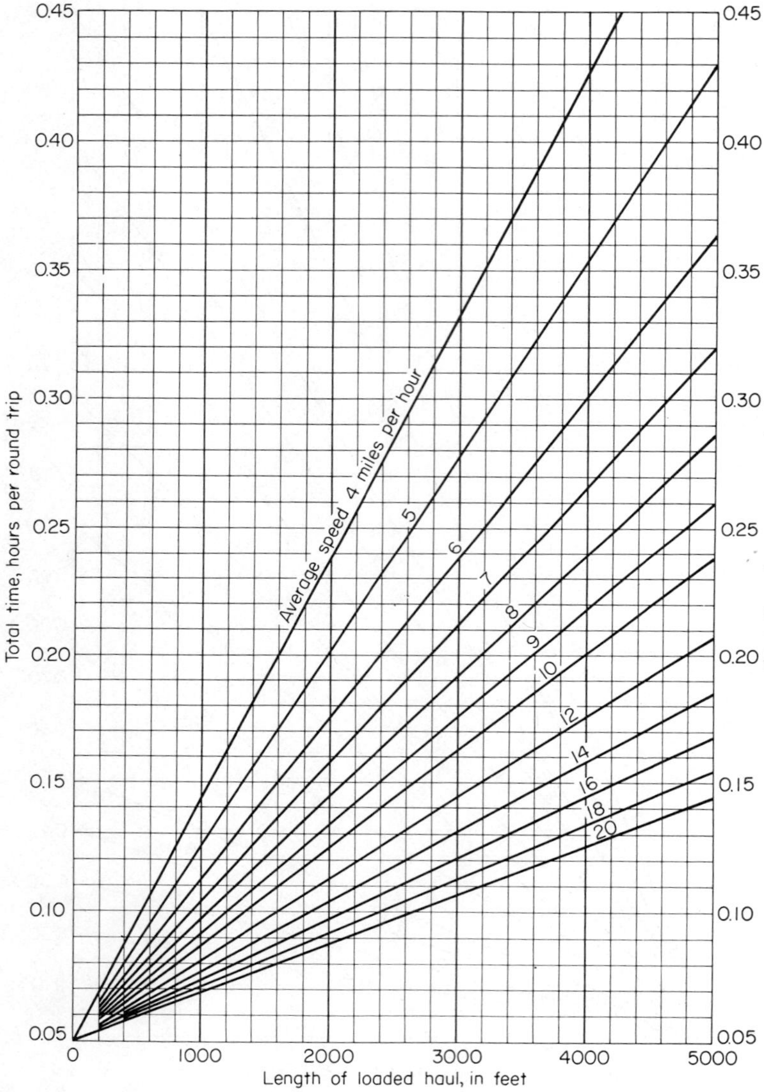

Diagram 3-4 Hours per round trip required for tractor-drawn scrapers with an allowance of 0.05 hr (3 min) per round trip for loading, unloading, turning, and minor delays.

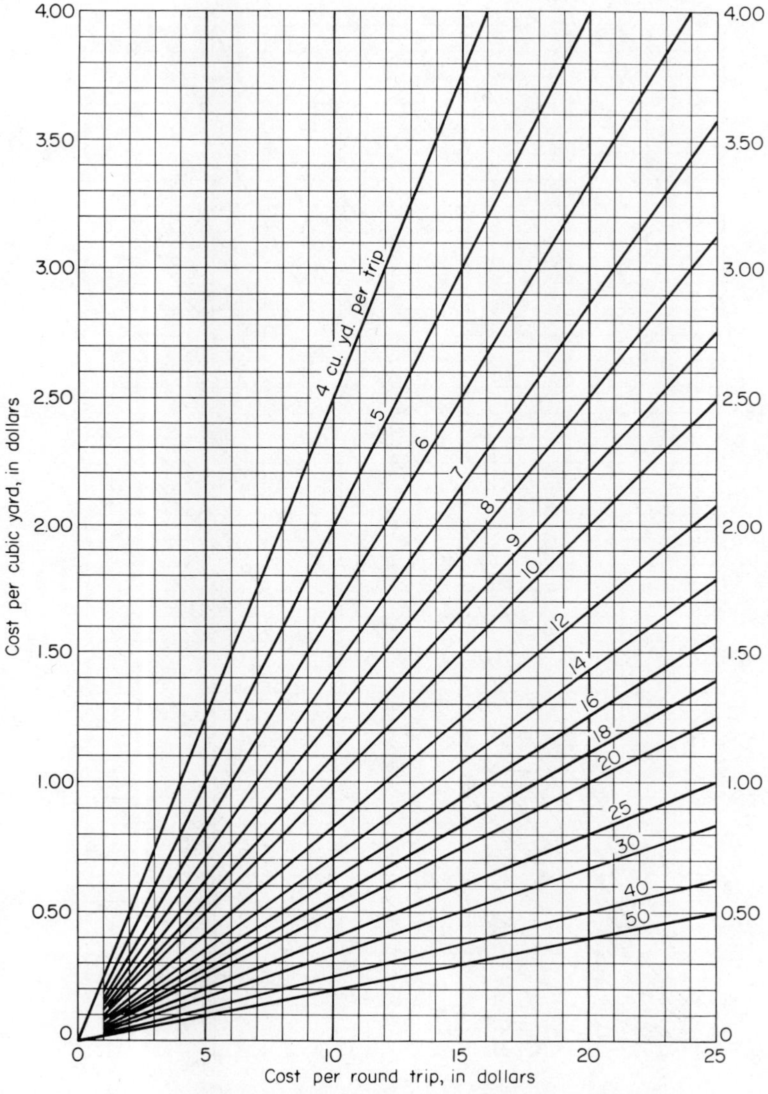

Diagram 3-5 Cost per cubic yard of excavating soil with tractor-drawn scrapers.

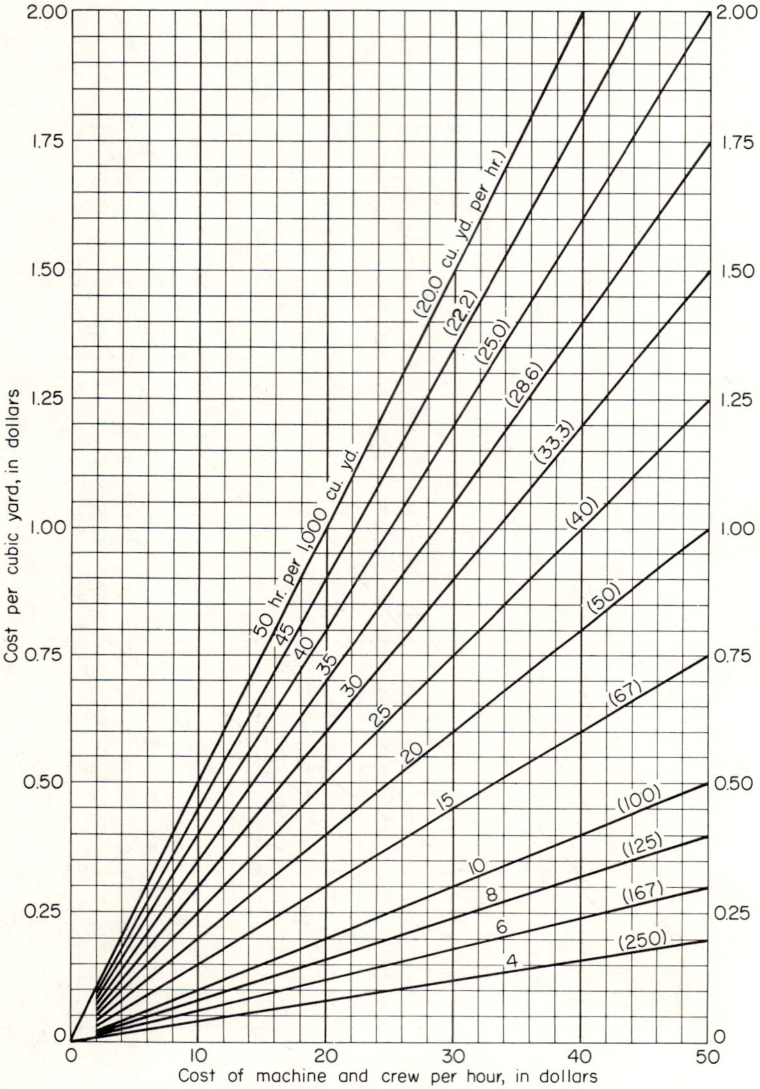

Diagram 3-6 Cost per cubic yard for digging and loading or spreading soil with power equipment.

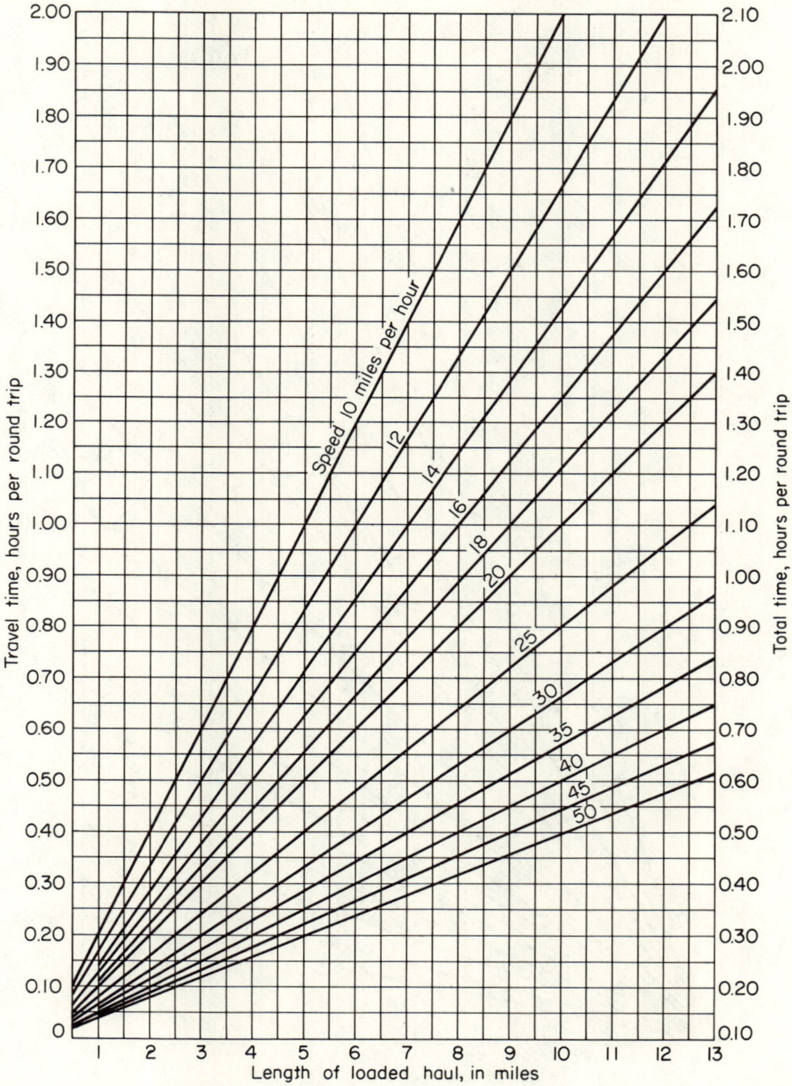

Diagram 3-7 Time in hours required for one round trip for transporting soil. Left-hand scale shows travel time only. Right-hand scale shows total time including 0.10 hr for loading, unloading, and minor delays.

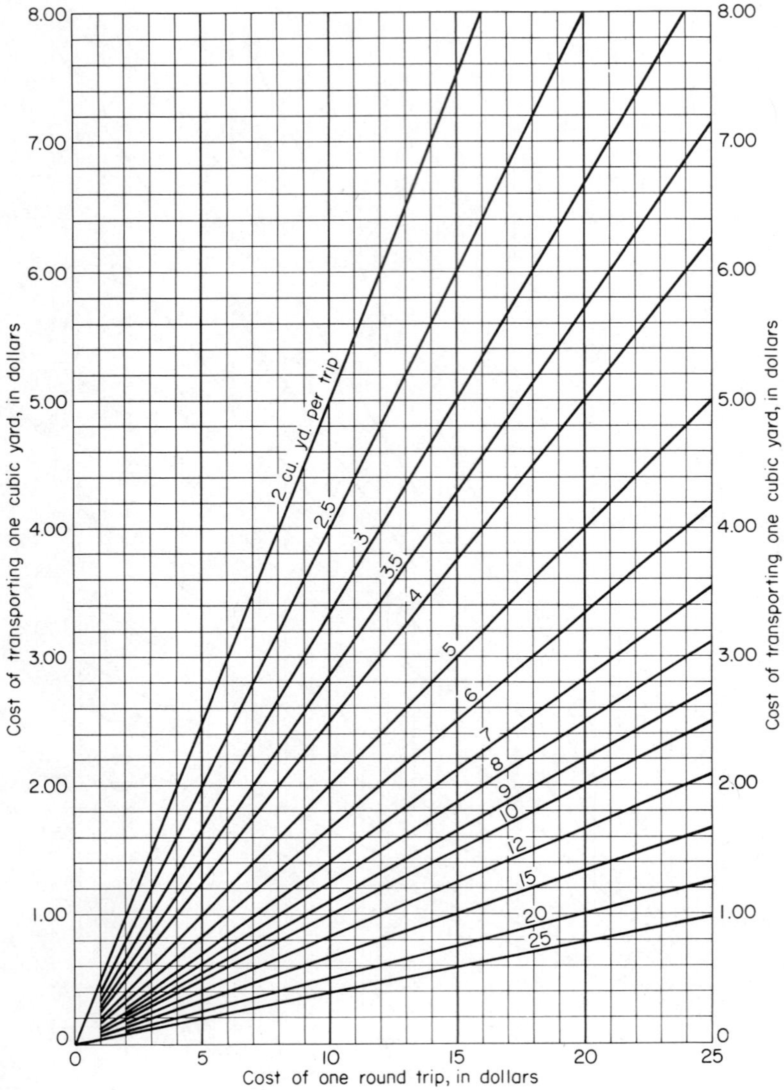

Diagram 3-8 Cost per cubic yard of transporting soil.

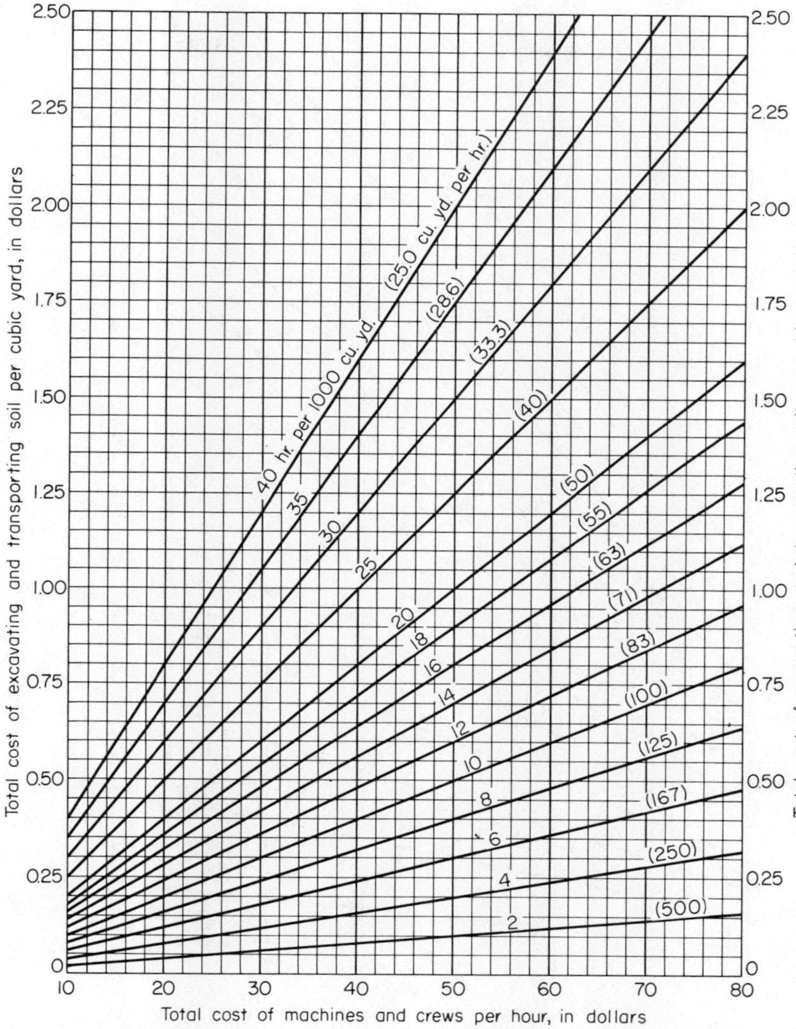

Diagram 3-9 Total cost per cubic yard of excavating and transporting soil with all power equipment.

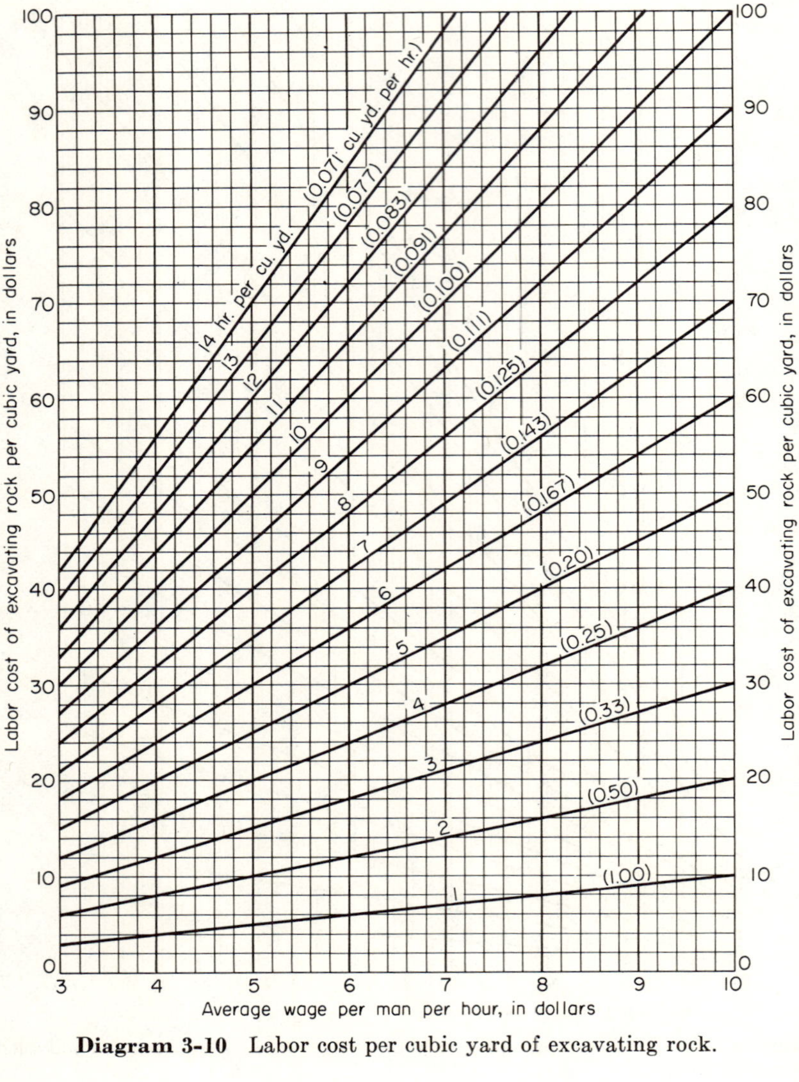

Diagram 3-10 Labor cost per cubic yard of excavating rock.

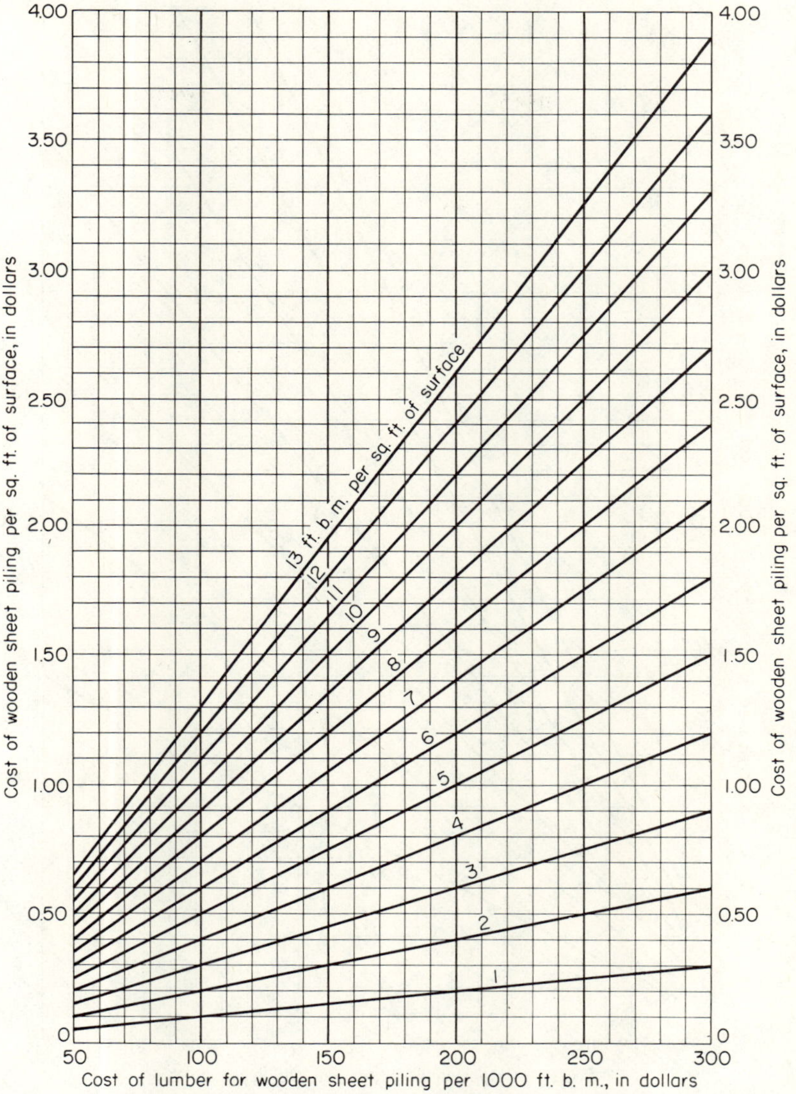

Diagram 4-1 Lumber cost of wooden sheet piling per square foot of surface.

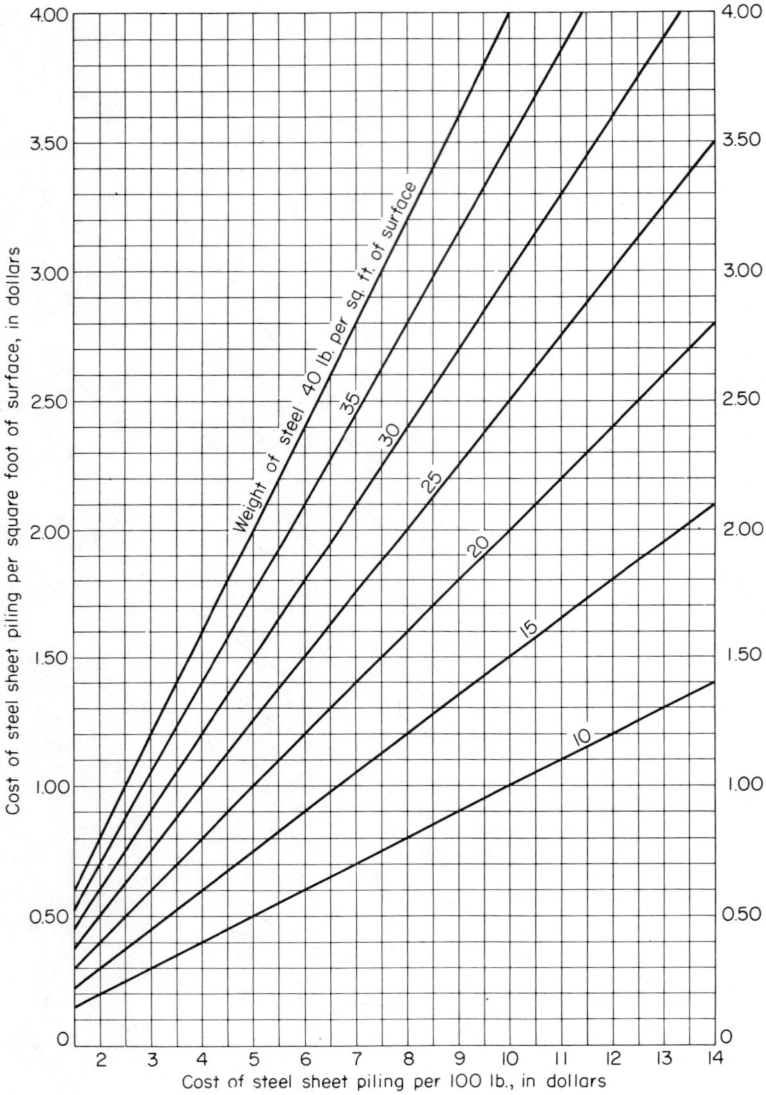

Diagram 4-2 Cost of steel sheet piling per square foot of surface.

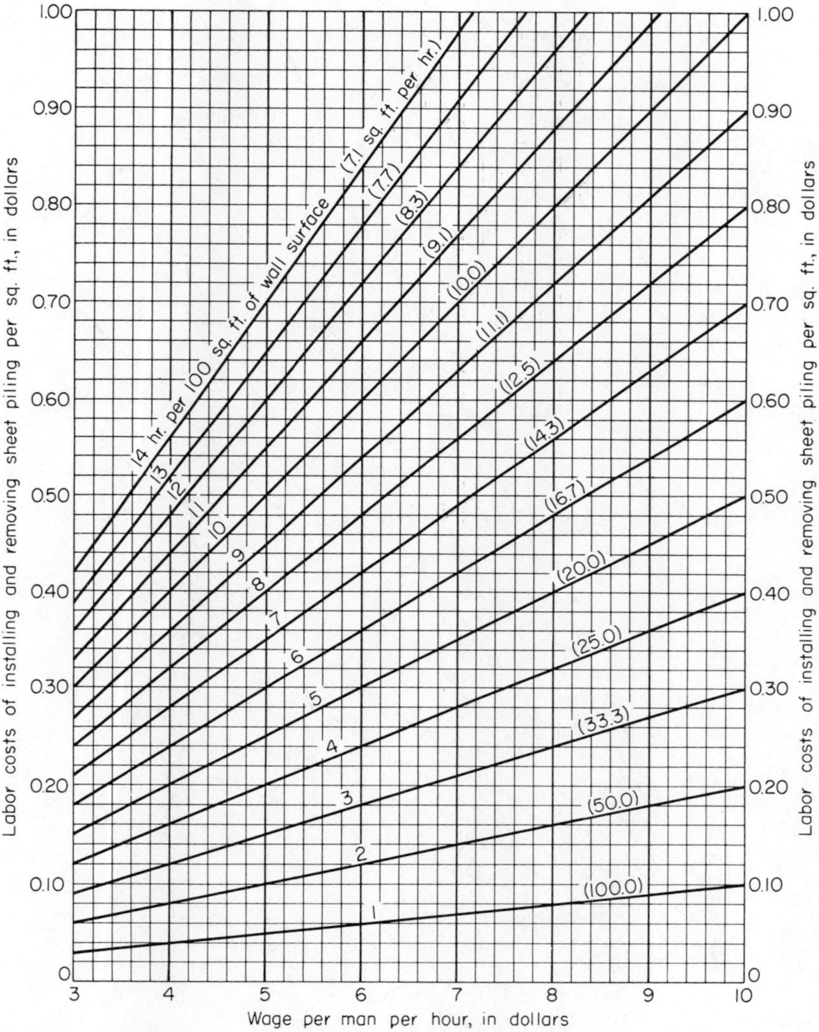

Diagram 4-3 Labor costs per square foot of surface of installing and removing sheet piling.

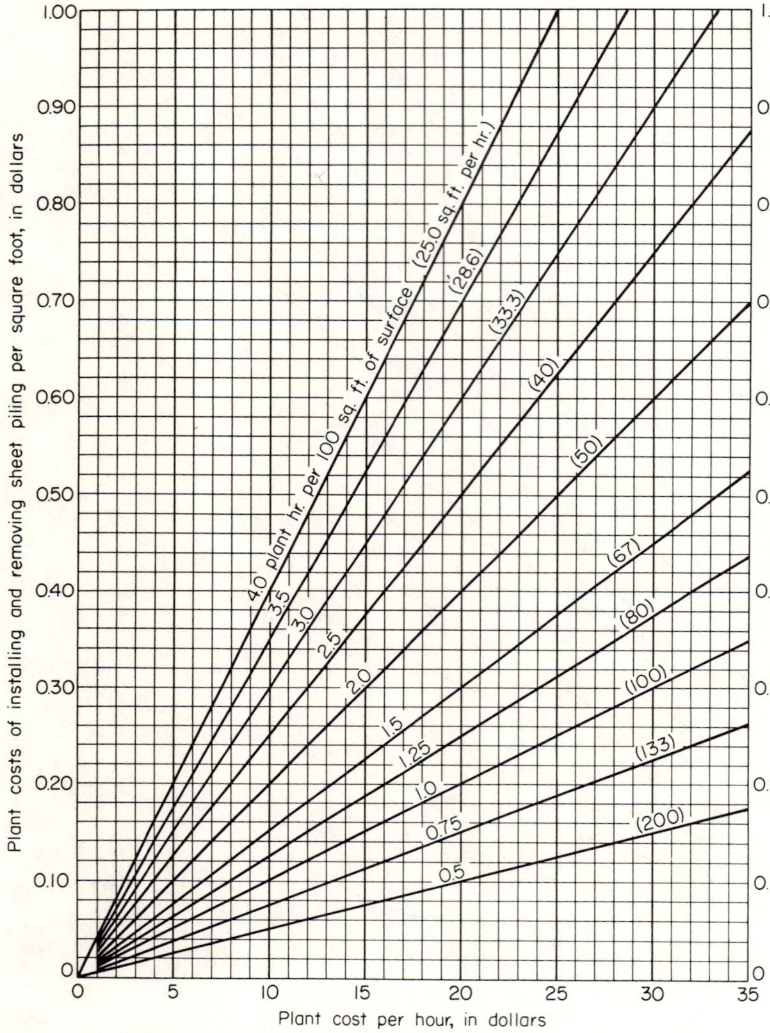

Diagram 4-4 Plant or equipment costs per square foot of surface for installing and removing sheet piling.

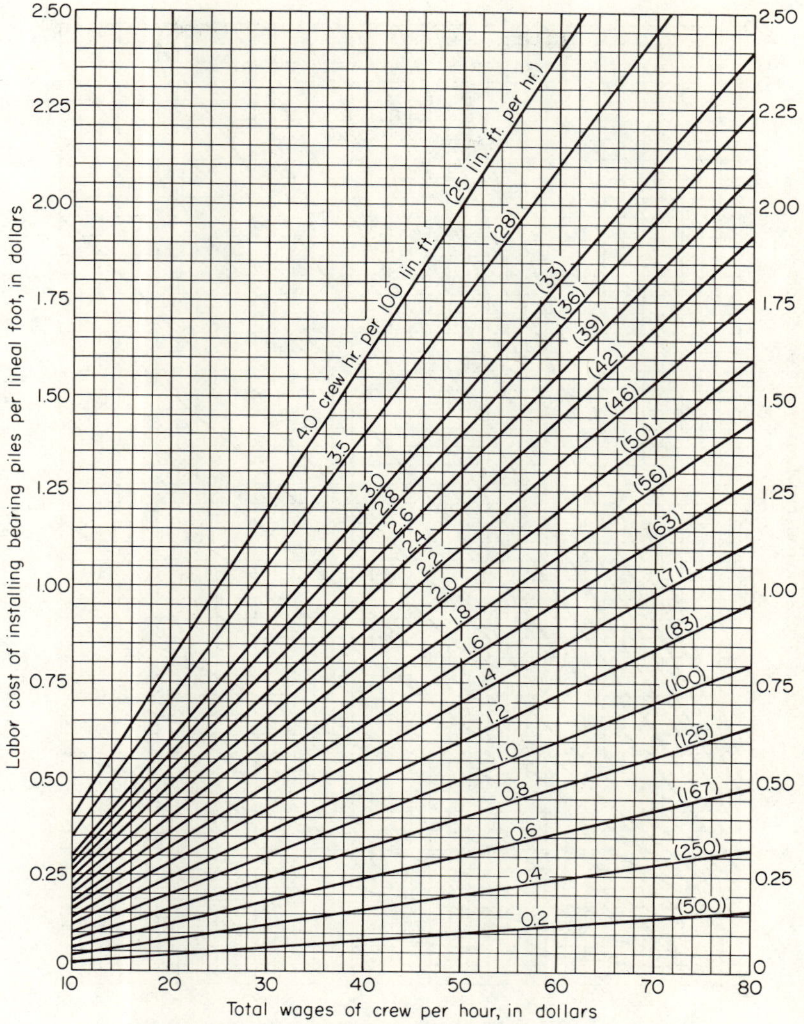

Diagram 4-5 Labor cost per lineal foot of installing bearing piles.

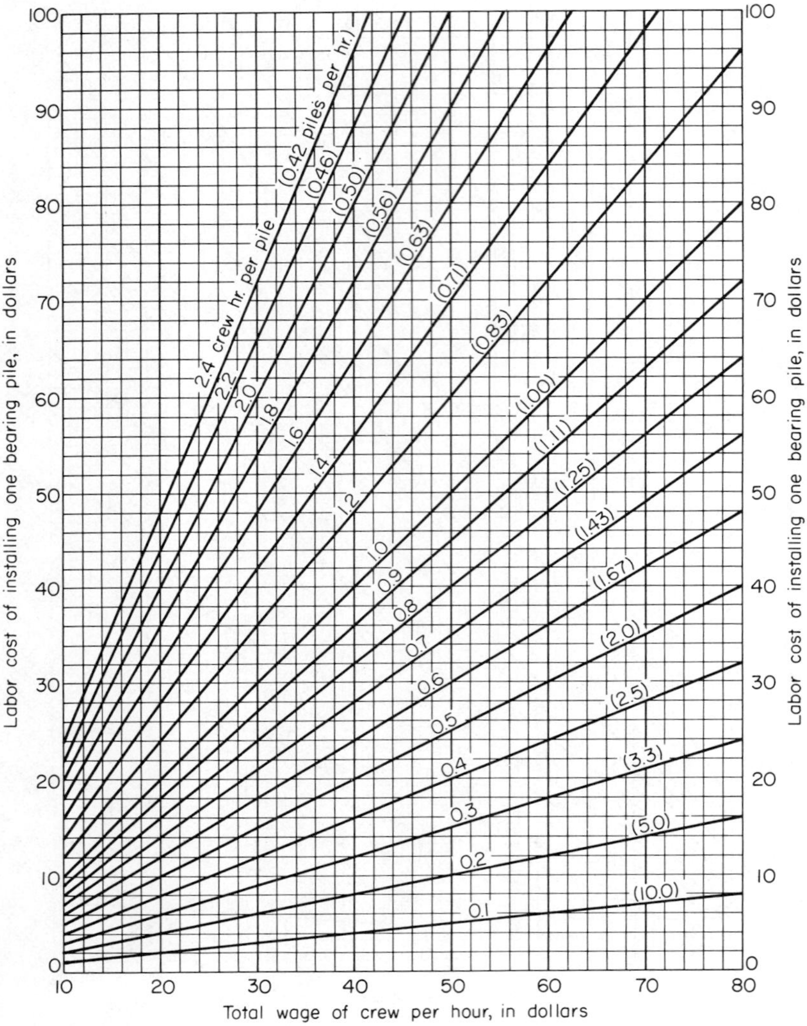

Diagram 4-6 Labor cost per pile of installing bearing piles.

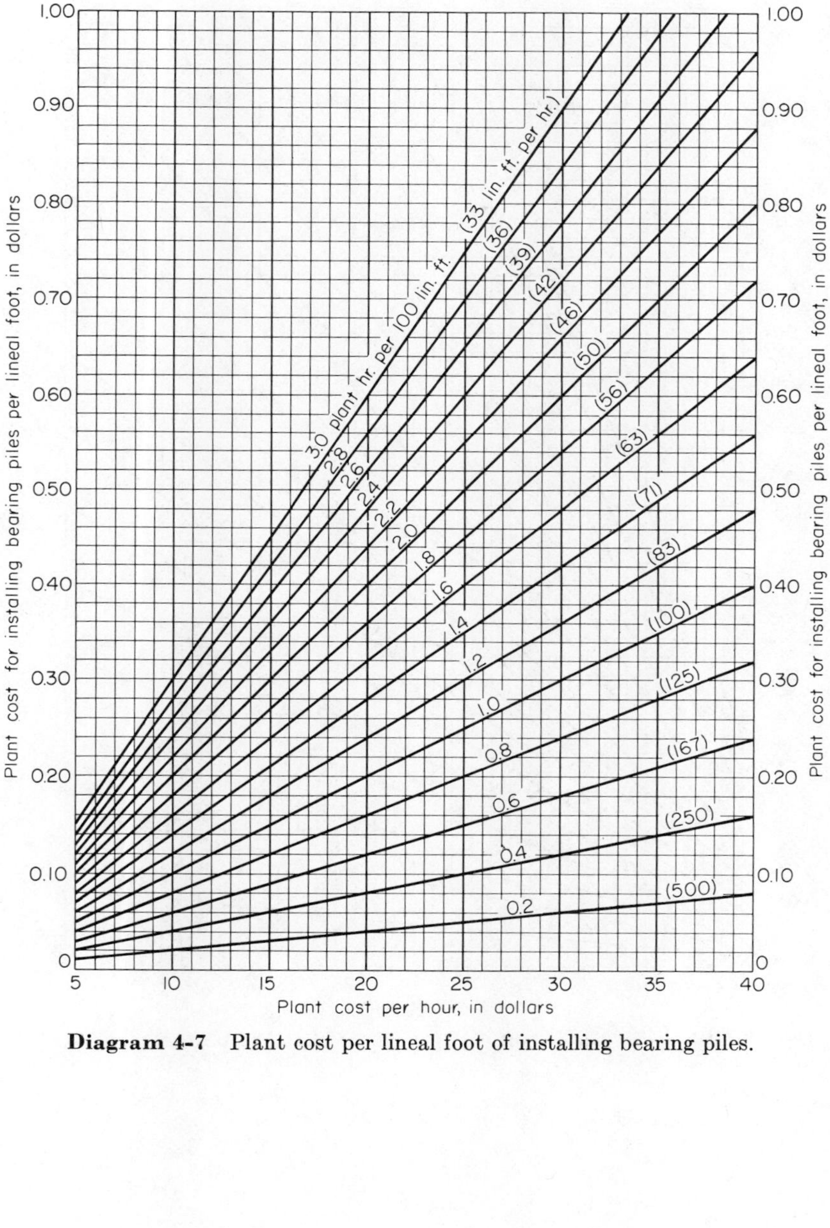

Diagram 4-7 Plant cost per lineal foot of installing bearing piles.

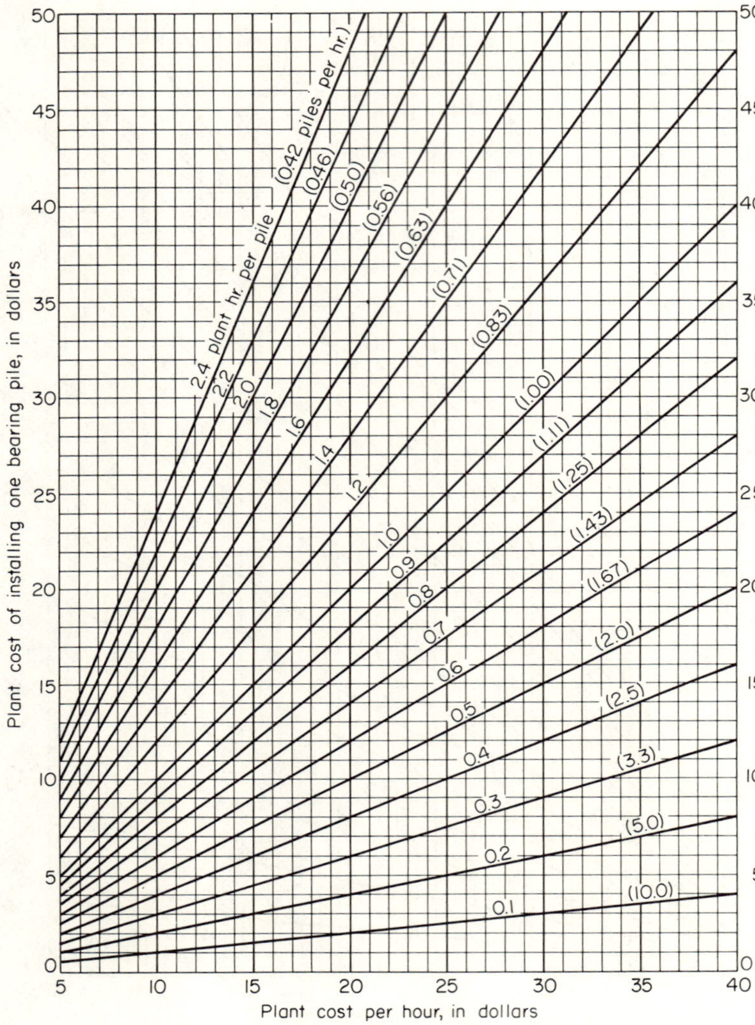

Diagram 4-8 Plant cost per pile of installing bearing piles.

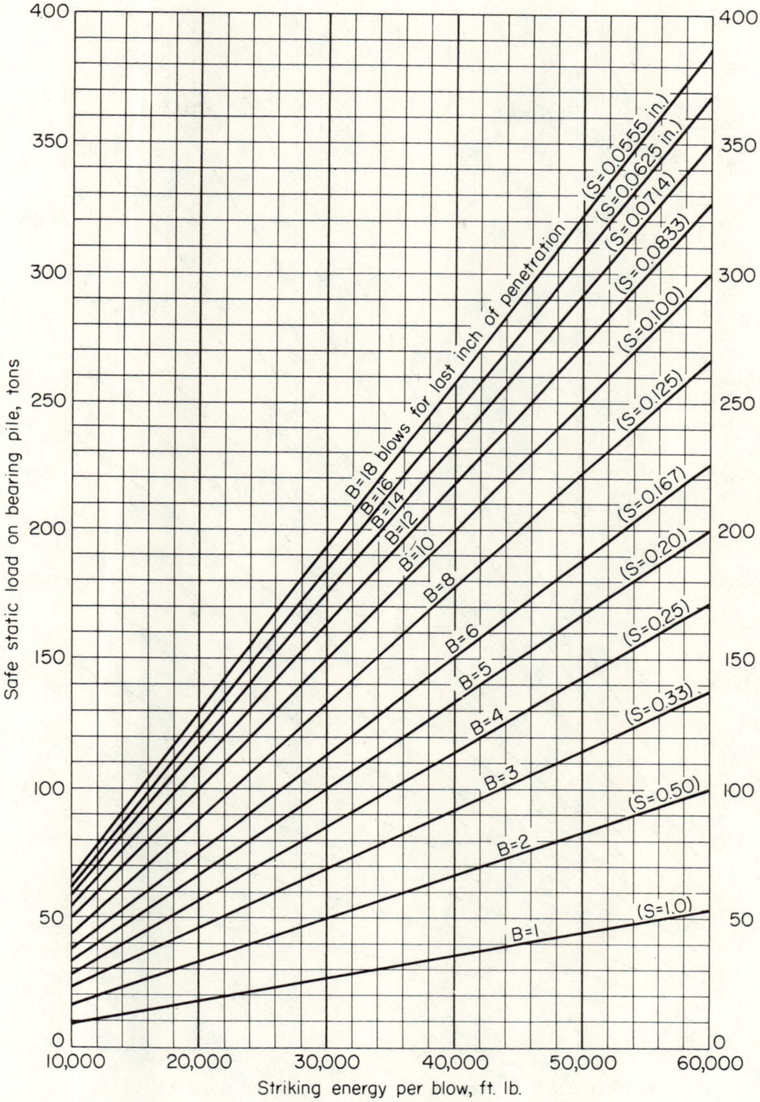

Diagram 4-9 Safe static load on bearing piles driven by power hammers. Based on the *Engineering News-Record* formula given in the article on Placing and Driving, of Chap. 4.

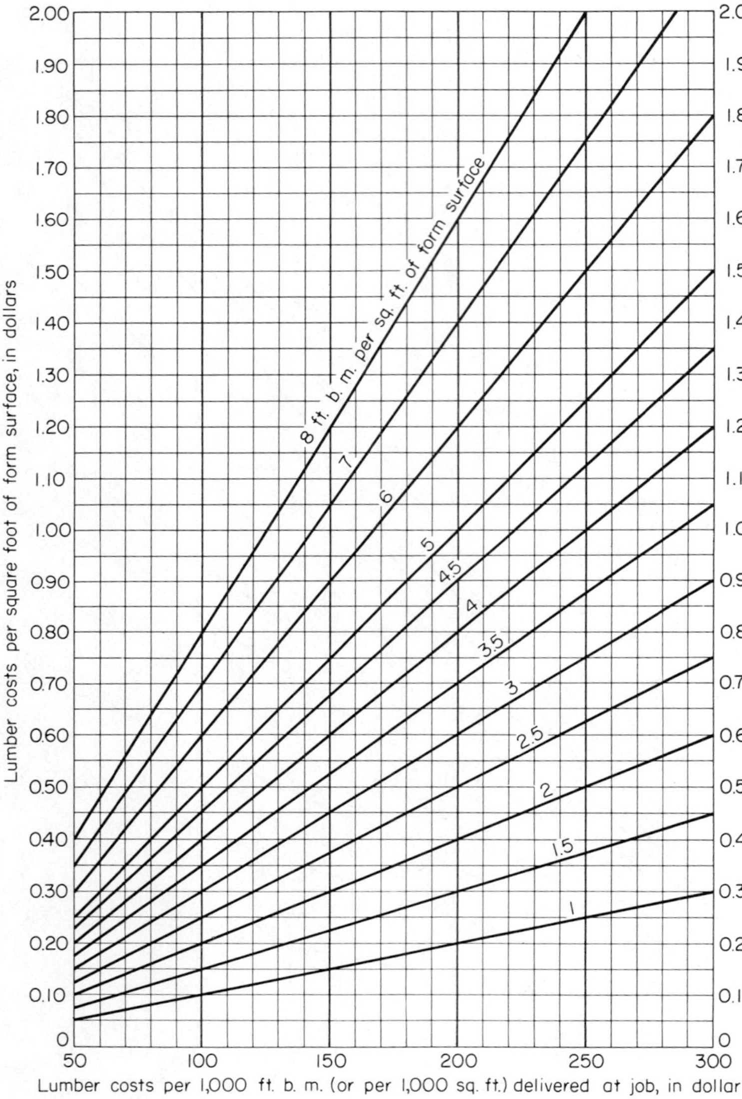

Diagram 5-1 Lumber costs for forms for concrete.

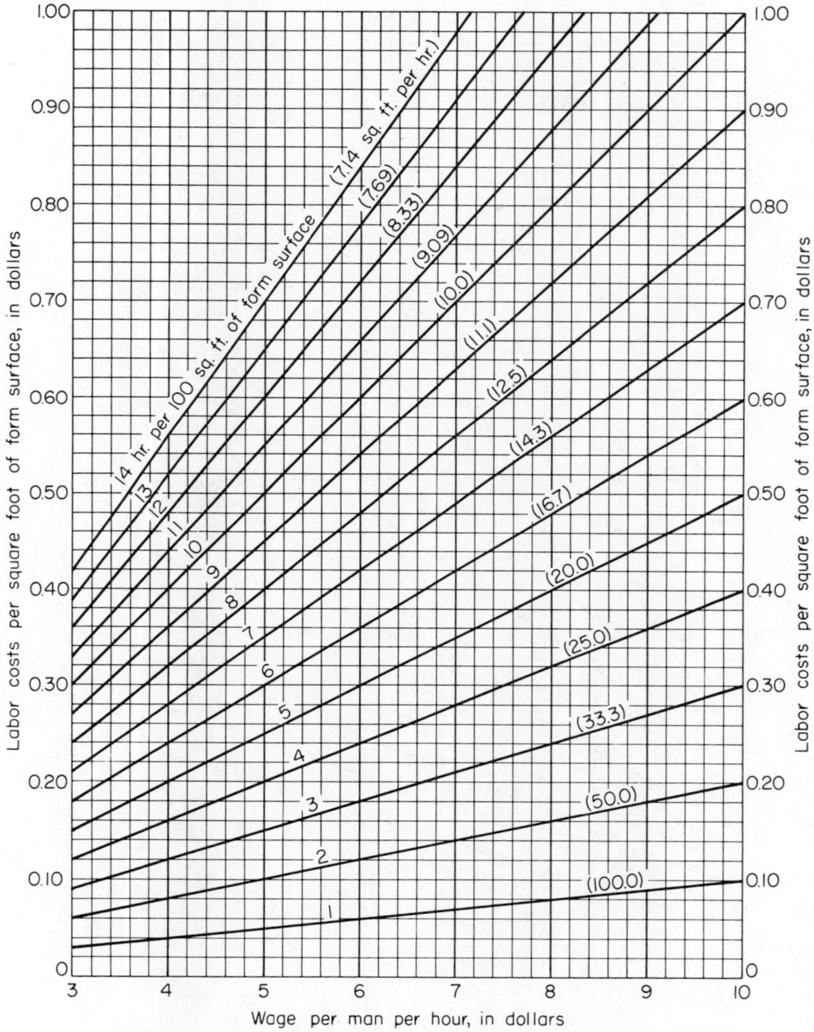

Diagram 5-2 Labor costs per square foot of form surface.

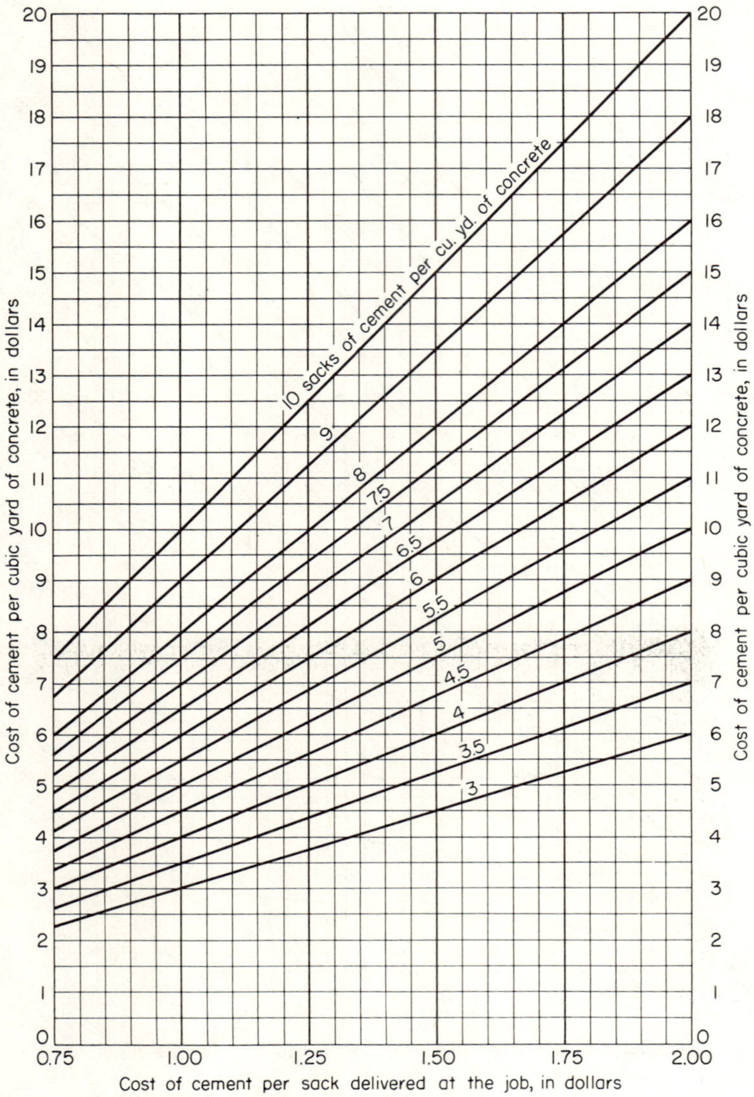

Diagram 5-3 Cost of cement per cubic yard of concrete.

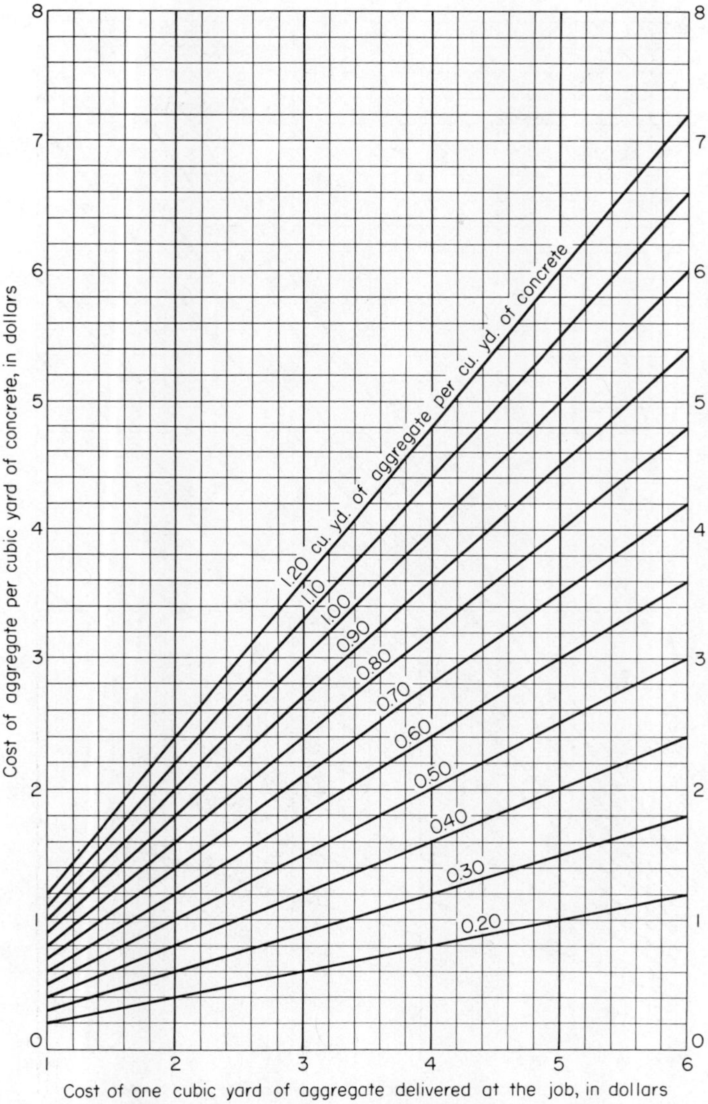

Diagram 5-4 Cost of aggregate per cubic yard of concrete. Aggregate measured and priced per cubic yard.

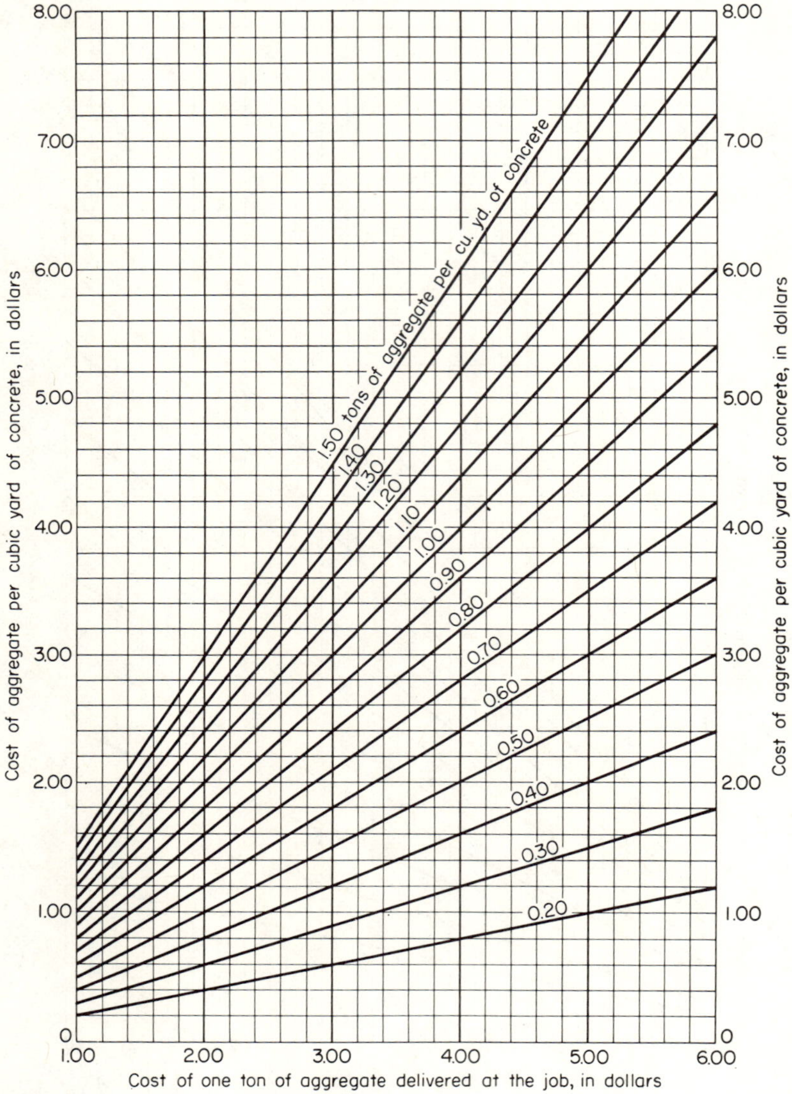

Diagram 5-5 Cost of aggregate per cubic yard of concrete. Aggregate measured and priced per ton.

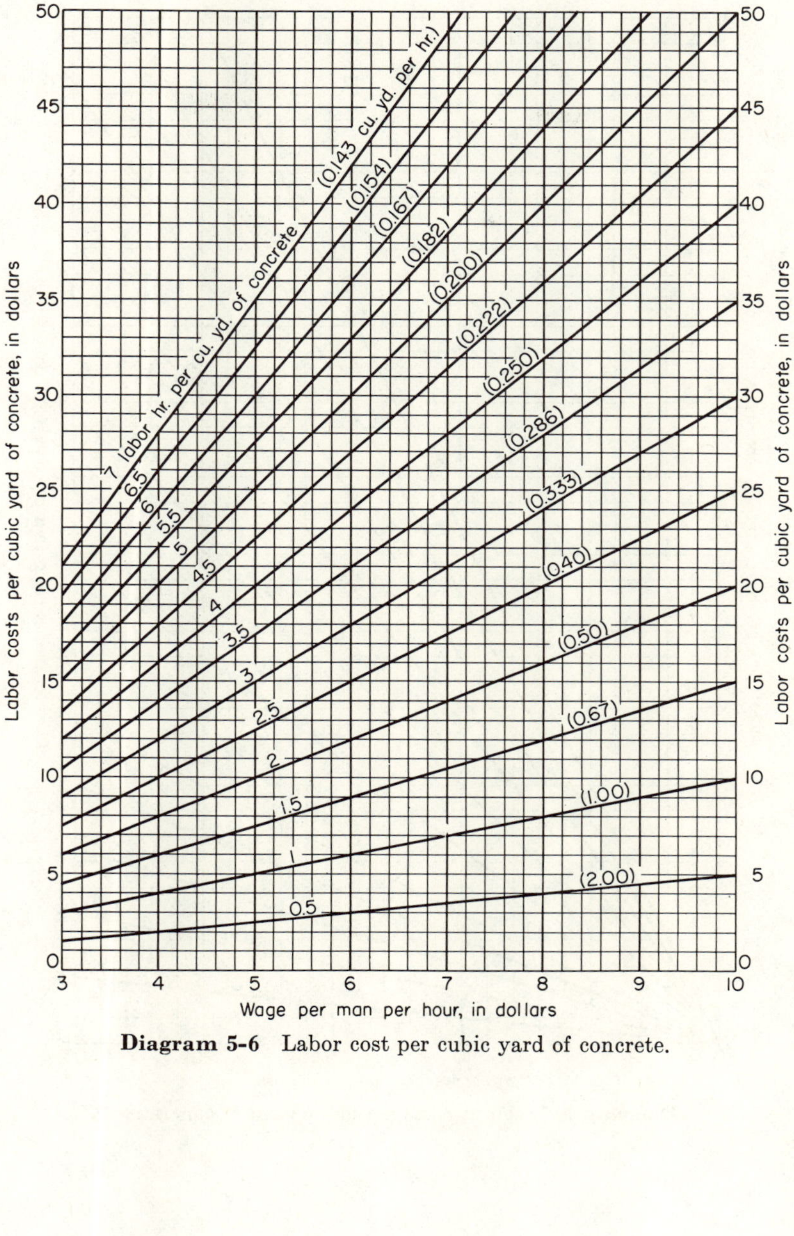

Diagram 5-6 Labor cost per cubic yard of concrete.

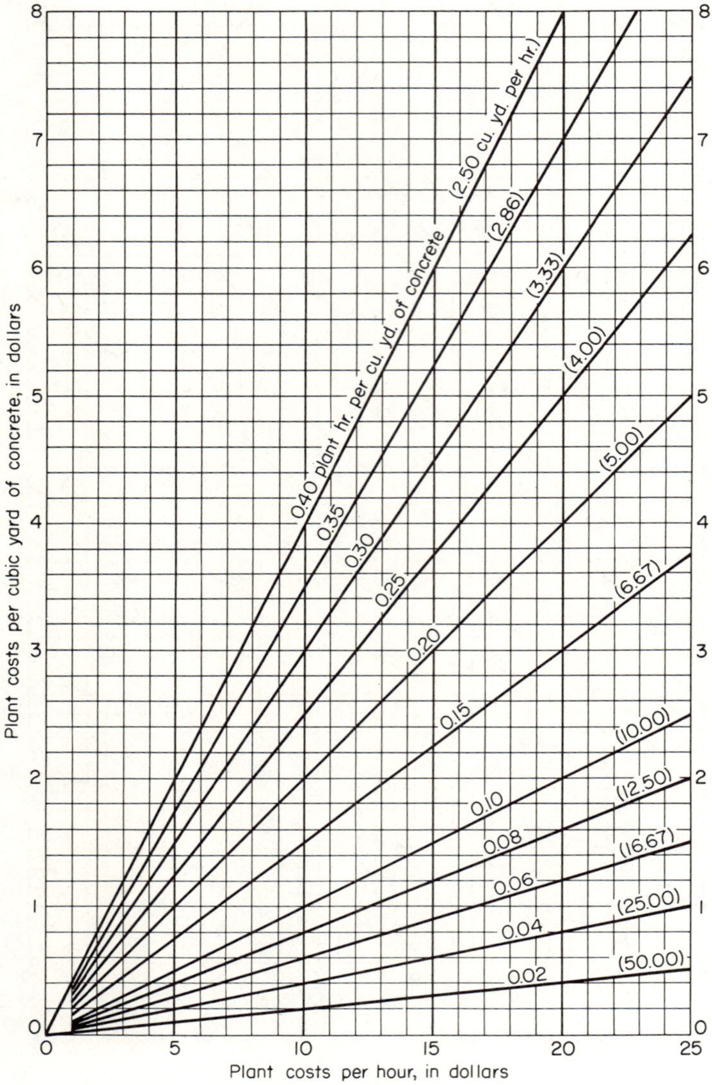

Diagram 5-7 Plant costs per cubic yard of concrete.

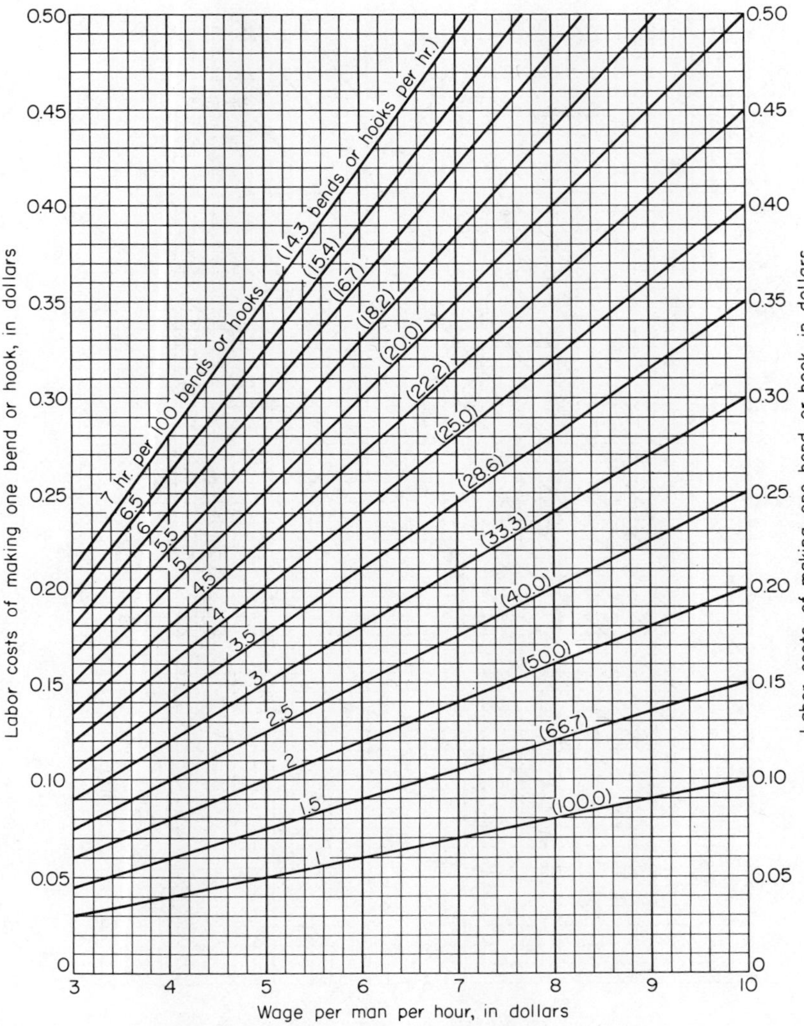

Diagram 5-8 Labor costs of making one-eighth or one-quarter bend or one hook.

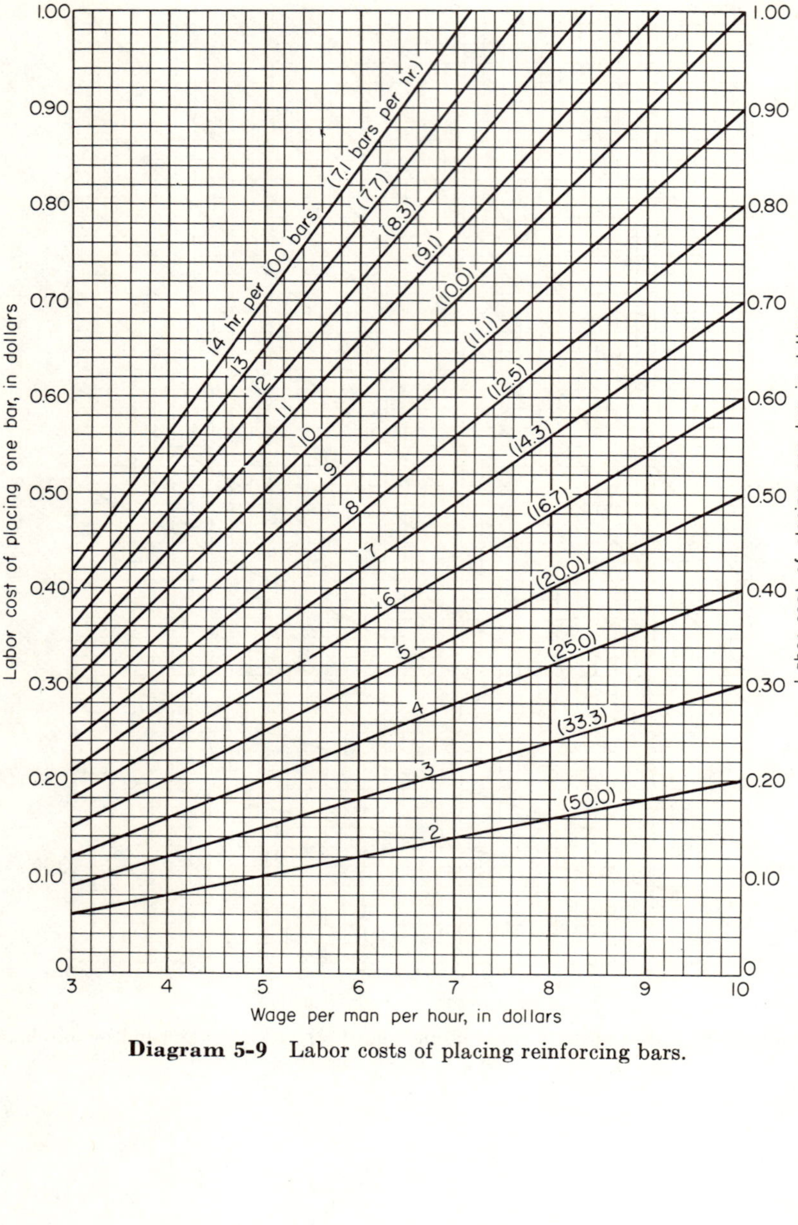

Diagram 5-9 Labor costs of placing reinforcing bars.

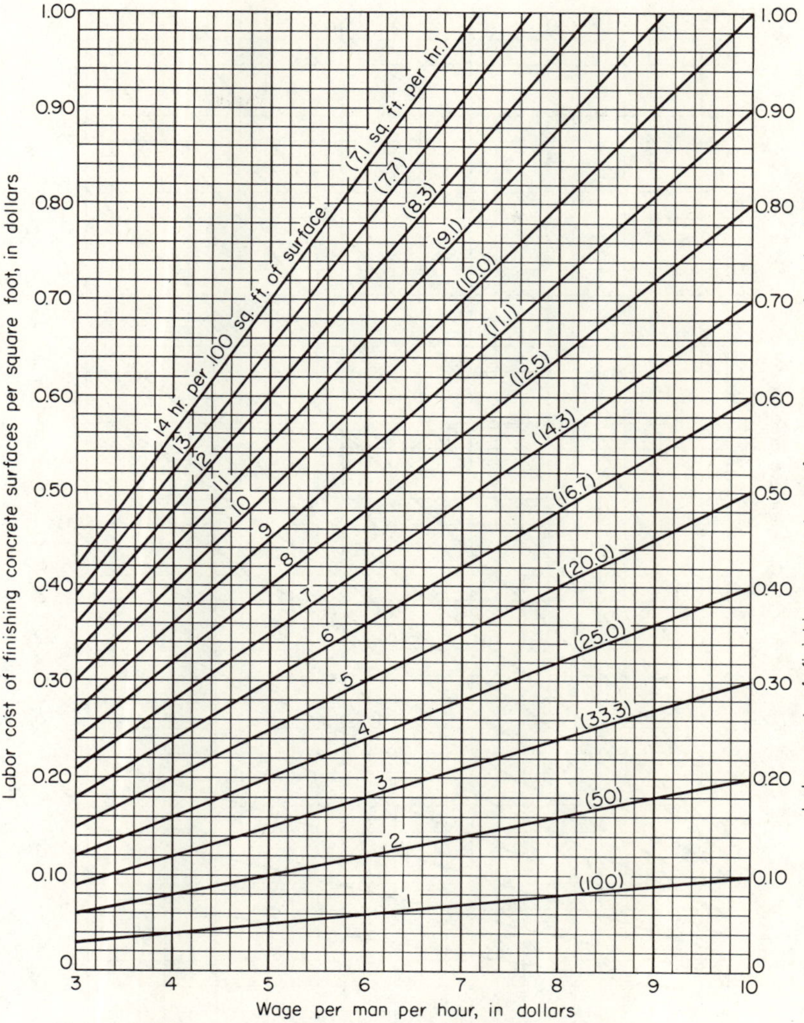

Diagram 5-10 Labor costs of finishing concrete surfaces.

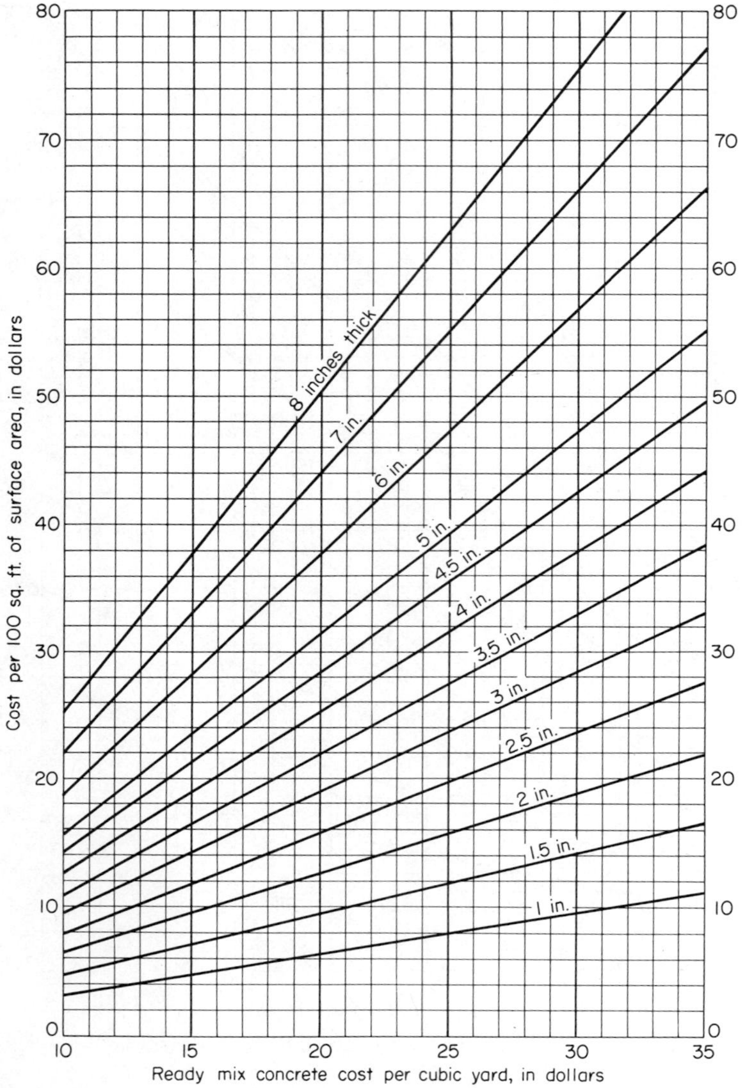

Diagram 5-11 Ready mix concrete cost per 100 sq ft of surface area, with an allowance of 2 percent for waste.

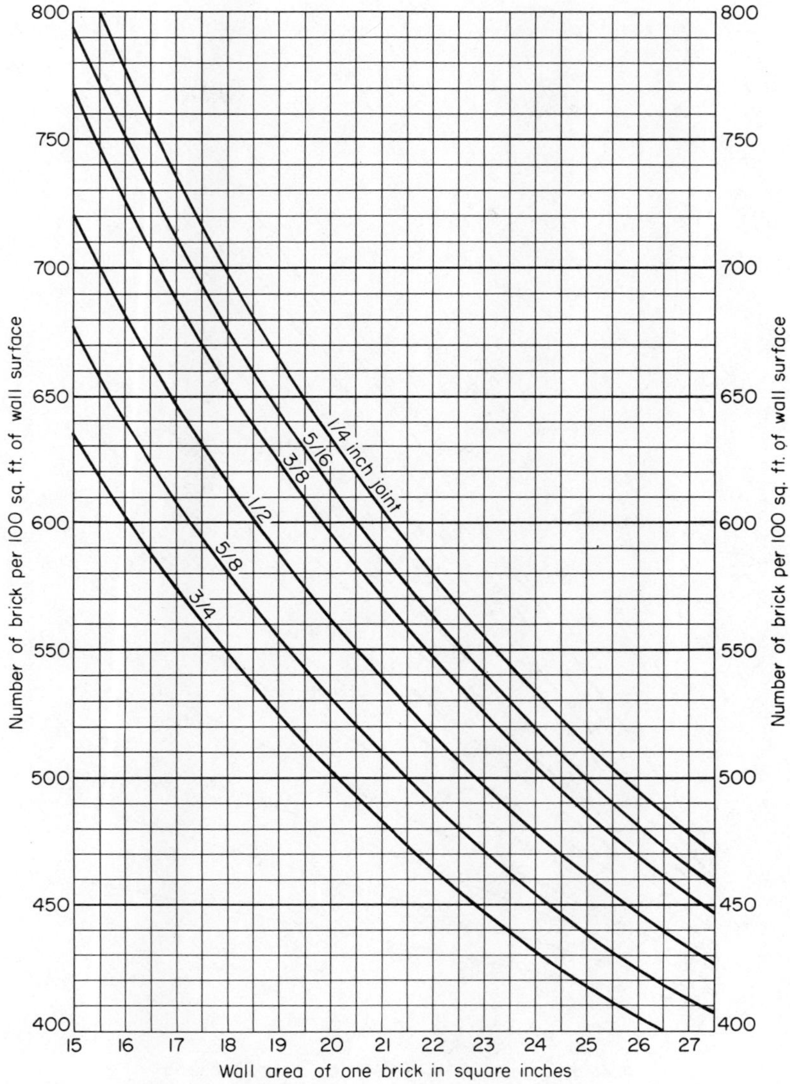

Diagram 6-1 Number of brick per 100 sq ft of wall surface. Varying thicknesses of joints.

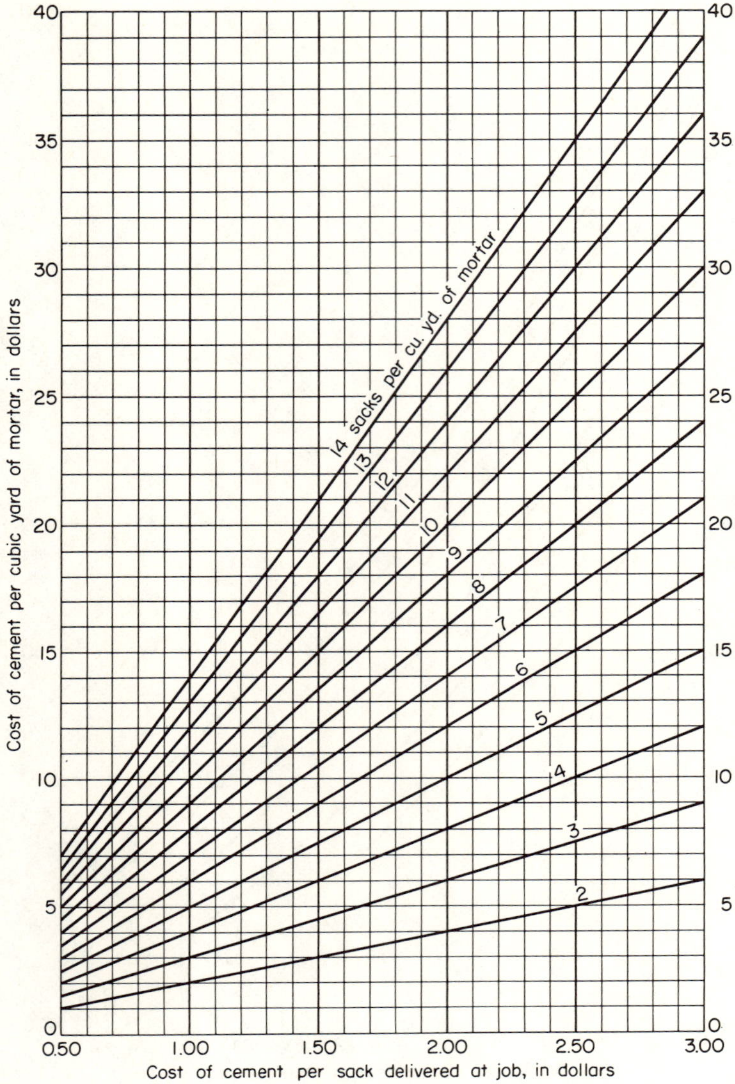

Diagram 6-2 Cost of cement per cubic yard of mortar.

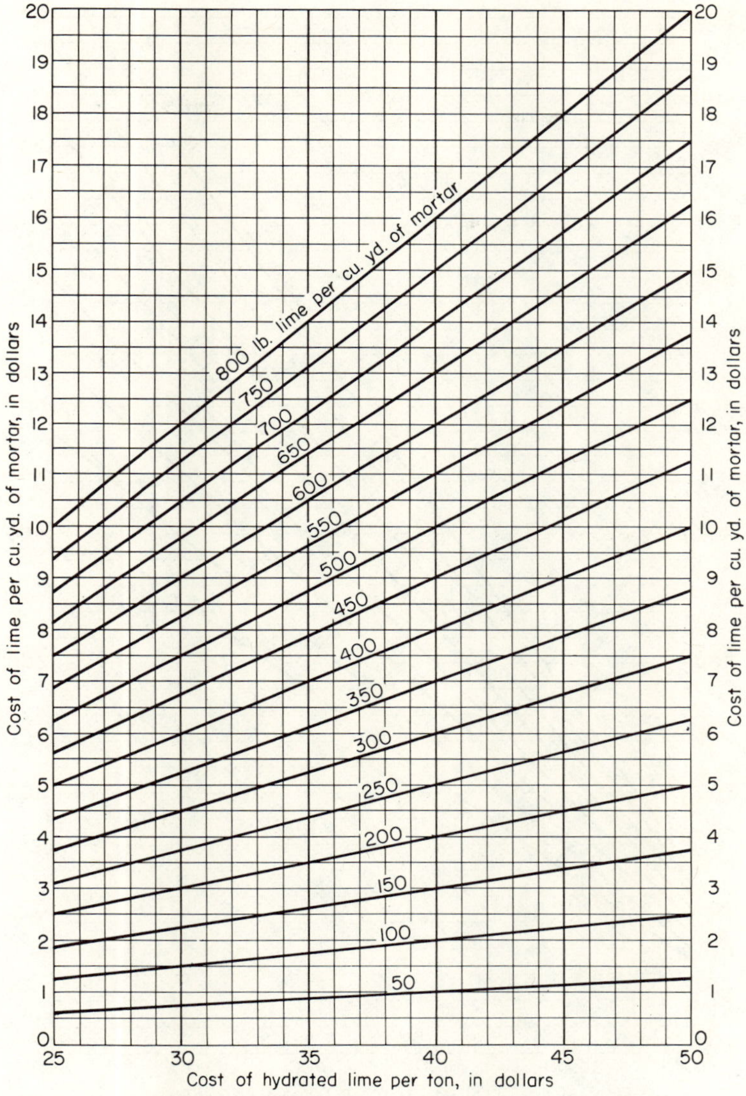

Diagram 6-3 Cost of lime per cubic yard of mortar.

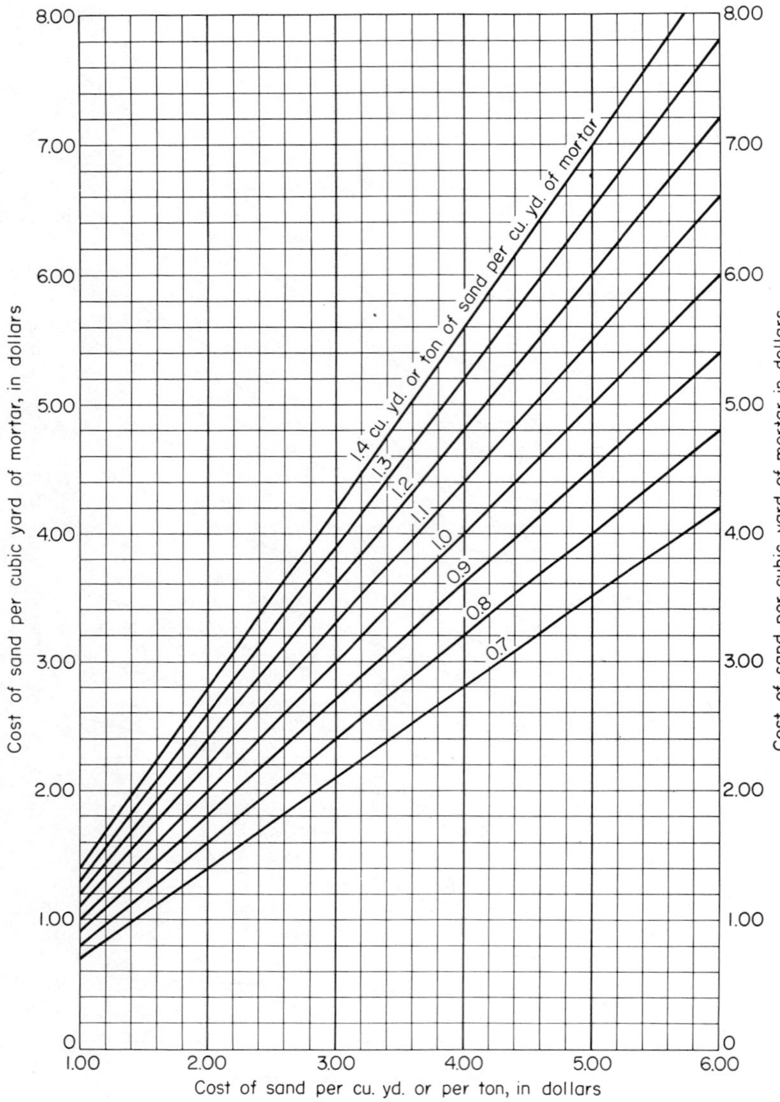

Diagram 6-4 Cost of sand per cubic yard of mortar.

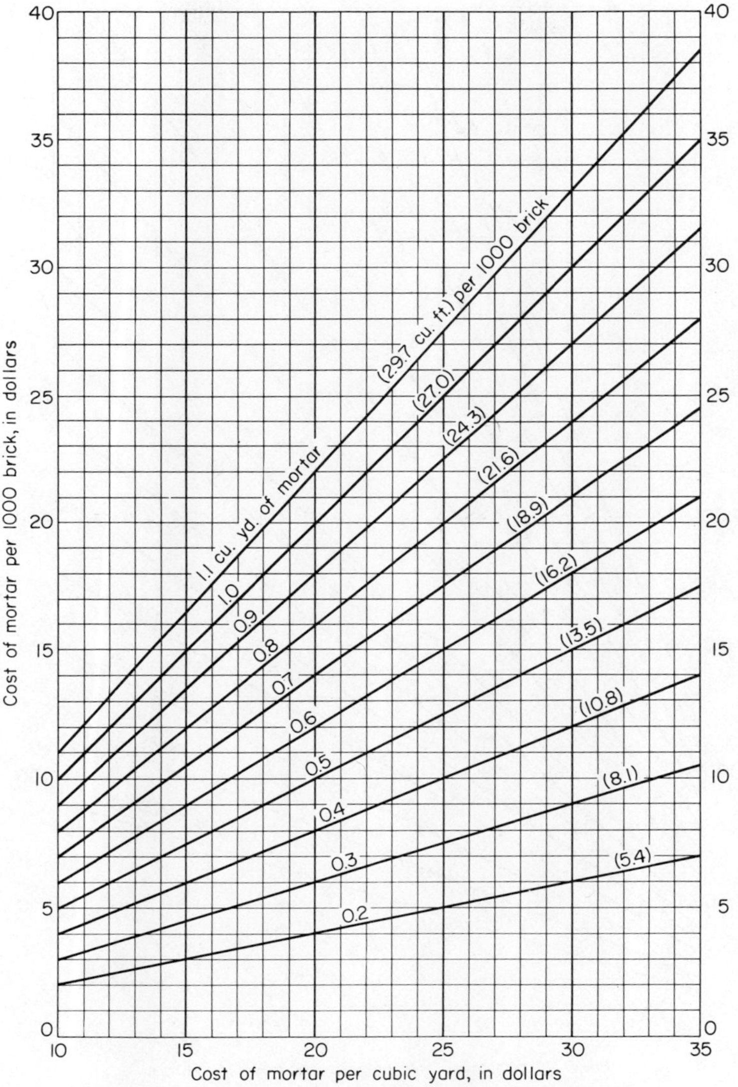

Diagram 6-5 Cost of mortar per 1,000 brick for brick masonry.

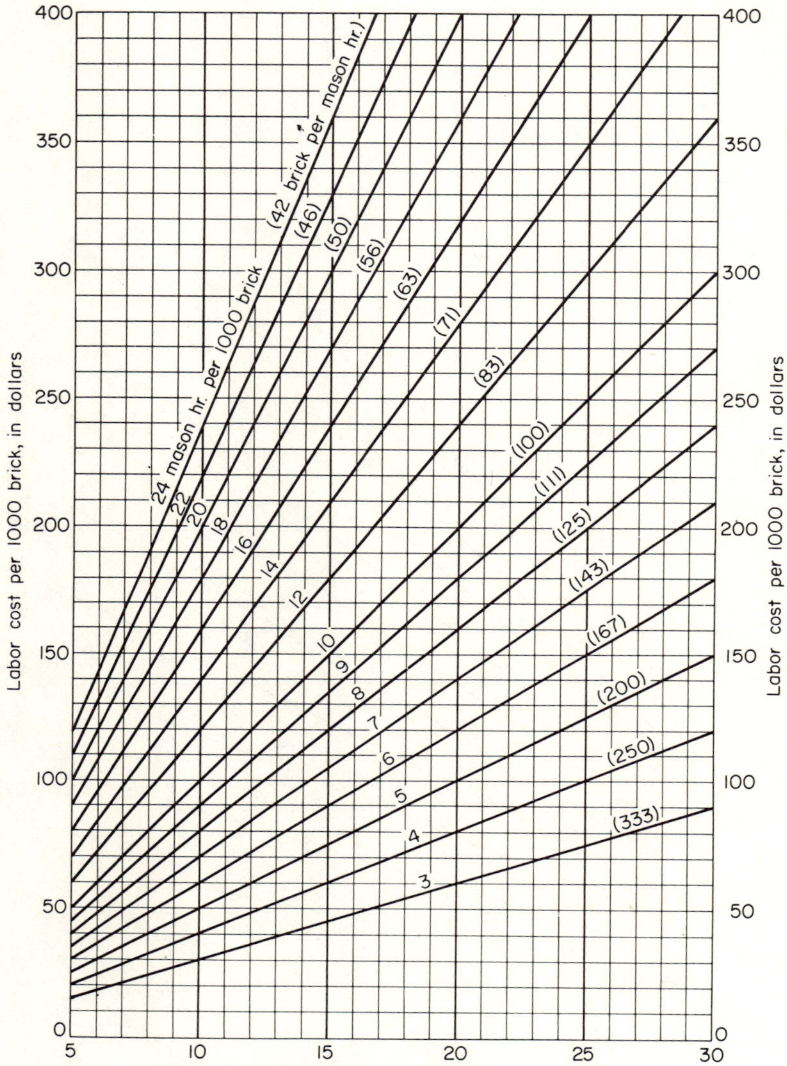

Diagram 6-6 Labor cost of laying brick masonry per 1,000 brick.

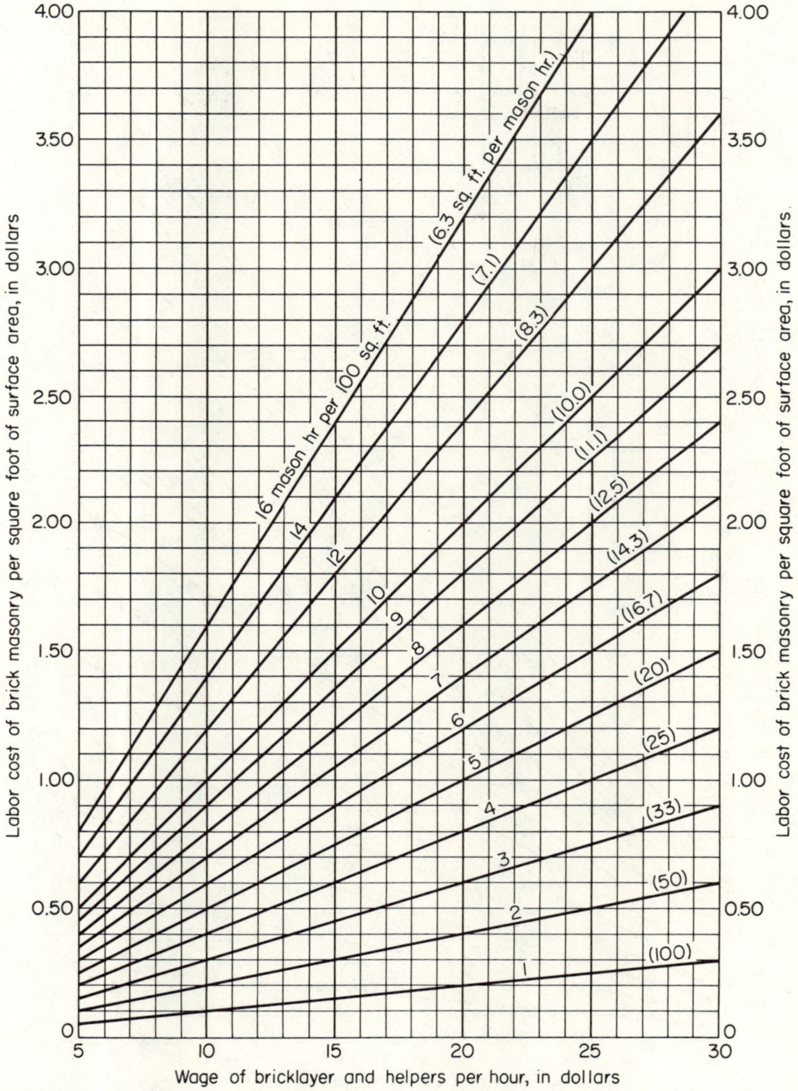

Diagram 6-7 Labor cost per square foot of surface area for brick masonry.

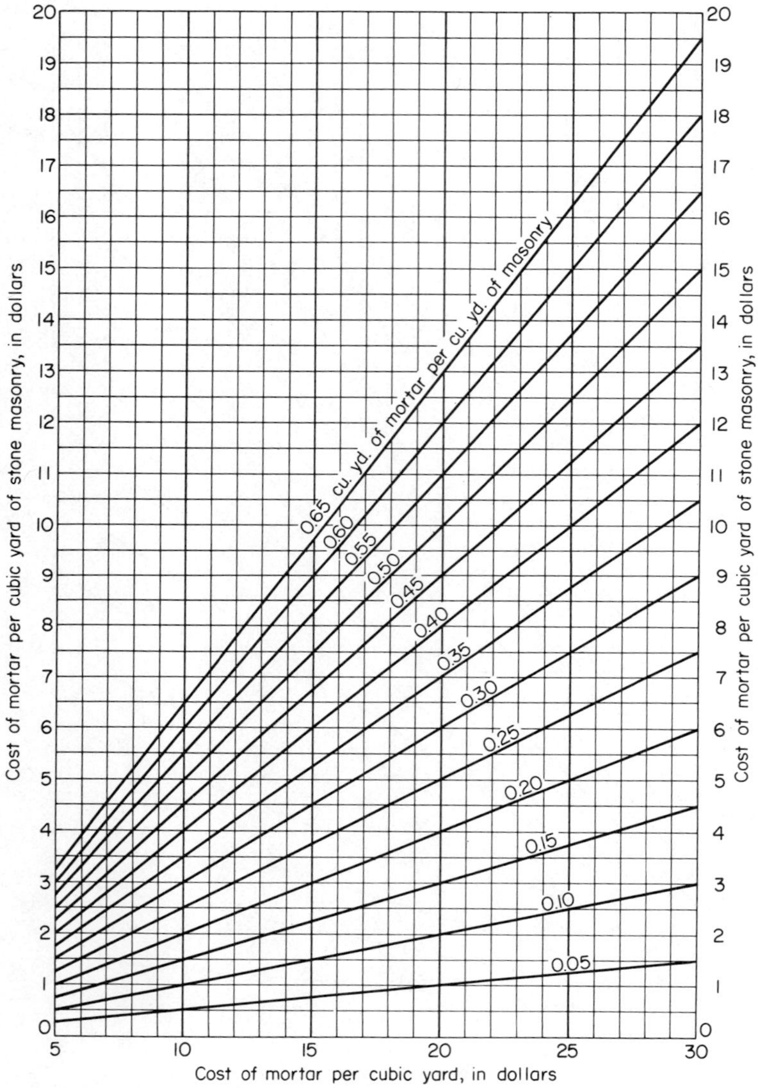

Diagram 6-8 Cost of mortar per cubic yard of stone masonry.

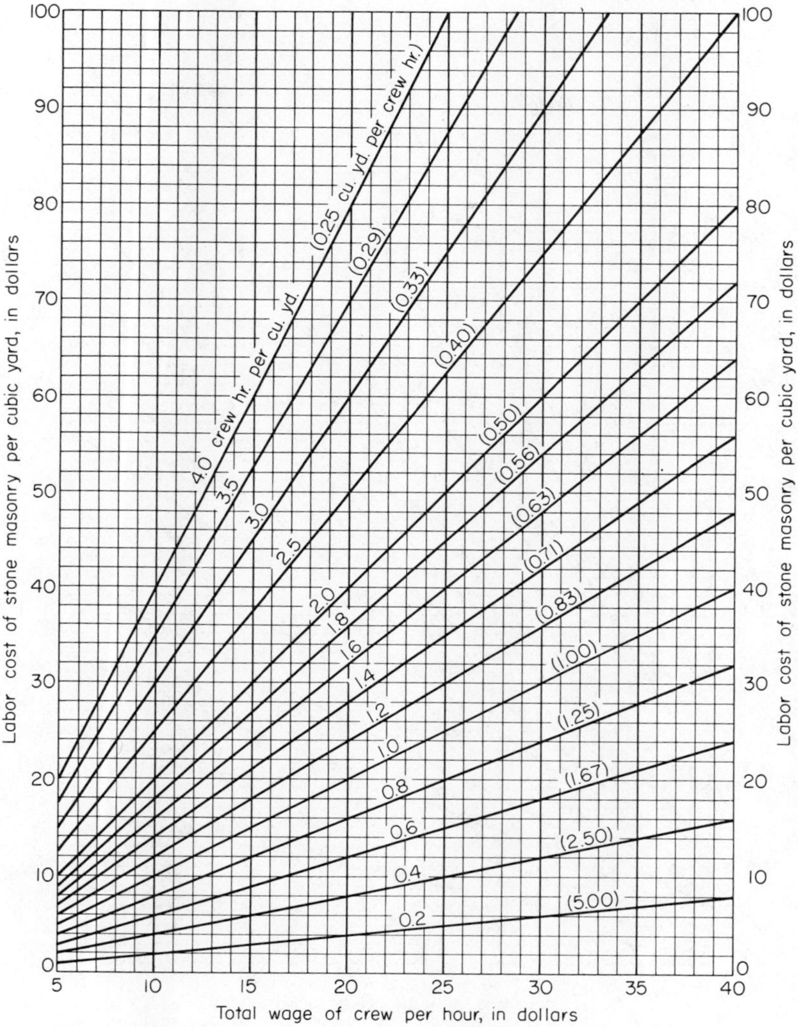

Diagram 6-9 Labor cost per cubic yard for laying stone masonry.

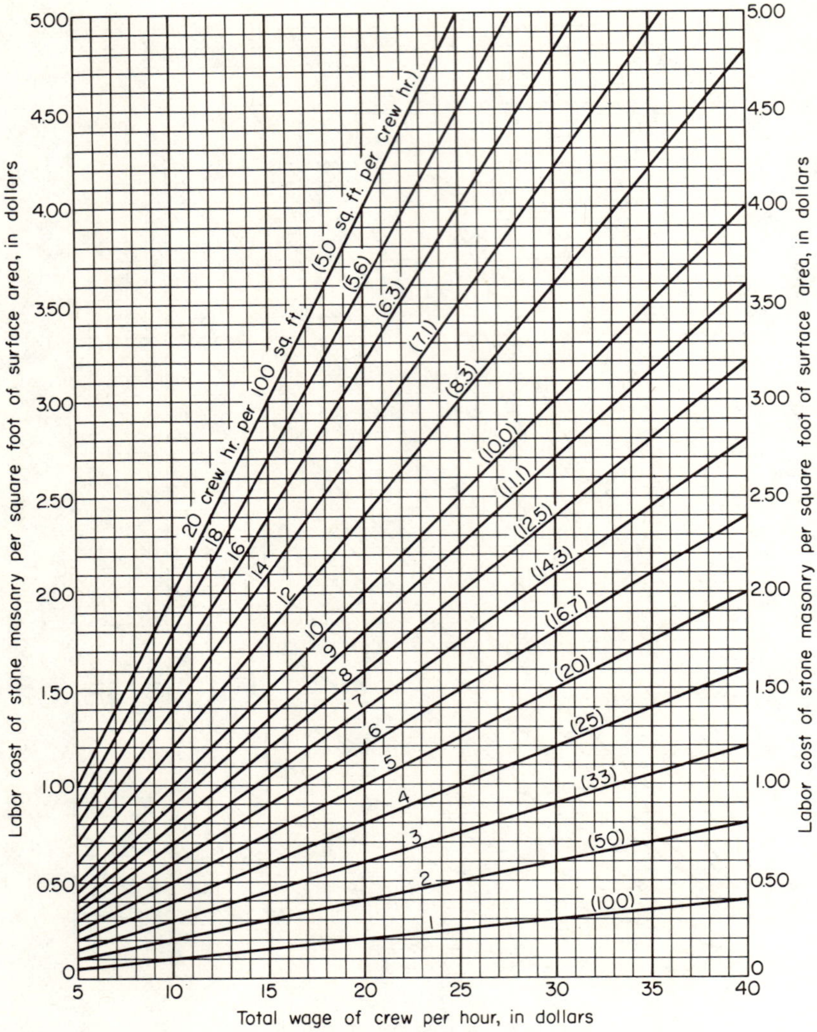

Diagram 6-10 Labor cost per square foot of surface area for stone masonry.

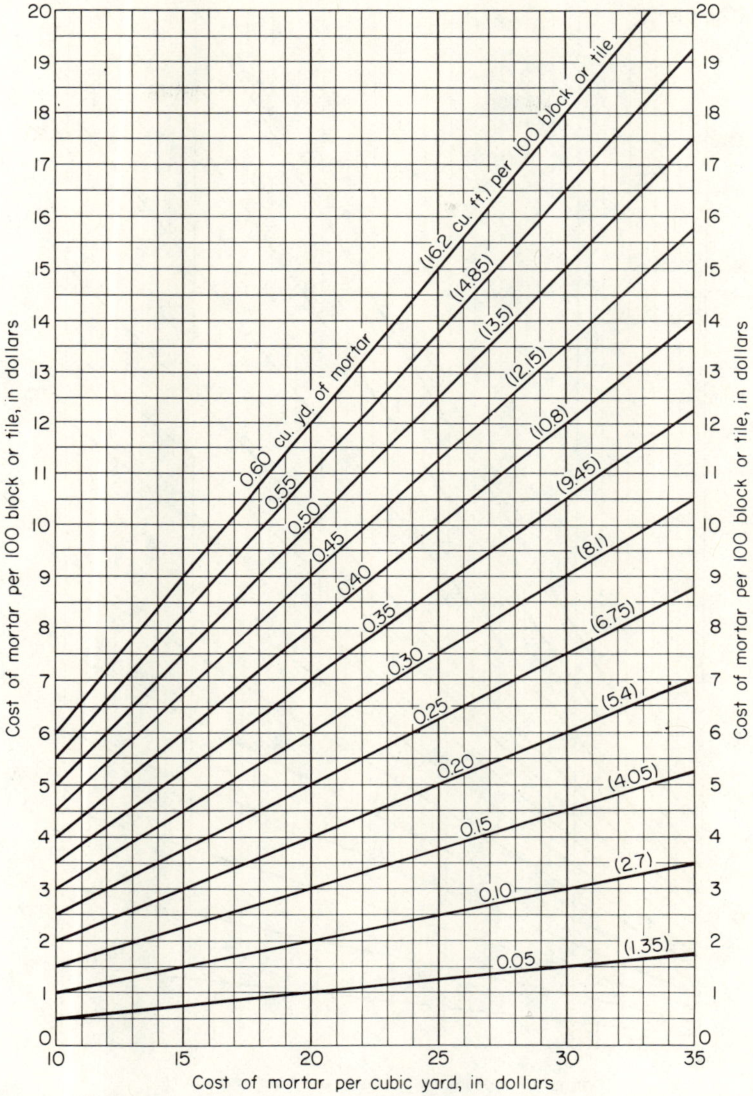

Diagram 6-11 Cost of mortar per 100 block or tile or per 100 cu ft of terra cotta masonry.

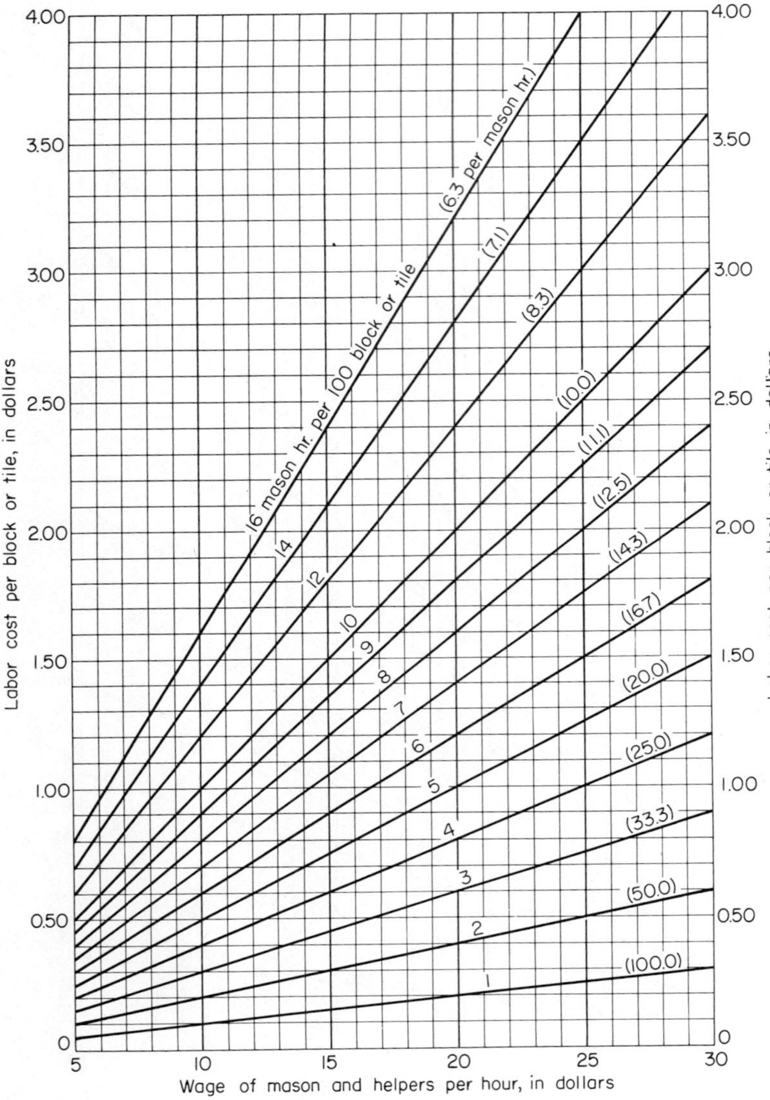

Diagram 6-12 Labor cost of laying one block or tile or 1 cu ft of architectural terra cotta.

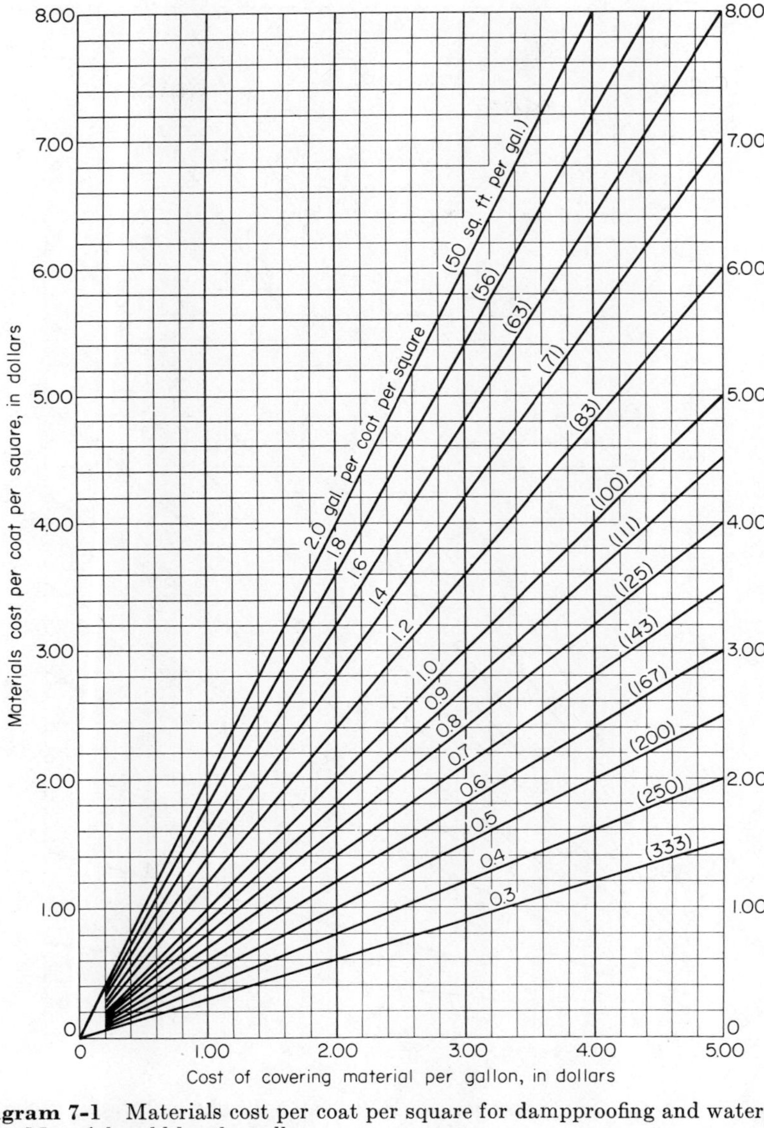

Diagram 7-1 Materials cost per coat per square for dampproofing and waterproofing. Materials sold by the gallon.

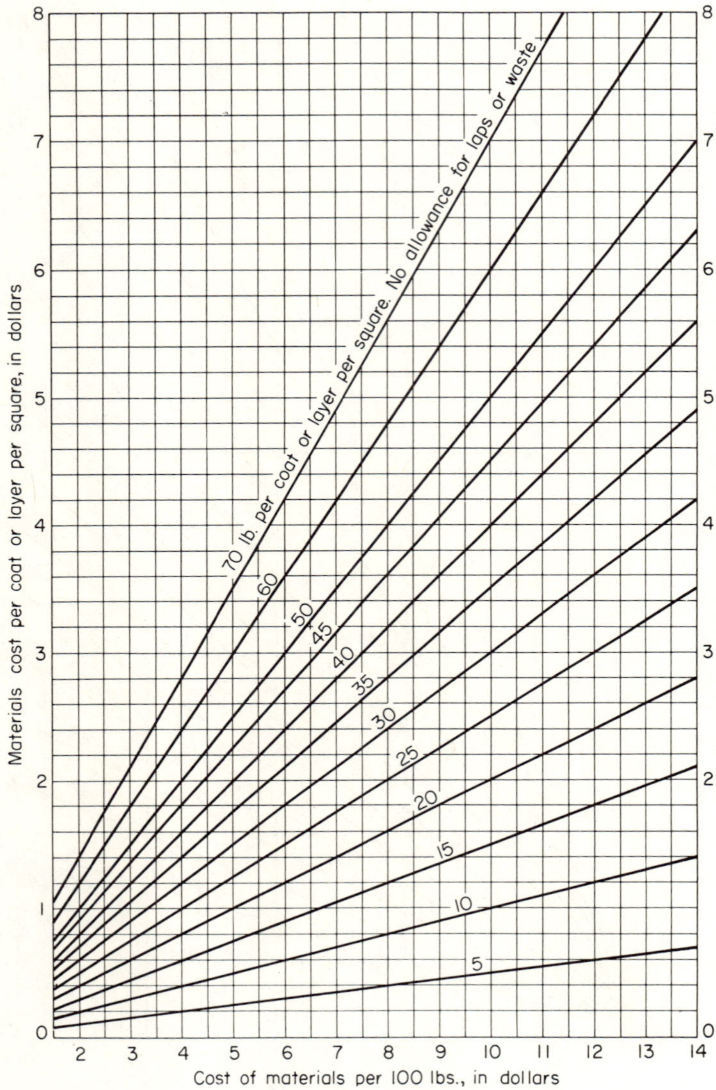

Diagram 7-2 Materials cost per square for dampproofing and waterproofing with no allowance for laps or waste. Materials sold by the 100 lb.

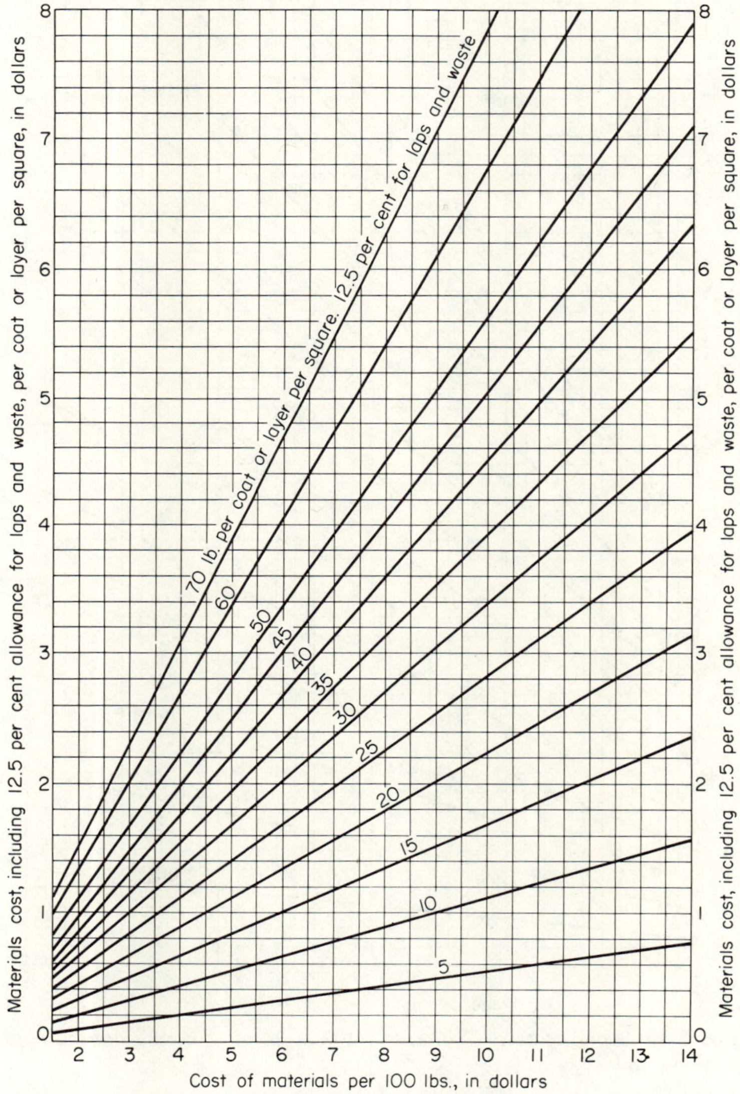

Diagram 7-3 Materials cost per square for dampproofing and waterproofing with an allowance of 12.5 percent for laps and waste. Materials sold by the 100 lb.

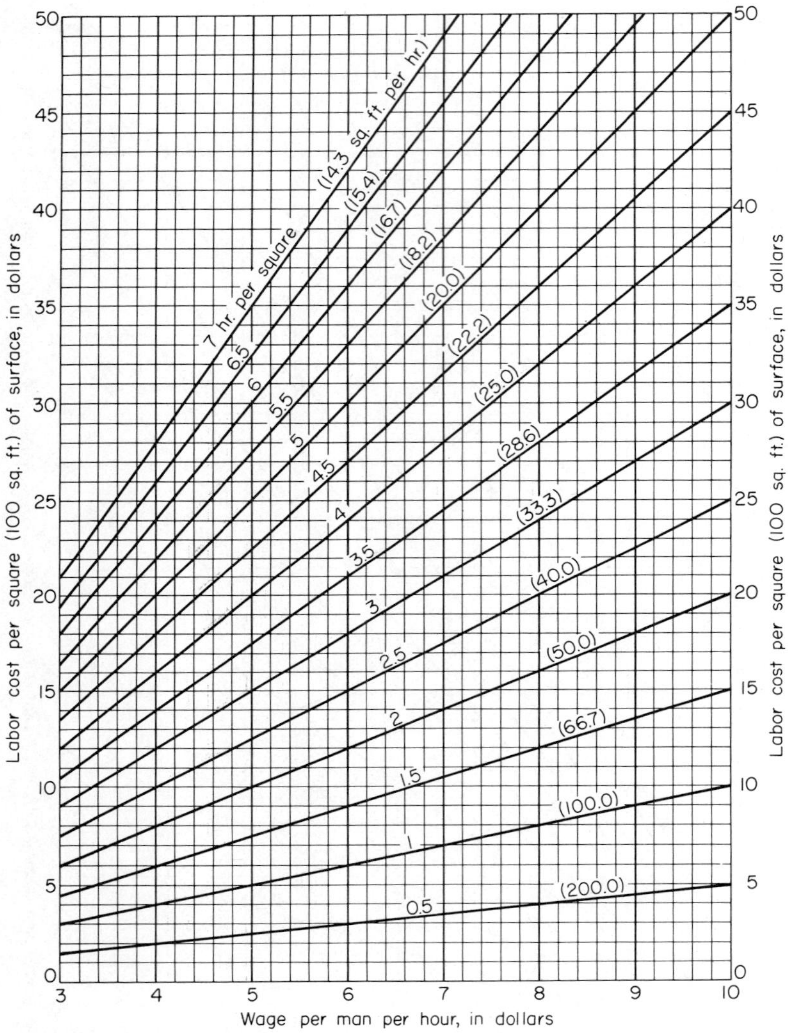

Diagram 7-4 Labor costs per square of 100 sq ft of applying dampproofing and waterproofing materials.

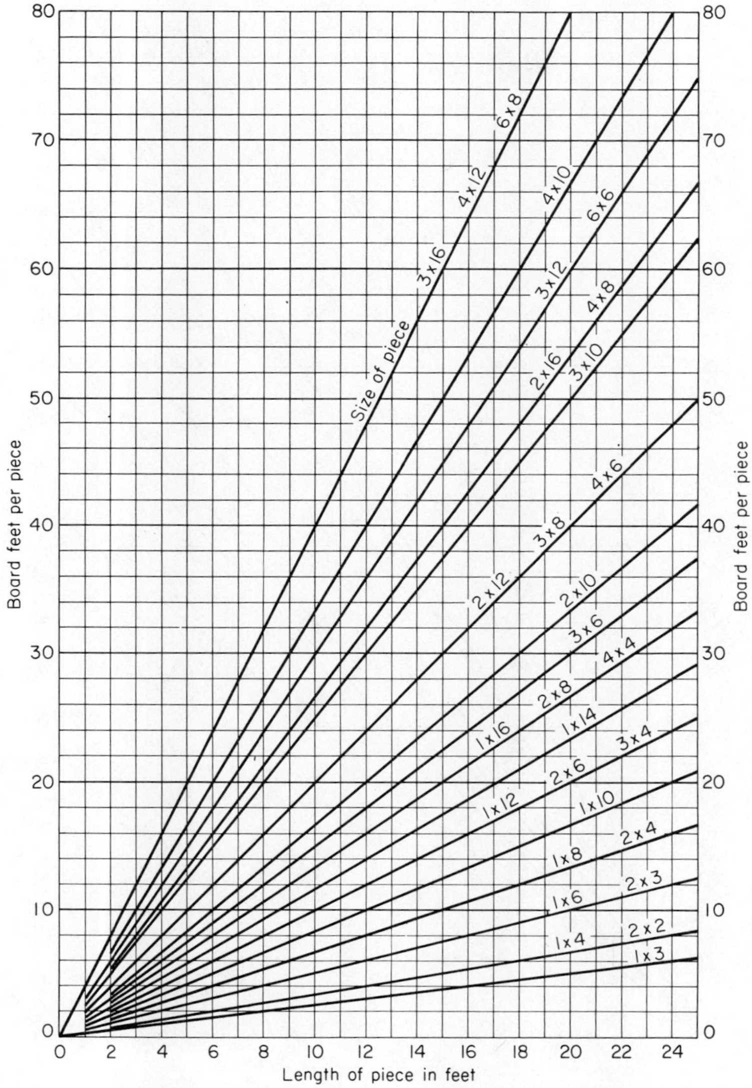

Diagram 8-1 Board feet in pieces of lumber varying from 3 to 48 sq in. in cross-sectional area.

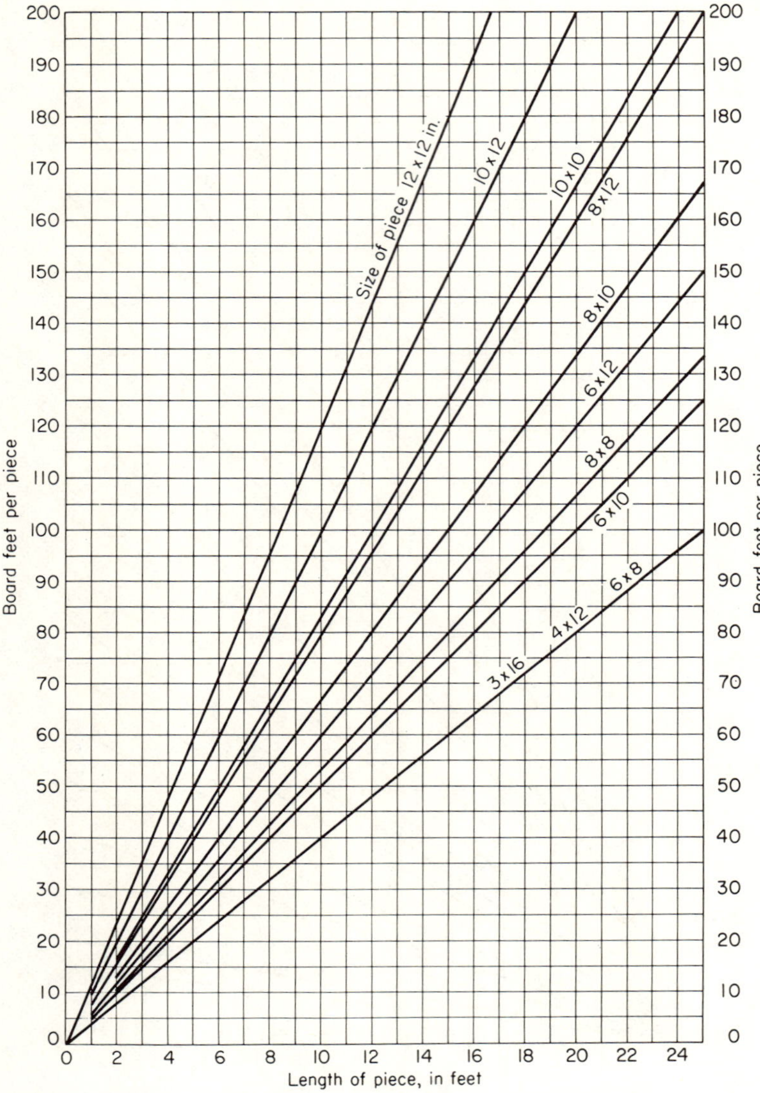

Diagram 8-2 Board feet in pieces of lumber varying from 48 to 144 sq in. in cross-sectional area.

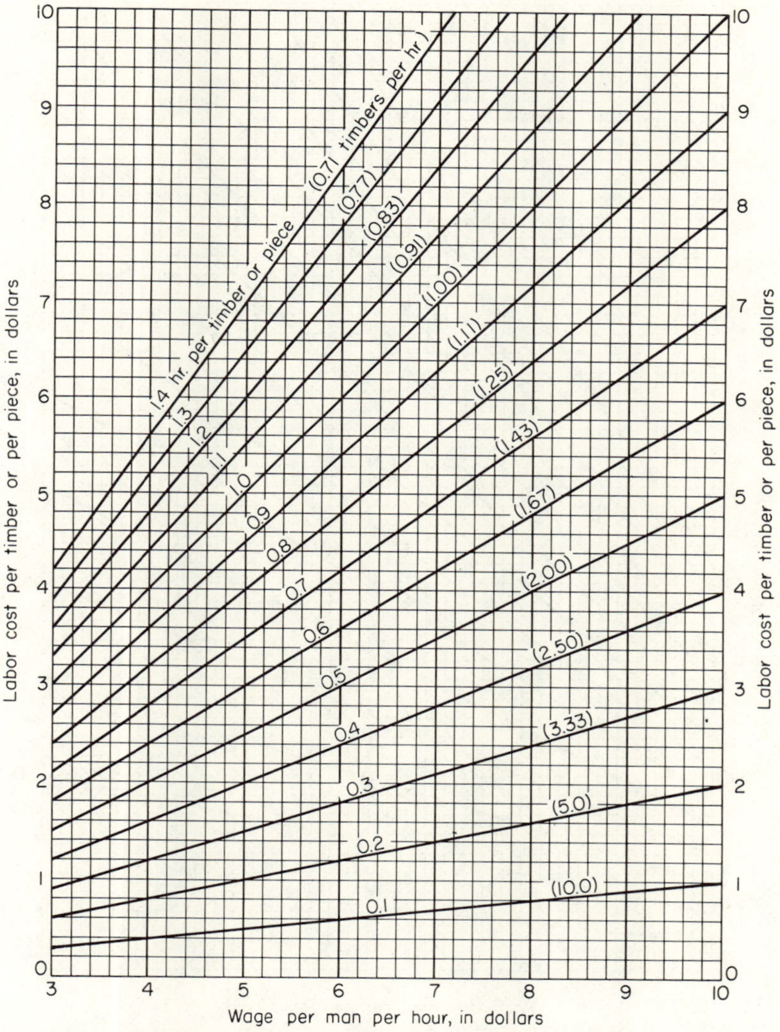

Diagram 8-3 Labor cost per timber or per piece in wood construction.

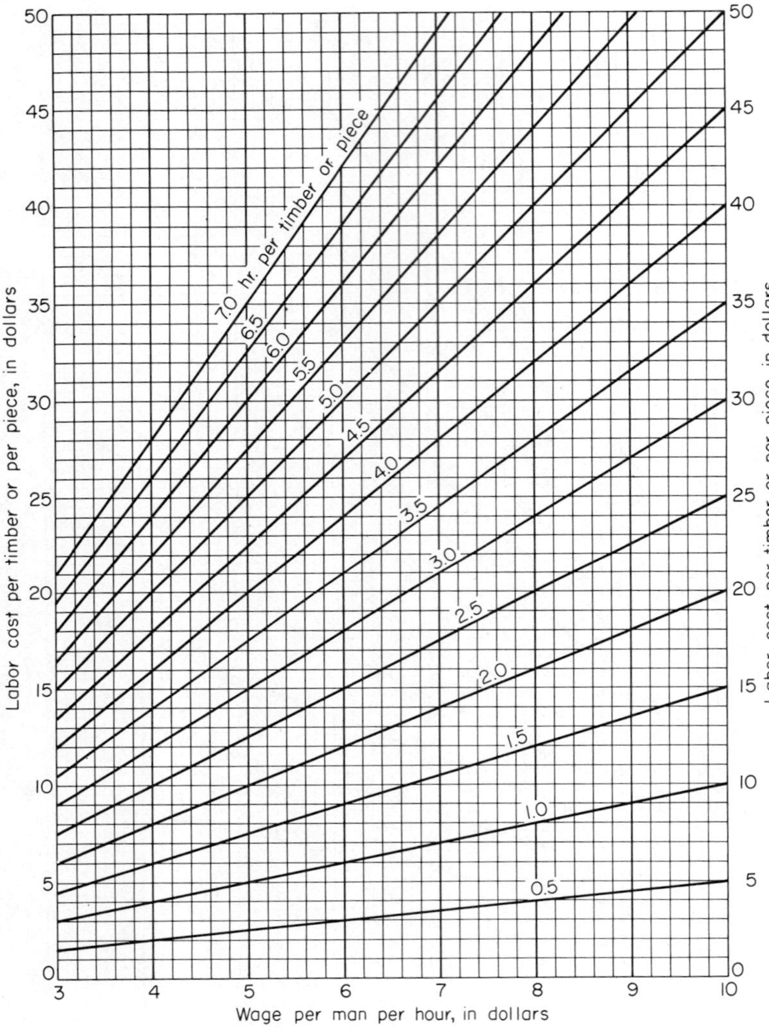

Diagram 8-4 Labor cost per timber or per piece in wood construction.

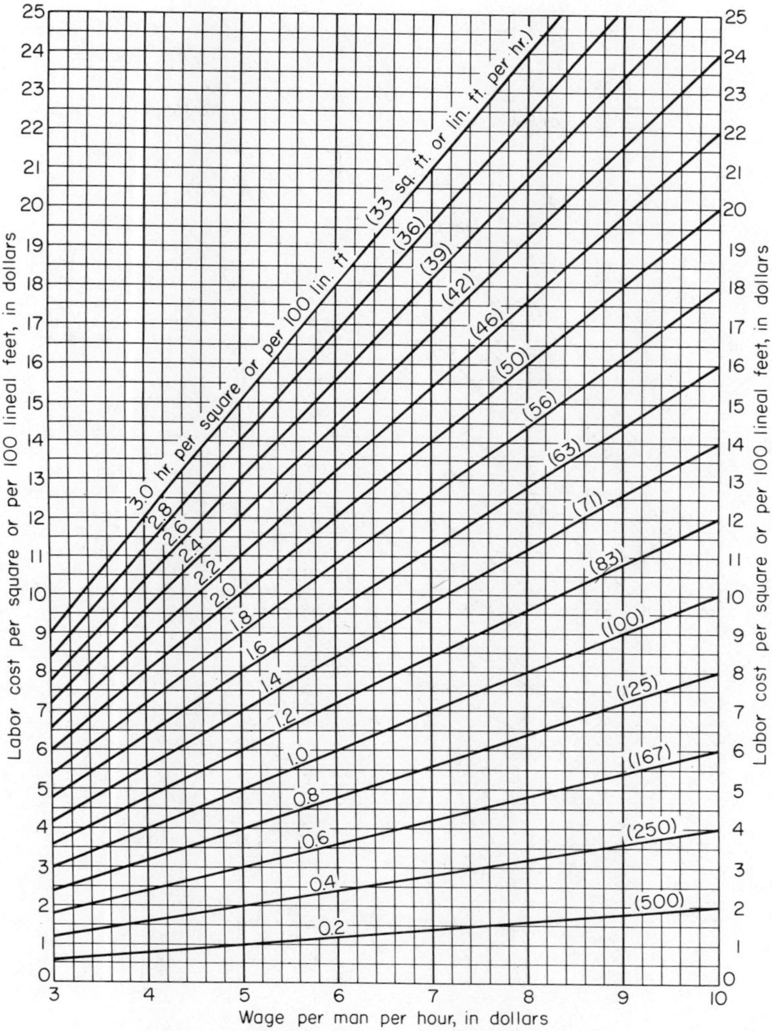

Diagram 8-5 Labor cost of wood construction per 100 lin ft or per square of 100 sq ft.

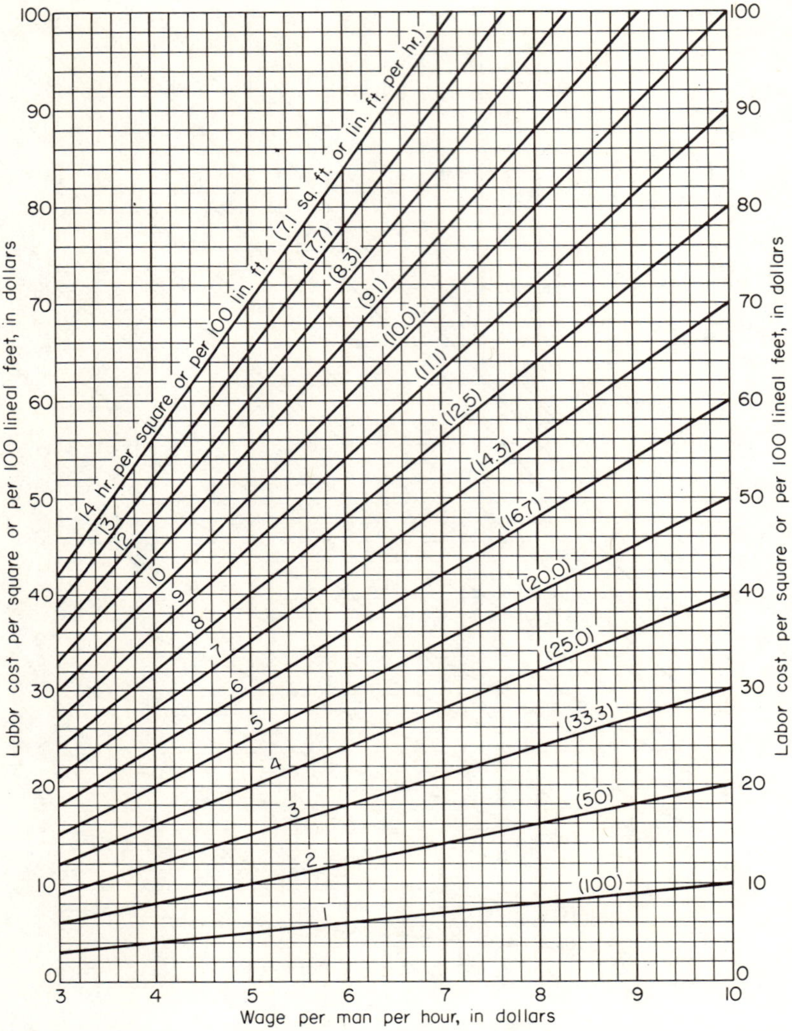

Diagram 8-6 Labor cost of wood construction per 100 lin ft or per square of 100 sq ft.

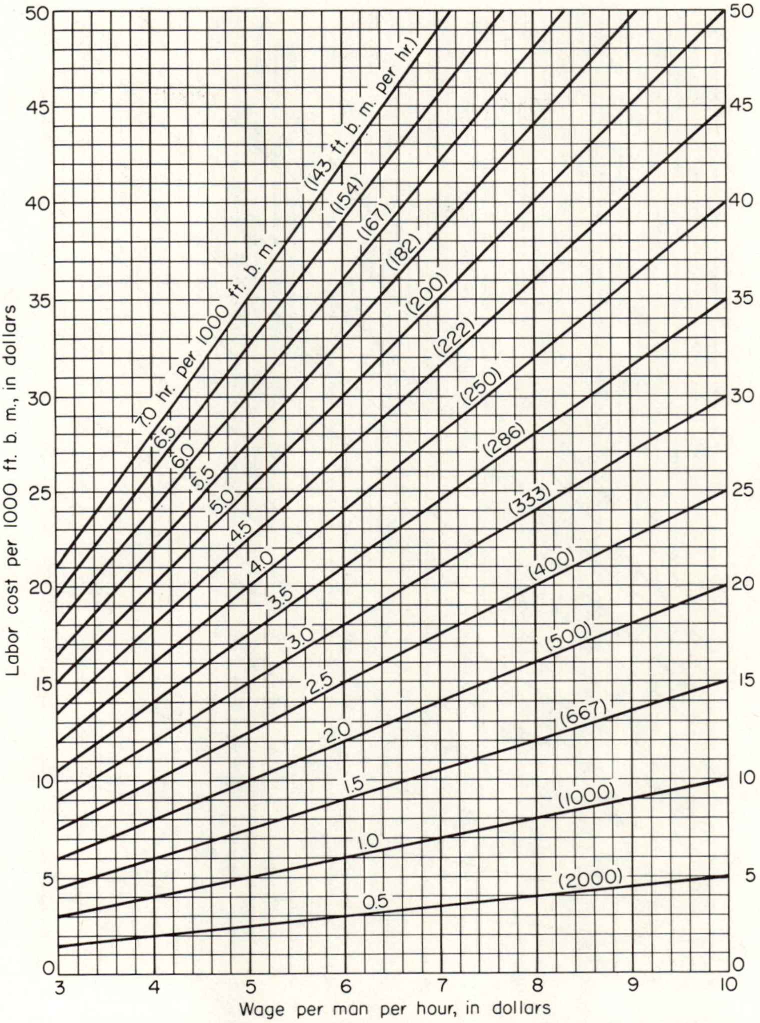

Diagram 8-7 Labor cost of wood construction per 1,000 fbm.

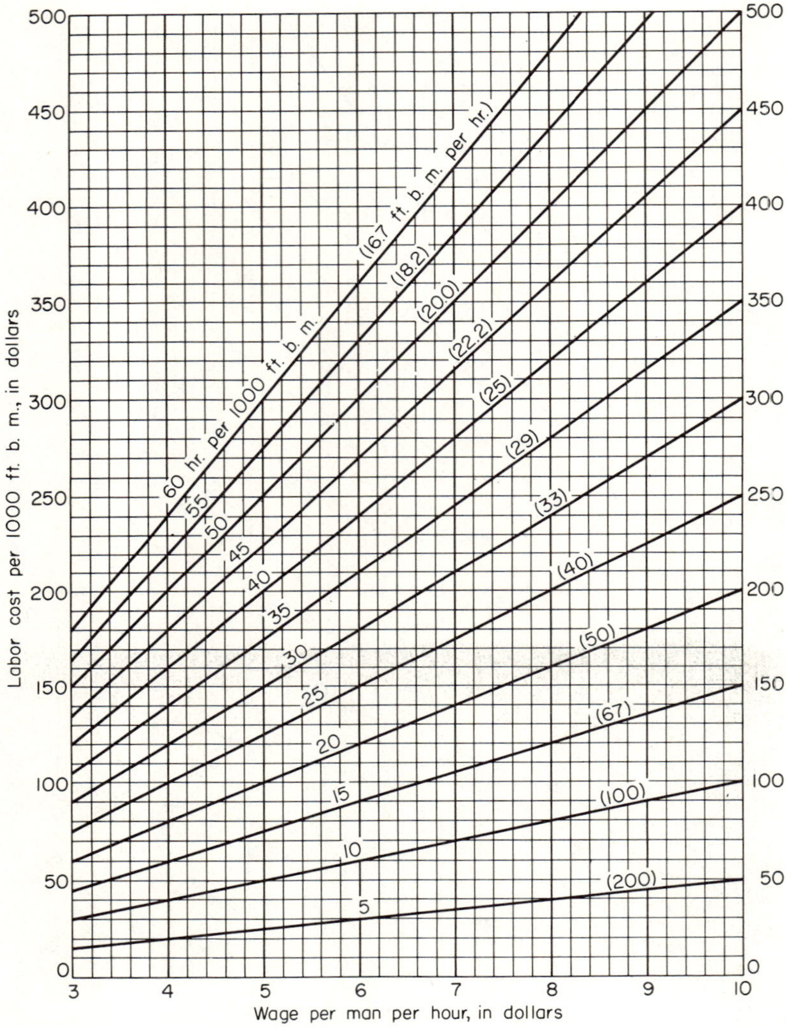

Diagram 8-8 Labor cost of wood construction per 1,000 fbm.

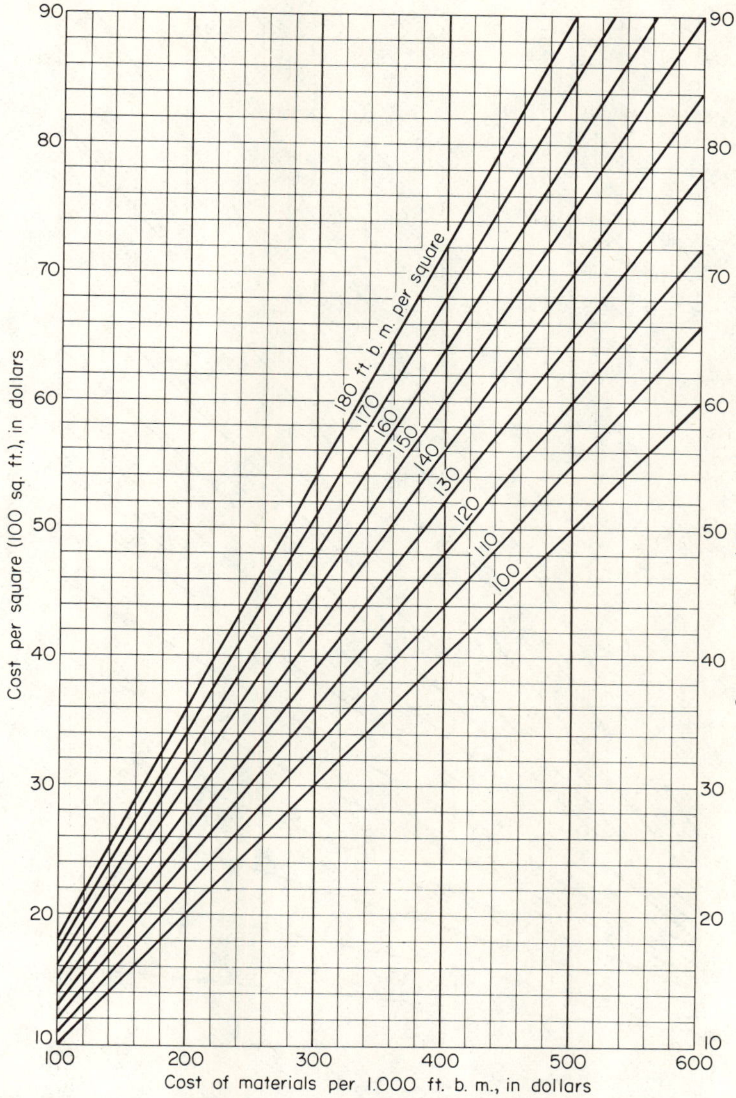

Diagram 8-9 Cost of wood materials per square of 100 sq ft with allowances for lap and waste.

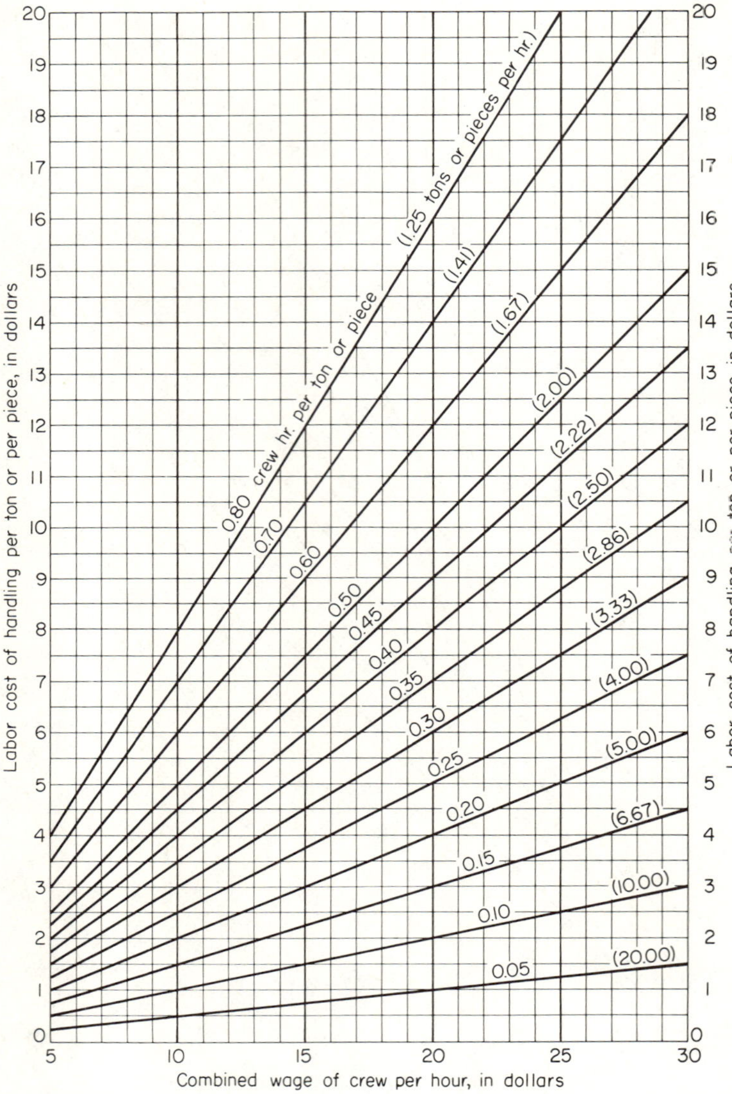

Diagram 9-1 Labor cost per ton or per piece of handling structural steel.

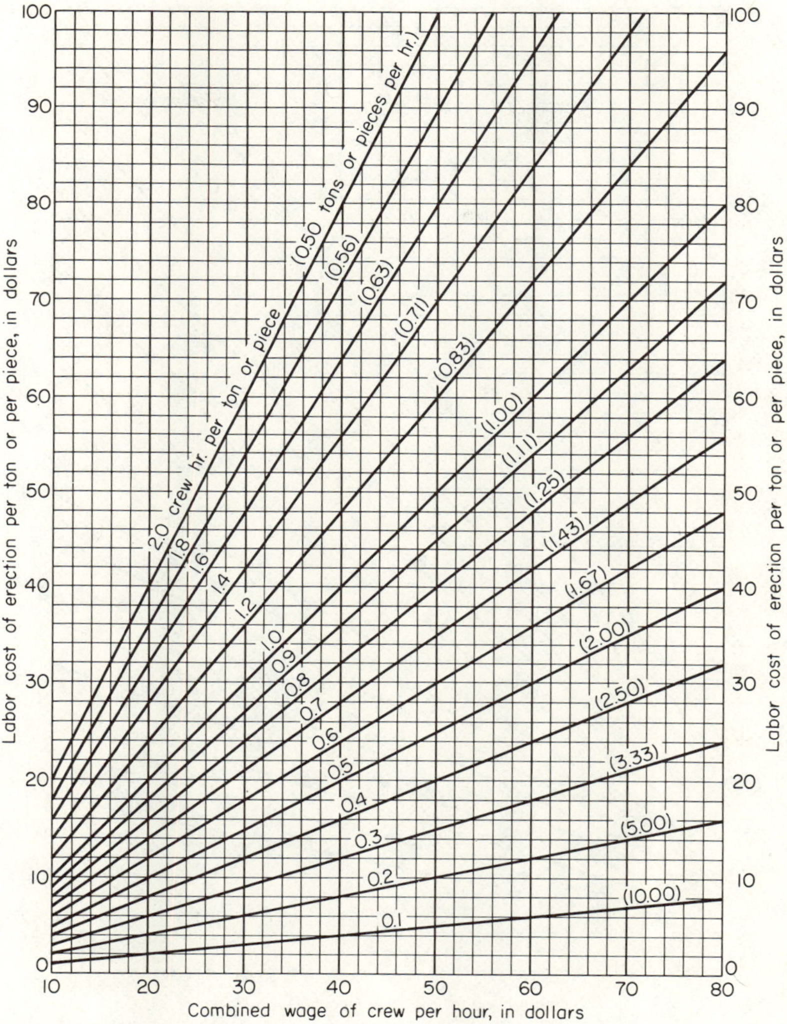

Diagram 9-2 Labor cost per ton or per piece of erecting structural steel for an erection crew of 4 to 10 men.

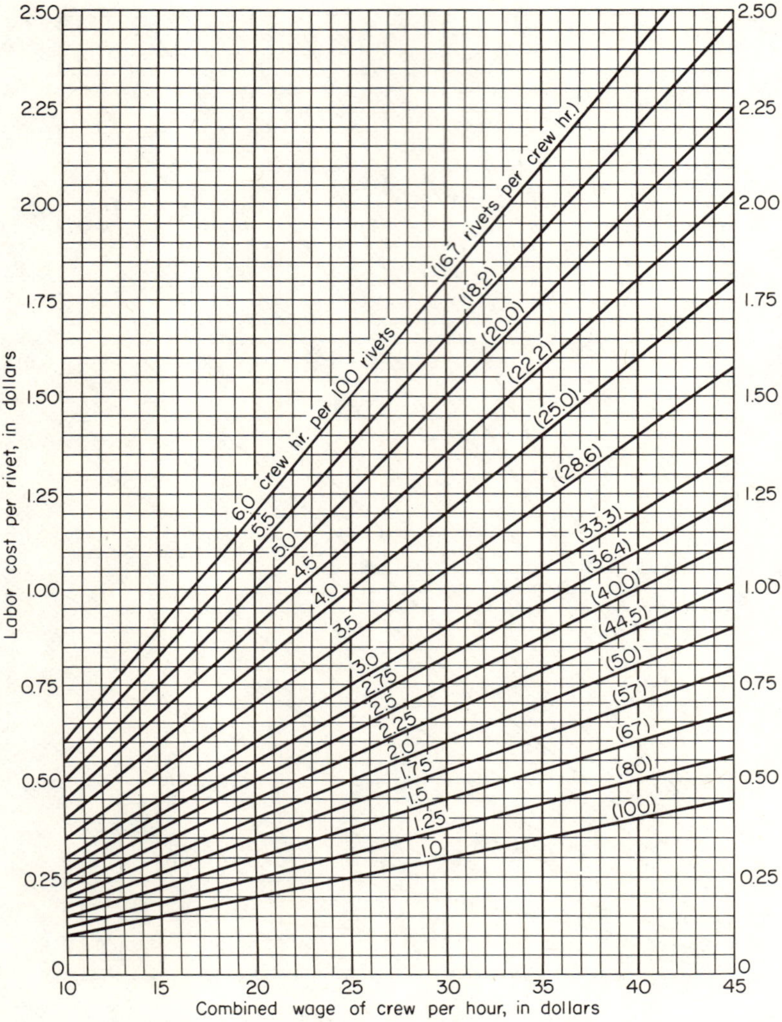

Diagram 9-3 Labor cost of riveting structural steel for a crew of 2 riveters, 1 heater, 1 catcher, and part time of a helper.

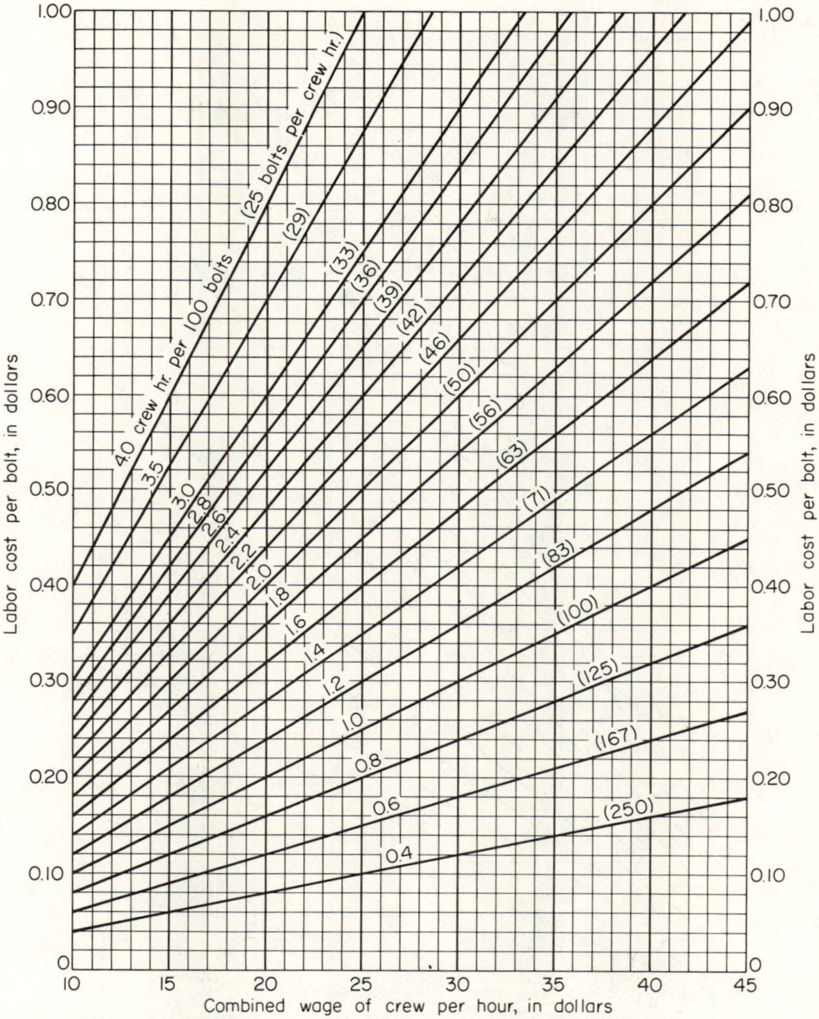

Diagram 9-4 Labor cost of bolting structural steel for a crew of 4 (or 2) bolters and part time of a helper.

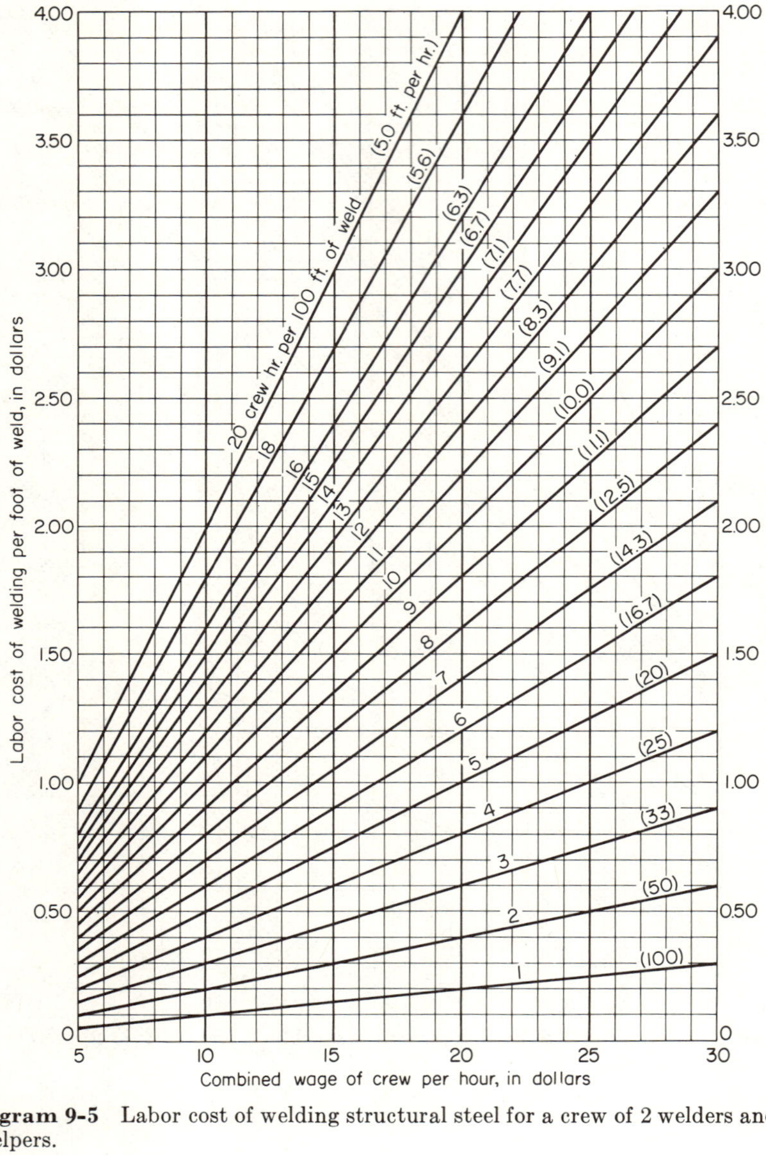

Diagram 9-5 Labor cost of welding structural steel for a crew of 2 welders and 1 or 2 helpers.

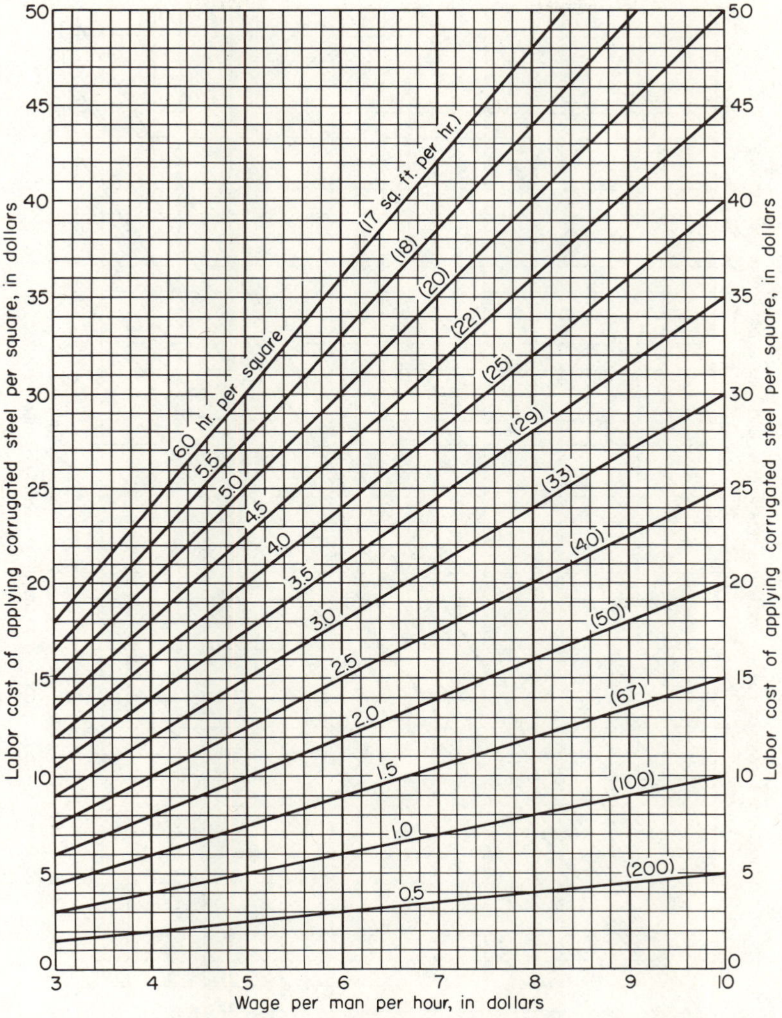

Diagram 9-6 Labor cost of applying corrugated steel per square of 100 sq ft.

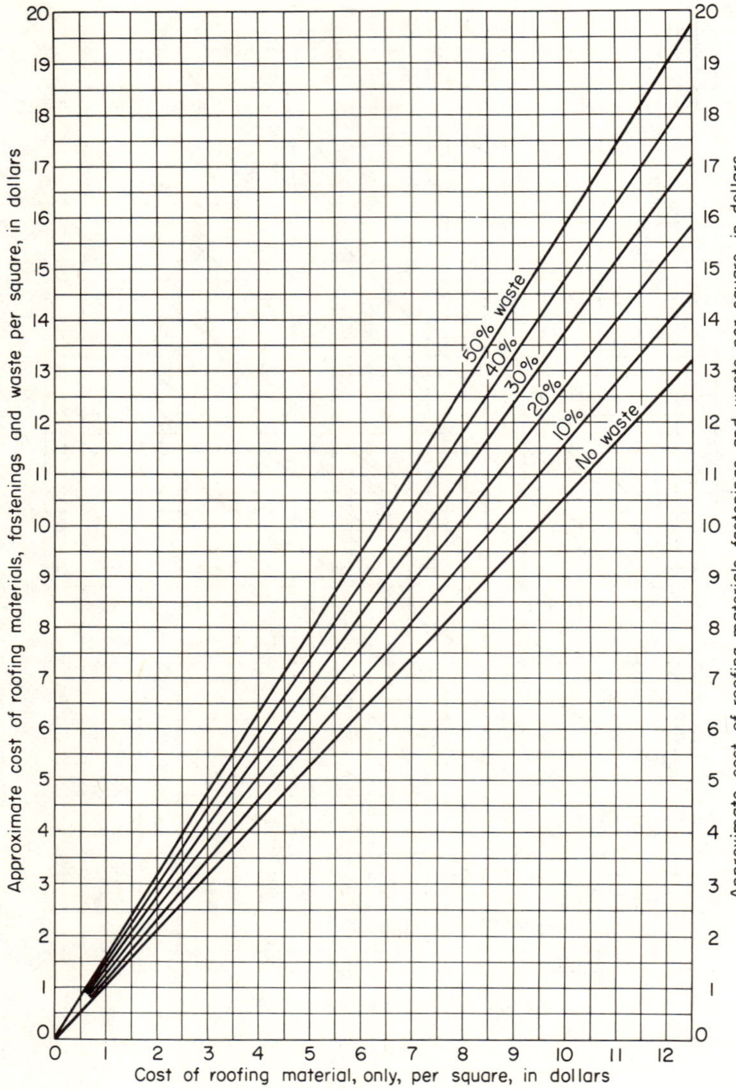

Diagram 10-1 Approximate cost of roofing materials per square with allowances for fastenings and waste.

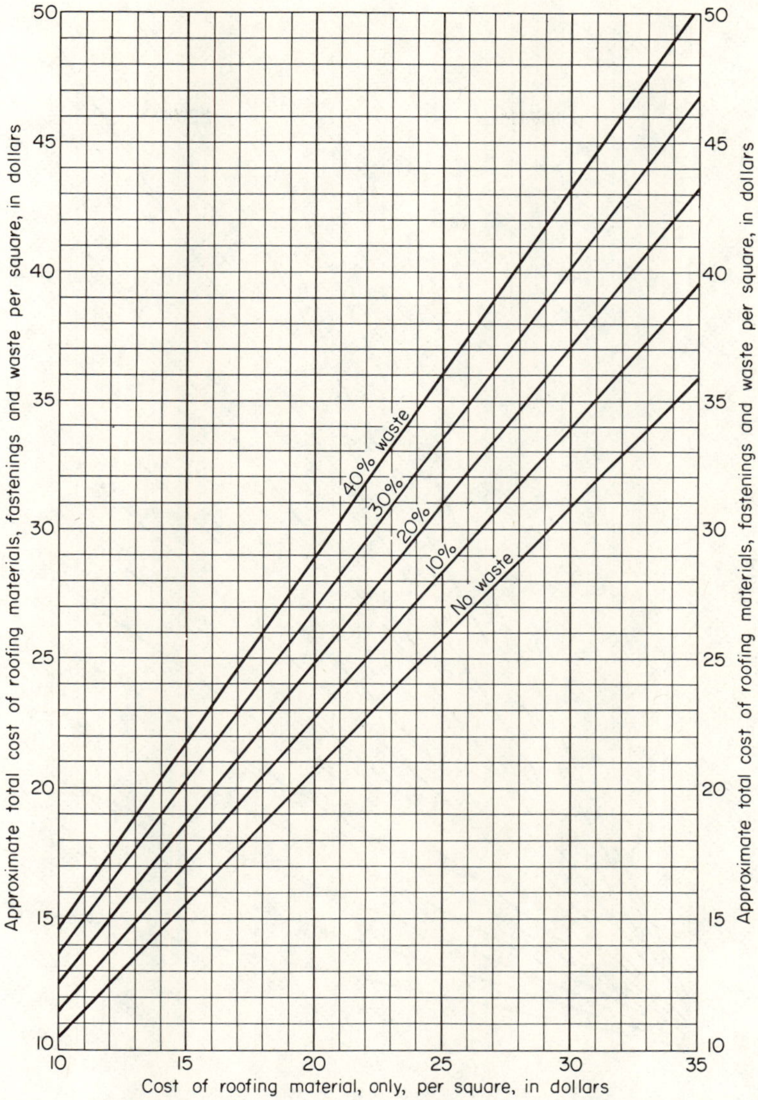

Diagram 10-2 Approximate cost of roofing materials per square with allowances for fastenings and waste.

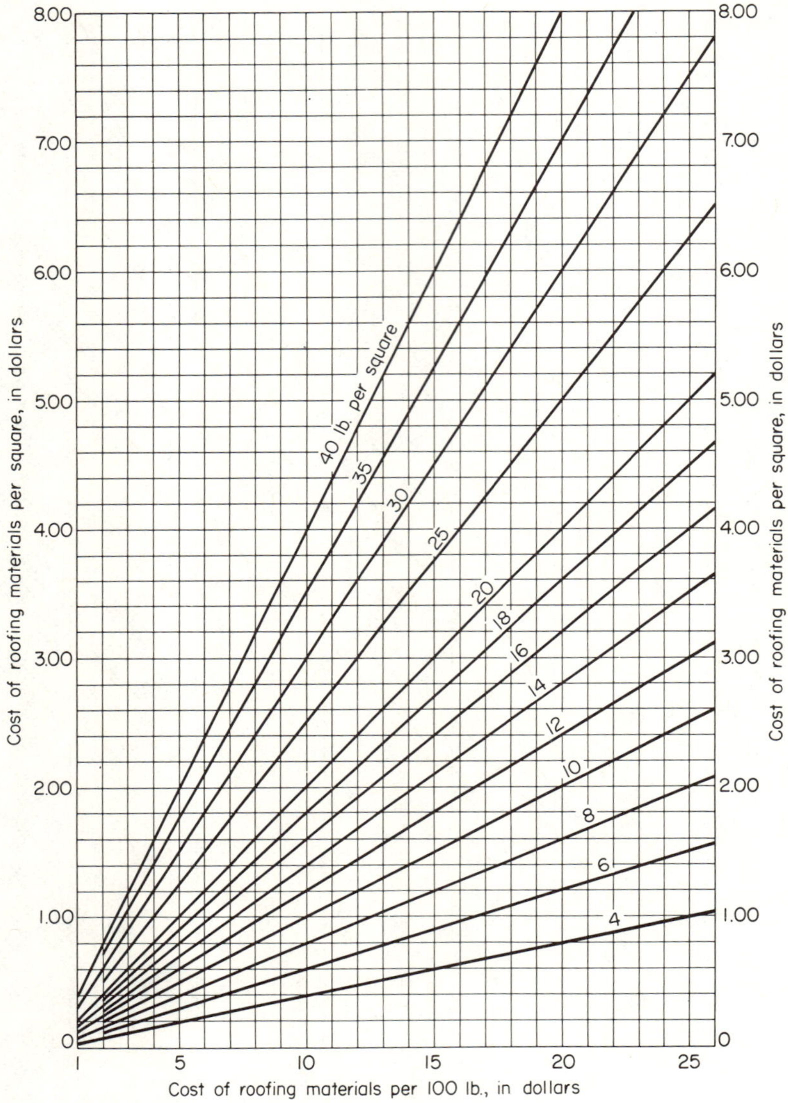

Diagram 10-3 Cost of roofing materials per square of roof area.

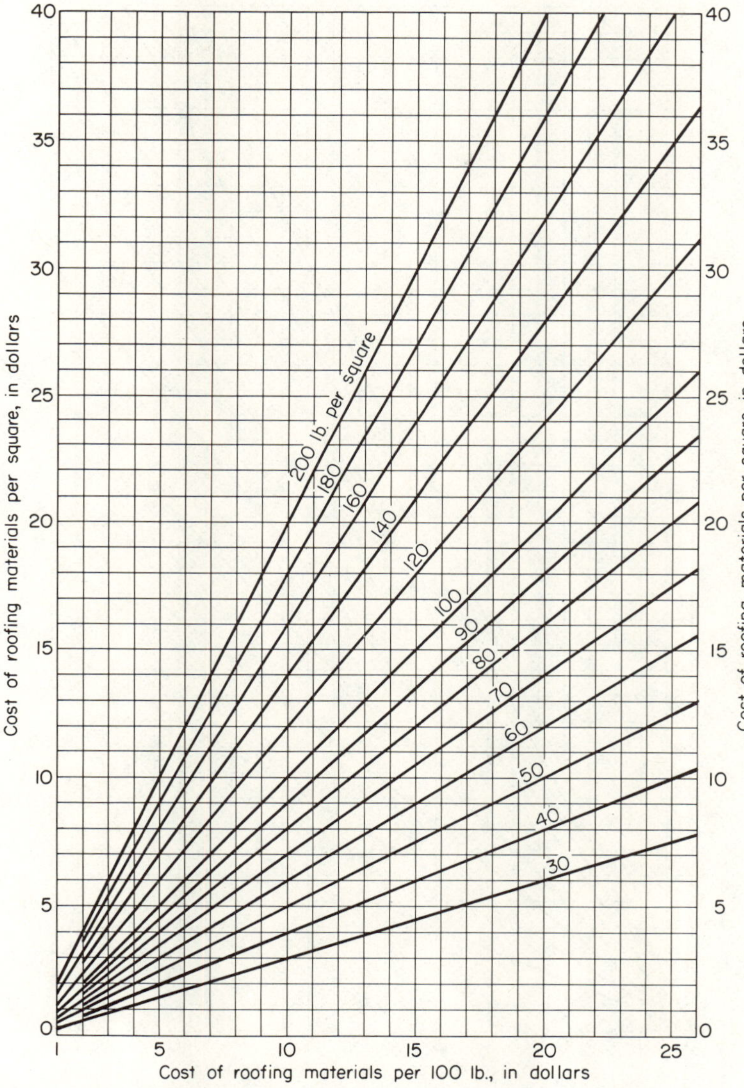

Diagram 10-4 Cost of roofing materials per square of roof area.

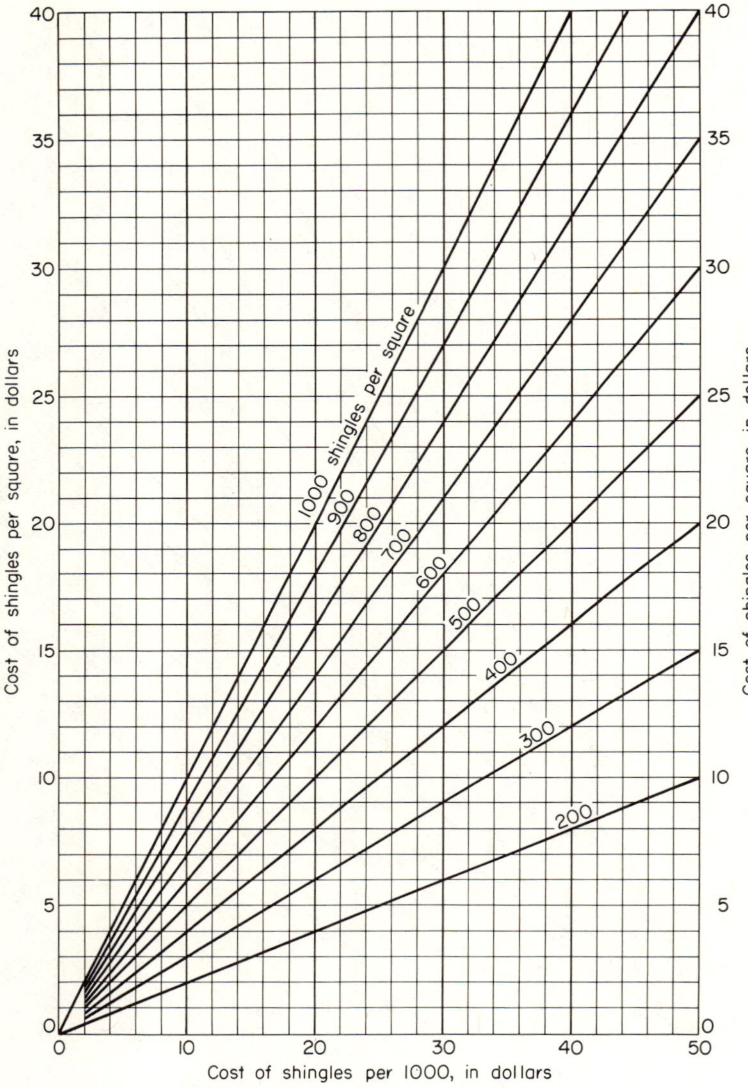

Diagram 10-5 Cost of shingles per square of roof area.

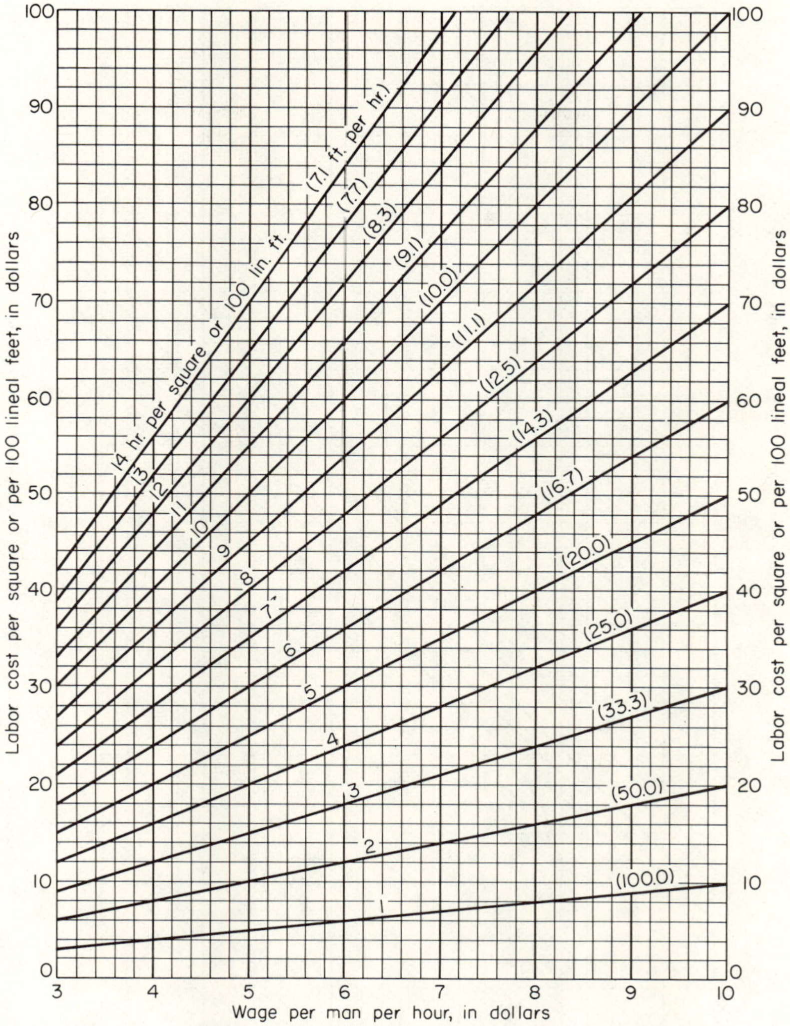

Diagram 10-6 Labor cost of roofing per square or per 100 lin ft.

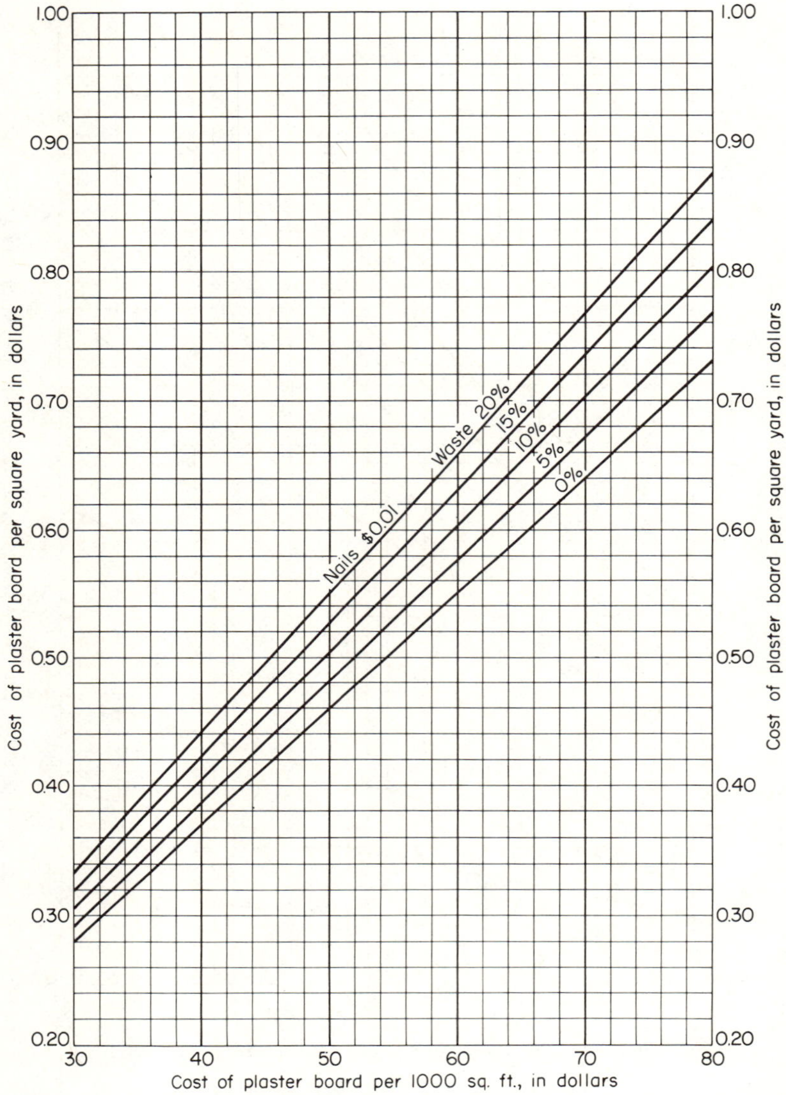

Diagram 11-1 Cost of plasterboard per square yard with an allowance of $0.01 per square yard for nails and varying allowances for waste.

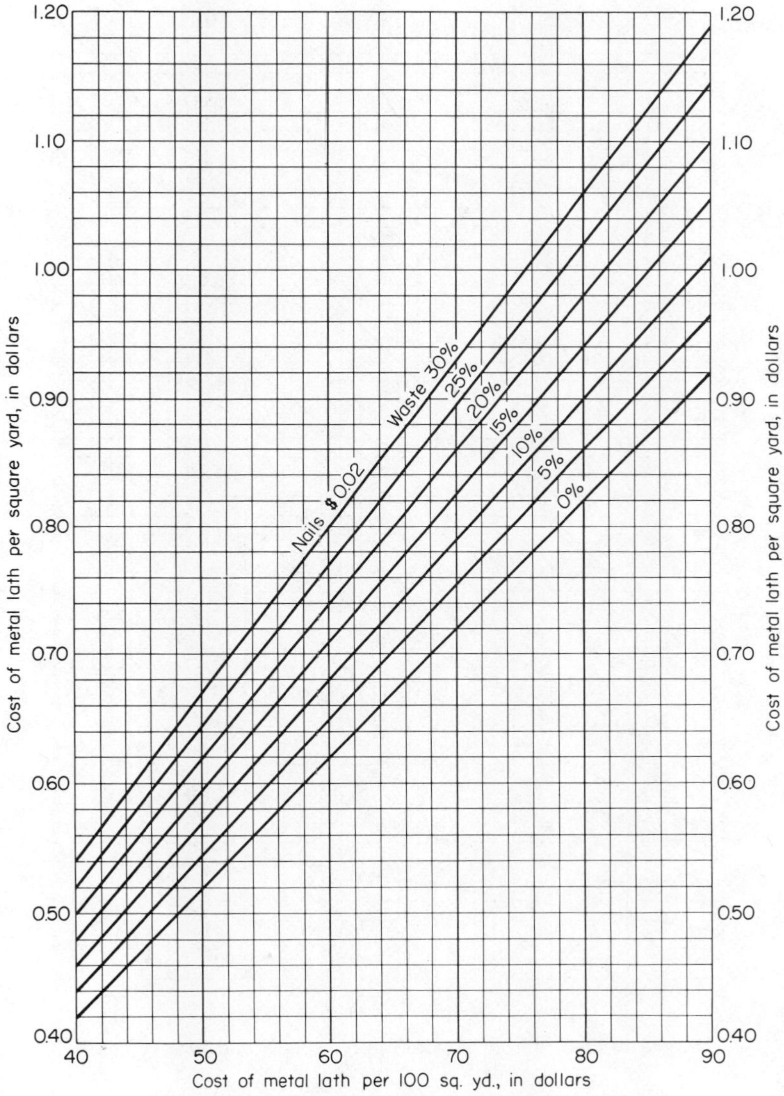

Diagram 11-2 Cost of metal lath per square yard with an allowance of $0.02 per square yard for nails and varying allowances for waste.

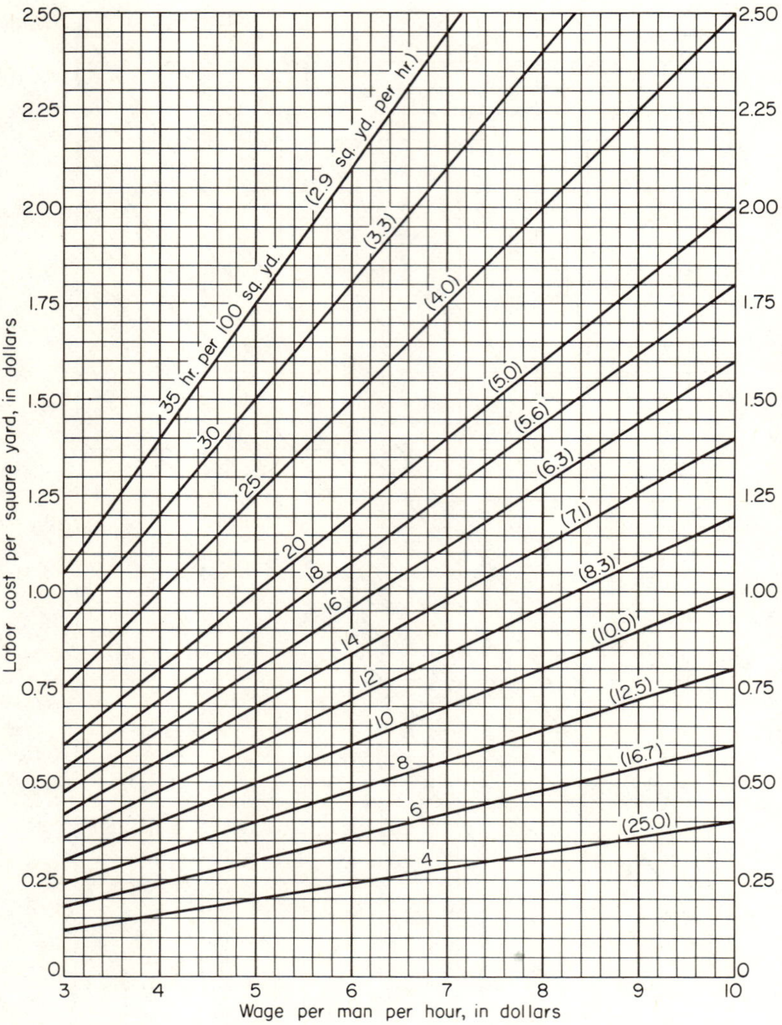

Diagram 11-3 Labor cost of applying lath and plaster board.

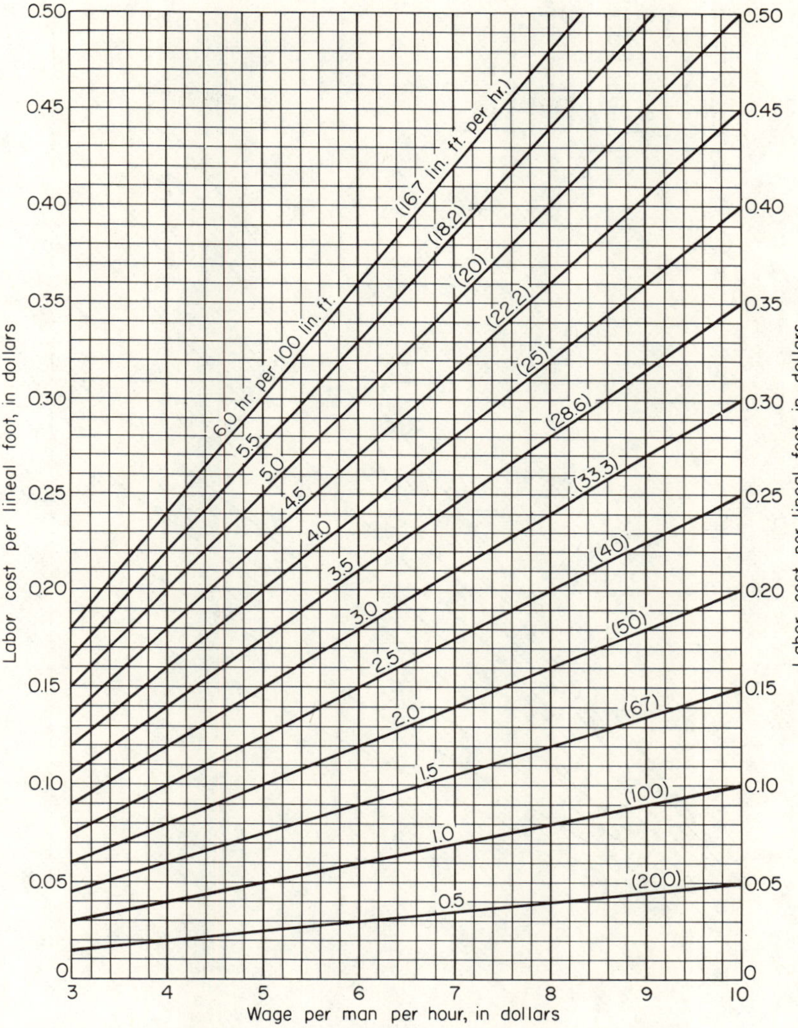

Diagram 11-4 Labor cost of applying furring, grounds, corners, corner bead, etc.

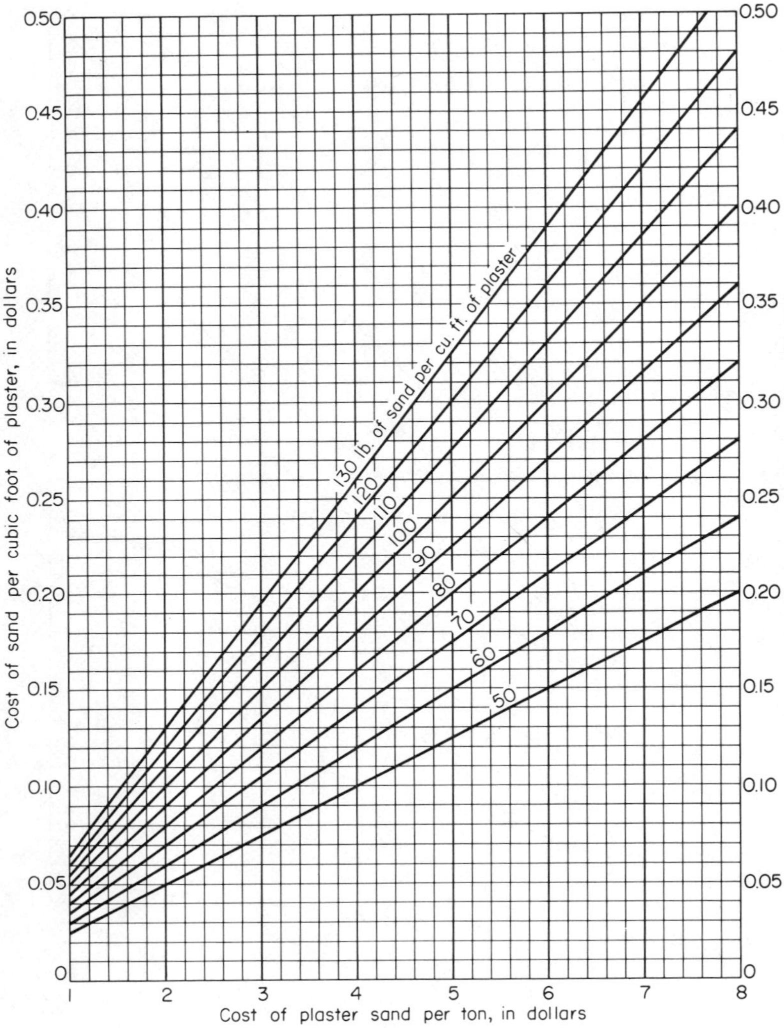

Diagram 11-5 Cost of sand per cubic foot of plaster.

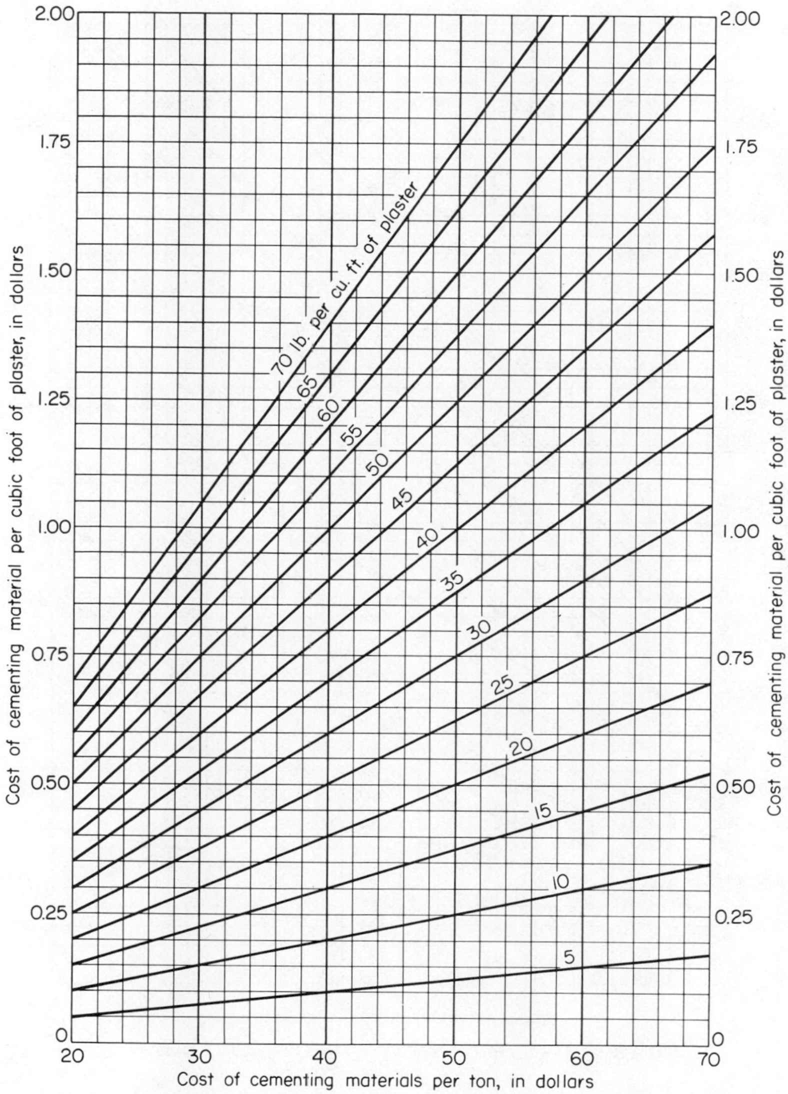

Diagram 11-6 Cost of cementing materials (lime, gypsum, plaster cement, etc.) per cubic foot of plaster.

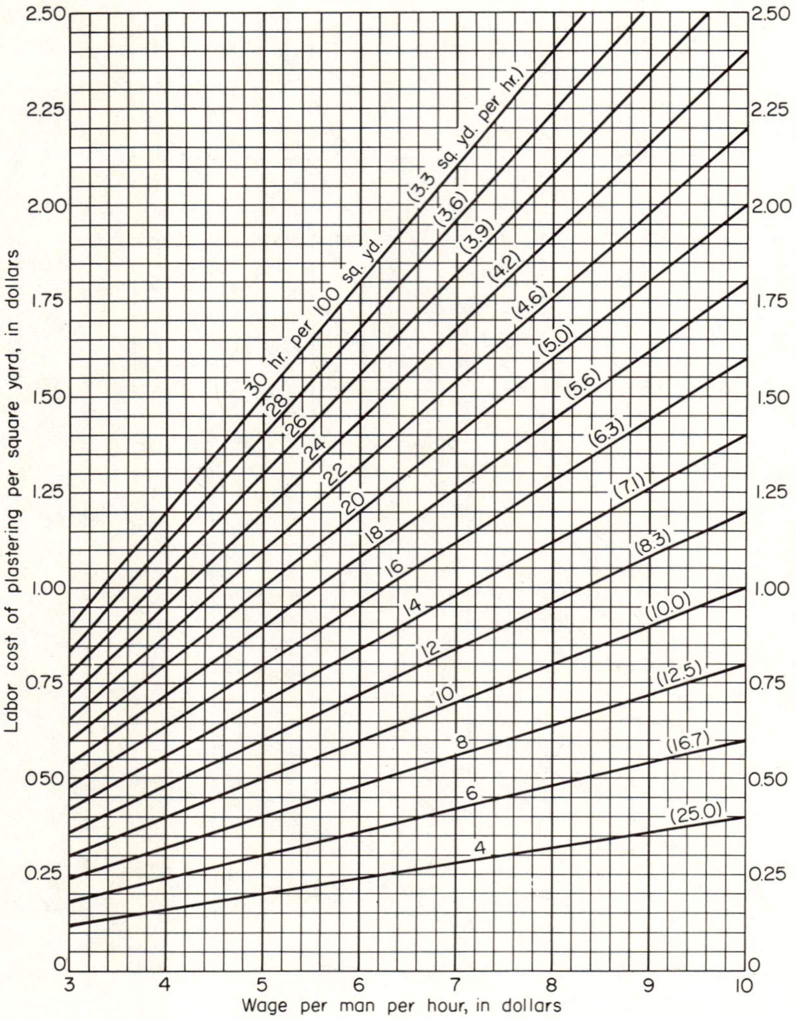

Diagram 11-7 Labor cost per square yard of plastering.

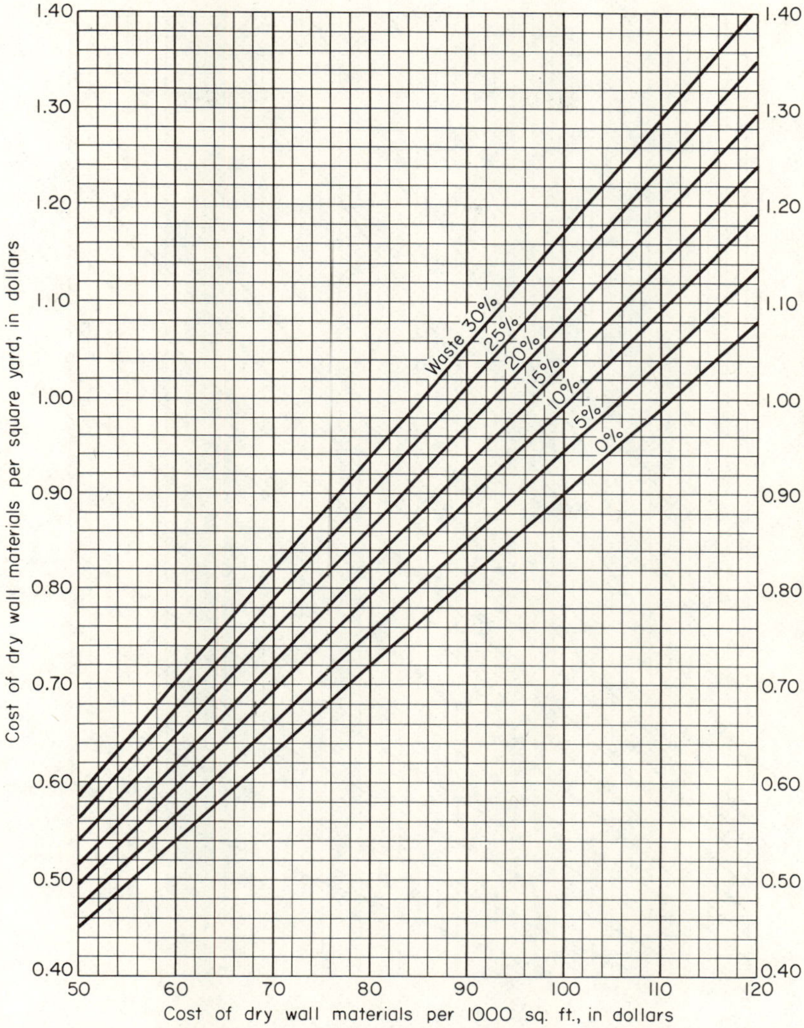

Diagram 11-8 Cost of drywall materials per square yard with allowances for waste.

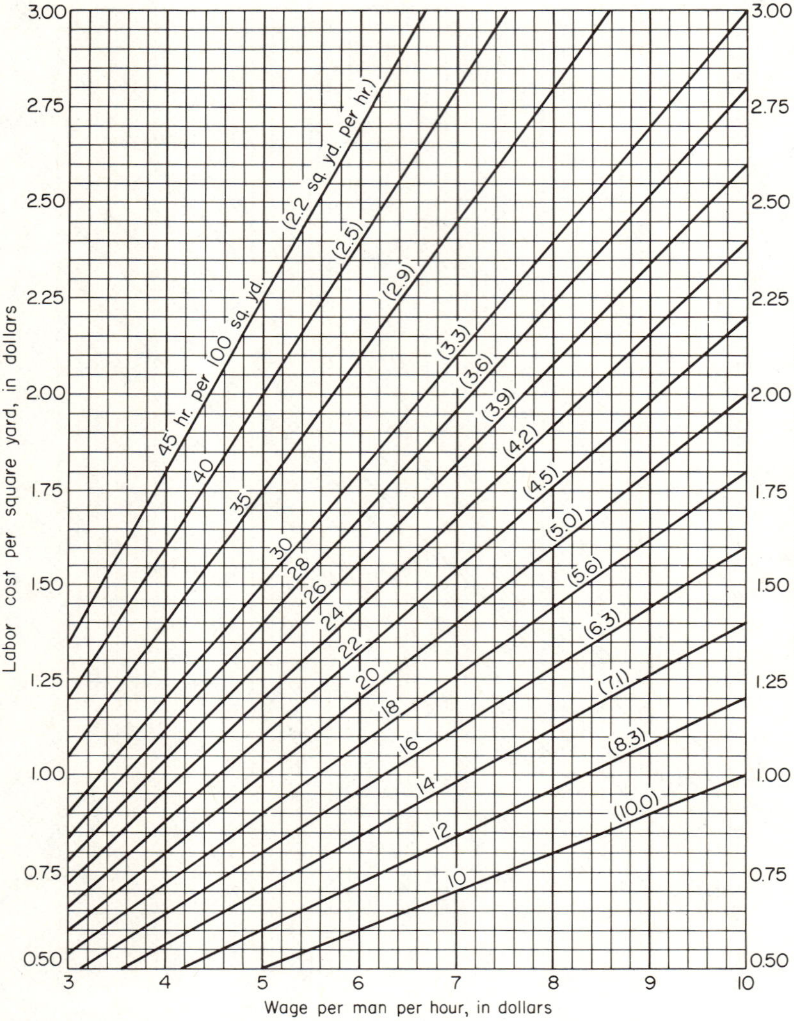

Diagram 11-9 Labor cost per square yard of applying drywall materials.

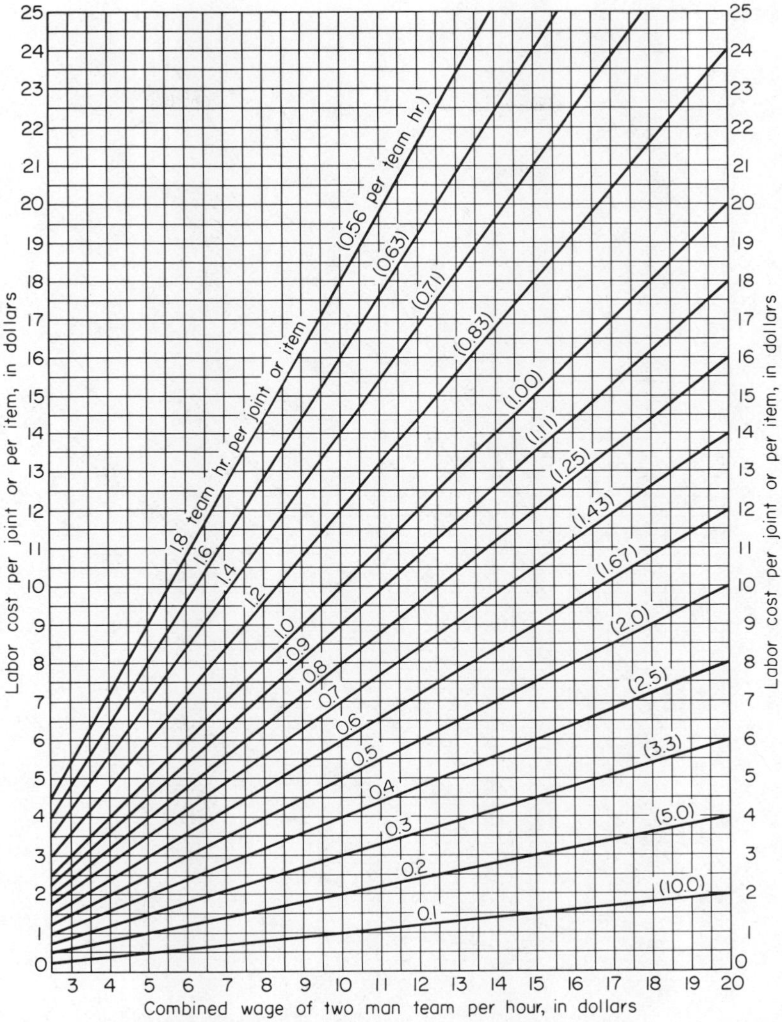

Diagram 12-1 Labor costs of installing pipe and other items in steam and hot-water heating systems.

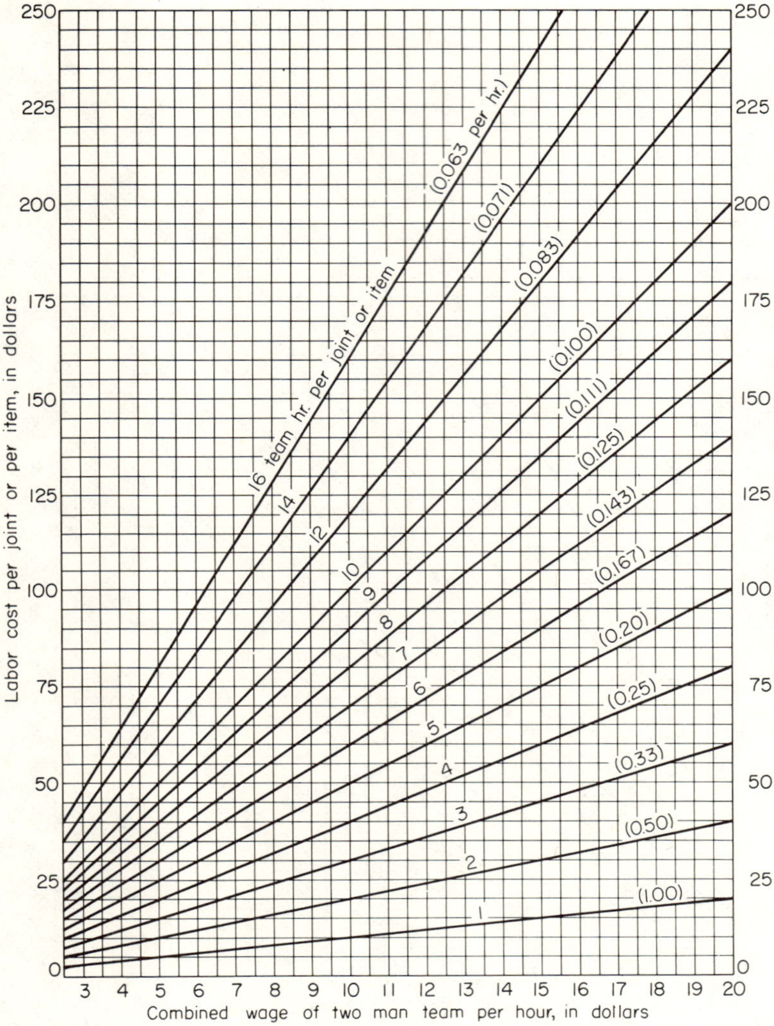

Diagram 12-2 Labor costs of installing pipe and other items in steam and hot-water heating systems.

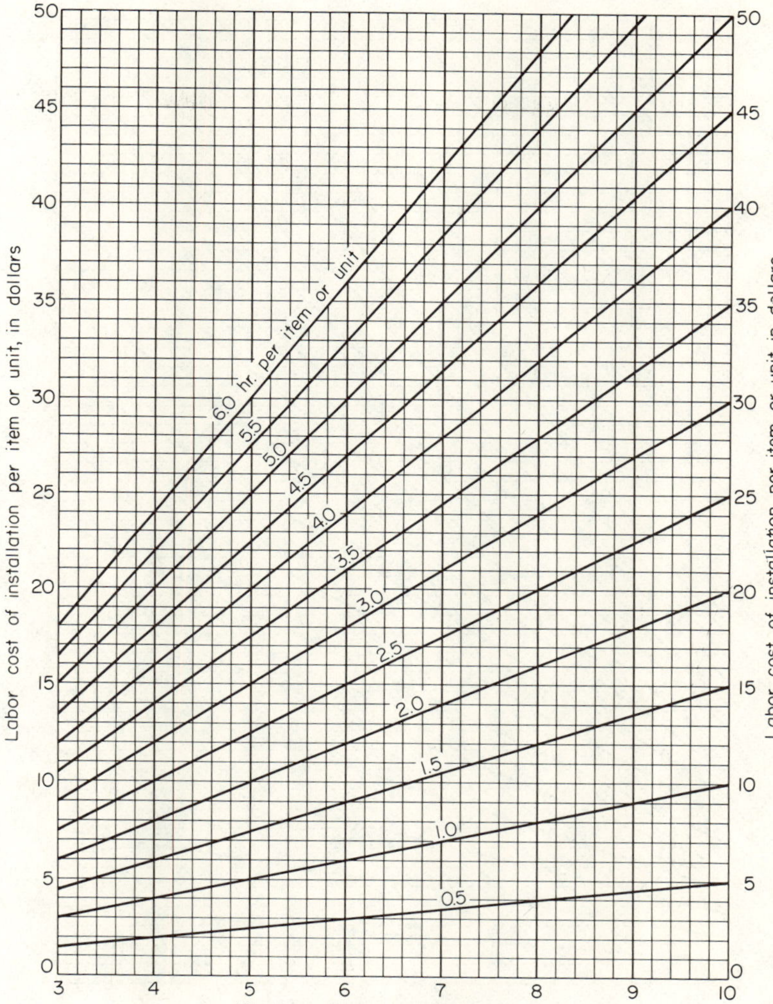

Diagram 12-3 Labor cost per item of installing warm-air heating and air conditioning systems.

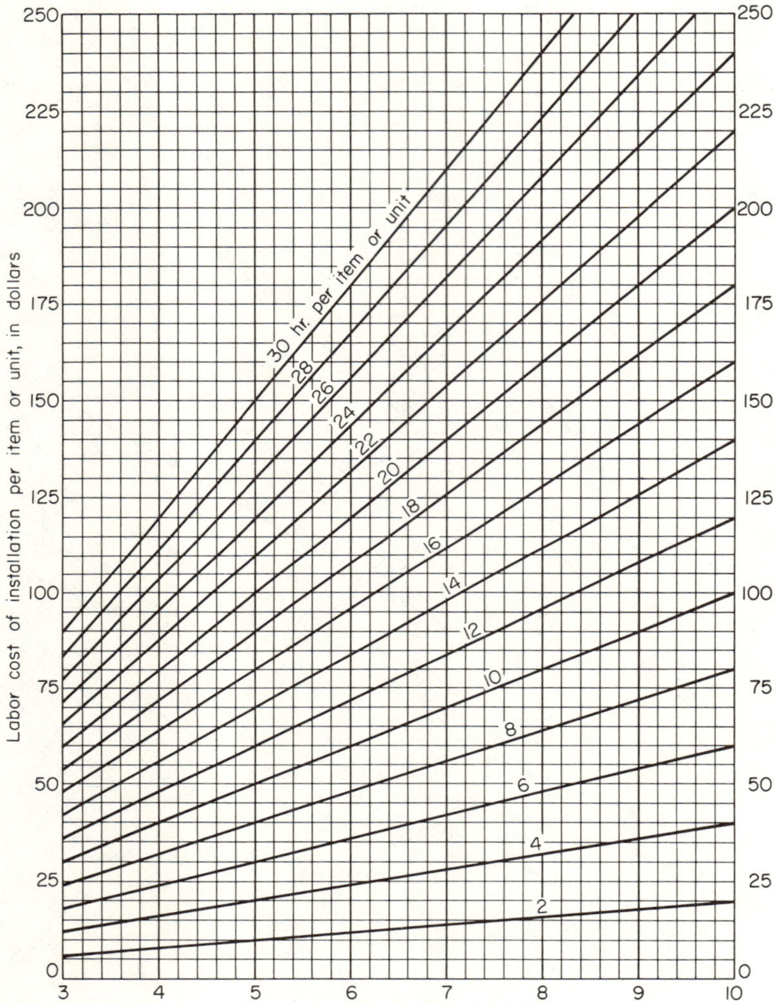

Diagram 12-4 Labor cost per item of installing warm-air heating and air conditioning systems.

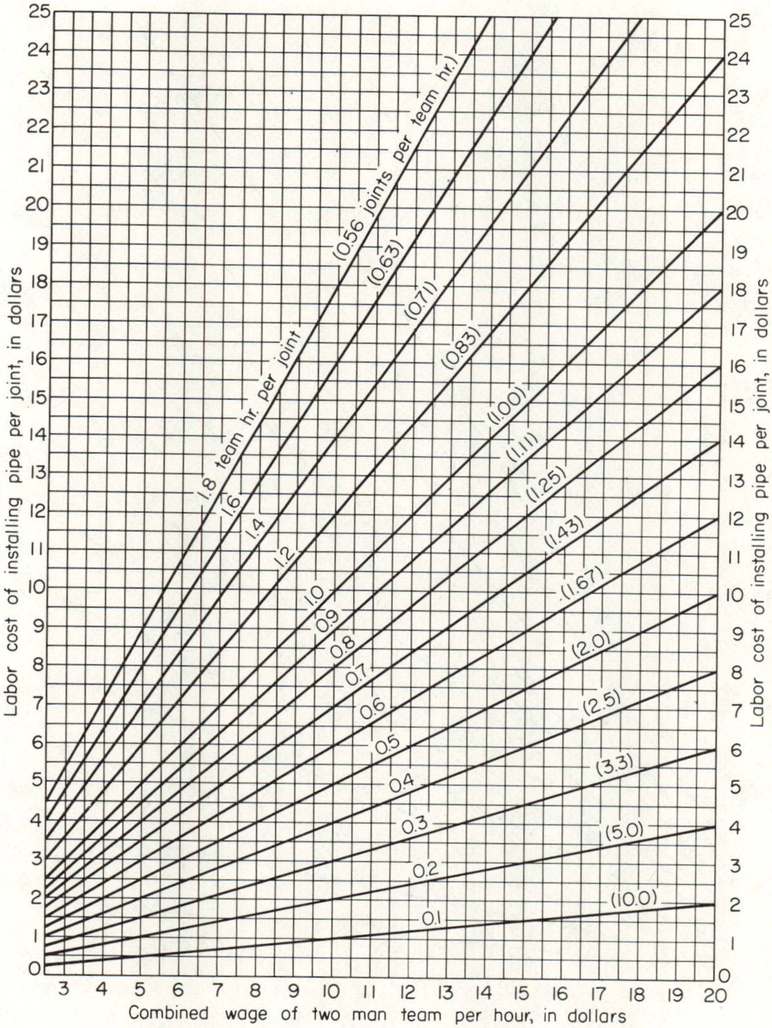

Diagram 13-1 Labor cost of installing pipe, per joint.

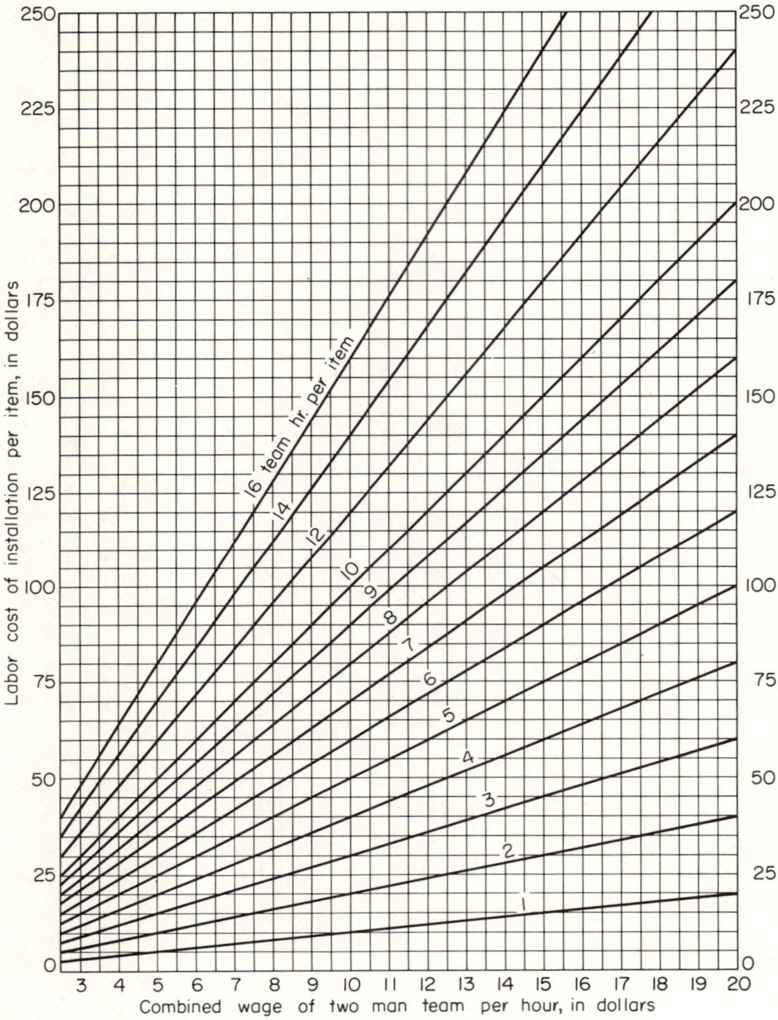

Diagram 13-2 Labor cost of installing rough plumbing work, per item.

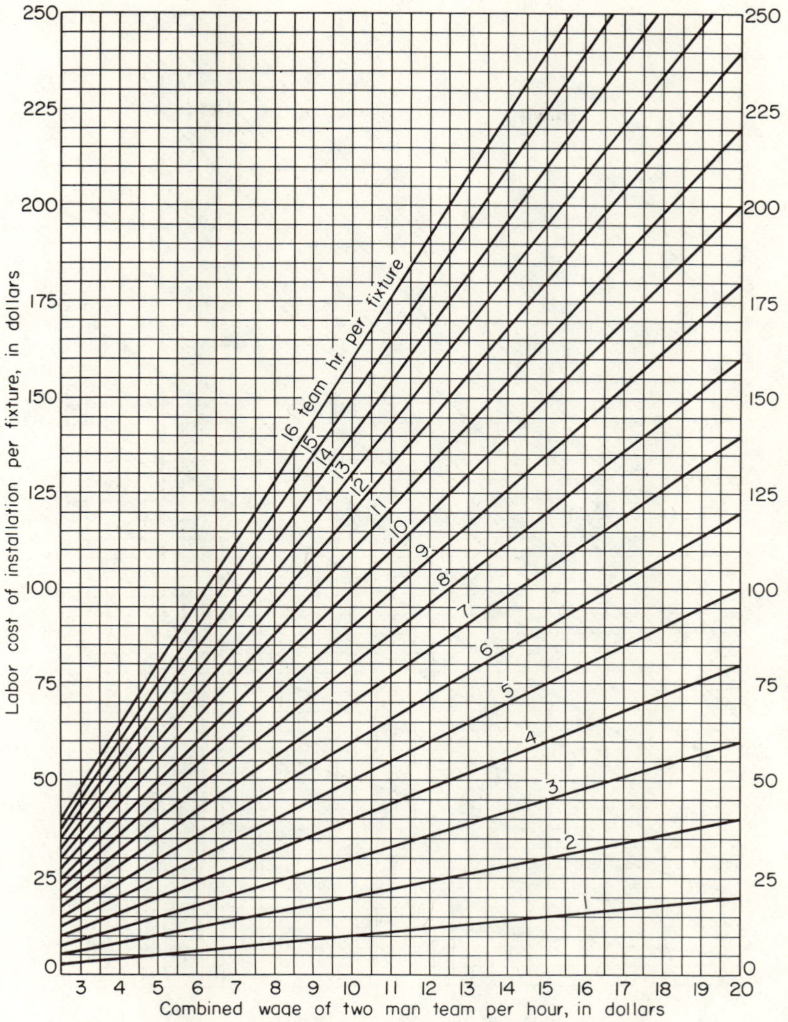

Diagram 13-3 Labor cost of installing finish plumbing fixtures.

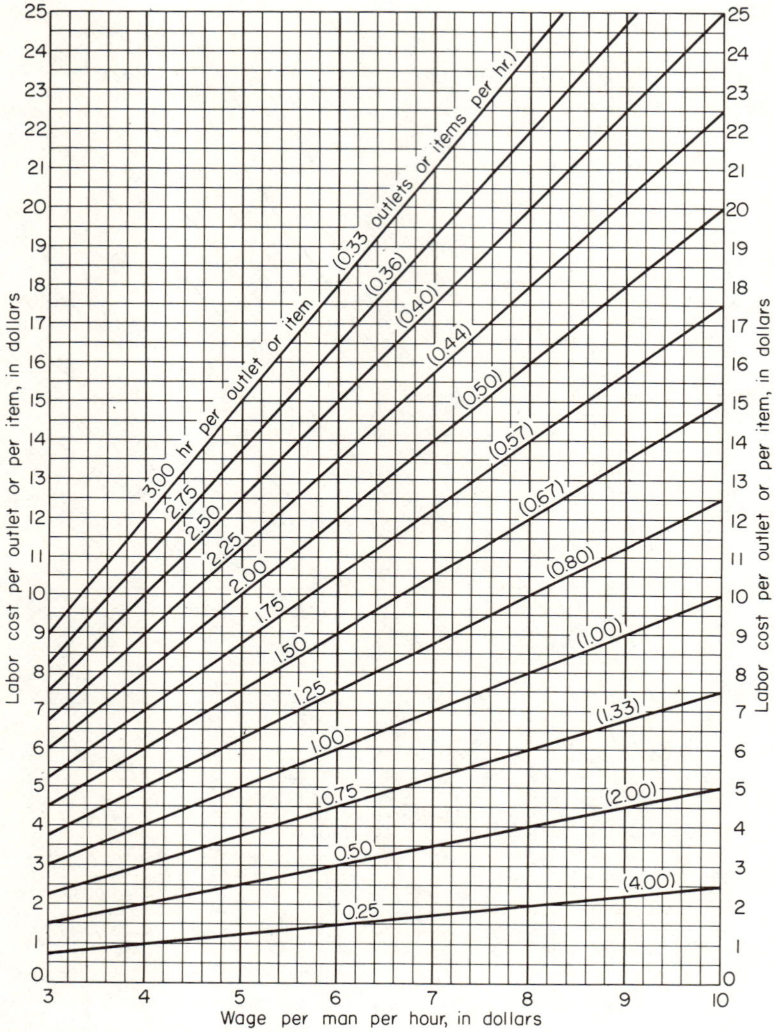

Diagram 14-1 Labor cost per outlet or per item for rough electrical work.

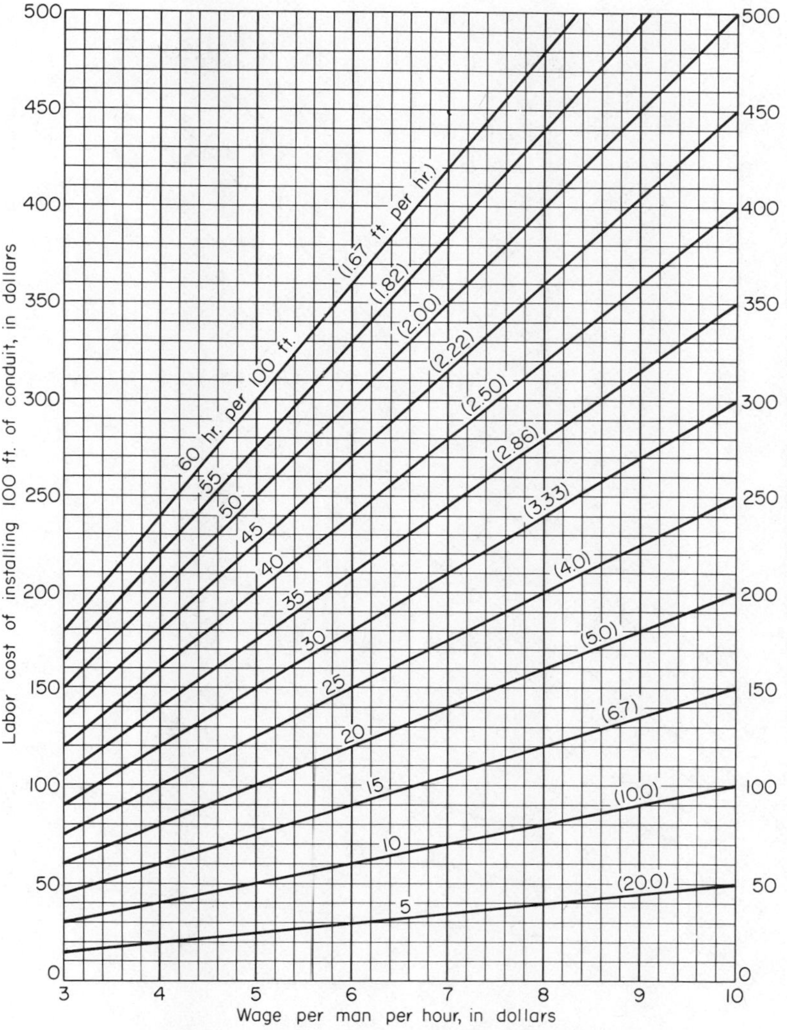

Diagram 14-2 Labor cost of installing electrical conduit.

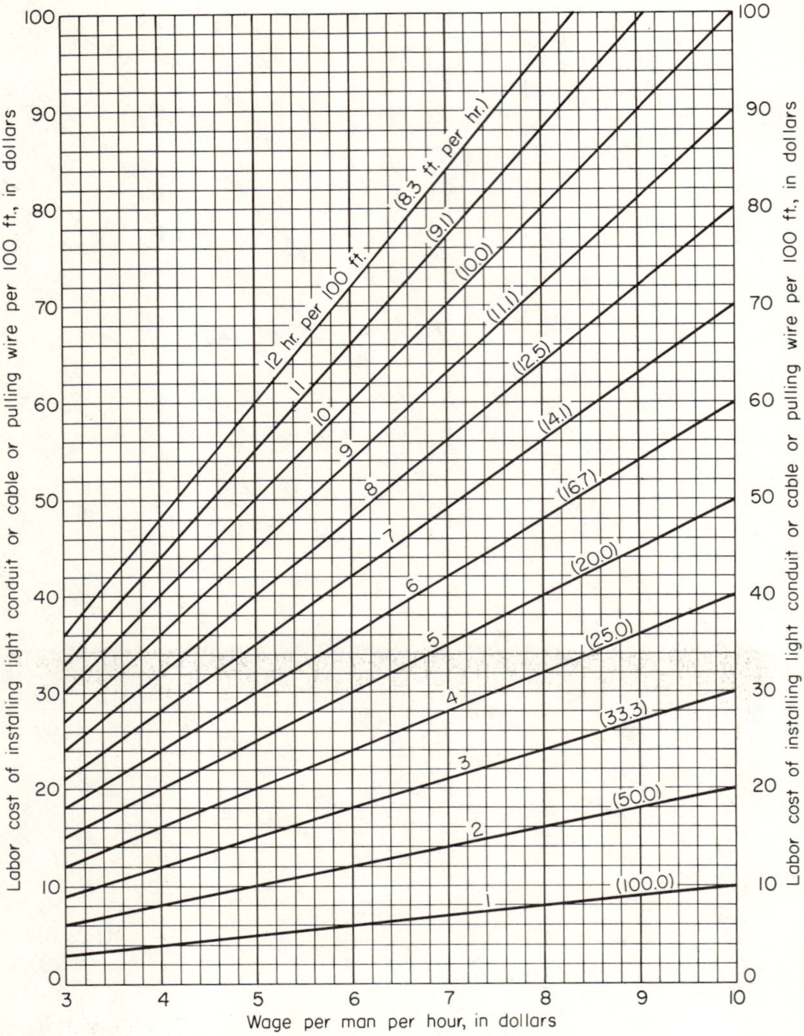

Diagram 14-3 Labor cost of pulling wire in conduit or of installing armored cable.

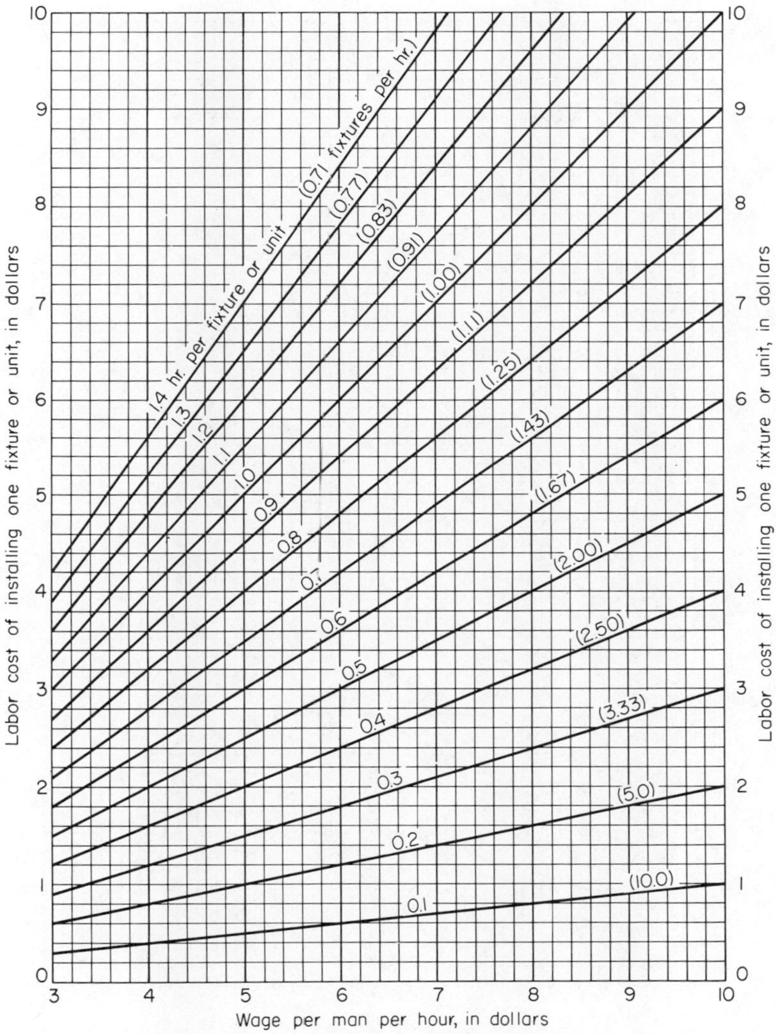

Diagram 14-4 Labor cost of installing electrical fixtures and other units.

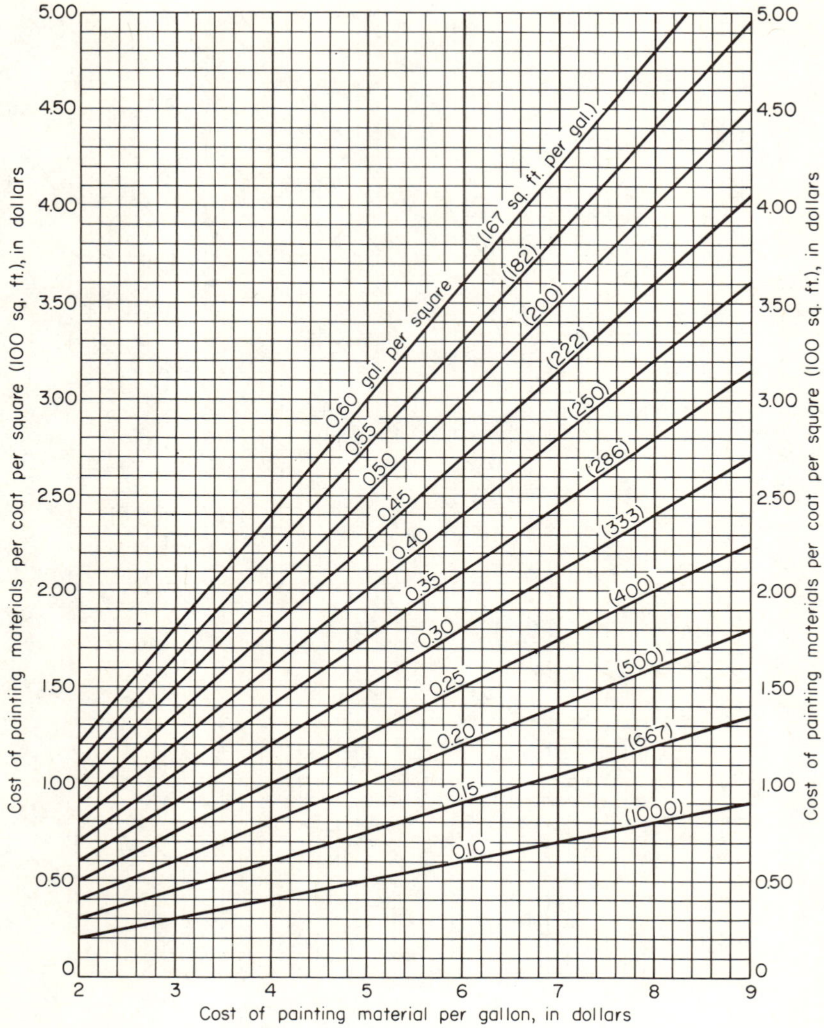

Diagram 15-1 Cost of painting materials per coat per square of 100 sq ft of surface.

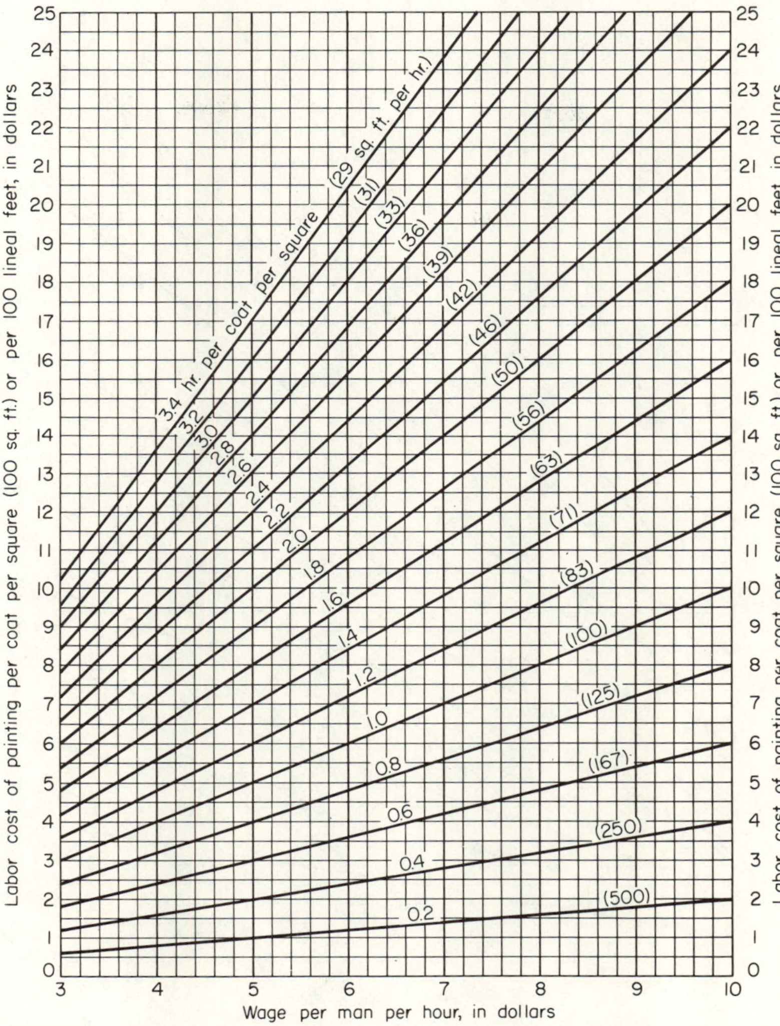

Diagram 15-2 Labor cost of painting per coat per square (100 sq ft) or per 100 lin ft.

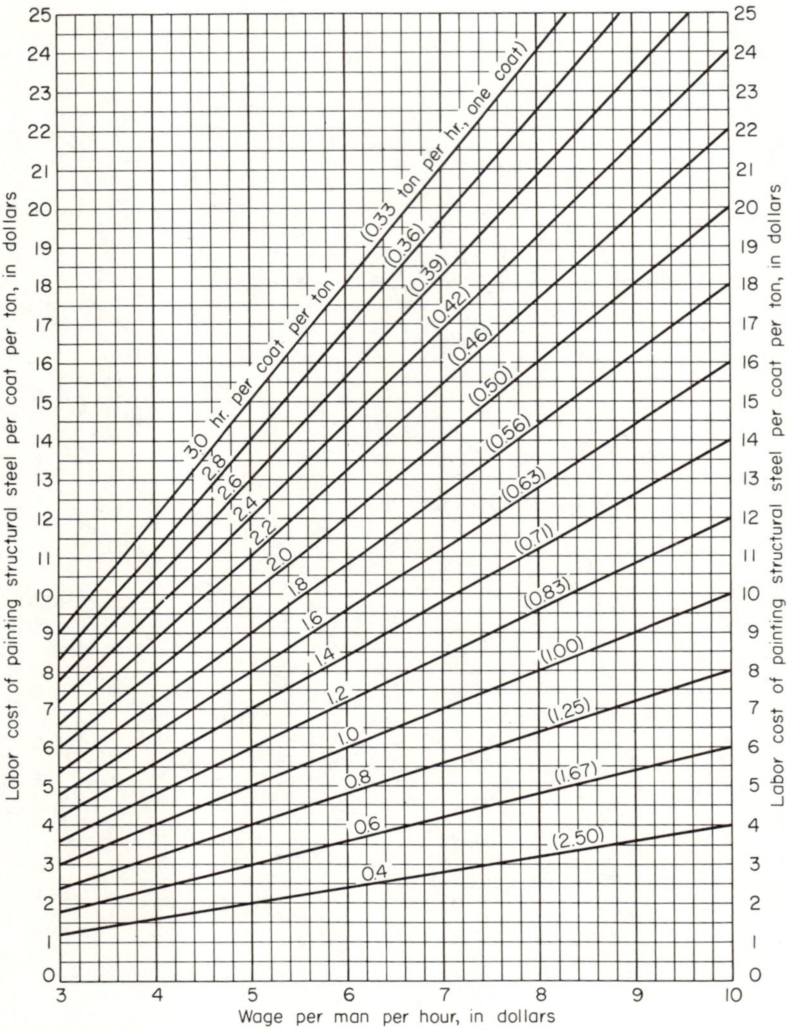

Diagram 15-3 Labor cost of painting structural steel per coat per ton.

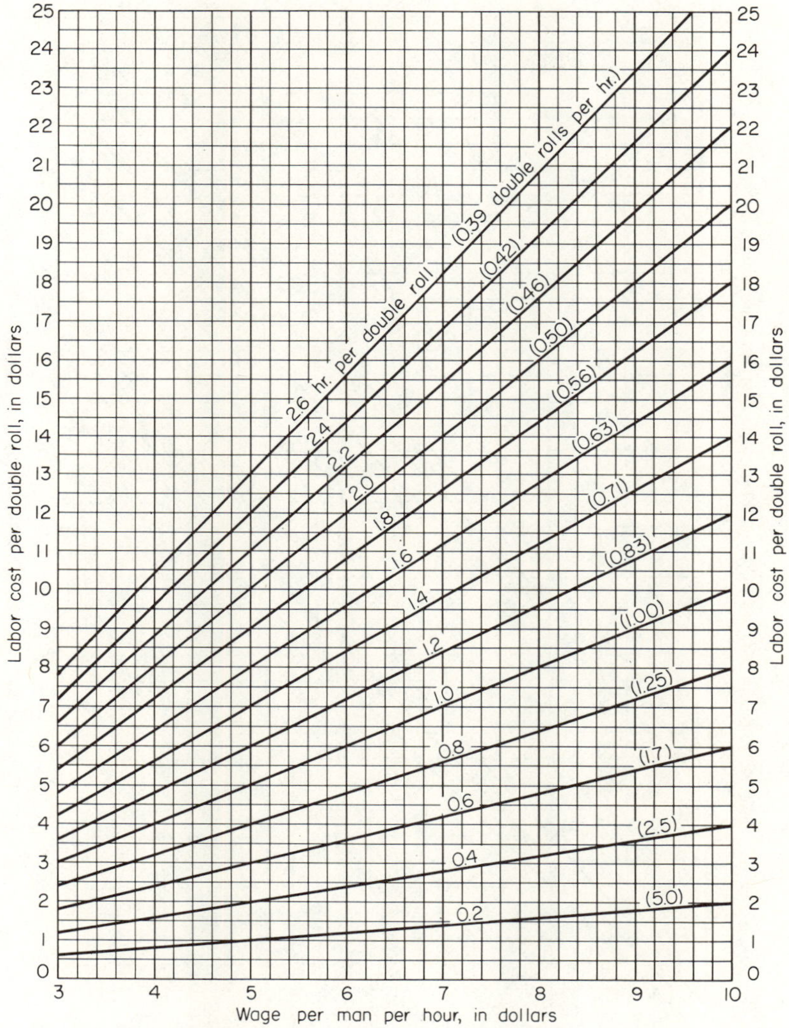

Diagram 15-4 Labor cost of papering per double roll. Cost per single roll is half of the cost per double roll.

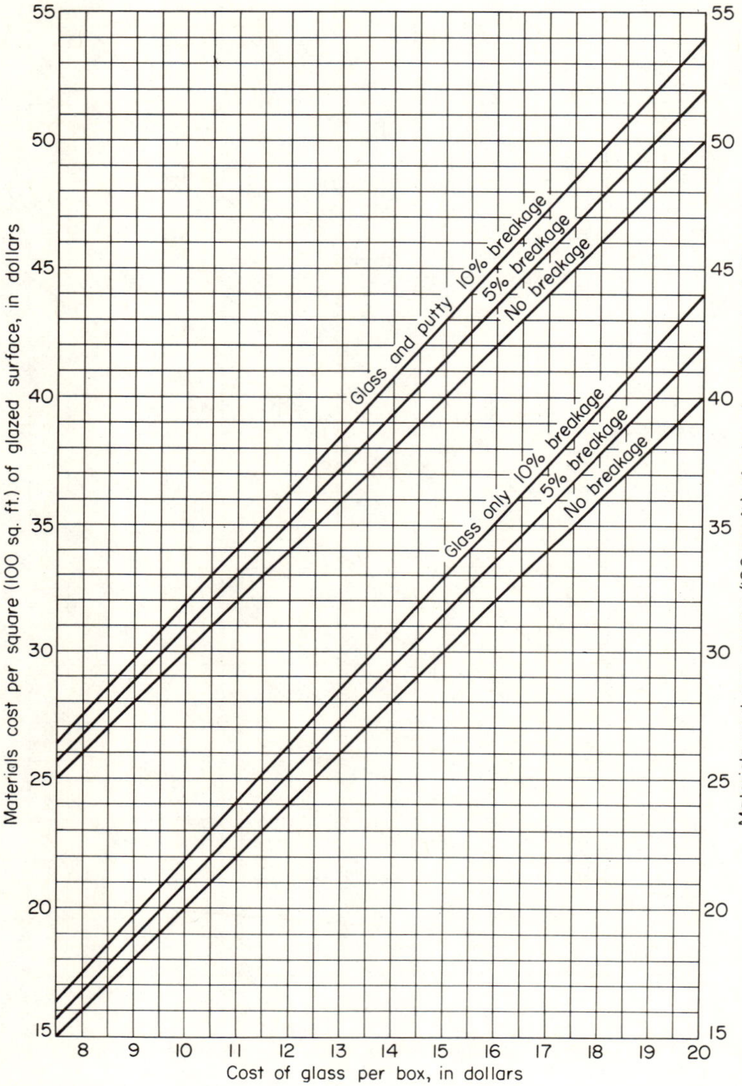

Diagram 15-5 Materials cost of glass for glazing with allowances for breakage and with or without an allowance of $5 per box of glass for putty.

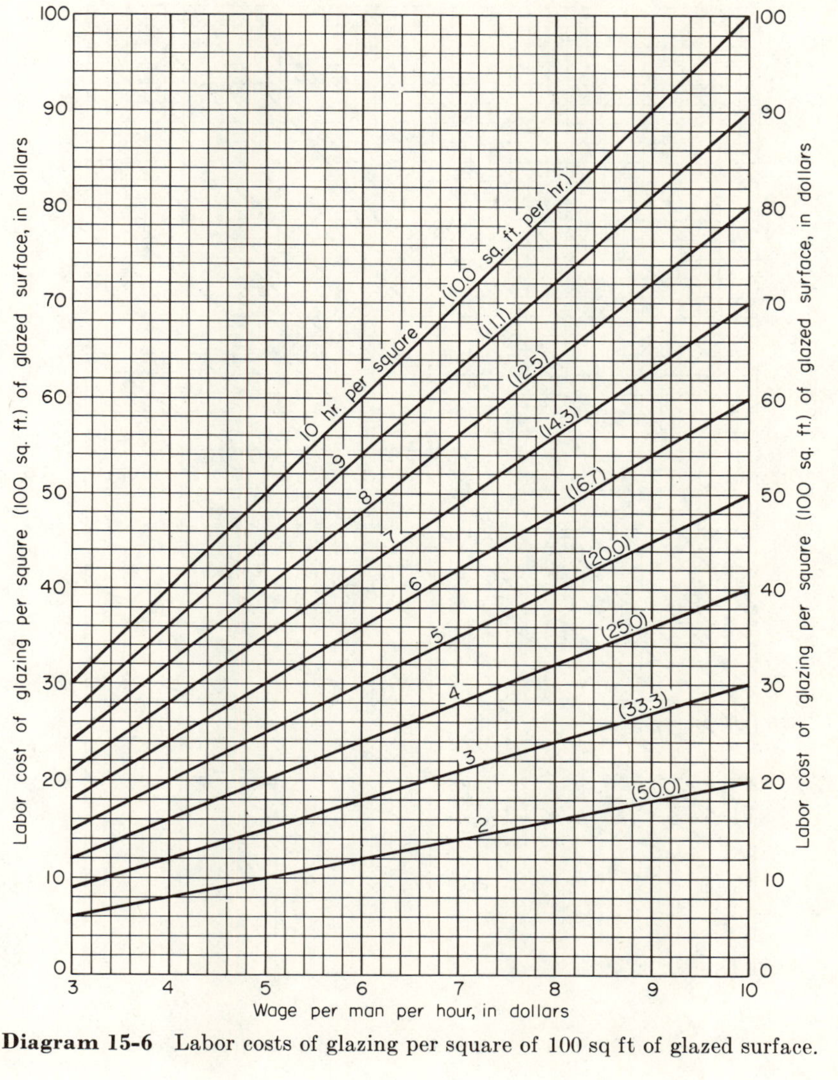

Diagram 15-6 Labor costs of glazing per square of 100 sq ft of glazed surface.

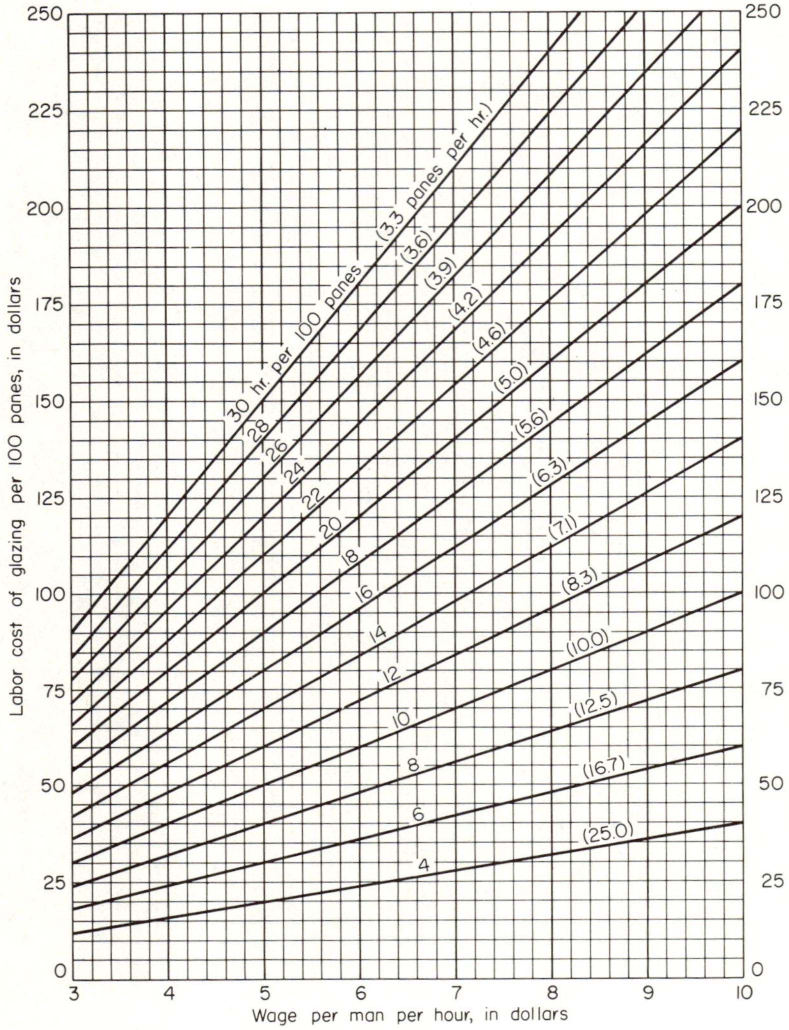

Diagram 15-7 Labor cost of glazing per 100 panes of glass.

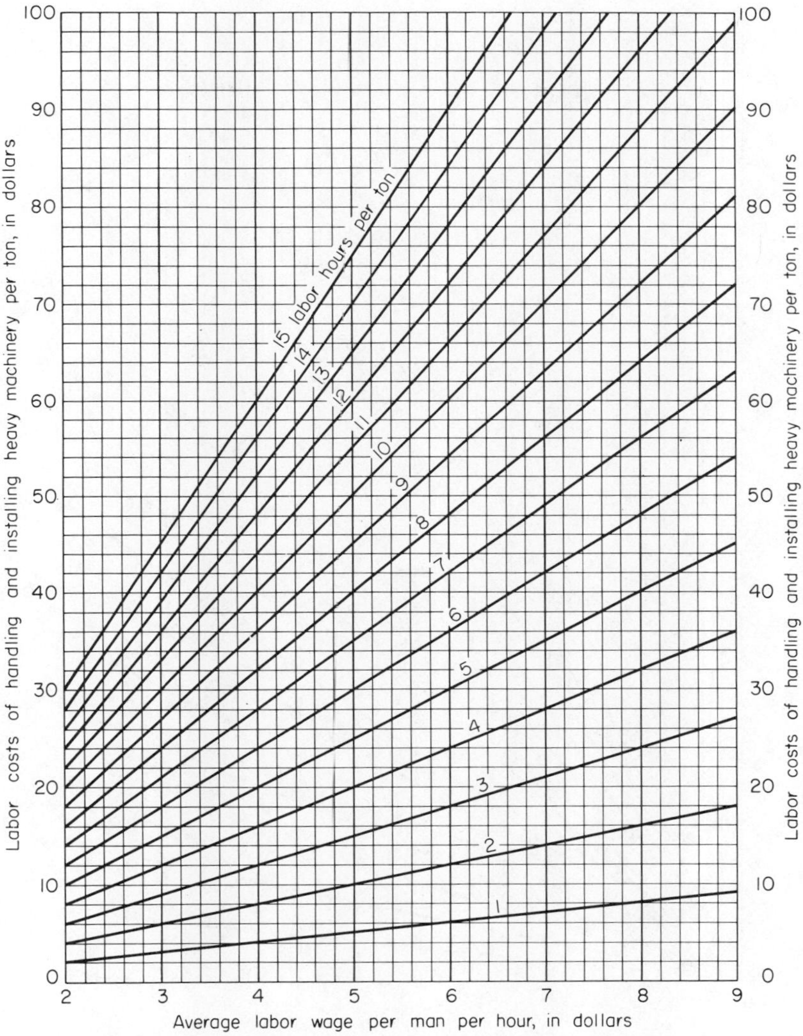

Diagram 16-1 Labor costs per ton of handling and installing heavy machinery.

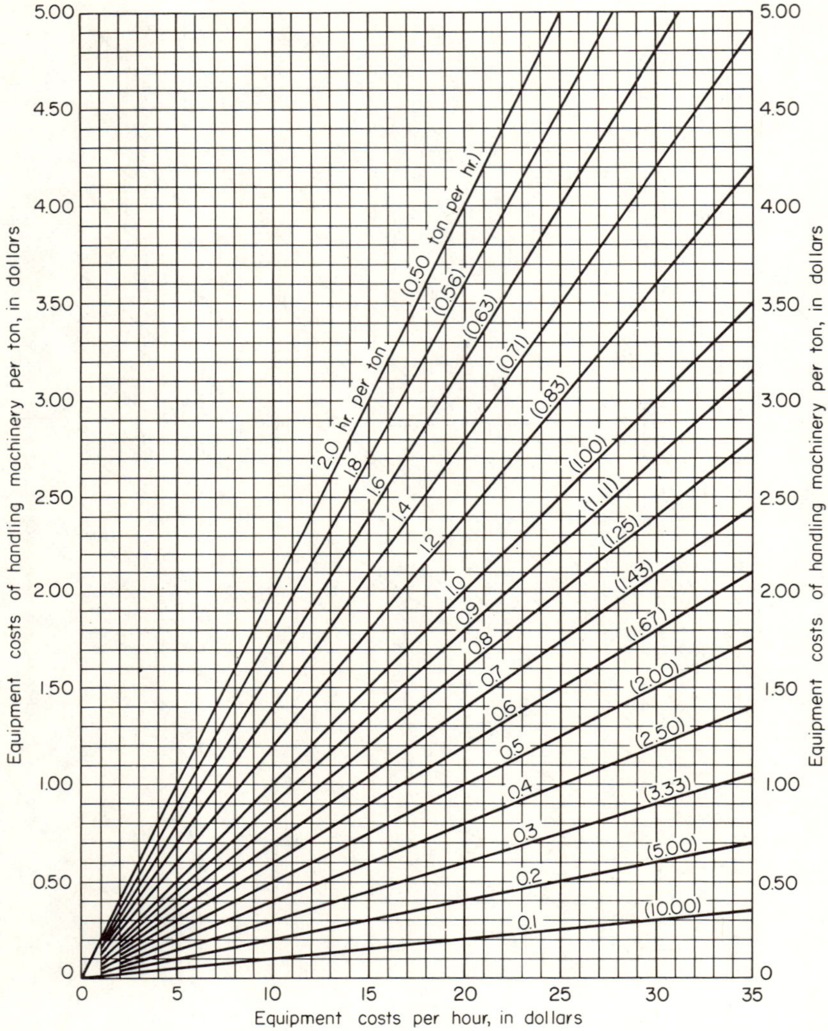

Diagram 16-2 Equipment costs of handling heavy machinery.

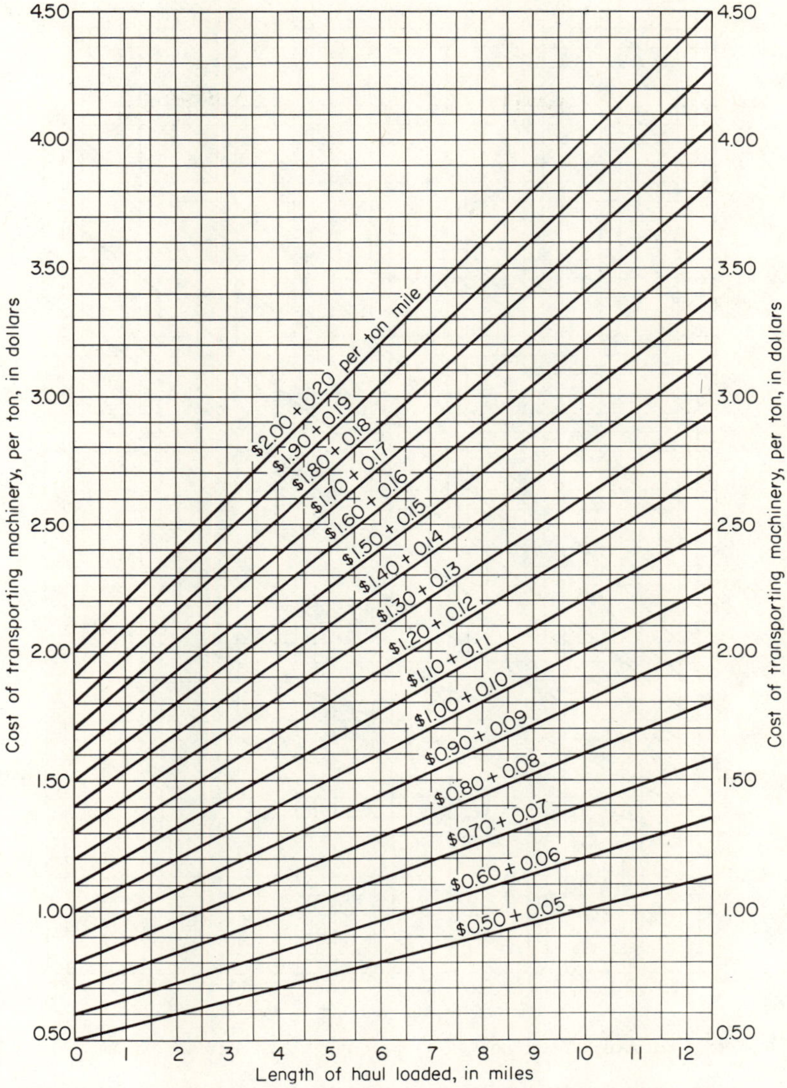

Diagram 16-3 Costs of transporting heavy machinery by autotruck or by autotruck and trailer.

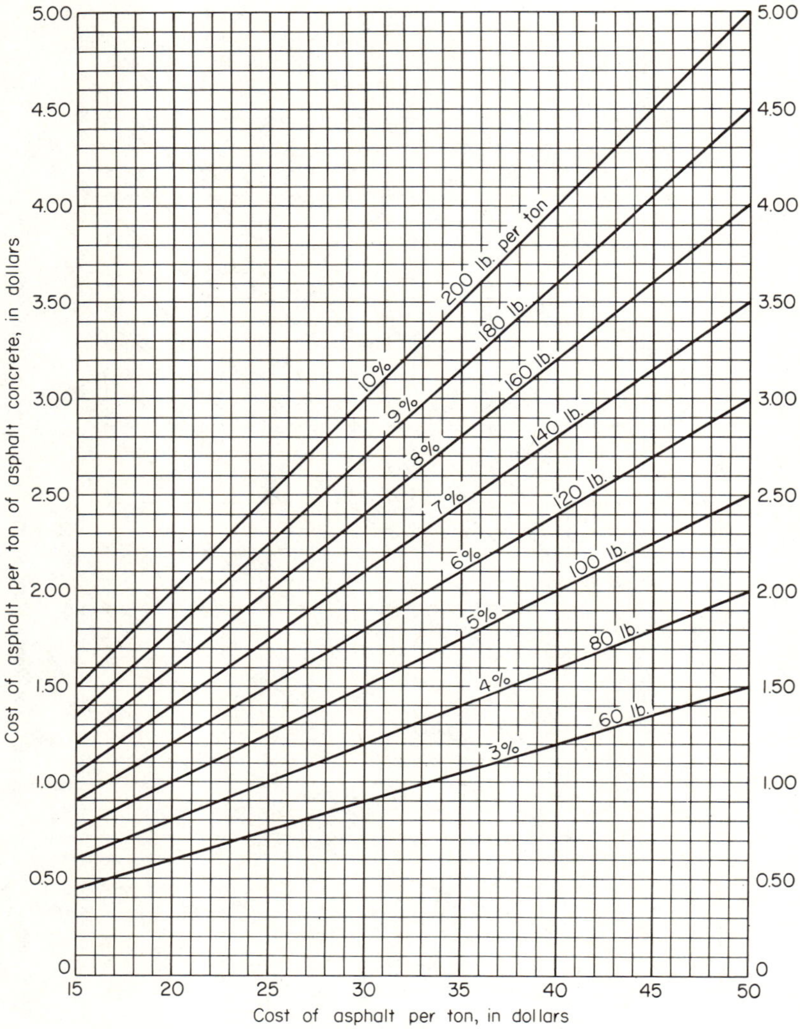

Diagram 17-1 Cost of asphalt per ton of asphalt concrete.

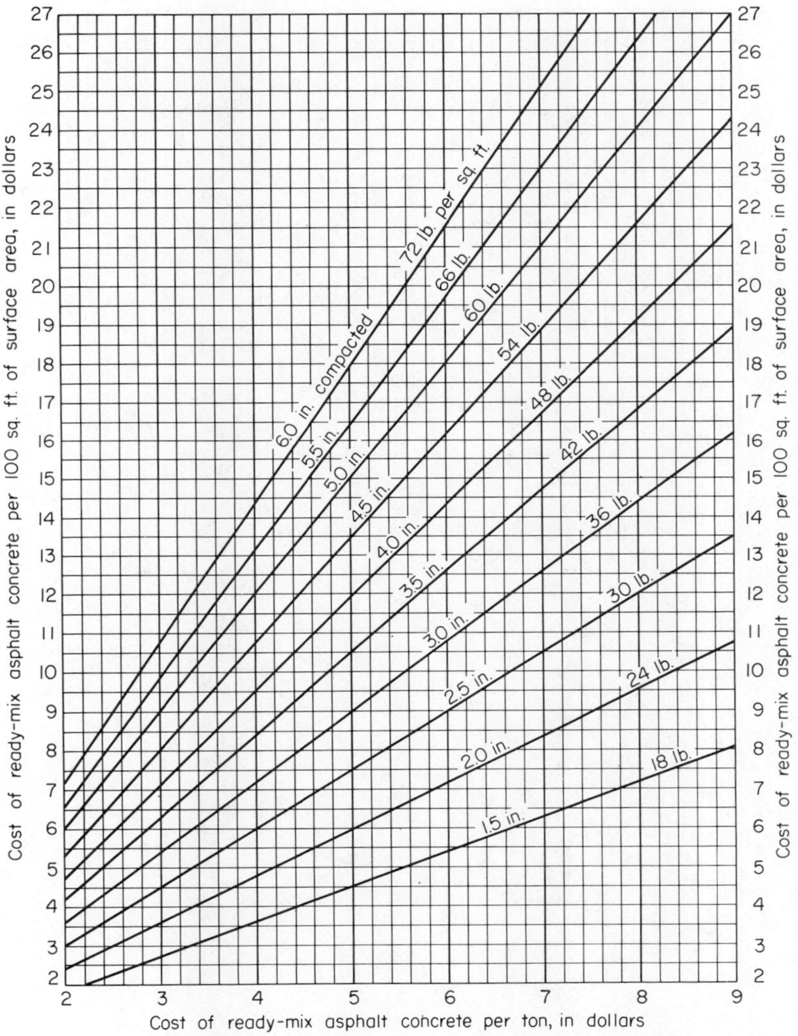

Diagram 17-2 Cost of ready-mix asphalt concrete for walks and drives.

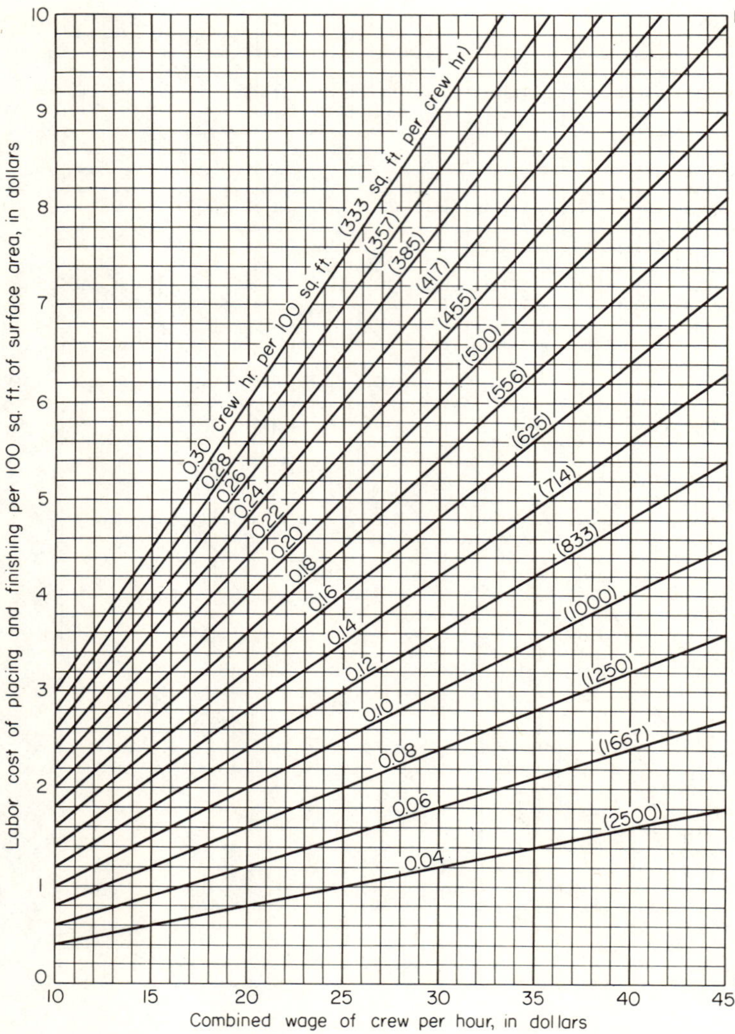

Diagram 17-3 Labor cost of placing and finishing asphalt-concrete walks and drives.

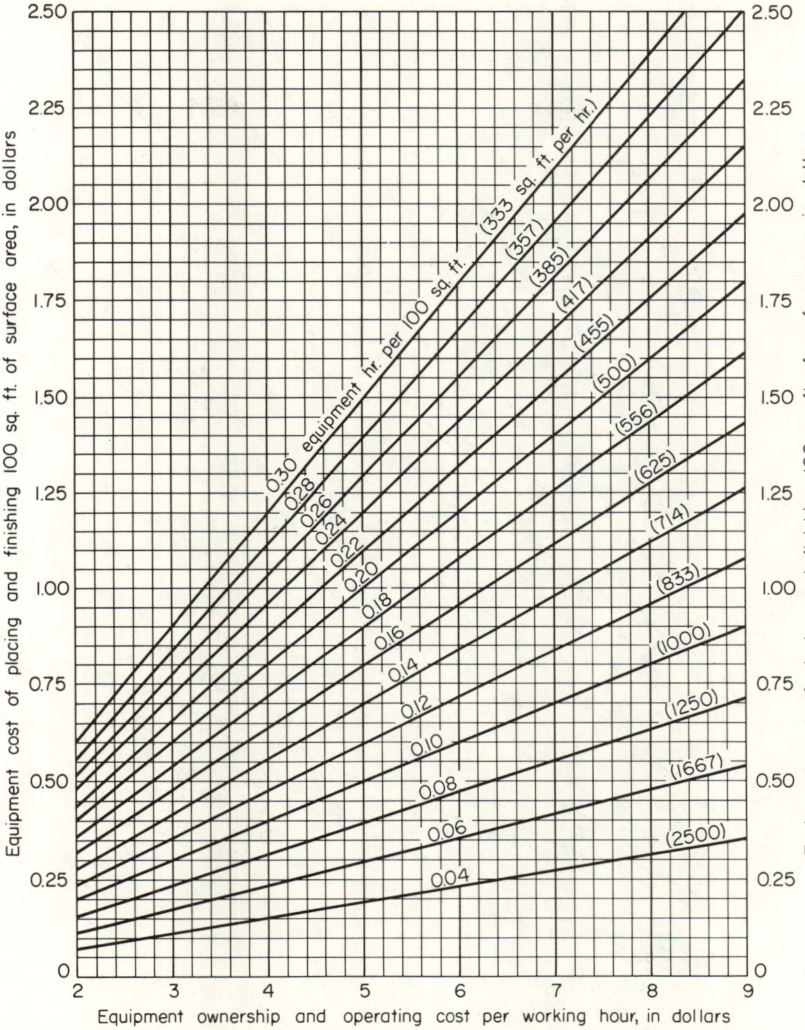

Diagram 17-4 Equipment cost of placing and finishing asphalt-concrete walks and drives.

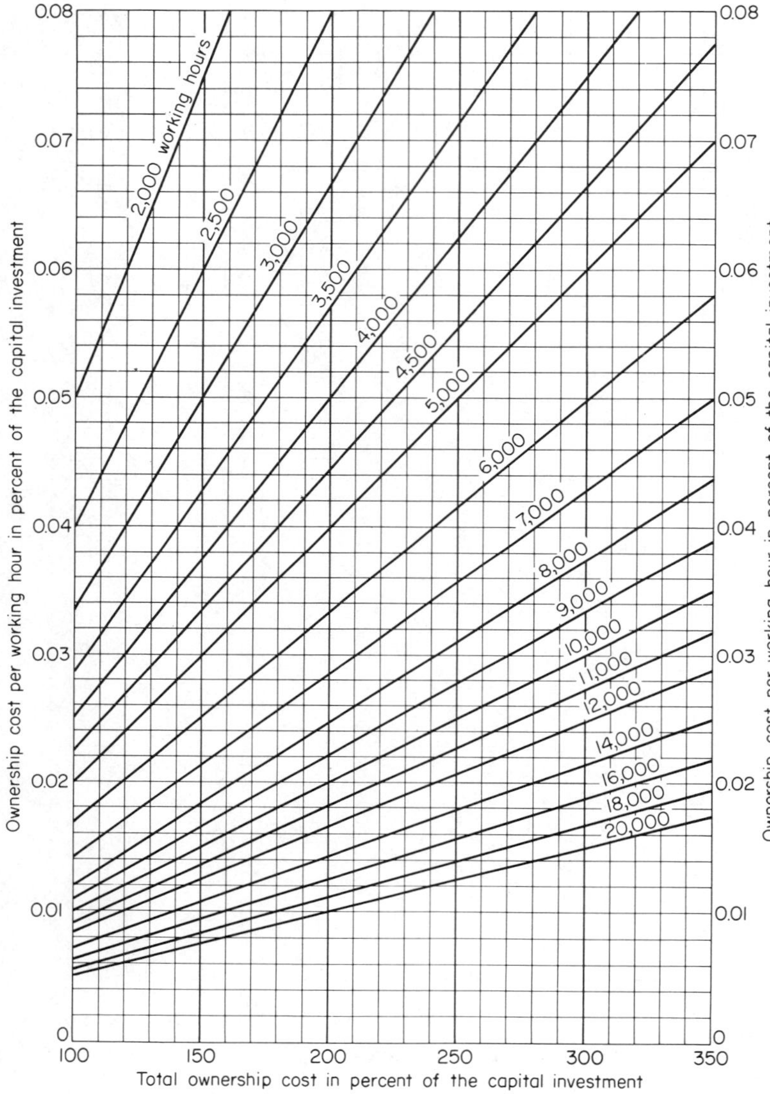

Diagram 18-1 Ownership cost per working hour in percent of the capital investment.

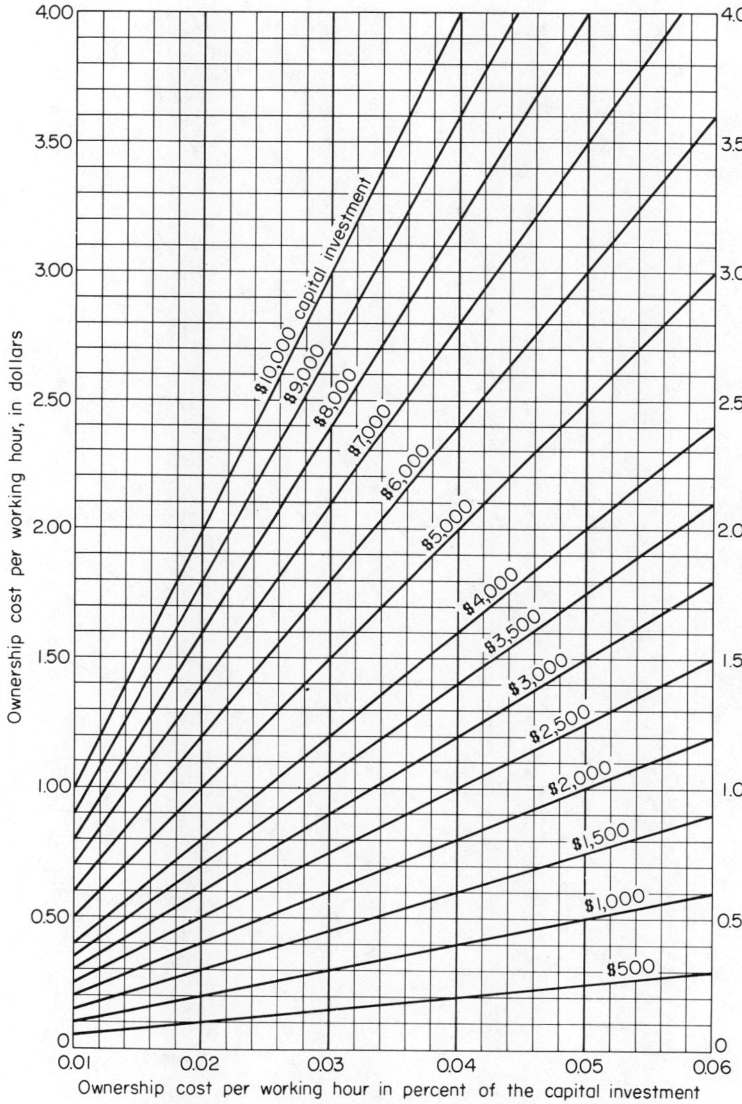

Diagram 18-2 Ownership cost per working hour in dollars.

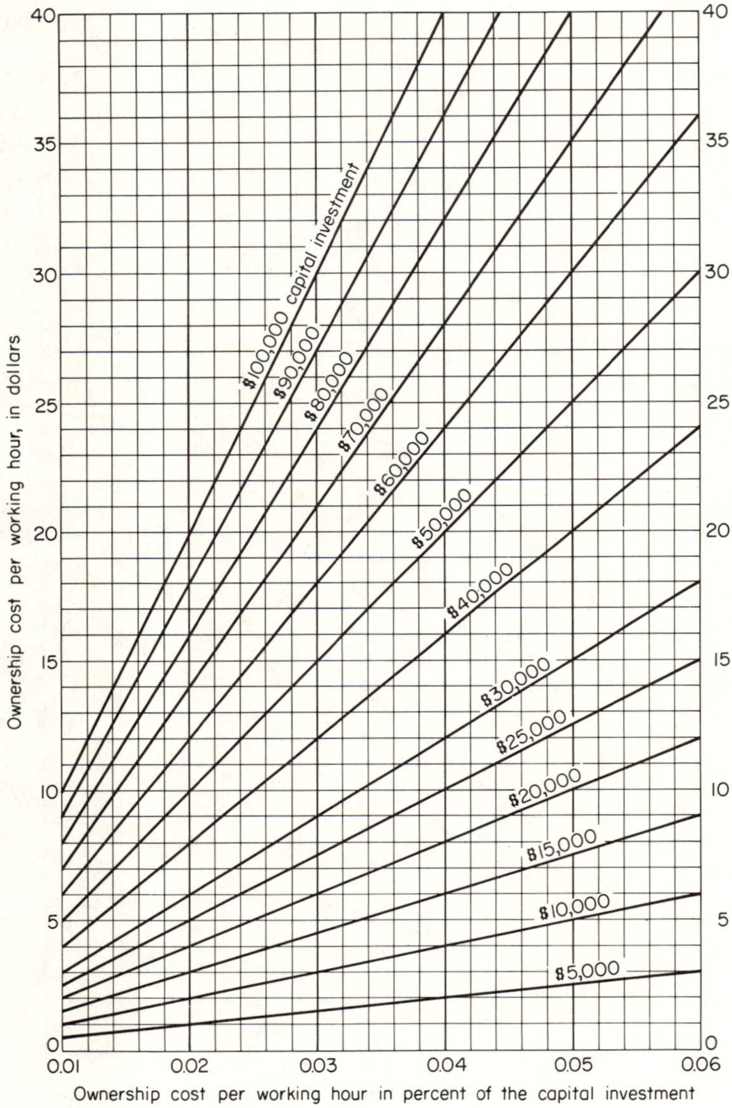

Diagram 18-3 Ownership cost per working hour in dollars.

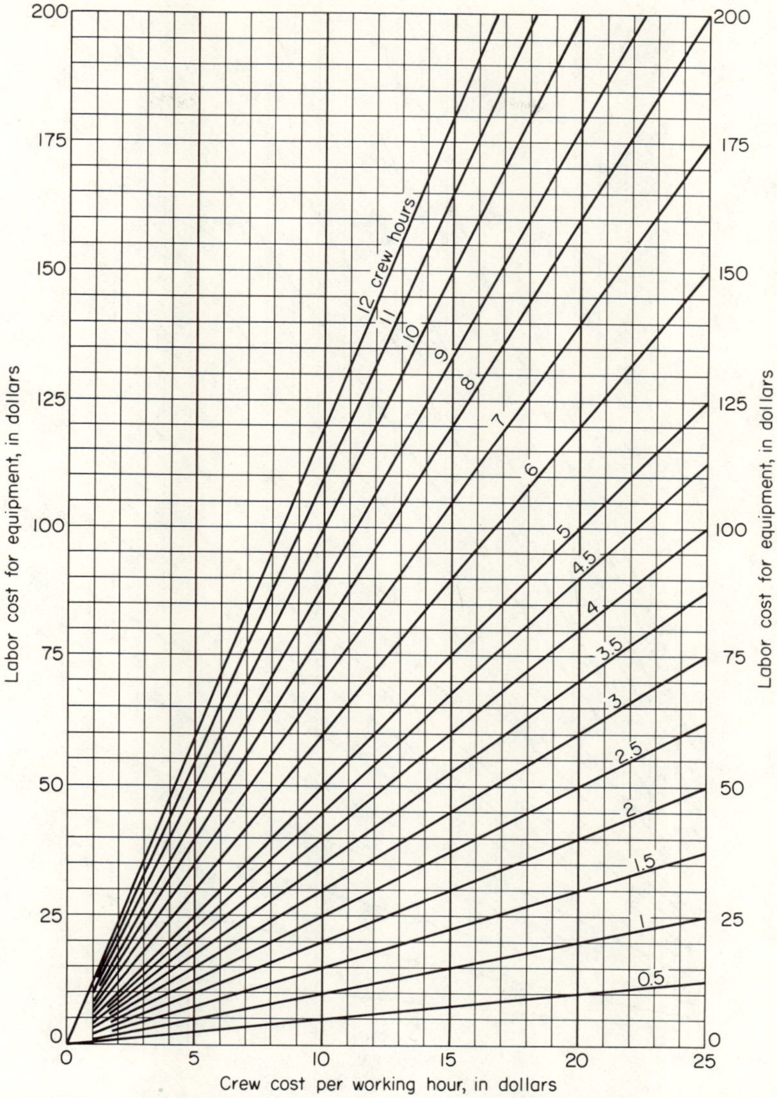

Diagram 18-4 Labor cost for equipment: loading, unloading, erecting, moving, dismantling, etc.

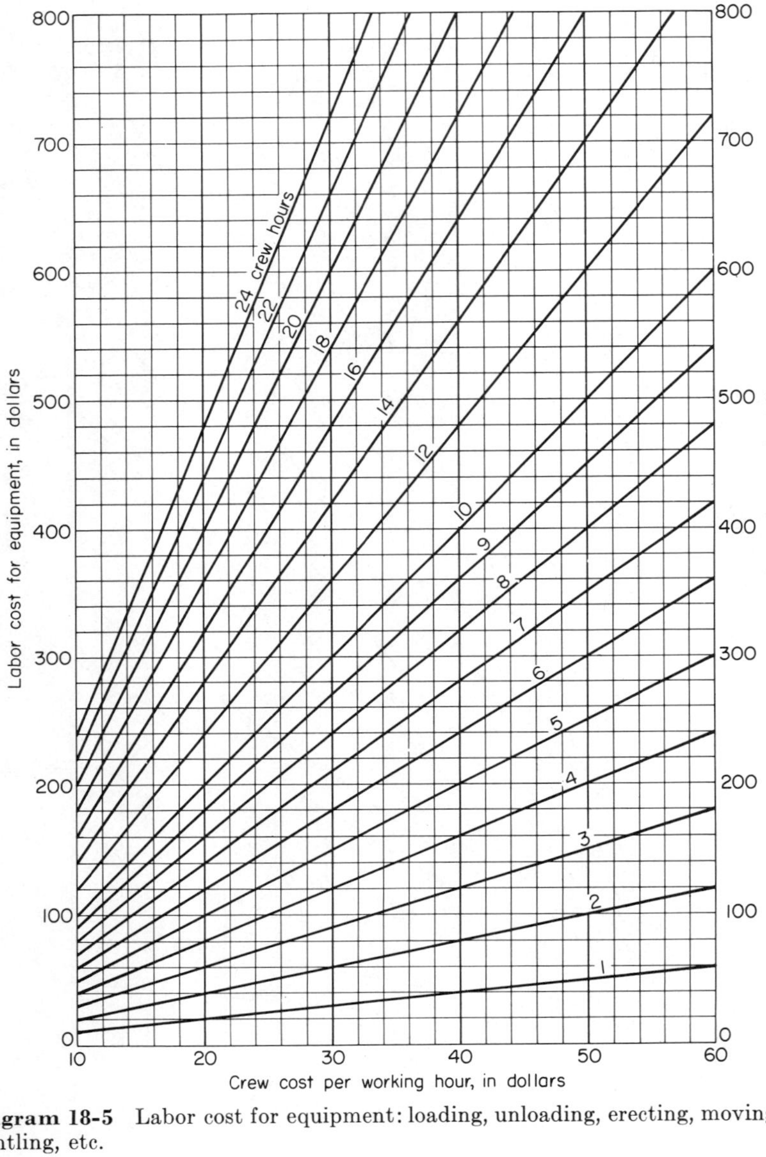

Diagram 18-5 Labor cost for equipment: loading, unloading, erecting, moving, dismantling, etc.

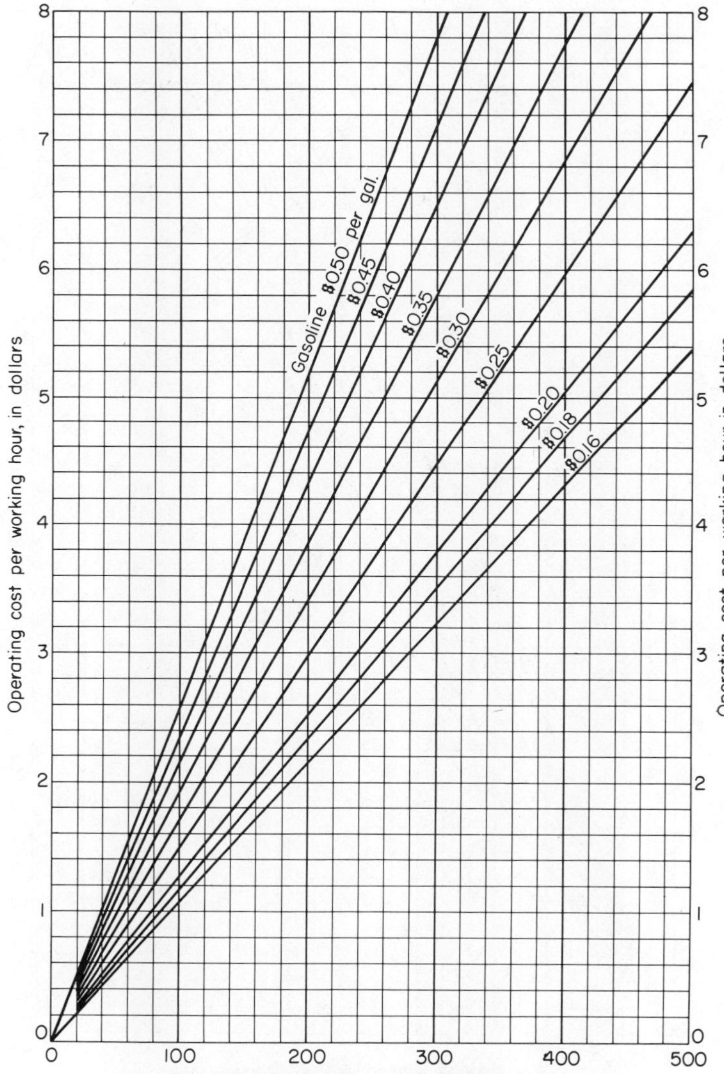

Diagram 18-6 Approximate operating cost per working hour for gasoline-powered equipment, assuming operation at 70 percent of rated brake horsepower and allowing $0.0035 per rated brake horsepower-hour for oil, grease, supplies, etc.

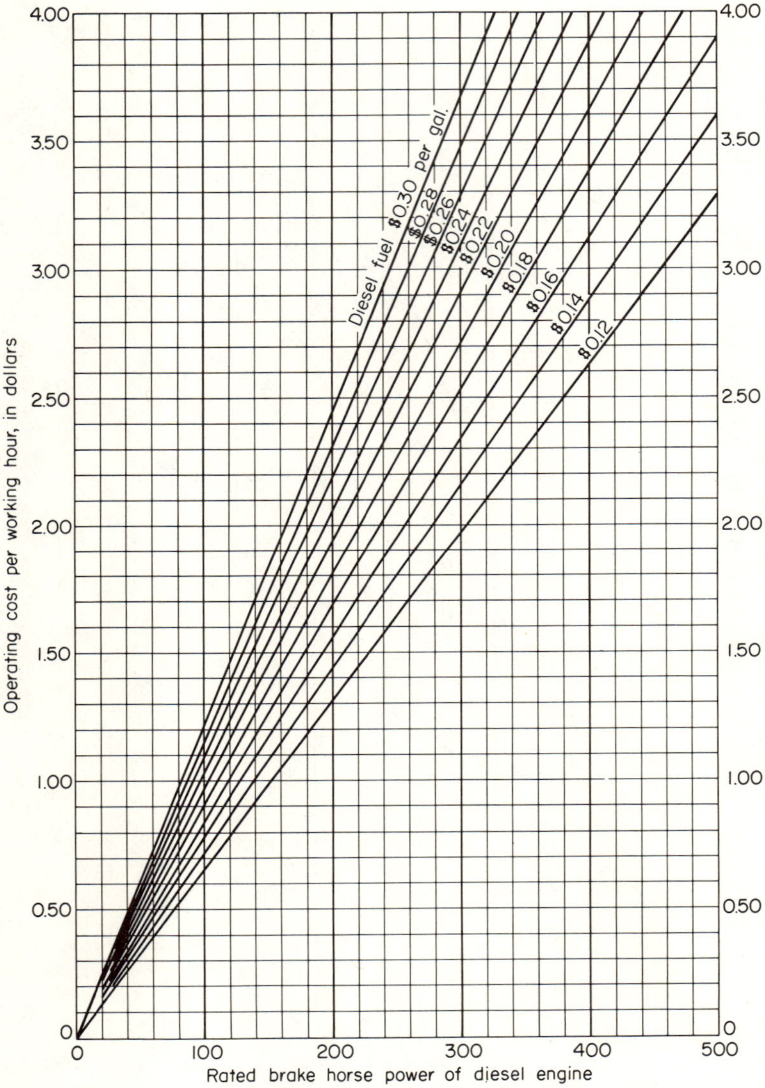

Diagram 18-7 Approximate operating cost per working hour for diesel-powered equipment, assuming operation at 70 percent of rated brake horsepower and allowing $0.0028 per rated brake horsepower-hour for oil, grease, supplies, etc.

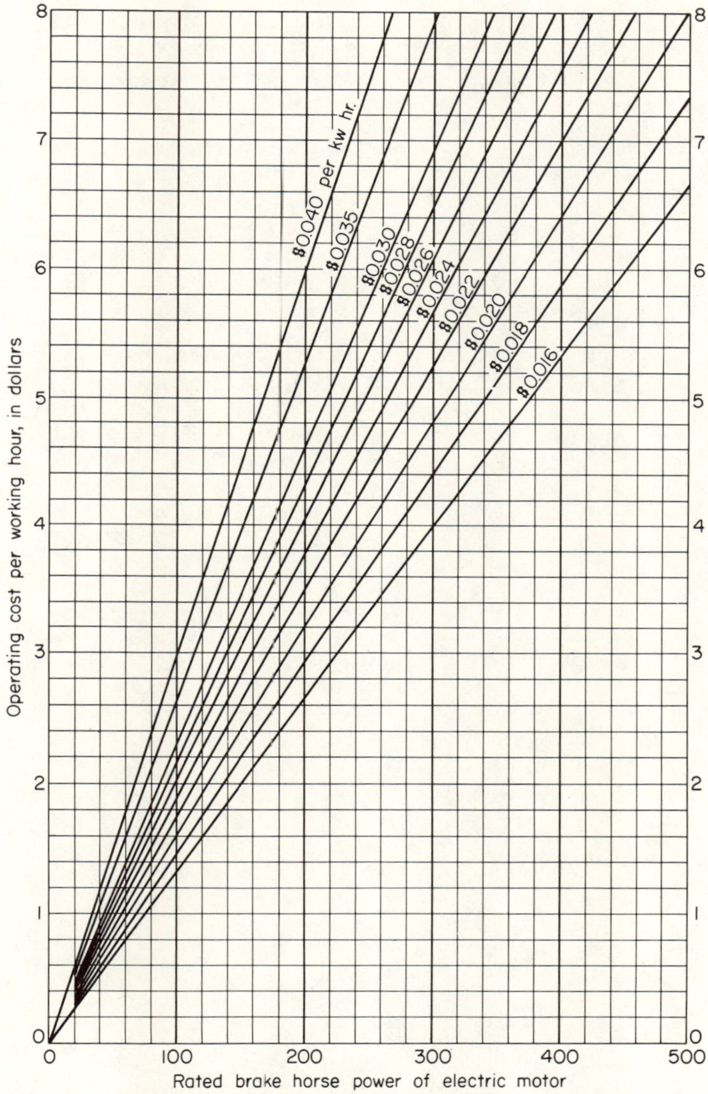

Diagram 18-8 Approximate operating cost per working hour for electric-motor-powered equipment, assuming that 1 kilowatthour of electricity is required for 1 brake horsepower-hour, assuming motor operation at 70 percent of rated brake horsepower, and allowing $0.0021 per rated brake horsepower-hour for oil, grease, supplies, etc.

Index